INTRODUCTION TO STOCHASTIC PROCESSES WITH R

INTRODUCTION TO STOCHASTIC PROCESSES WITH R

ROBERT P. DOBROW

WILEY

Published by John Wiley & Sons, Inc., Hoboken, New Jersey
Published simultaneously in Canada

Library of Congress Cataloging-in-Publication Data:

Dobrow, Robert P., author.
Introduction to stochastic processes with R / Robert P. Dobrow.
pages cm
Includes bibliographical references and index.
ISBN 978-1-118-74065-1 (cloth)
1. Stochastic processes. 2. R (Computer program language) I. Title.
QC20.7.S8D63 2016
519.2′302855133–dc23
2015032706

Set in 10/12pt, Times-Roman by SPi Global, Chennai, India.

Printed in Singapore

M WEP220846 230823

To my family

CONTENTS

PREFACE

> The last thing one discovers in composing a work is what to put first.
>
> —Blaise Pascal

The intended audience for this book are students who like probability. With that prerequisite, I am confident that you will love stochastic processes.

Stochastic, or random, processes is the dynamic side of probability. What differential equations is to calculus, stochastic processes is to probability. The material appeals to those who like applications and to those who like theory. It is both excellent preparation for future study, as well as a *terminal* course, in the sense that we do not have to tell students to wait until the next class or the next year before seeing the good stuff. This *is* the good stuff! Stochastic processes, as a branch of probability, speaks in the language of rolling dice, flipping coins, and gambling games, but in the service of applications as varied as the spread of infectious diseases, the evolution of genetic sequences, models for climate change, and the growth of the World Wide Web.

The book assumes that the reader has taken a calculus-based probability course and is familiar with matrix algebra. Conditional probability and conditional expectation, which are essential tools, are offered in the introductory chapter, but may be skimmed over depending upon students' background. Some topics assume a greater knowledge of linear algebra than basic matrices (such as eigenvalues and eigenvectors) but these are optional, and relevant sections are starred. The book does not assume background in combinatorics, differential equations, or real analysis. Necessary mathematics is introduced as needed.

A focus of this book is the use of simulation. I have chosen to use the popular statistical freeware `R`, which is an accessible interactive computing environment. The use of simulation, important in its own right for applied work and mathematical research, is a powerful pedagogical tool for making theoretical concepts come alive with practical, hands-on demonstrations. It is not necessary to use R in order to use this book; code and script files are supplemental. However, the software is easy—and fun—to learn, and there is a tutorial and exercises in an appendix for bringing students up to speed.

The book contains more than enough material for a standard one-semester course. Several topics may lend themselves to individual or group projects, such as card shuffling, perfect sampling (coupling from the past), queueing theory, stochastic calculus, martingales, and stochastic differential equations. Such specialized material is contained in starred sections.

An undergraduate textbook poses many challenges. I have struggled with trying to find the right balance between theory and application, between conceptual understanding and formal proof. There are, of course, some things that cannot be said. Continuous-time processes, in particular, require advanced mathematics based on measure theory to be made precise. Where these subjects are presented I have emphasized intuition over rigor.

Following is a synopsis of the book's nine chapters.

Chapter 1 introduces stochastic and deterministic models, the generic features of stochastic processes, and simulation. This is essential material. The second part of the chapter treats conditional probability and conditional expectation, which can be reviewed at a fast pace.

The main features of discrete-time Markov chains are covered in Chapters 2 and 3. Many examples of Markov chains are introduced, and some of them are referenced throughout the book. Numerical and simulation-based methods motivate the discussion of limiting behavior. In addition to basic computations, topics include stationary distributions, ergodic and absorbing chains, time reversibility, and the strong Markov property. Several important limit theorems are discussed in detail, with proofs given at the end of the chapter. Instructors may choose to limit how much time is spent on proofs.

Branching processes are the topic of Chapter 4. Although branching processes are Markov chains, the methods of analysis are different enough to warrant a separate chapter. Probability-generating functions are introduced, and do not assume prior exposure.

The focus of Chapter 5 is Markov chain Monte Carlo, a relatively new topic but one with exponentially growing application. Instructors will find many subjects to pick and choose. Several case studies make for excellent classroom material, in particular (i) a randomized method for decoding text, from Diaconis (2009), and (ii) an application that combines ecology and counting matrices with fixed row and column totals, based on Cobb and Chen (2003). Other topics include coupling from the past, card shuffling, and rates of convergence of Markov chains.

Chapter 6 is devoted to the Poisson process. The approach emphasizes three alternate definitions and characterizations, based on the (i) counting process, (ii) arrival process, and (iii) infinitesimal description. Additional topics are spatial processes, nonhomogeneous Poisson processes, embedding, and arrival time paradoxes.

Continuous-time Markov chains are discussed in Chapter 7. For continuous-time stochastic processes, here and in Chapter 8, there is an emphasis on intuition, examples, and applications. In addition to basic material, there are sections on queueing theory (with Little's formula), absorbing processes, and Poisson subordination.

Brownian motion is the topic of Chapter 8. The material is more challenging. Topics include the invariance principle, transformations, Gaussian processes, martingales, and the optional stopping theorem. Examples include scoring in basketball and animal tracking. The Black–Scholes options pricing formula is derived.

Chapter 9 is a gentle introduction to stochastic calculus. *Gentle* means no measure theory, sigma fields, or filtrations, but an emphasis on examples and applications. I decided to include this material because of its growing popularity and application. Stochastic differential equations are introduced. Simulation and numerical methods help make the topic accessible.

Book appendices include (i) getting started with `R`, with exercises, (ii) probability review, with short sections on the main discrete and continuous probability distributions, (iii) summary table of common probability distributions, and (iv) matrix algebra review. Resources for students include a suite of `R` functions and script files for generating many of the processes from the book.

The book contains more than 200 examples, and about 600 end-of-chapter exercises. Short solutions to most odd-numbered exercises are given at the end of the book. A web site www.people.carleton.edu/rdobrow/stochbook is established. It contains errata and relevant files. All the `R` code and script files used in the book are available at this site. A solutions manual with detailed solutions to all exercises is available for instructors.

Much of this book is a reflection of my experience teaching the course over the past 10 years. Here is a suggested one-semester syllabus, which I have used.

1. Introduction and review—1.1, 1.2, 1.3 (quickly skim 1.4 and 1.5)
2. One-day introduction to `R`—Appendix A
3. Markov chains—All of chapters 2 and 3
4. Branching processes—Chapter 4
5. MCMC—5.1, 5.2
6. Poisson process—6.1, 6.2, 6.4, 6.5, 6.8
7. Continuous-time Markov chains—7.1, 7.2, 7.3, 7.4
8. Brownian motion—8.1, 8.2, 8.3, 8.4, 8.5, 8.7

If instructors have questions on syllabus, homework assignments, exams, or projects, I am happy to share resources and experiences teaching this most rewarding course.

Stochastic Processes is a great mathematical adventure. Bon voyage!

ACKNOWLEDGMENTS

I am indebted to family, friends, students, and colleagues who encouraged, supported, and inspired this project. My past students are all responsible, in some measure, for this work. The students of my Spring 2015 Introduction to Stochastic Processes class at Carleton College field-tested the first draft. Those who found errors in the text, and thus earned extra credit points (some quite ample), were Taeyoung Choi, Elsa Cristofaro, Graham Earley, Michelle Marinello, Il Shan Ng, Risako Owan, John Pedtke, and Maggie Sauer. Also, Edward Heo, Jack Hessle, Alex Trautman, and Zach Wood-Doughty contributed to case studies and examples, `R` code, and aesthetic design.

Professors Bill Peterson at Middlebury College, and Michele Joynor at East Tennessee State University, gave helpful feedback based on field-testing the manuscript in their classes. Others who offered suggestions and encouragement include David Aldous, Laura Chihara, Hosam Mahmoud, Jack O'Brien, Jacob Spear, Jeff Rosenthal, and John Wierman.

Carleton College and the Department of Mathematics and Statistics were enormously supportive. Thanks to Sue Jandro, the Department's Administrative Assistant, and Mike Tie, the Department's Technical Director, for help throughout the past 2 years.

The staff at Wiley, including Steve Quigley, Amy Hendrickson, Jon Gurstelle, and Sari Friedman, provided encouragement and assistance in preparing this book.

I am ever grateful to my family, who gave loving support. My wife Angel was always enthusiastic and gave me that final nudge to "go for it" when I was still in the procrastination stage. My sons Tom and Joe read many sections, including the preface and introduction, with an eagle's eye to style and readability. Special thanks to my son, Danny, and family friend Kellen Kirchberg, for their beautiful design of the front cover.

LIST OF SYMBOLS AND NOTATION

Such is the advantage of a well constructed language that its simplified notation often becomes the source of profound theories.

—Pierre-Simon Laplace

$\emptyset$	empty set
$A \cup B$	union of sets A and B
$A \cap B$	intersection of sets A and B
$A \subseteq B$	A is a subset of B
$x \in A$	x is an element of set A
$\binom{n}{k}$	binomial coefficient $\frac{n!}{k!(n-k)!}$
$\lfloor x \rfloor$	floor of x
$\lceil x \rceil$	ceiling of x
$\parallel \boldsymbol{x} \parallel$	length of vector $\boldsymbol{x}$
$\boldsymbol{A}^{-1}$	matrix inverse of $\boldsymbol{A}$
$\boldsymbol{A}^T$	matrix transpose of $\boldsymbol{A}$
$e^{\boldsymbol{A}}$	matrix exponential of $\boldsymbol{A}$
B_t	standard Brownian motion
$\text{Corr}(X, Y)$	correlation of X and Y
$\text{Cov}(X, Y)$	covariance of X and Y
$d(i)$	period of state i
$\deg(v)$	degree of vertex v in a graph
e	extinction probability (for branching process)

$E(X)$	expectation of X
$E(X^k)$	kth moment of X
$E(X\|Y)$	conditional expectation of X given Y
f_j	probability of eventual return to j
$f_X(x)$	probability density function of X
$f(x, y)$	joint probability density function of X and Y
$f_{X\|Y}(x\|y)$	conditional density function of X given $Y = y$
$G_X(s)$	probability generating function of X
$\boldsymbol{I}$	identity matrix
I_A	indicator function of the event A
$k(s, t)$	covariance function
$K_t(x, y)$	Markov transition kernel
λ_*	second largest eigenvalue
$m_X(t)$	moment generating function of X
N_t	Poisson process
N_A	spatial Poisson process ($A \subseteq \mathbb{R}^d$)
ω	simple outcome of a sample space
Ω	sample space
$\boldsymbol{\pi}$	stationary distribution of Markov chain
$\boldsymbol{P}$	transition matrix for Markov chain
$\boldsymbol{P}(t)$	transition function for continuous-time Markov chain
$\boldsymbol{Q}$	generator matrix
$\mathbb{R}$	real numbers
$\mathbb{R}^2$	two-dimensional Euclidean space
$\mathbb{R}^n$	n-dimensional Euclidean space
$\mathcal{S}$	state space for Markov chain
$SD(X)$	standard deviation of X
T_a	first hitting time of a
$U_{(1)}, \ldots, U_{(n)}$	order statistics
$v(t)$	total variation distance
$Var(X)$	variance of X
$Var(X\|Y)$	conditional variance of X given Y

NOTATION CONVENTIONS

Matrices are represented by bold, capital letters, for example, $\boldsymbol{M}, \boldsymbol{P}$.

Vectors are represented by bold, lowercase letters, for example, $\boldsymbol{\alpha}$, $\boldsymbol{\lambda}$, $\boldsymbol{\pi}$. When vectors are used to represent discrete probability distributions they are row vectors.

ABBREVIATIONS

cdf	cumulative distribution function
gcd	greatest common divisor
i.i.d.	independent and identically distributed
MCMC	Markov chain Monte Carlo
mgf	moment-generating function
pdf	probability density function
pgf	probability generating function
pmf	probability mass function
sde	stochastic differential equation
slln	strong law of large numbers
wlln	weak law of large numbers

ABOUT THE COMPANION WEBSITE

This book is accompanied by a companion website:
http://www.people.carleton.edu/~rdobrow/stochbook/

The website includes:

- Solutions manual available to instructors.
- R script files
- Errata

1

INTRODUCTION AND REVIEW

We demand rigidly defined areas of doubt and uncertainty!
–Douglas Adams, *The Hitchhiker's Guide to the Galaxy*

1.1 DETERMINISTIC AND STOCHASTIC MODELS

Probability theory, the mathematical science of uncertainty, plays an ever growing role in how we understand the world around us—whether it is the climate of the planet, the spread of an infectious disease, or the results of the latest news poll.

The word "stochastic" comes from the Greek *stokhazesthai*, which means to aim at, or guess at. A stochastic process, also called a random process, is simply one in which outcomes are uncertain. By contrast, in a deterministic system there is no randomness. In a deterministic system, the same output is always produced from a given input.

Functions and differential equations are typically used to describe deterministic processes. Random variables and probability distributions are the building blocks for stochastic systems.

Consider a simple exponential growth model. The number of bacteria that grows in a culture until its food source is exhausted exhibits exponential growth. A common

Introduction to Stochastic Processes with R, First Edition. Robert P. Dobrow.

deterministic growth model is to assert that the population of bacteria grows at a fixed rate, say 20% per minute. Let $y(t)$ denote the number of bacteria present after t minutes. As the growth rate is proportional to population size, the model is described by the differential equation

$$\frac{dy}{dt} = (0.20)y.$$

The equation is solved by the exponential function

$$y(t) = y_0 e^{(0.20)t},$$

where $y_0 = y(0)$ is the initial size of the population.

As the model is deterministic, bacteria growth is described by a function, and no randomness is involved. For instance, if there are four bacteria present initially, then after 15 minutes, the model asserts that the number of bacteria present is

$$y(15) = 4e^{(0.20)15} = 80.3421 \approx 80.$$

The deterministic model does not address the uncertainty present in the reproduction rate of individual organisms. Such uncertainty can be captured by using a stochastic framework where the times until bacteria reproduce are modeled by random variables. A simple stochastic growth model is to assume that the times until individual bacteria reproduce are independent exponential random variables, in this case with rate parameter 0.20. In many biological processes, the exponential distribution is a common choice for modeling the times of *births* and *deaths*.

In the deterministic model, when the population size is n, the number of bacteria increases by $(0.20)n$ in 1 minute. Similarly, for the stochastic model, after n bacteria arise the time until the next bacteria reproduces has an exponential probability distribution with rate $(0.20)n$ per minute. (The stochastic process here is called a *birth process*, which is introduced in Chapter 7.)

While the outcome of a deterministic system is fixed, the outcome of a stochastic process is uncertain. See Figure 1.1 to compare the graph of the deterministic exponential growth function with several possible outcomes of the stochastic process.

The dynamics of a stochastic process are described by random variables and probability distributions. In the deterministic growth model, one can say with certainty how many bacteria are present after t minutes. For the stochastic model, questions of interest might include:

- What is the *average* number of bacteria present at time t?
- What is the *probability* that the number of bacteria will exceed some threshold after t minutes?
- What is the *distribution* of the time it takes for the number of bacteria to double in size?

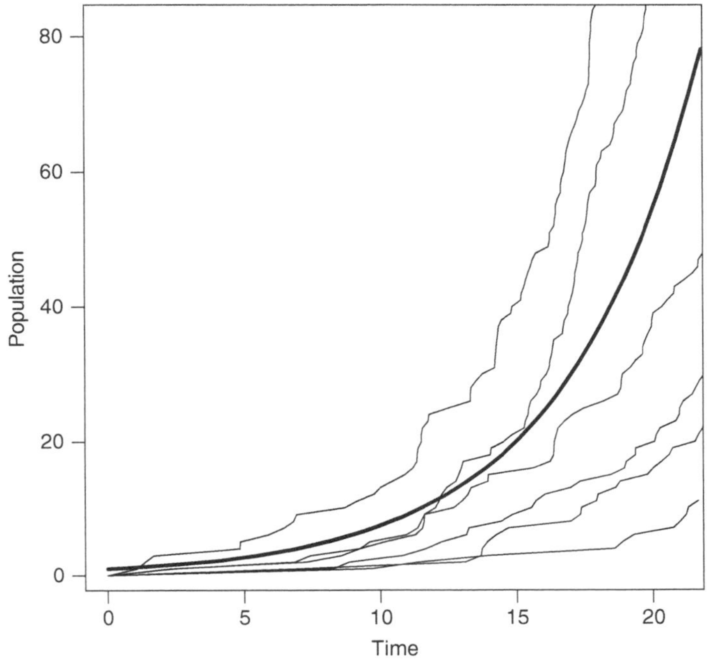

Figure 1.1 Growth of a bacteria population. The deterministic exponential growth curve (dark line) is plotted against six realizations of the stochastic process.

In more sophisticated stochastic growth models, which allow for births and deaths, one might be interested in the likelihood that the population goes extinct, or reaches a long-term equilibrium.

In all cases, conclusions are framed using probability with the goal of quantifying the uncertainty in the system.

Example 1.1 (PageRank) The power of internet search engines lies in their ability to respond to a user's query with an ordered list of web sites ranked by importance and relevance. The heart of Google's search engine is the PageRank algorithm, which assigns an *importance value* to each web page, called its *page rank*. The algorithm is remarkable given the massiveness of the web with over one trillion web pages, and is an impressive achievement of mathematics, particularly linear algebra.

Although the actual PageRank algorithm is complex with many technical (and secret) details, the page rank of a particular web page is easily described by means of a stochastic model. Consider a hypothetical web surfer who travels across the internet moving from page to page at random. When the surfer is on a particular web page, they pick one of the available hypertext links on that page uniformly at random and then move to that page.

The model can be described as a random walk by the web surfer on a giant graph called the *webgraph*. In the webgraph, vertices (nodes) are web pages. Vertex x is joined to vertex y by a directed edge if there is a hypertext link on page x that leads to page y. When the surfer is at vertex x, they choose an edge leading away from x uniformly at random from the set of available edges, and move to the vertex which that edge points to.

The random surfer model is an example of a more general stochastic process called *random walk on a graph*.

Imagine that the web surfer has been randomly walking across the web for a long, long time. What is the probability that the surfer will be at, say, page x? To make this more precise, let p_x^k denote the probability that the surfer is at page x after k steps. The long-term probability of being at page x is defined as $\lim_{k\to\infty} p_x^k$.

This long-term probability is precisely the page rank of page x. Intuitively, the long-term probability of being at a particular page will tend to be higher for pages with more incoming links and smaller for pages with few links, and is a measure of the importance, or popularity, of a page. The PageRank algorithm can be understood as an assignment of probabilities to each site on the web.

Figure 1.2 shows a simplified network of five pages. The numbers under each vertex label are the long-term probabilities of reaching that vertex, and thus the page rank assigned to that page.

Many stochastic processes can be expressed as random walks on graphs in discrete time, or as the limit of such walks in continuous time. These models will play a central role in this book. ■

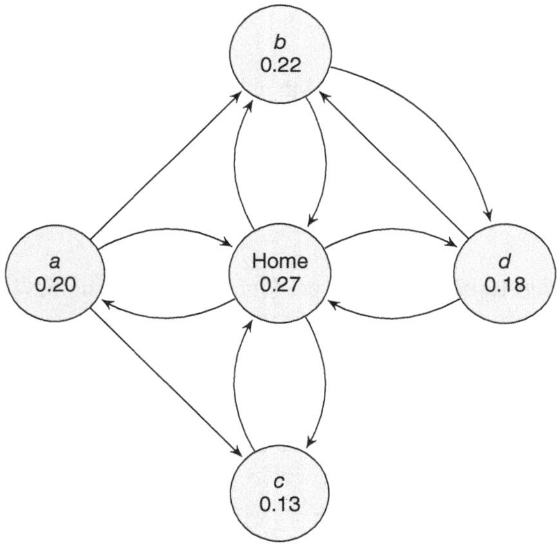

Figure 1.2 Five-page webgraph. Vertex labels show long-term probabilities of reaching each page.

Example 1.2 (Spread of infectious diseases) Models for the spread of infectious diseases and the development of epidemics are of interest to health scientists, epidemiologists, biologists, and public health officials. Stochastic models are relevant because of the randomness inherent in person-to-person contacts and population fluctuations.

The SIR (Susceptible–Infected–Removed) model is a basic framework, which has been applied to the spread of measles and other childhood diseases. At time t, let S_t represent the number of people susceptible to a disease, I_t the number infected, and R_t the number recovered and henceforth immune from infection. Individuals in the population transition from being susceptible to possibly infected to recovered ($S \to I \to R$).

The deterministic SIR model is derived by a system of three nonlinear differential equations, which model interactions and the rate of change in each subgroup.

A stochastic SIR model in discrete time was introduced in the 1920s by medical researchers Lowell Reed and Wade Frost from Johns Hopkins University. In the Reed–Frost model, when a susceptible individual comes in contact with someone who is infected there is a fixed probability z that they will be infected.

Assume that each susceptible person is in contact with all those who are infected. Let p be the probability that a susceptible individual is infected at time t. This is equal to 1 minus the probability that the person is not infected at time t, which occurs if they are not infected by any of the already infected persons, of which there are I_t. This gives

$$p = 1 - (1 - z)^{I_t}.$$

Disease evolution is modeled in discrete time, where one time unit is the incubation period—also the recovery time—of the disease.

The model can be described with a coin-flipping analogy. To find I_{t+1}, the number of individuals infected at time $t + 1$, flip S_t coins (one for each susceptible), where the probability of heads for each coin is the infection probability p. Then, the number of newly infected individuals is the number of coins that land heads.

The number of heads in n independent coin flips with heads probability p has a binomial distribution with parameters n and p. In other words, I_{t+1} has a binomial distribution with $n = S_t$ and $p = 1 - (1 - z)^{I_t}$.

Having found the number of infected individuals at time $t + 1$, the number of susceptible persons decreases by the number of those infected. That is,

$$S_{t+1} = S_t - I_{t+1}.$$

Although the Reed–Frost model is not easy to analyze exactly, it is straightforward to simulate on a computer. The graphs in Figure 1.3 were obtained by simulating the process assuming an initial population of 3 infected and 400 susceptible individuals, with individual infection probability $z = 0.004$. The number of those infected is plotted over 20 time units. ■

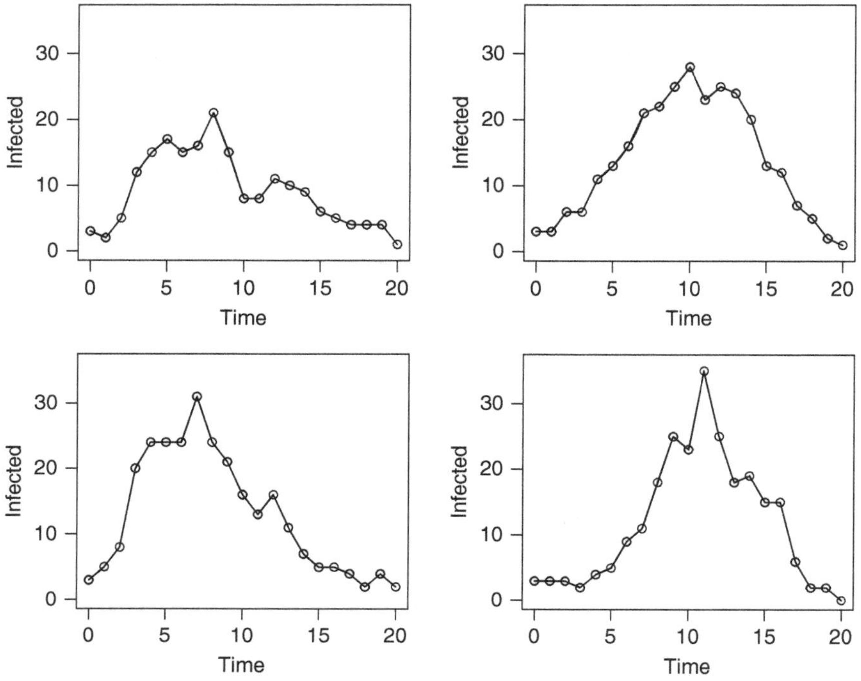

Figure 1.3 Four outcomes of the Reed–Frost epidemic model.

1.2 WHAT IS A STOCHASTIC PROCESS?

In its most general expression, a stochastic process is simply a collection of random variables $\{X_t, t \in I\}$. The index t often represents time, and the set I is the *index set* of the process. The most common index sets are $I = \{0, 1, 2, \ldots\}$, representing discrete time, and $I = [0, \infty)$, representing continuous time. Discrete-time stochastic processes are sequences of random variables. Continuous-time processes are uncountable collections of random variables.

The random variables of a stochastic process take values in a common *state space* $\mathcal{S}$, either discrete or continuous. A stochastic process is specified by its index and state spaces, and by the dependency relations among its random variables.

Stochastic Process

A stochastic process is a collection of random variables $\{X_t, t \in I\}$. The set I is the *index set* of the process. The random variables are defined on a common *state space* $\mathcal{S}$.

Example 1.3 (Monopoly) The popular board game *Monopoly* can be modeled as a stochastic process. Let $X_0, X_1, X_2 \ldots$ represent the successive board positions of an individual player. That is, X_k is the player's board position after k plays.

The state space is $\{1, \ldots, 40\}$ denoting the 40 squares of a Monopoly board—from Go to Boardwalk. The index set is $\{0, 1, 2, \ldots\}$ Both the index set and state space are discrete.

An interesting study is to rank the squares of the board in increasing order of probability. Which squares are most likely to be landed on?

Using Markov chain methods (discussed in Chapter 2), Stewart (1996) shows that the most landed-on square is Jail. The next most frequented square is Illinois Avenue, followed by Go, whereas the least frequented location on the board is the third Chance square from Go. ■

Example 1.4 (Discrete time, continuous state space) An air-monitoring station in southern California records oxidant concentration levels every hour in order to monitor smog pollution. If it is assumed that hourly concentration levels are governed by some random mechanism, then the station's data can be considered a realization of a stochastic process $X_0, X_1, X_2, \ldots$, where X_k is the oxidant concentration level at the kth hour. The time variable is discrete. Since concentration levels take a continuum of values, the state space is continuous. ■

Example 1.5 (Continuous time, discrete state space) Danny receives text messages at random times day and night. Let X_t be the number of texts he receives up to time t. Then, $\{X_t, t \in [0, \infty)\}$ is a continuous-time stochastic process with discrete state space $\{0, 1, 2, \ldots\}$.

This is an example of an *arrival process*. If we assume that the times between Danny's texts are independent and identically distributed (i.i.d.) exponential random variables, we obtain a *Poisson process*. The Poisson process arises in myriad settings involving random *arrivals*. Examples include the number of births each day on a maternity ward, the decay of a radioactive substance, and the occurrences of oil spills in a harbor. ■

Example 1.6 (Random walk and gambler's ruin) A random walker starts at the origin on the integer line. At each discrete unit of time the walker moves either right or left, with respective probabilities p and $1 - p$. This describes a *simple random walk* in one dimension.

A stochastic process is built as follows. Let $X_1, X_2, \ldots$ be a sequence of i.i.d. random variables with

$$X_k = \begin{cases} +1, & \text{with probability } p, \\ -1, & \text{with probability } 1 - p, \end{cases}$$

for $k \geq 1$. Set

$$S_n = X_1 + \cdots + X_n, \text{ for } n \geq 1,$$

with $S_0 = 0$. Then, S_n is the random walk's position after n steps. The sequence $S_0, S_1, S_2, \ldots$ is a discrete-time stochastic process whose state space is $\mathbb{Z}$, the set of all integers.

Consider a gambler who has an initial stake of k dollars, and repeatedly wagers \$1 on a game for which the probability of winning is p and the probability of losing is $1 - p$. The gambler's successive fortunes is a simple random walk started at k.

Assume that the gambler decides to stop when their fortune reaches \$$n$ $(n > k)$, or drops to 0, whichever comes first. What is the probability that the gambler is eventually ruined? This is the classic gambler's ruin problem, first discussed by mathematicians Blaise Pascal and Pierre Fermat in 1656.

See Figure 1.4 for simulations of gambler's ruin with $k = 20$, $n = 60$, and $p = 1/2$. Observe that four of the nine outcomes result in the gambler's ruin before 1,000 plays. In the next section, it is shown that the probability of eventual ruin is $(n - k)/n = (60 - 20)/60 = 2/3$. ■

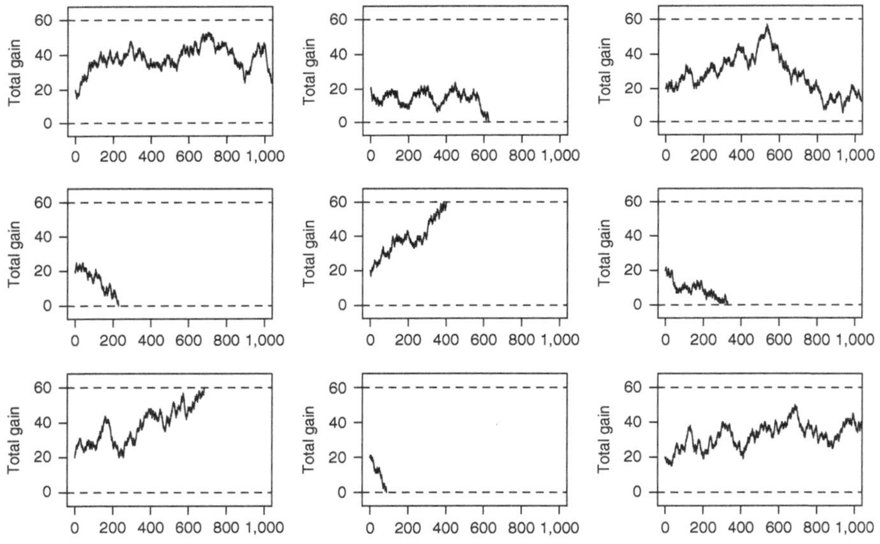

Figure 1.4 Random walk and gambler's ruin.

Example 1.7 (Brownian motion) Brownian motion is a continuous-time, continuous state space stochastic process. The name also refers to a physical process, first studied by the botanist Robert Brown in 1827. Brown observed the seemingly erratic, zigzag motion of tiny particles ejected from pollen grains suspended in water. He gave a detailed study of the phenomenon but could not explain its cause. In 1905, Albert Einstein showed that the motion was the result of water molecules bombarding the particles.

The mathematical process known as Brownian motion arises as the *limiting process* of a discrete-time random walk. This is obtained by *speeding up* the walk, letting the interval between discrete steps tend to 0. The process is used as a model for many phenomena that exhibit "erratic, zigzag motion," such as stock prices, the growth of crystals, and signal noise.

Brownian motion has remarkable properties, which are explored in Chapter 8. Paths of the process are continuous everywhere, yet differentiable nowhere. Figure 1.5 shows simulations of two-dimensional Brownian motion. For this case, the index set is $[0, \infty)$ and the state space is $\mathbb{R}^2$. ∎

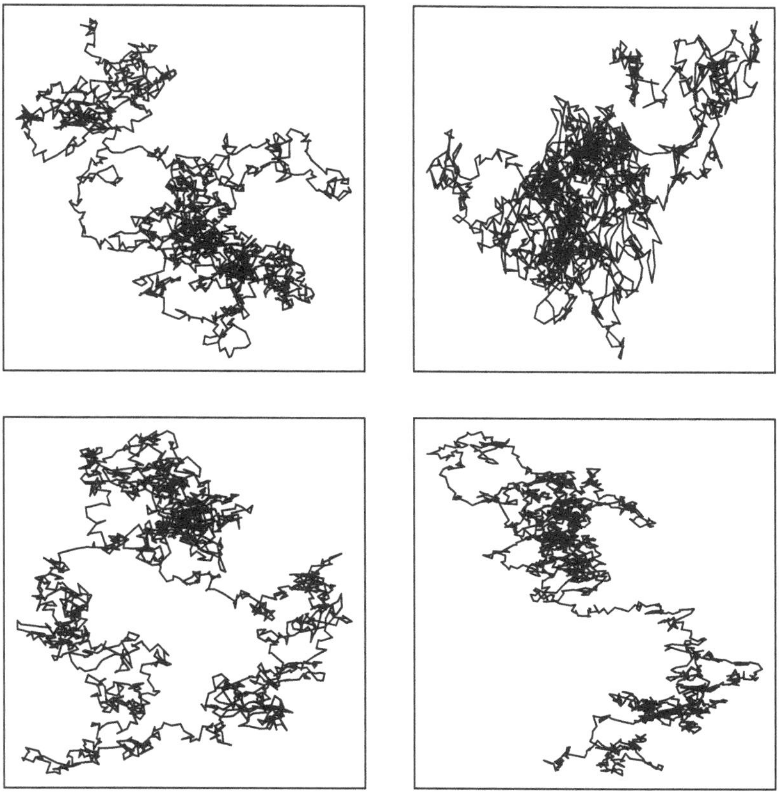

Figure 1.5 Simulations of two-dimensional Brownian motion.

1.3 MONTE CARLO SIMULATION

Advancements in modern computing have revolutionized the study of stochastic systems, allowing for the visualization and simulation of increasingly complex models.

At the heart of the many simulation techniques developed to generate random variables and stochastic processes lies the Monte Carlo method. Given a random experiment and event A, a Monte Carlo estimate of $P(A)$ is obtained by repeating the random experiment many times and taking the proportion of trials in which A occurs as an approximation for $P(A)$.

The name Monte Carlo evidently has its origins in the fact that the mathematician Stanislaw Ulam, who developed the method in 1946, had an uncle who regularly gambled at the Monte Carlo casino in Monaco.

Monte Carlo simulation is intuitive and matches up with our sense of how probabilities *should* behave. The relative frequency interpretation of probability says that the probability of an event is the long-term proportion of times that the event occurs in repeated trials. It is justified theoretically by the strong law of large numbers.

Consider repeated independent trials of a random experiment. Define the sequence $X_1, X_2, \ldots$, where

$$X_k = \begin{cases} 1, & \text{if } A \text{ occurs on the } k\text{th trial,} \\ 0, & \text{if } A \text{ does not occur on the } k\text{th trial,} \end{cases}$$

for $k \geq 1$. Then, $(X_1 + \cdots + X_n)/n$ is the proportion of n trials in which A occurs. The X_k are identically distributed with common mean $E(X_k) = P(A)$.

By the strong law of large numbers,

$$\lim_{n\to\infty} \frac{X_1 + \cdots + X_n}{n} = P(A), \text{ with probability 1.} \tag{1.1}$$

For large n, the Monte Carlo estimate of $P(A)$ is

$$P(A) \approx \frac{X_1 + \cdots + X_n}{n}.$$

In this book, we use the software package R for simulation. R is a flexible and interactive environment. We often use R to illustrate the result of an exact, theoretical calculation with numerical verification. The easy-to-learn software allows the user to see the impact of varying parameters and assumptions of the model. For example, in the Reed–Frost epidemic model of Example 1.2, it is interesting to see how small changes in the infection probability affect the duration and intensity of the epidemic. See the R script file **ReedFrost.R** and Exercise 1.36 to explore this question.

If you have not used R before, work through the exercises in the introductory tutorial in Appendix A: Getting Started with R.

1.4 CONDITIONAL PROBABILITY

The simplest stochastic process is a sequence of i.i.d. random variables. Such sequences are often used to model random samples in statistics. However, most real-world systems exhibit some type of dependency between variables, and an independent sequence is often an unrealistic model.

Thus, the study of stochastic processes really begins with conditional probability—conditional distributions and conditional expectation. These will become essential tools for all that follows.

Starting with a random experiment, the sample space Ω is the set of all possible outcomes. An *event* is a subset of the sample space. For events A and B, the *conditional probability of A given B* is

$$P(A|B) = \frac{P(A \cap B)}{P(B)},$$

defined for $P(B) > 0$. Events A and B are *independent* if $P(A|B) = P(A)$. Equivalently, A and B are independent if

$$P(A \cap B) = P(A)P(B).$$

Events that are not independent are said to be *dependent*.

For many problems where the goal is to find $P(A)$, partial information and dependencies between events in the sample space are brought to bear. If the sample space can be partitioned into a collection of disjoint events $B_1, \ldots, B_k$, then A can be expressed as the disjoint union

$$A = (A \cap B_1) \cup \cdots \cup (A \cap B_k).$$

If conditional probabilities of the form $P(A|B_i)$ are known, then the law of total probability can be used to find $P(A)$.

Law of Total Probability

Let $B_1, \ldots, B_k$ be a sequence of events that partition the sample space. That is, the B_i are mutually exclusive (disjoint) and their union is equal to Ω. Then, for any event A,

$$P(A) = \sum_{i=1}^{k} P(A \cap B_i) = \sum_{i=1}^{k} P(A|B_i)P(B_i).$$

Example 1.8 According to the Howard Hughes Medical Institute, about 7% of men and 0.4% of women are colorblind—either cannot distinguish red from green or see red and green differently from most people. In the United States, about 49% of the population is male and 51% female. A person is selected at random. What is the probability they are colorblind?

Solution Let C, M, and F denote the events that a random person is colorblind, male, and female, respectively. By the law of total probability,

$$\begin{aligned} P(C) &= P(C|M)P(M) + P(C|F)P(F) \\ &= (0.07)(0.49) + (0.004)(0.51) = 0.03634. \end{aligned}$$

■

Using the law of total probability in this way is called *conditioning*. Here, we find the *total probability* of being colorblind by conditioning on sex.

Example 1.9 In a standard deck of cards, the probability that the suit of a random card is hearts is $13/52 = 1/4$. Assume that a standard deck has one card missing. A card is picked from the deck. Find the probability that it is a heart.

Solution Assume that the missing card can be any of the 52 cards picked uniformly at random. Let M denote the event that the missing card is a heart, with the complement M^c the event that the missing card is not a heart. Let H denote the event that the card that is picked from the deck is a heart. By the law of total probability,

$$\begin{aligned} P(H) &= P(H|M)P(M) + P(H|M^c)P(M^c) \\ &= \left(\frac{12}{51}\right)\frac{1}{4} + \left(\frac{13}{51}\right)\frac{3}{4} = \frac{1}{4}. \end{aligned}$$

The result can also be obtained by appealing to symmetry. Since all cards are equally likely, and all four suits are equally likely, the argument by symmetry gives that the desired probability is $1/4$. ∎

Example 1.10 (Gambler's ruin) The gambler's ruin problem was introduced in Example 1.6. A gambler starts with k dollars. On each play a fair coin is tossed and the gambler wins \$1 if heads occurs, or loses \$1 if tails occurs. The gambler stops when he reaches $\$n$ $(n > k)$ or loses all his money. Find the probability that the gambler will eventually lose.

Solution We make two observations, which are made more precise in later chapters. First, the gambler will eventually stop playing, either by reaching n or by reaching 0. One might argue that the gambler could play forever. However, it can be shown that that event occurs with probability 0. Second, assume that after, say, 100 wagers, the gambler's capital returns to $\$k$. Then, the probability of eventually winning $\$n$ is the same as it was initially. The memoryless character of the process means that the probability of winning $\$n$ or losing all his money only depends on how much capital the gambler has, and not on how many previous wagers the gambler made.

Let p_k denote the probability of reaching n when the gambler's fortune is k. What is the gambler's status if heads is tossed? Their fortune increases to $k+1$ and the probability of winning is the same as it would be if the gambler had started the game with $k+1$. Similarly, if tails is tossed and the gambler's fortune decreases to $k-1$. Hence,

$$p_k = p_{k+1}\left(\frac{1}{2}\right) + p_{k-1}\left(\frac{1}{2}\right),$$

or

$$p_{k+1} - p_k = p_k - p_{k-1}, \quad \text{for} \quad k = 1,\ldots,n-1, \tag{1.2}$$

with $p_0 = 0$ and $p_n = 1$. Unwinding the recurrence gives

$$p_k - p_{k-1} = p_{k-1} - p_{k-2} = p_{k-2} - p_{k-3} = \cdots = p_1 - p_0 = p_1,$$

for $k = 1, \ldots, n$. We have that $p_2 - p_1 = p_1$, giving $p_2 = 2p_1$. Also, $p_3 - p_2 = p_3 - 2p_1 = p_1$, giving $p_3 = 3p_1$. More generally, $p_k = kp_1$, for $k = 1, \ldots, n$.

Sum Equation (1.2) over suitable k to obtain

$$\sum_{k=1}^{n-1}(p_{k+1} - p_k) = \sum_{k=1}^{n-1}(p_k - p_{k-1}).$$

Both sums telescope to

$$p_n - p_1 = p_{n-1} - p_0,$$

which gives $1 - p_1 = p_{n-1} = (n-1)p_1$, so $p_1 = 1/n$. Thus,

$$p_k = kp_1 = \frac{k}{n}, \text{ for } k = 0, \ldots, n.$$

The probability that the gambler eventually wins \$$n$ is k/n. Hence, the probability of the gambler's ruin is $(n-k)/n$. ■

R : Simulating Gambler's Ruin

The file **gamblersruin.R** contains the function `gamble(k,n,p)`, which simulates the gambler's ruin process. At each wager, the gambler wins with probability p, and loses with probability $1 - p$. The gambler's initial stake is \$$k$. The function `gamble` returns 1, if the gambler is eventually ruined, or 0, if the gambler gains \$$n$.

In the simulation the function is called 1,000 times, creating a list of 1,000 ruins and wins, which are represented by 1s and 0s. The mean of the list gives the proportion of 1s, which estimates the probability of eventual ruin.

```
> k <- 20
> n <- 60
> p <- 1/2
> trials <- 1000
> simlist <- replicate(trials, gamble(k,n,p))
> mean(simlist) # Estimate of probability of ruin
[1] 0.664
# Exact probability of ruin is 2/3
```

Sometimes, we need to find a conditional probability of the form $P(B|A)$, but what is given in the problem are *reverse* probabilities of the form $P(A|B)$ and $P(A|B^c)$. Bayes' rule provides a method for *inverting* the conditional probability.

Bayes' Rule

For events A and B,

$$P(B|A) = \frac{P(A|B)P(B)}{P(A|B)P(B) + P(A|B^c)P(B^c)}.$$

Bayes' rule is a consequence of the law of total probability and the definition of conditional probability, as

$$P(B|A) = \frac{P(A \cap B)}{P(A)} = \frac{P(A|B)P(B)}{P(A)}.$$

For any event B, the events B and B^c partition the sample space. Given a countable sequence of events $B_1, B_2, \ldots$, which partition the sample space, a more general form of Bayes' rule is

$$P(B_i|A) = \frac{P(A|B_i)P(B_i)}{\sum_j P(A|B_j)P(B_j)}, \quad \text{for} \quad i = 1, 2, \ldots$$

Example 1.11 The use of polygraphs (lie detectors) is controversial, and many scientists feel that they should be banned. On the contrary, some polygraph advocates claim that they are mostly accurate. In 1998, the Supreme Court (*United States v. Sheffer*) supported the right of state and federal governments to bar polygraph evidence in court.

Assume that one person in a company of 100 employees is a thief. To find the thief the company will administer a polygraph test to all its employees. The lie detector has the property that if a subject is a liar, there is a 95% probability that the polygraph will detect that they are lying. However, if the subject is telling the truth, there is a 10% chance the polygraph will report a *false positive* and assert that the subject is lying.

Assume that a random employee is given the polygraph test and asked whether they are the thief. The employee says "no," and the lie detector reports that they are lying. Find the probability that the employee is in fact lying.

Solution Let L denote the event that the employee is a liar. Let D denote the event that the lie detector reports that the employee is a liar. The desired probability is $P(L|D)$. By Bayes' rule,

$$\begin{aligned} P(L|D) &= \frac{P(D|L)P(L)}{P(D|L)P(L) + P(D|L^c)P(L^c)} \\ &= \frac{(0.95)(0.01)}{(0.95)(0.01) + (0.10)(0.99)} = 0.088. \end{aligned}$$

There is less than a 10% chance that the employee is in fact the thief!

Many people, when first given this problem and asked to guess the answer, choose a probability closer to 90%. The mistake is a consequence of confusing the conditional probabilities $P(L|D)$ and $P(D|L)$, the probability that the individual is a liar, given that the polygraph says they are, with the probability that the polygraph says they are lying, given that they are a liar. Since the population of truth tellers is relatively big, the number of false positives—truth tellers whom the lie detector falsely records as being a liar—is also significant. In this case, about 10% of 99, or about 10 employees will be false positives. Assuming that the lie detector correctly identifies the thief as lying, there will be about 11 employees who are identified as liars by the polygraph. The probability that one of them chosen at random is in fact the thief is only about 1/11. ■

Conditional Distribution

The *distribution* of a random variable X refers to the set of values of X and their corresponding probabilities. The distribution of a random variable is specified with either a probability mass function (pmf), if X is discrete, or a probability density function (pdf), if X is continuous.

For more than one random variable, there is a joint distribution, specified by either a joint pmf or a joint pdf.

If X and Y are discrete random variables, their joint pmf is $P(X = x, Y = y)$, considered a function of x and y. If X and Y are continuous, the joint density function $f(x, y)$ satisfies

$$P(X \leq x, Y \leq y) = \int_{-\infty}^{x} \int_{-\infty}^{y} f(s, t)\, dt\, ds,$$

for all $x, y \in \mathbb{R}$.

For jointly distributed random variables X and Y, the *conditional distribution of Y given $X = x$* is specified by either a conditional pmf or a conditional pdf.

Discrete Case

The *conditional pmf of Y given $X = x$* is

$$P(Y = y|X = x) = \frac{P(X = x, Y = y)}{P(X = x)},$$

defined when $P(X = x) > 0$. The conditional pmf is a function of y, where x is treated as fixed.

Example 1.12 Max chooses an integer X uniformly at random between 1 and 100. If $X = x$, Mary then chooses an integer Y uniformly at random between 1 and x. Find the conditional pmf of Y given $X = x$.

Solution By the structure of this two-stage random experiment, the conditional distribution of Y given $X = x$ is uniform on $\{1,\ldots,x\}$. Thus, the conditional pmf is

$$P(Y = y|X = x) = \frac{1}{x}, \text{ for } y = 1,\ldots,x.$$

■

Note that the conditional pmf *is* a probability function. For fixed x, the probabilities $P(Y = y|X = x)$ are nonnegative and sum to 1, as

$$\sum_y P(Y = y|X = x) = \sum_y \frac{P(X = x, Y = y)}{P(X = x)} = \frac{P(X = x)}{P(X = x)} = 1.$$

Example 1.13 The joint pmf of X and Y is

$$P(X = x, Y = y) = \frac{x + y}{18}, \quad \text{for } x, y = 0, 1, 2.$$

Find the conditional pmf of Y given $X = x$.

Solution The marginal distribution of X is

$$P(X = x) = \sum_{y=0}^{2} P(X = x, Y = y) = \frac{x}{18} + \frac{x+1}{18} + \frac{x+2}{18} = \frac{x+1}{6},$$

for $x = 0, 1, 2$. The conditional pmf is

$$P(Y = y|X = x) = \frac{P(X = x, Y = y)}{P(X = x)} = \frac{(x + y)/18}{(x + 1)/6} = \frac{x + y}{3(x + 1)},$$

for $y = 0, 1, 2$. ■

Example 1.14 A bag contains 2 red, 3 blue, and 4 white balls. Three balls are picked from the bag (sampling without replacement). Let B be the number of blue balls picked. Let R be the number of red balls picked. Find the conditional pmf of B given $R = 1$.

Solution We have

$$\begin{aligned} P(B = b|R = 1) &= \frac{P(B = b, R = 1)}{P(R = 1)} \\ &= \frac{\binom{3}{b}\binom{2}{1}\binom{4}{3-b-1}\Big/\binom{9}{3}}{\binom{2}{1}\binom{7}{2}\Big/\binom{9}{3}} = \frac{2\binom{3}{b}\binom{4}{2-b}}{42} \end{aligned}$$

$$= \frac{\binom{3}{b}\binom{4}{2-b}}{21} = \begin{cases} 2/7, & \text{if } b = 0, \\ 4/7, & \text{if } b = 1, \\ 1/7, & \text{if } b = 2. \end{cases}$$

■

Continuous Case

For continuous random variables X and Y, the *conditional density function of* Y *given* $X = x$ is

$$f_{Y|X}(y|x) = \frac{f(x, y)}{f_X(x)},$$

where f_X is the marginal density function of X. The conditional density is a function of y, where x is treated as fixed.

Conditional densities are used to compute conditional probabilities. For $R \subseteq \mathbb{R}$,

$$P(Y \in R|X = x) = \int_R f_{Y|X}(y|x)\, dy.$$

Example 1.15 Random variables X and Y have joint density

$$f(x, y) = e^{-x}, \text{for } 0 < y < x < \infty.$$

Find $P(Y < 2|X = 5)$.

Solution The desired probability is

$$P(Y < 2|X = 5) = \int_0^2 f_{Y|X}(y|5)\, dy.$$

To find the conditional density function $f_{Y|X}(y|x)$, find the marginal density

$$f_X(x) = \int_{-\infty}^{\infty} f(x, y)\, dy = \int_0^x e^{-x}\, dy = xe^{-x}, \quad \text{for} \quad x > 0.$$

This gives

$$f_{Y|X}(y|x) = \frac{f(x, y)}{f_X(x)} = \frac{e^{-x}}{xe^{-x}} = \frac{1}{x}, \quad \text{for} \quad 0 < y < x.$$

The conditional distribution of Y given $X = x$ is uniform on $(0, x)$. Hence,

$$P(Y < 2|X = 5) = \int_0^2 f_{Y|X}(y|5)\, dy = \int_0^2 \frac{1}{5}\, dy = \frac{2}{5}.$$

■

Example 1.16 Tom picks a real number X uniformly distributed on $(0, 1)$. Tom shows his number x to Marisa who then picks a number Y uniformly distributed on $(0, x)$. Find (i) the conditional distribution of Y given $X = x$; (ii) the joint distribution of X and Y; and (iii) the marginal density of Y.

Solution

(i) The conditional distribution of Y given $X = x$ is given directly in the statement of the problem. The distribution is uniform on $(0, x)$. Thus,

$$f_{Y|X}(y|x) = \frac{1}{x}, \quad \text{for} \quad 0 < y < x.$$

(ii) For the joint density,

$$f(x, y) = f_{Y|X}(y|x)f_X(x) = \frac{1}{x}(1) = \frac{1}{x}, \quad \text{for} \quad 0 < y < x < 1.$$

(iii) To find the marginal density of Y, integrate out the x variable in the joint density function. This gives

$$f_Y(y) = \int_{-\infty}^{\infty} f(x, y)\, dx = \int_y^1 \frac{1}{x}\, dx = -\ln\ y, \quad \text{for} \quad 0 < y < 1.$$

■

1.5 CONDITIONAL EXPECTATION

A conditional expectation is an expectation computed with respect to a conditional probability distribution. Write $E(Y|X = x)$ for the *conditional expectation of Y given $X = x$*.

Conditional Expectation of Y given $X = x$

$$E(Y|X = x) = \begin{cases} \sum_y yP(Y = y|X = x), & \text{discrete,} \\ \int_{-\infty}^{\infty} yf_{Y|X}(y|x)\, dy, & \text{continuous.} \end{cases}$$

Most important is that $E(Y|X = x)$ is a function of x.

Example 1.17 A school cafeteria has two registers. Let X and Y denote the number of students in line at the first and second registers, respectively. The joint pmf of X and Y is specified by the following joint distribution table.

		Y: 0	1	2	3	4
	0	0.15	0.14	0.03	0	0
	1	0.14	0.12	0.06	0.01	0
X	2	0.03	0.06	0.10	0.03	0.02
	3	0	0.01	0.03	0.02	0.01
	4	0	0	0.02	0.01	0.01

Find the expected number of people in line at the second register if there is one person at the first register.

Solution The problem asks for $E(Y|X = 1)$. We have

$$E(Y|X=1) = \sum_{y=0}^{4} yP(Y=y|X=1) = \sum_{y=0}^{4} y\frac{P(X=1, Y=y)}{P(X=1)}.$$

The marginal probability $P(X = 1)$ is obtained by summing over the $X = 1$ row of the table. That is, $P(X = 1) = 0.14 + 0.12 + 0.06 + 0.01 + 0 = 0.33$. Hence,

$$\begin{aligned} E(Y|X=1) &= \frac{1}{0.33}\sum_{y=0}^{4} yP(X=1, Y=y) \\ &= \frac{1}{0.33}[0(0.14) + 1(0.12) + 2(0.06) + 3(0.01) + 4(0)] \\ &= 0.818. \end{aligned}$$

■

Example 1.18 Let X and Y be independent Poisson random variables with respective parameters λ and μ. Find the conditional expectation of Y given $X + Y = n$.

Solution First find the conditional pmf of Y given $X + Y = n$. We use the fact that the sum of independent Poisson random variables has a Poisson distribution whose parameter is the sum of the individual parameters. That is, $X + Y$ has a Poisson distribution with parameter $\lambda + \mu$. This gives

$$\begin{aligned} P(Y=y|X+Y=n) &= \frac{P(Y=y, X+Y=n)}{P(X+Y=n)} = \frac{P(Y=y, X=n-y)}{P(X+Y=n)} \\ &= \frac{P(Y=y)P(X=n-y)}{P(X+Y=n)} \\ &= \frac{(e^{-\mu}\mu^y/y!)(e^{-\lambda}\lambda^{n-y}/(n-y)!)}{e^{-(\lambda+\mu)}(\lambda+\mu)^n/n!} \end{aligned}$$

$$= \frac{n!}{y!(n-y)!}\left(\frac{\mu^y \lambda^{n-y}}{(\lambda+\mu)^n}\right)$$

$$= \binom{n}{y}\left(\frac{\mu}{\lambda+\mu}\right)^y\left(\frac{\lambda}{\lambda+\mu}\right)^{n-y},$$

for $y = 0,\ldots,n$. The form of the conditional pmf shows that the conditional distribution is binomial with parameters n and $p = \mu/(\lambda+\mu)$. The desired conditional expectation is the mean of this binomial distribution. That is,

$$E(Y|X+Y=n) = np = \frac{n\mu}{\lambda+\mu}.$$

■

Example 1.19 Assume that X and Y have joint density

$$f(x,y) = \frac{2}{xy}, \text{ for } 1 < y < x < e.$$

Find $E(Y|X=x)$.

Solution The marginal density of X is

$$f_X(x) = \int_1^x \frac{2}{xy}\, dy = \frac{2\ln x}{x}, \text{ for } 1 < x < e.$$

The conditional density of Y given $X = x$ is

$$f_{Y|X}(y|x) = \frac{f(x,y)}{f_X(x)} = \frac{2/(xy)}{2\ln x/x} = \frac{1}{y\ln x}, \text{ for } 1 < y < x,$$

with conditional expectation

$$E(Y|X=x) = \int_1^x y f_{Y|X}(y|x)\, dy = \int_1^x \frac{y}{y\ln x}\, dy = \frac{x-1}{\ln x}.$$

■

Key properties of conditional expectation follow.

Properties of Conditional Expectation

1. (Linearity) For constants a and b and random variables X, Y, and Z,

$$E(aY+bZ|X=x) = aE(Y|X=x) + bE(Z|X=x).$$

2. If g is a function,

$$E(g(Y)|X = x) = \begin{cases} \sum_y g(y)P(Y = y|X = x), & \text{discrete,} \\ \int_{-\infty}^{\infty} g(y)f_{Y|X}(y|x)\, dy, & \text{continuous.} \end{cases}$$

3. (Independence) If X and Y are independent,

$$E(Y|X = x) = E(Y).$$

4. If $Y = g(X)$ is a function of X,

$$E(Y|X = x) = g(x).$$

Proof. Properties 1 and 2 are consequences of the fact that conditional expectation *is* an expectation and thus retains all the properties, such as linearity, of the regular expectation. For a proof of property 2, which is sometimes called *the law of the unconscious statistician*, see Dobrow (2013).

For the independence property in the discrete case, if X and Y are independent,

$$E(Y|X = x) = \sum_y yP(Y = y|X = x) = \sum_y yP(Y = y) = E(Y).$$

The continuous case is similar. For property 4,

$$E(Y|X = x) = E(g(X)|X = x) = E(g(x)|X = x) = g(x).$$

■

Example 1.20 Consider random variables X, Y, and U, where U is uniformly distributed on $(0, 1)$. Find the conditional expectation

$$E\left(UX^2 + (1 - U)Y^2|U = u\right).$$

Solution By linearity of conditional expectation,

$$\begin{aligned} E\left(UX^2 + (1 - U)Y^2|U = u\right) &= E\left(uX^2 + (1 - u)Y^2|U = u\right) \\ &= uE\left(X^2|U = u\right) + (1 - u)E\left(Y^2|U = u\right). \end{aligned}$$

If X and Y are also independent of U, the latter expression reduces to

$$uE\left(X^2\right) + (1 - u)E\left(Y^2\right).$$

■

Extending the scope of conditional expectation, we show how to condition on a general event. Given an event A, the *indicator* for A is the 0–1 random variable defined as

$$I_A = \begin{cases} 1, & \text{if } A \text{ occurs,} \\ 0, & \text{if } A \text{ does not occur.} \end{cases}$$

Conditional Expectation Given an Event

Let A be an event such that $P(A) > 0$. The *conditional expectation of Y given A* is

$$E(Y|A) = \frac{E(YI_A)}{P(A)}.$$

For the discrete case, the formula gives

$$E(Y|A) = \frac{1}{P(A)} \sum_y yP(\{Y = y\} \cap A) = \sum_y yP(Y = y|A).$$

Let $A_1, \ldots, A_k$ be a sequence of events that partition the sample space. Observe that

$$I_{A_1} + \cdots + I_{A_k} = 1,$$

since every outcome $\omega \in \Omega$ is contained in exactly one of the A_is. It follows that

$$Y = \sum_{i=1}^{k} YI_{A_i}.$$

Taking expectations gives

$$E(Y) = \sum_{i=1}^{k} E\left(YI_{A_i}\right) = \sum_{i=1}^{k} \left(\frac{E\left(YI_{A_i}\right)}{P(A)}\right) P(A) = \sum_{i=1}^{k} E(Y|A_i)P(A_i).$$

The result is known as the law of total expectation. Note the similarity with the law of total probability.

The law of total expectation is often used with partitioning events $\{X = x\}$. This gives

$$E(Y) = \sum_x E(Y|X = x)P(X = x).$$

In summary, here are two forms of the law of total expectation.

Law of Total Expectation

Let Y be a random variable. Let $A_1, \ldots, A_k$ be a sequence of events that partition the sample space. Then,

$$E(Y) = \sum_{i=1}^{k} E(Y|A_i)P(A_i).$$

If X and Y are jointly distributed random variables,

$$E(Y) = \sum_x E(Y|X = x)P(X = x).$$

Example 1.21 A fair coin is flipped repeatedly. Find the expected number of flips needed to get two heads in a row.

Solution Let Y be the number of flips needed. Consider three events: (i) T, the first flip is tails; (ii) HT, the first flip is heads and the second flip is tails; and (iii) HH, the first two flips are heads. The events T, HT, HH partition the sample space. By the law of total expectation,

$$\begin{aligned} E(Y) &= E(Y|T)P(T) + E(Y|HT)P(HT) + E(Y|HH)P(HH) \\ &= E(Y|T)\frac{1}{2} + E(Y|HT)\frac{1}{4} + (2)\frac{1}{4}. \end{aligned}$$

Consider $E(Y|T)$. Assume that the first flip is tails. Since successive flips are independent, after the first tails we *start over* waiting for two heads in a row. Since one flip has been used, it follows that $E(Y|T) = 1 + E(Y)$. Similarly, after first heads and then tails we start over again, having used up two coin tosses. Thus, $E(Y|HT) = 2 + E(Y)$. This gives

$$E(Y) = (1 + E(Y))\frac{1}{2} + (2 + E(Y))\frac{1}{4} + (2)\frac{1}{4} = E(Y)\frac{3}{4} + \frac{3}{2}.$$

Solving for $E(Y)$ gives $E(Y)(1/4) = 3/2$, or $E(Y) = 6$. ■

Example 1.22 Every day Bob goes to the pizza shop, orders a slice of pizza, and picks a topping—pepper, pepperoni, pineapple, prawns, or prosciutto—uniformly at random. On the day that Bob first picks pineapple, find the expected number of prior days in which he picked pepperoni.

Solution Let Y be the number of days, before the day Bob first picked pineapple, in which he picks pepperoni. Let X be the number of days needed for Bob to first pick pineapple. Then, X has a geometric distribution with parameter $1/5$.

If $X = x$, then on the first $x - 1$ days pineapple was not picked. And for each of these days, given that pineapple was not picked, there was a $1/4$ chance of picking pepperoni. The conditional distribution of Y given $X = x$ is binomial with parameters $x - 1$ and $1/4$. Thus, $E[Y|X = x] = (x - 1)/4$, and

$$E(Y) = \sum_{x=1}^{\infty} E(Y|X = x)P(X = x)$$

$$= \sum_{x=1}^{\infty} \left(\frac{x-1}{4}\right)\left(\frac{4}{5}\right)^{x-1}\frac{1}{5}$$
$$= \left(\frac{1}{4}\sum_{x=1}^{\infty} x\left(\frac{4}{5}\right)^{x-1}\frac{1}{5}\right) - \left(\frac{1}{4}\sum_{x=1}^{\infty}\left(\frac{4}{5}\right)^{x-1}\frac{1}{5}\right)$$
$$= \frac{1}{4}E(X) - \frac{1}{4} = \frac{5}{4} - \frac{1}{4} = 1.$$

■

R : Simulation of Bob's Pizza Probability

```
> trials <- 10000  # simulation repeated 10,000 times
> simlist <- numeric(trials)
> toppings <- c("pepper","pepperoni","pineapple",
   "prawns","prosciutto")
> for (i in 1:trials) {
> pineapple <- 0
> pepperoni <- 0 #counts pepperonis before pineapple
> while (pineapple == 0) {
     # pick toppings until pineapple is selected
   pick <- sample(toppings,1)
   if (pick == "pepperoni") pepperoni <-pepperoni + 1
   if (pick == "pineapple") pineapple <- 1    }
> simlist[i] <- pepperoni }
> mean(simlist)
[1] 0.9966
```

The next example illustrates conditional expectation given an event in the continuous case.

Example 1.23 The time that Joe spends talking on the phone is exponentially distributed with mean 5 minutes. What is the expected length of his phone call if he talks for more than 2 minutes?

Solution Let Y be the length of Joe's phone call. With $A = \{Y > 2\}$, the desired conditional expectation is

$$E(Y|A) = E(Y|Y > 2) = \frac{1}{P(Y > 2)} \int_2^{\infty} y\frac{1}{5}e^{-y/5}\, dy$$
$$= \left(\frac{1}{e^{-2/5}}\right) 7e^{-2/5} = 7 \text{ minutes}.$$

Note that the solution can be obtained using the memoryless property of the exponential distribution. The conditional distribution of Y given $Y > 2$ is equal to the distribution of $2 + Z$, where Z has the same distribution has Y. Thus,

$$E(Y|Y > 2) = E(2 + Z) = 2 + E(Z) = 2 + E(Y) = 2 + 5 = 7.$$

■

Conditioning on a Random Variable

From conditioning on an event, we introduce the notion of conditioning on a random variable, a powerful tool for computing conditional expectations and probabilities.

Recall that if X is a random variable and g is a function, then $Y = g(X)$ is a random variable, which is a function of X. The conditional expectation $E(Y|X = x)$ is a function of x. We apply this function to the random variable X and define a new random variable called the *conditional expectation of Y given X*, written $E(Y|X)$. The defining properties of $E(Y|X)$ are given here.

Conditional Expectation of *Y* given *X*

The conditional expectation $E(Y|X)$ has three defining properties.

1. $E(Y|X)$ is a random variable.
2. $E(Y|X)$ is a function of X.
3. $E(Y|X)$ is equal to $E(Y|X = x)$ whenever $X = x$. That is, if

$$E(Y|X = x) = g(x), \text{ for all } x,$$

then $E(Y|X) = g(X)$.

Example 1.24 Let X be uniformly distributed on $(0, 1)$. If $X = x$, a second number Y is picked uniformly on $(0, x)$. Find $E(Y|X)$.

Solution For this two-stage random experiment, the conditional distribution of Y given $X = x$ is uniform on $(0, x)$, for $0 < x < 1$. It follows that $E(Y|X = x) = x/2$. Since this is true for all x, $E(Y|X) = X/2$. ■

It may seem that the difference between $E(Y|X)$ and $E(Y|X = x)$ is just a matter of notation, with capital letters replacing lowercase letters. However, as much as they look the same, there is a fundamental difference. The conditional expectation $E(Y|X = x)$ is a function of x. Its domain is a set of real numbers. The deterministic function can be evaluated and graphed. For instance, in the last example $E(Y|X = x) = x/2$ is a linear function of x with slope 1/2.

On the contrary, $E(Y|X)$ is a random variable. As such, it has a probability distribution. Since $E(Y|X)$ is a random variable with some probability distribution, it makes sense to take *its* expectation with respect to that distribution. The expectation of a conditional expectation may seem pretty far out. But it leads to one of the most important results in probability.

Law of Total Expectation

For random variables X and Y,

$$E(Y) = E(E(Y|X)).$$

We prove this result for the discrete case, and leave the continuous case as an exercise for the reader.

Proof.

$$\begin{aligned}
E(E(Y|X)) &= \sum_x E(Y|X = x)P(X = x) \\
&= \sum_x \left(\sum_y yP(Y = y|X = x) \right) P(X = x) \\
&= \sum_y y \sum_x P(Y = y|X = x)P(X = x) \\
&= \sum_y y \sum_x P(X = x, Y = y) \\
&= \sum_y yP(Y = y) = E(Y).
\end{aligned}$$

■

Example 1.25 Angel will harvest N tomatoes in her garden, where N has a Poisson distribution with parameter λ. Each tomato is checked for defects. The chance that a tomato has defects is p. Defects are independent from tomato to tomato. Find the expected number of tomatoes with defects.

Solution Let X be the number of tomatoes with defects. The conditional distribution of X given $N = n$ is binomial with parameters n and p. This gives $E(X|N = n) = np$. Since this is true for all n, $E(X|N) = Np$. By the law of total expectation,

$$E(X) = E(E(X|N)) = E(Np) = pE(N) = p\lambda.$$

■

Example 1.26 Ellen's insurance will pay for a medical expense subject to a \$100 deductible. Assume that the amount of the expense is exponentially distributed with mean \$500. Find the expectation and standard deviation of the payout.

Solution Let M be the amount of the medical expense and let X be the insurance company's payout. Then,

$$X = \begin{cases} M - 100, & \text{if } M > 100, \\ 0, & \text{if } M \le 100, \end{cases}$$

where M is exponentially distributed with parameter $1/500$. To find the expected payment, apply the law of total expectation, giving

$$\begin{aligned} E(X) = E(E(X|M)) &= \int_0^\infty E(X|M = m)\lambda e^{-\lambda m}\, dm \\ &= \int_{100}^\infty E(M - 100|M = m)\frac{1}{500}e^{-m/500}\, dm \\ &= \int_{100}^\infty (m - 100)\frac{1}{500}e^{-m/500}\, dm \\ &= 500e^{-100/500} = \$409.365. \end{aligned}$$

For the standard deviation, first find

$$\begin{aligned} E\left(X^2\right) = E\left(E\left(X^2|M\right)\right) &= \int_0^\infty E\left(X^2|M = m\right)\lambda e^{-\lambda m}\, dm \\ &= \int_{100}^\infty E\left((M - 100)^2|M = m\right)\frac{1}{500}e^{-m/500}\, dm \\ &= \int_{100}^\infty (m - 100)^2\frac{1}{500}e^{-m/500}\, dm \\ &= 500000e^{-1/5} = 409365. \end{aligned}$$

This gives

$$\begin{aligned} SD(X) &= \sqrt{Var(X)} = \sqrt{E(X^2) - E(X)^2} \\ &= \sqrt{409365 - (409.365)^2} = \$491.72. \end{aligned}$$

■

R : Simulation of Ellen's Payout

```
> trials <- 100000
> simlist <- numeric(trials)
> for (i in 1:trials) {
> expense <- rexp(1,1/500)
> payout <- max(0, expense-100)
```

```
> simlist[i] <- payout}
> mean(simlist)
[1] 410.0308
> sd(simlist)
[1] 493.5457
```

Example 1.27 (Random sum of random variables) A stochastic model for the cost of damage from traffic accidents is given by Van der Lann and Louter (1986). Let X_k be the amount of damage from an individual's kth traffic accident. Assume $X_1, X_2, \ldots$ is an i.i.d. sequence with mean μ. The number of accidents N for an individual driver is a random variable with mean λ. It is assumed that the number of accidents is independent of the amount of damages for each accident. That is, N is independent of the X_k. For an individual driver, find the total cost of damages.

Solution Let T be the total cost of damages. Then,

$$T = X_1 + \cdots + X_N = \sum_{k=1}^{N} X_k.$$

The number of summands is random. The random variable T is a random sum of random variables. By the law of total expectation $E(T) = E(E(T|N))$. To find $E(T|N)$, consider

$$\begin{aligned} E(T|N = n) &= E\left(\sum_{k=1}^{N} X_k | N = n \right) = E\left(\sum_{k=1}^{n} X_k | N = n \right) \\ &= E\left(\sum_{k=1}^{n} X_k \right) = \sum_{k=1}^{n} E(X_k) = n\mu, \end{aligned}$$

where the third equality is because N is independent of the X_k. Since the final equality holds for all n, $E(T|N) = N\mu$. By the law of total expectation,

$$E(T) = E(E(T|N)) = E(N\mu) = \mu E(N) = \mu\lambda.$$

The result is intuitive. The expected total cost is the product of the expected number of accidents and the expected cost per accident.

Note that it would have been incorrect to write

$$E\left(\sum_{k=1}^{N} X_k \right) = \sum_{k=1}^{N} E(X_k).$$

Linearity of expectation does not apply here because the number of summands is random, not fixed. Indeed, this equation does not even make sense as the left-hand side is a fixed number (the expectation of a random variable), while the right-hand side is a random variable. ■

Computing Probabilities by Conditioning

For an event A, let I_A be the indicator for A. Then,

$$E(I_A) = (1)P(A) + (0)P(A^c) = P(A).$$

From this simple fact, one sees that probabilities can always be expressed as expectations. As such, the law of total expectation can be used when computing probabilities. In particular, if X is a discrete random variable,

$$\begin{aligned} P(A) = E(I_A) &= E(E(I_A|X)) \\ &= \sum_x E(I_A|X = x)P(X = x) \\ &= \sum_x [(1)P(I_A = 1|X = x)P(X = x) + (0)P(I_A = 0|X = x)P(X = x)] \\ &= \sum_x P(A|X = x)P(X = x). \end{aligned}$$

We have recaptured the law of total probability, where the conditioning events are $\{X = x\}$ for all x.

If X is continuous with density function f_X,

$$P(A) = \int_{-\infty}^{\infty} E(I_A|X = x)f_X(x)\, dx = \int_{-\infty}^{\infty} P(A|X = x)f_X(x)\, dx,$$

which gives the continuous version of the law of total probability.

Example 1.28 Max arrives to class at time X. Mary arrives at time Y. Assume that X and Y have exponential distributions with respective parameters λ and μ. If arrival times are independent, find the probability that Mary arrives before Max.

Solution Let $A = \{Y < X\}$ be the event that Mary arrives to class before Max. By conditioning on Max's arrival time,

$$\begin{aligned} P(A) = P(Y < X) &= \int_{-\infty}^{\infty} P(Y < X|X = x)f_X(x)\, dx \\ &= \int_0^{\infty} P(Y < x|X = x)\lambda e^{-\lambda x}\, dx \\ &= \int_0^{\infty} P(Y < x)\lambda e^{-\lambda x}\, dx \end{aligned}$$

$$= \int_0^\infty (1 - e^{-\mu x})\lambda e^{-\lambda x}\,dx$$

$$= 1 - \int_0^\infty \lambda e^{-(\lambda+\mu)x}\,dx$$

$$= 1 - \frac{\lambda}{\lambda+\mu} = \frac{\mu}{\lambda+\mu}.$$

The fourth equality is by independence of X and Y. ■

Example 1.29 (Sums of independent random variables) Assume that X and Y are independent continuous random variables with density functions f_X and f_Y, respectively. (i) Find the density function of $X + Y$. (ii) Use part (i) to find the density of the sum of two independent standard normal random variables.

Solution

(i) Conditioning on Y,

$$P(X + Y \le t) = \int_{-\infty}^\infty P(X + Y \le t | Y = y) f_Y(y)\,dy$$

$$= \int_{-\infty}^\infty P(X \le t - y | Y = y) f_Y(y)\,dy$$

$$= \int_{-\infty}^\infty P(X \le t - y) f_Y(y)\,dy.$$

Differentiating with respect to t gives

$$f_{X+Y}(t) = \int_{-\infty}^\infty f_X(t - y) f_Y(y)\,dy. \tag{1.3}$$

Equation (1.3) is known as a *convolution* formula.

(ii) For X and Y independent standard normal random variables, by Equation (1.3), $X + Y$ has density

$$f_{X+Y}(t) = \int_{-\infty}^\infty \frac{1}{\sqrt{2\pi}} e^{-(t-y)^2/2} \frac{1}{\sqrt{2\pi}} e^{-y^2/2}\,dy$$

$$= \frac{1}{\sqrt{4\pi}} e^{-t^2/4} \int_{-\infty}^\infty \frac{1}{\sqrt{2\pi(1/2)}} e^{-(y-t/2)^2/2(1/2)}\,dy \tag{1.4}$$

$$= \frac{1}{\sqrt{4\pi}} e^{-t^2/4},$$

which is the density of a normal distribution with mean 0 and variance 2. The last equality is because the integrand in Equation (1.4) is the density of a normal distribution with mean $t/2$ and variance 1/2, and thus integrates to 1. ■

Conditional Variance

Analogous to conditional expectation, the conditional variance is a variance taken with respect to a conditional distribution. Given random variables X and Y, let $\mu_x = E(Y|X = x)$. Then, the conditional variance $Var(Y|X = x)$ is defined as

$$Var(Y|X = x) = E\left((Y - \mu_x)^2|X = x\right)$$

$$= \begin{cases} \sum_y (y - \mu_x)^2 P(Y = y|X = x), & \text{discrete,} \\ \int_{-\infty}^{\infty} (y - \mu_x)^2 f_{Y|X}(y|x)\, dy, & \text{continuous.} \end{cases}$$

Compare with the regular variance formula

$$Var(Y) = E\left((Y - \mu)^2\right),$$

where $\mu = E(Y)$. Observe that the conditional expectation $E(Y|X = x)$ takes the place of the unconditional expectation $E(Y)$ in the regular variance formula.

Example 1.30 Let N be a positive, integer-valued random variable. If $N = n$, flip n coins, each of which has heads probability p. Let Y be the number of coins which come up heads. Find $Var(Y|N = n)$.

Solution The conditional distribution of Y given $N = n$ is binomial with parameters n and p. From the properties of the binomial distribution,

$$Var(Y|N = n) = np(1 - p).$$

■

Properties of the variance transfer to the conditional variance.

Properties of Conditional Variance

1. $$Var(Y|X = x) = E\left(Y^2|X = x\right) - (E(Y|X = x))^2.$$

2. For constants a and b,

$$Var(aY + b|X = x) = a^2 Var(Y|X = x).$$

As with the development of conditional expectation, define the *conditional variance* $Var(Y|X)$ as the random variable which is a function of X and which takes the value $Var(Y|X = x)$ when $X = x$.

The law of total variance shows how to find the variance of a random variable by conditioning.

Law of Total Variance

$$Var(Y) = E(Var(Y|X)) + Var(E(Y|X)).$$

The proof is easier than you might think. We have that

$$\begin{aligned} E(Var(Y|X)) &= E\left(E\left(Y^2|X\right) - (E(Y|X))^2\right) \\ &= E\left(E\left(Y^2|X\right)\right) - E\left((E(Y|X))^2\right) \\ &= E\left(Y^2\right) - E\left((E(Y|X))^2\right). \end{aligned}$$

And

$$\begin{aligned} Var(E(Y|X)) &= E\left((E(Y|X))^2\right) - (E(E(Y|X)))^2 \\ &= E\left((E(Y|X))^2\right) - (E(Y))^2. \end{aligned}$$

Thus,

$$\begin{aligned} &E(Var(Y|X)) + Var(E(Y|X)) \\ &\quad = \left(E(Y^2) - E\left((E(Y|X)^2)\right)\right) + \left(E\left((E(Y|X))^2\right) - (E(Y))^2\right) \\ &\quad = E\left(Y^2\right) - (E(Y))^2 = Var(Y). \end{aligned}$$

Example 1.31 A number X is uniformly distributed on $(0, 1)$. If $X = x$, then Y is picked uniformly on $(0, x)$. Find the variance of Y.

Solution The conditional distribution of Y given $X = x$ is uniform on $(0, x)$. From properties of the uniform distribution,

$$E(Y|X = x) = \frac{x}{2} \quad \text{and} \quad Var(Y|X = x) = \frac{x^2}{12}.$$

This gives $E(Y|X) = X/2$ and $Var(Y|X) = X^2/12$. By the law of total variance,

$$Var(Y) = E(Var(Y|X)) + Var(E(Y|X)) = E\left(\frac{X^2}{12}\right) + Var\left(\frac{X}{2}\right)$$

$$= \frac{1}{12}E\left(X^2\right) + \frac{1}{4}Var(X) = \frac{1}{12}\left(\frac{1}{3}\right) + \frac{1}{4}\left(\frac{1}{12}\right)$$
$$= \frac{7}{144} = 0.04861.$$

■

R : Simulation of *Var*(*Y*)

The structure of this two-stage random experiment makes it especially easy to simulate in R .

```
> trials <- 100000
> simlist <- replicate(trials,runif(1,0,runif(1)))
> var(simlist)
[1] 0.04840338
```

Example 1.32 (Variance of a random sum of random variables) Assume that $X_1, X_2, \ldots$ is an i.i.d. sequence with common mean μ_X and variance σ_X^2. Let N be a positive, integer-valued random variable that is independent of the X_i with mean μ_N and variance σ_N^2. Let $T = X_1 + \cdots + X_N$. Find the variance of T.

Solution Random sums of random variables were introduced in Example 1.27 where the expectation $E(T) = \mu_X\mu_N$ was found using the law of total expectation. By the law of total variance,

$$Var(T) = Var\left(\sum_{k=1}^{N} X_k\right) = E\left(Var\left(\sum_{k=1}^{N} X_k|N\right)\right) + Var\left(E\left(\sum_{k=1}^{N} X_k|N\right)\right).$$

We have that

$$Var\left(\sum_{k=1}^{N} X_k|N = n\right) = Var\left(\sum_{k=1}^{n} X_k|N = n\right)$$
$$= Var\left(\sum_{k=1}^{n} X_k\right) = \sum_{k=1}^{n} Var(X_k)$$
$$= n\sigma_X^2.$$

The second equality is because N is independent of the X_k. The third equality is because the X_k are independent. This gives

$$Var\left(\sum_{k=1}^{N} X_k|N\right) = N\sigma_X^2.$$

From results for conditional expectation,

$$E\left(\sum_{k=1}^{N} X_k | N\right) = NE(X_1) = N\mu_X.$$

This gives

$$\begin{aligned} Var(T) &= E\left(Var\left(\sum_{k=1}^{N} X_k | N\right)\right) + Var\left(E\left(\sum_{k=1}^{N} X_k | N\right)\right) \\ &= E\left(N\sigma_X^2\right) + Var(N\mu_X) \\ &= \sigma_X^2 E(N) + \mu_X^2 Var(N) \\ &= \sigma_X^2 \mu_N + \mu_X^2 \sigma_N^2. \end{aligned}$$

■

Random sums of independent random variables will arise in several contexts. Results are summarized here.

Random Sums of Random Variables

Let $X_1, X_2, \ldots$ be an i.i.d. sequence of random variables with common mean μ_X and variance σ_X^2. Let N be a positive, integer-valued random variable independent of the X_i with $E(N) = \mu_N$ and $Var(N) = \sigma_N^2$. Let $T = \sum_{i=1}^{N} X_i$. Then,

$$E(T) = \mu_X \mu_N \text{ and } Var(T) = \sigma_X^2 \mu_N + \sigma_N^2 \mu_X^2.$$

EXERCISES

1.1 For the following scenarios identify a stochastic process $\{X_t, t \in I\}$, describing (i) X_t in context, (ii) state space, and (iii) index set. State whether the state space and index set are discrete or continuous.

(a) From day to day the weather in International Falls, Minnesota is either rain, clear, or snow.

Solution: Let X_t denote the weather on day t. Discrete state space is $S = \{\text{Rain, Clear, Snow}\}$. *Discrete index set is* $I = \{0, 1, 2, \cdots\}$.

(b) At the end of each year, a 4-year college student either advances in grade, repeats their grade, or drops out.

(c) Seismologists record daily earthquake magnitudes in Chile. The largest recorded earthquake in history was the Valdivia, Chile earthquake on May 22, 1960, which had a magnitude of 9.5 on the Richter scale.

(d) Data are kept on the circumferences of trees in an arboretum. The arboretum covers a two square-mile area.

(e) Starting Monday morning at 9 a.m., as students arrive to class, the teacher records student arrival times. The class has 30 students and lasts for 60 minutes.

(f) A card player shuffles a standard deck of cards by the following method: the top card of the deck is placed somewhere in the deck at random. The player does this 100 times to mix up the deck.

1.2 A regional insurance company insures homeowners against flood damage. Half of their policyholders are in Florida, 30% in Louisiana, and 20% in Texas. Company actuaries give the estimates in Table 1.1 for the probability that a policyholder will file a claim for flood damage over the next year.

(a) Find the probability that a random policyholder will file a claim for flood damage next year.

(b) A claim was filed. Find the probability that the policyholder is from Texas.

TABLE 1.1 Probability of Claim for Flood Damage

Florida	Louisiana	Texas
0.03	0.015	0.02

1.3 Let $B_1, \ldots, B_k$ be a partition of the sample space. For events A and C, prove the *law of total probability for conditional probability*

$$P(A|C) = \sum_{i=1}^{k} P(A|B_i \cap C)P(B_i|C).$$

1.4 See Exercise 1.2. Among all policyholders who live within five miles of the Atlantic Ocean, 75% live in Florida, 20% live in Louisiana, and 5% live in Texas. For those who live close to the ocean the probabilities of filing a claim increase, as given in Table 1.2.

Assume that a policyholder lives within five miles of the Atlantic coast. Use the law of total probability for conditional probability in Exercise 1.3 to find the chance they will file a claim for flood damage next year.

1.5 Two fair, six-sided dice are rolled. Let X_1, X_2 be the outcomes of the first and second die, respectively.

TABLE 1.2 Probability of Claim for Those Within Five Miles of Atlantic Coast

Florida	Louisiana	Texas
0.10	0.06	0.06

(a) Find the conditional distribution of X_2 given that $X_1 + X_2 = 7$.

(b) Find the conditional distribution of X_2 given that $X_1 + X_2 = 8$.

1.6 Bob has n coins in his pocket. One is two-headed, the rest are fair. A coin is picked at random, flipped, and shows heads. Find the probability that the coin is two-headed.

1.7 A rat is trapped in a maze with three doors and some hidden cheese. If the rat takes door one, he will wander around the maze for 2 minutes and return to where he started. If he takes door two, he will wander around the maze for 3 minutes and return to where he started. If he takes door three, he will find the cheese after 1 minute. If the rat returns to where he started he immediately picks a door to pass through. The rat picks each door uniformly at random. How long, on average, will the rat wander before finding the cheese?

1.8 A bag contains 1 red, 3 green, and 5 yellow balls. A sample of four balls is picked. Let G be the number of green balls in the sample. Let Y be the number of yellow balls in the sample.

(a) Find the conditional probability mass function of G given $Y = 2$ assuming the sample is picked without replacement.

(b) Find the conditional probability mass function of G given $Y = 2$ assuming the sample is picked with replacement.

1.9 Assume that X is uniformly distributed on $\{1, 2, 3, 4\}$. If $X = x$, then Y is uniformly distributed on $\{1, \ldots, x\}$. Find

(a) $P(Y = 2|X = 2)$

(b) $P(Y = 2)$

(c) $P(X = 2|Y = 2)$

(d) $P(X = 2)$

(e) $P(X = 2, Y = 2)$.

1.10 A die is rolled until a 3 occurs. By conditioning on the outcome of the first roll, find the probability that an even number of rolls is needed.

1.11 Consider the gambler's ruin process where at each wager, the gambler wins with probability p and loses with probability $q = 1 - p$. The gambler stops when reaching \$$n$ or losing all their money. If the gambler starts with \$$k$, with $0 < k < n$, find the probability of eventual ruin. See Example 1.10.

1.12 In n rolls of a fair die, let X be the number of times 1 is rolled, and Y the number of times 2 is rolled. Find the conditional distribution of X given $Y = y$.

1.13 Random variables X and Y have joint density

$$f(x, y) = 3y, \quad \text{for } 0 < x < y < 1.$$

(a) Find the conditional density of Y given $X = x$.

(b) Find the conditional density of X given $Y = y$. Describe the conditional distribution.

1.14 Random variables X and Y have joint density function

$$f(x, y) = 4e^{-2x}, \quad \text{for } 0 < y < x < \infty.$$

(a) Find the conditional density of X given $Y = y$.

(b) Find the conditional density of Y given $X = x$. Describe the conditional distribution.

1.15 Let X and Y be uniformly distributed on the disk of radius 1 centered at the origin. Find the conditional distribution of Y given $X = x$.

1.16 A poker hand consists of five cards drawn from a standard 52-card deck. Find the expected number of aces in a poker hand given that the first card drawn is an ace.

1.17 Let X be a Poisson random variable with $\lambda = 3$. Find $E(X|X > 2)$.

1.18 From the definition of conditional expectation given an event, show that

$$E(I_B|A) = P(B|A).$$

1.19 See Example 1.21. Find the variance of the number of flips needed to get two heads in a row.

1.20 A fair coin is flipped repeatedly.

(a) Find the expected number of flips needed to get three heads in a row.

(b) Find the expected number of flips needed to get k heads in a row.

1.21 Let T be a nonnegative, continuous random variable. Show

$$E(T) = \int_0^\infty P(T > t)\, dt.$$

1.22 Find $E(Y|X)$ when (X, Y) is uniformly distributed on the following regions.

(a) The rectangle $[a, b] \times [c, d]$.

(b) The triangle with vertices $(0, 0), (1, 0), (1, 1)$.

(c) The disc of radius 1 centered at the origin.

1.23 Let $X_1, X_2, \ldots$ be an i.i.d. sequence of random variables with common mean μ. Let $S_n = X_1 + \cdots + X_n$, for $n \geq 1$.

(a) Find $E(S_m|S_n)$, for $m \leq n$.

(b) Find $E(S_m|S_n)$ for $m > n$.

1.24 Prove the law of total expectation $E(Y) = E(E(Y|X))$ for the continuous case.

1.25 Let X and Y be independent exponential random variables with respective parameters 1 and 2. Find $P(X/Y < 3)$ by conditioning.

1.26 The density of X is $f(x) = xe^{-x}$, for $x > 0$. Given $X = x$, Y is uniformly distributed on $(0, x)$. Find $P(Y < 2)$ by conditioning on X.

1.27 A restaurant receives N customers per day, where N is a random variable with mean 200 and standard deviation 40. The amount spent by each customer is normally distributed with mean \$15 and standard deviation \$3. The amounts that customers spend are independent of each other and independent of N. Find the mean and standard deviation of the total amount spent at the restaurant per day.

1.28 On any day, the number of accidents on the highway has a Poisson distribution with parameter Λ. The parameter Λ varies from day to day and is itself a random variable. Find the mean and variance of the number of accidents per day when Λ is uniformly distributed on $(0, 3)$.

1.29 If X and Y are independent, does $Var(Y|X) = Var(Y)$?

1.30 Assume that $Y = g(X)$ is a function of X. Find simple expressions for

(a) $E(Y|X)$.

(b) $Var(Y|X)$.

1.31 Consider a sequence of i.i.d. Bernoulli trials with success parameter p. Let X be the number of trials needed until the first success occurs. Then, X has a geometric distribution with parameter p. Find the variance of X by conditioning on the first trial.

1.32 R: Simulate flipping three fair coins and counting the number of heads X.

(a) Use your simulation to estimate $P(X = 1)$ and $E(X)$.

(b) Modify the above to allow for a biased coin where $P(\text{Heads}) = 3/4$.

1.33 R: Cards are drawn from a standard deck, with replacement, until an ace appears. Simulate the mean and variance of the number of cards required.

1.34 R: The time until a bus arrives has an exponential distribution with mean 30 minutes.

(a) Use the command `rexp()` to simulate the probability that the bus arrives in the first 20 minutes.

(b) Use the command `pexp()` to compare with the exact probability.

1.35 R: See the script file **gamblersruin.R**. A gambler starts with a \$60 initial stake.

(a) The gambler wins, and loses, each round with probability $p = 0.50$. Simulate the probability the gambler wins \$100 before he loses everything.

(b) The gambler wins each round with probability $p = 0.51$. Simulate the probability the gambler wins \$100 before he loses everything.

1.36 R: See Example 1.2 and the script file **ReedFrost.R**. Observe the effect on the course of the disease by changing the initial values for the number of people susceptible and infected. How does increasing the number of infected people affect the duration of the disease?

1.37 R: Simulate the results of Exercise 1.28. Estimate the mean and variance of the number of accidents per day.

2

MARKOV CHAINS: FIRST STEPS

> Let us finish the article and the whole book with a good example of dependent trials, which approximately can be considered as a simple chain.
>
> –Andrei Andreyevich Markov

2.1 INTRODUCTION

Consider a game with a playing board consisting of squares numbered 1–10 arranged in a circle. (Think of miniature Monopoly.) A player starts at square 1. At each turn, the player rolls a die and moves around the board by the number of spaces shown on the face of the die. The player keeps moving around and around the board according to the roll of the die. (Granted, this is not a very exciting game.)

Let X_k be the number of squares the player lands on after k moves, with $X_0 = 1$. Assume that the player successively rolls 2, 1, and 4. The first four positions are

$$(X_0, X_1, X_2, X_3) = (1, 3, 4, 8).$$

Given this information, what can be said about the player's next location X_4? Even though we know the player's full past history of moves, the only information relevant

Introduction to Stochastic Processes with R, First Edition. Robert P. Dobrow.

for predicting their future position is their most recent location X_3. Since $X_3 = 8$, then necessarily $X_4 \in \{9, 10, 1, 2, 3, 4\}$, with equal probability. Formally,

$$P(X_4 = j | X_0 = 1, X_1 = 3, X_2 = 4, X_3 = 8) = P(X_4 = j | X_3 = 8) = \frac{1}{6},$$

for $j = 9, 10, 1, 2, 3, 4$. Given the player's most recent location X_3, their future position X_4 is independent of past history X_0, X_1, X_2.

The sequence of player's locations $X_0, X_1, \ldots$ is a stochastic process called a *Markov chain*. The game illustrates the essential property of a Markov chain: the future, given the present, is independent of the past.

Markov Chain

Let S be a discrete set. A Markov chain is a sequence of random variables $X_0, X_1, \ldots$ taking values in S with the property that

$$\begin{aligned} P(X_{n+1} &= j | X_0 = x_0, \ldots, X_{n-1} = x_{n-1}, X_n = i) \\ &= P(X_{n+1} = j | X_n = i), \end{aligned} \tag{2.1}$$

for all $x_0, \ldots, x_{n-1}, i, j \in S$, and $n \geq 0$. The set S is the *state space* of the Markov chain.

We often use descriptive language to describe the evolution of a Markov chain. For instance, if $X_n = i$, we say that the chain *visits* state i, or *hits* i, at *time* n.

A Markov chain is *time-homogeneous* if the probabilities in Equation (2.1) do not depend on n. That is,

$$P(X_{n+1} = j | X_n = i) = P(X_1 = j | X_0 = i), \tag{2.2}$$

for all $n \geq 0$. Unless stated otherwise, the Markov chains in this book are time-homogeneous.

Since the probabilities in Equation (2.2) only depend on i and j, they can be arranged in a matrix $\boldsymbol{P}$, whose ijth entry is $P_{ij} = P(X_1 = j | X_0 = i)$. This is the *transition matrix*, or *Markov matrix*, which contains the one-step transition probabilities of moving from state to state.

If the state space has k elements, then the transition matrix is a square $k \times k$ matrix. If the state space is countably infinite, the transition matrix is infinite.

For the simple board game Markov chain, the sample space is

$$S = \{1, 2, 3, 4, 5, 6, 7, 8, 9, 10\},$$

with transition matrix

$$
P = \begin{array}{c} \\ 1 \\ 2 \\ 3 \\ 4 \\ 5 \\ 6 \\ 7 \\ 8 \\ 9 \\ 10 \end{array}
\begin{array}{c}
\begin{array}{cccccccccc} 1 & 2 & 3 & 4 & 5 & 6 & 7 & 8 & 9 & 10 \end{array} \\
\begin{pmatrix}
0 & 1/6 & 1/6 & 1/6 & 1/6 & 1/6 & 1/6 & 0 & 0 & 0 \\
0 & 0 & 1/6 & 1/6 & 1/6 & 1/6 & 1/6 & 1/6 & 0 & 0 \\
0 & 0 & 0 & 1/6 & 1/6 & 1/6 & 1/6 & 1/6 & 1/6 & 0 \\
0 & 0 & 0 & 0 & 1/6 & 1/6 & 1/6 & 1/6 & 1/6 & 1/6 \\
1/6 & 0 & 0 & 0 & 0 & 1/6 & 1/6 & 1/6 & 1/6 & 1/6 \\
1/6 & 1/6 & 0 & 0 & 0 & 0 & 1/6 & 1/6 & 1/6 & 1/6 \\
1/6 & 1/6 & 1/6 & 0 & 0 & 0 & 0 & 1/6 & 1/6 & 1/6 \\
1/6 & 1/6 & 1/6 & 1/6 & 0 & 0 & 0 & 0 & 1/6 & 1/6 \\
1/6 & 1/6 & 1/6 & 1/6 & 1/6 & 0 & 0 & 0 & 0 & 1/6 \\
1/6 & 1/6 & 1/6 & 1/6 & 16 & 1/6 & 0 & 0 & 0 & 0
\end{pmatrix}
\end{array}.
$$

The entries of every Markov transition matrix $\boldsymbol{P}$ are nonnegative, and each row sums to 1, as

$$\sum_j P_{ij} = \sum_j P(X_1 = j | X_0 = i) = \sum_j \frac{P(X_1 = j, X_0 = i)}{P(X_0 = i)} = \frac{P(X_0 = i)}{P(X_0 = i)} = 1,$$

for all rows i. A nonnegative matrix whose rows sum to 1 is called a *stochastic matrix.*

Stochastic Matrix

A stochastic matrix is a square matrix $\boldsymbol{P}$, which satisfies

1. $P_{ij} \geq 0$ for all i, j.
2. For each row i,

$$\sum_j P_{ij} = 1.$$

2.2 MARKOV CHAIN CORNUCOPIA

Following is a taste of the wide range of applications of Markov chains. Many of these examples are referenced throughout the book.

Example 2.1 (Heads you win) Successive coin flips are the very model of independent events. Yet in a fascinating study of how people actually flip coins, Diaconis (2007) shows that vigorously flipped coins are ever so slightly biased to come up the same way they started. "For natural flips," Diaconis asserts, "the chance of coming up as started is about 0.51."

In other words, successive coin flips are not independent. They can be described by a Markov chain with transition matrix

$$P = \begin{array}{c} \\ H \\ T \end{array} \begin{array}{c} \begin{array}{cc} H & T \end{array} \\ \begin{pmatrix} 0.51 & 0.49 \\ 0.49 & 0.51 \end{pmatrix} \end{array}.$$

■

Note on Notation

In Example 2.1, the state space is $S = \{H, T\}$, and the transition matrix is labeled with row and column identifiers. For many Markov chains, such as this, random variables take values in a discrete state space whose elements are not necessarily numbers.

For matrix entries, we can use suitable labels to identify states. For instance, for this example $P_{HH} = P_{TT} = 0.51$ and $P_{HT} = P_{TH} = 0.49$.

Example 2.2 (Poetry and dependent sequences) Andrei Andreyevich Markov, the Russian mathematician who introduced Markov chains over 100 years ago, first applied them in the analysis of the poem *Eugene Onegin* by Alexander Pushkin. In the first 20,000 letters of the poem, Markov counted (by hand!) 8,638 vowels and 11,362 consonants. He also tallied pairs of successive letters. Of the 8,638 pairs that start with vowels, 1,104 pairs are vowel–vowel. Of the 11,362 pairs that start with consonants, 3,827 are consonant–consonant. Markov treated the succession of letter types as a random sequence. The resulting transition matrix is

$$P = \begin{array}{c} \\ \text{v} \\ \text{c} \end{array} \begin{array}{c} \begin{array}{cc} \text{vowel} & \text{consonant} \end{array} \\ \begin{pmatrix} 1,104/8,638 & 7,534/8,638 \\ 7,535/11,362 & 3,827/11,362 \end{pmatrix} \end{array} = \begin{pmatrix} 0.175 & 0.825 \\ 0.526 & 0.474 \end{pmatrix}.$$

Markov showed that the succession of letter types was not an independent sequence. For instance, if letter types were independent, the probability of two successive consonants would be $(11,362/20,000)^2 = 0.323$, whereas from Pushkin's poem the probability is $P_{cc} = 0.474$.

Markov's work was a polemic against a now obscure mathematician who argued that the law of large numbers only applied to independent sequences. Markov disproved the claim by showing that the Pushkin letter sequence was a dependent sequence for which the law of large numbers applied. ■

Example 2.3 (Chained to the weather) Some winter days in Minnesota it seems like the snow will never stop. A Minnesotan's view of winter might be described by the following transition matrix for a weather Markov chain, where r, s, and c denote rain, snow, and clear, respectively.

$$
\boldsymbol{P} = \begin{matrix} & r & s & c \\ r & 0.2 & 0.6 & 0.2 \\ s & 0.1 & 0.8 & 0.1 \\ c & 0.1 & 0.6 & 0.3 \end{matrix}.
$$

For this model, no matter what the weather today, there is always at least a 60% chance that it will snow tomorrow. ■

Example 2.4 (I.i.d. sequence) An independent and identically distributed sequence of random variables is trivially a Markov chain. Assume that $X_0, X_1, \ldots$ is an i.i.d. sequence that takes values in $\{1, \ldots, k\}$ with

$$
P(X_n = j) = p_j, \text{ for } j = 1, \ldots, k, \quad \text{and} \quad n \geq 0,
$$

where $p_1 + \cdots + p_k = 1$. By independence,

$$
P(X_1 = j | X_0 = i) = P(X_1 = j) = p_j.
$$

The transition matrix is

$$
\boldsymbol{P} = \begin{matrix} & 1 & 2 & \cdots & k \\ 1 & p_1 & p_2 & \cdots & p_k \\ 2 & p_1 & p_2 & \cdots & p_k \\ \vdots & \vdots & \vdots & & \vdots \\ k & p_1 & p_2 & \cdots & p_k \end{matrix}.
$$

The matrix for an i.i.d. sequence has identical rows as the next state of the chain is independent of, and has the same distribution as, the present state. ■

Example 2.5 (Gambler's ruin) Gambler's ruin was introduced in Chapter 1. In each round of a gambling game a player either wins \$1, with probability p, or loses \$1, with probability $1 - p$. The gambler starts with \$$k$. The game stops when the player either loses all their money, or gains a total of \$$n$ ($n > k$).

The gambler's successive fortunes form a Markov chain on $\{0, 1, \ldots, n\}$ with $X_0 = k$ and transition matrix given by

$$
P_{ij} = \begin{cases} p, & \text{if } j = i + 1, \ 0 < i < n, \\ 1 - p, & \text{if } j = i - 1, \ 0 < i < n, \\ 1, & \text{if } i = j = 0, \ \text{ or } i = j = n, \\ 0, & \text{otherwise.} \end{cases}
$$

Here is the transition matrix with $n = 6$ and $p = 1/3$.

$$P = \begin{array}{c} \\ 0 \\ 1 \\ 2 \\ 3 \\ 4 \\ 5 \\ 6 \end{array}\begin{array}{c} \begin{array}{ccccccc} 0 & 1 & 2 & 3 & 4 & 5 & 6 \end{array} \\ \begin{pmatrix} 1 & 0 & 0 & 0 & 0 & 0 & 0 \\ 2/3 & 0 & 1/3 & 0 & 0 & 0 & 0 \\ 0 & 2/3 & 0 & 1/3 & 0 & 0 & 0 \\ 0 & 0 & 2/3 & 0 & 1/3 & 0 & 0 \\ 0 & 0 & 0 & 2/3 & 0 & 1/3 & 0 \\ 0 & 0 & 0 & 0 & 2/3 & 0 & 1/3 \\ 0 & 0 & 0 & 0 & 0 & 0 & 1 \end{pmatrix} \end{array}.$$

Gambler's ruin is an example of *simple random walk with absorbing boundaries.* Since $P_{00} = P_{nn} = 1$, when the chain reaches 0 or n, it stays there forever. ■

Example 2.6 (Wright–Fisher model) The Wright–Fisher model describes the evolution of a fixed population of k genes. Genes can be one of two types, called alleles: A or a. Let X_n denote the number of A alleles in the population at time n, where time is measured by generations. Under the model, the number of A alleles at time $n + 1$ is obtained by drawing with replacement from the gene population at time n. Thus, conditional on there being i alleles of type A at time n, the number of A alleles at time $n + 1$ has a binomial distribution with parameters k and $p = i/k$. This gives a Markov chain with transition matrix defined by

$$P_{ij} = \binom{k}{j}\left(\frac{i}{k}\right)^j\left(1 - \frac{i}{k}\right)^{k-j}, \quad \text{for } 0 \le i,j \le k.$$

Observe that $P_{00} = P_{kk} = 1$. As the chain progresses, the population is eventually made up of all a alleles (state 0) or all A alleles (state k). A question of interest is what is the probability that the population evolves to the all-A state? ■

Example 2.7 (Squash) Squash is a popular racket sport played by two or four players in a four-walled court. In the international scoring system, the first player to score nine points is the winner. However, if the game is tied at 8-8, the player who reaches 8 points first has two options: (i) to play to 9 points (set one) or to play to 10 points (set two). Set one means that the next player to score wins the game. Set two means that the first player to score two points wins the game. Points are only scored by the player who is serving. A player who wins a rally serves the next rally. Thus, if the game is tied 8-8, the player who is not serving decides. Should they choose set one or set two? This endgame play is modeled by a Markov chain in Broadie and Joneja (1993). The two players are called A and B and a score of xy means that A has scored x points and B has scored y points. The states of the chain are defined by the score and the server. The authors let p be the probability that A wins a rally given that A is serving, and q be the probability that A wins a rally given that B is serving. Assumptions are that p and q are constant over time and independent of the current score. Following is the transition matrix of the Markov chain for the set two options. The chain is used to solve for the optimal strategy.

$$
P = \begin{array}{c} \\ 88B \\ 88A \\ 89B \\ 89A \\ 98B \\ 98A \\ 99B \\ 99A \\ A\text{ loses} \\ A\text{ wins} \end{array}
\begin{array}{c}
\begin{array}{cccccccccc} 88B & 88A & 89B & 89A & 98B & 98A & 99B & 99A & A\text{ loses} & A\text{ wins} \end{array} \\
\left(\begin{array}{cccccccccc}
0 & q & 1-q & 0 & 0 & 0 & 0 & 0 & 0 & 0 \\
1-p & 0 & 0 & 0 & 0 & p & 0 & 0 & 0 & 0 \\
0 & 0 & 0 & q & 0 & 0 & 0 & 0 & 1-q & 0 \\
0 & 0 & 1-p & 0 & 0 & 0 & 0 & p & 0 & 0 \\
0 & 0 & 0 & 0 & 0 & q & 1-q & 0 & 0 & 0 \\
0 & 0 & 0 & 0 & 1-p & 0 & 0 & 0 & 0 & p \\
0 & 0 & 0 & 0 & 0 & 0 & 0 & q & 1-q & 0 \\
0 & 0 & 0 & 0 & 0 & 0 & 1-p & 0 & 0 & p \\
0 & 0 & 0 & 0 & 0 & 0 & 0 & 0 & 1 & 0 \\
0 & 0 & 0 & 0 & 0 & 0 & 0 & 0 & 0 & 1
\end{array}\right)
\end{array}.
$$

■

Example 2.8 (Random walk on a graph) A *graph* is a set of vertices and a set of edges. Two vertices are *neighbors* if there is an edge joining them. The *degree* of vertex v is the number of neighbors of v. For the graph in Figure 2.1, $\deg(a) = 1$, $\deg(b) = \deg(c) = \deg(d) = 4$, $\deg(e) = 3$, and $\deg(f) = 2$.

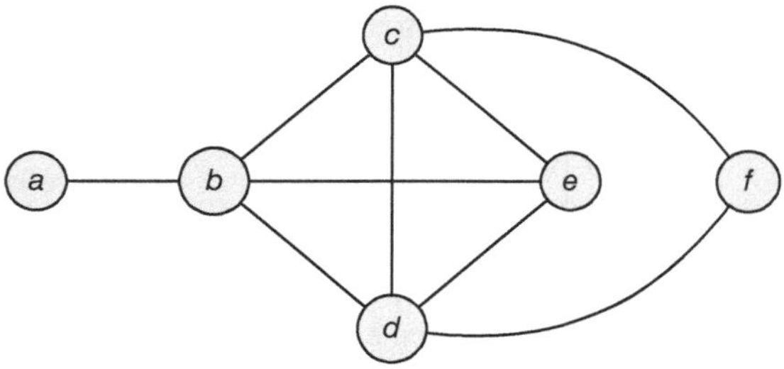

Figure 2.1 Graph on six vertices.

Imagine the vertices as lily pads on a pond. A frog is sitting on one lily pad. At each discrete unit of time, the frog hops to a neighboring lily pad chosen uniformly at random. For instance, if the frog is on lily pad c, it jumps to b, d, e, or f with probability 1/4 each. If the frog is on f, it jumps to c or d with probability 1/2 each. If the frog is on a, it always jumps to b.

Let X_n be the frog's location after n hops. The sequence $X_0, X_1, \ldots$ is a Markov chain. Given a graph G such a process is called *simple random walk on G*.

For vertices i and j, write $i \sim j$ if i and j are neighbors. The one-step transition probabilities are

$$
P(X_1 = j | X_0 = i) = \begin{cases} \dfrac{1}{\deg(i)}, & \text{if } i \sim j, \\ 0, & \text{otherwise.} \end{cases}
$$

The transition matrix for simple random walk on the graph in Figure 2.1 is

$$P = \begin{array}{c} \\ a \\ b \\ c \\ d \\ e \\ f \end{array} \begin{array}{c} \begin{array}{cccccc} a & b & c & d & e & f \end{array} \\ \begin{pmatrix} 0 & 1 & 0 & 0 & 0 & 0 \\ 1/4 & 0 & 1/4 & 1/4 & 1/4 & 0 \\ 0 & 1/4 & 0 & 1/4 & 1/4 & 1/4 \\ 0 & 1/4 & 1/4 & 0 & 1/4 & 1/4 \\ 0 & 1/3 & 1/3 & 1/3 & 0 & 0 \\ 0 & 0 & 1/2 & 1/2 & 0 & 0 \end{pmatrix} \end{array}.$$

Of particular interest is the long-term behavior of the random walk. What can be said of the frog's position after it has been hopping for a long time?

(a) The *cycle graph* on nine vertices is shown in Figure 2.2. Simple random walk on the cycle moves left or right with probability 1/2. Each vertex has degree two. The transition matrix is defined using clock arithmetic. For a cycle with k vertices,

$$P_{ij} = \begin{cases} 1/2, & \text{if } j \equiv i \pm 1 \mod k, \\ 0, & \text{otherwise.} \end{cases}$$

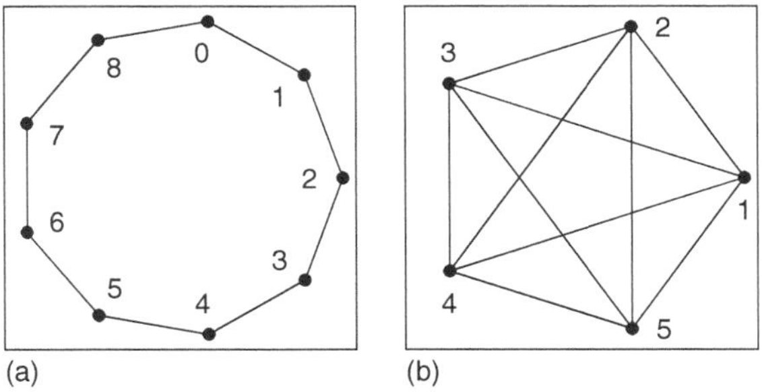

Figure 2.2 (a) Cycle graph on nine vertices. (b) Complete graph on five vertices.

(b) In the *complete graph*, every pair of vertices is joined by an edge. The complete graph on five vertices is shown in Figure 2.2. The complete graph on k vertices has $\binom{k}{2}$ edges. Each vertex has degree $k-1$. The entries of the transition matrix are

$$P_{ij} = \begin{cases} 1/(k-1), & \text{if } i \neq j \\ 0, & \text{if } i = j. \end{cases}$$

(c) The *k-hypercube graph* has vertex set consisting of all k-element sequences of 0s and 1s. Two vertices (sequences) are connected by an edge if they differ in exactly one coordinate. The graph has 2^k vertices and $k2^{k-1}$ edges. Each vertex has degree k. See Figure 2.3.

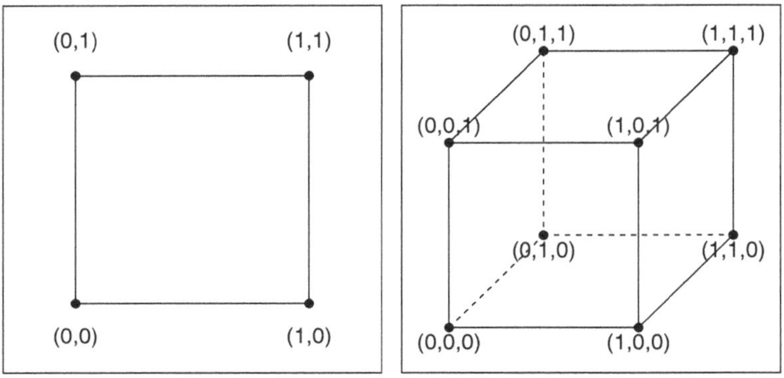

Figure 2.3 The k-hypercube graphs for $k = 2$ and $k = 3$.

Random walk on the k-hypercube can be described as follows. Assume that the walk is at a particular k-element 0–1 sequence. The next sequence of the walk is found by picking one of the k coordinates uniformly at random and *flipping the bit* at that coordinate. That is, switch from 0 to 1, or from 1 to 0. Here is the transition matrix for random walk on the 3-hypercube graph.

$$P = \begin{array}{c} \\ 000 \\ 100 \\ 010 \\ 110 \\ 001 \\ 101 \\ 011 \\ 111 \end{array} \begin{array}{c} \begin{array}{cccccccc} 000 & 100 & 010 & 110 & 001 & 101 & 011 & 111 \end{array} \\ \begin{pmatrix} 0 & 1/3 & 1/3 & 0 & 1/3 & 0 & 0 & 0 \\ 1/3 & 0 & 0 & 1/3 & 0 & 1/3 & 0 & 0 \\ 1/3 & 0 & 0 & 1/3 & 0 & 0 & 1/3 & 0 \\ 0 & 1/3 & 1/3 & 0 & 0 & 0 & 0 & 1/3 \\ 1/3 & 0 & 0 & 0 & 0 & 1/3 & 1/3 & 0 \\ 0 & 1/3 & 0 & 0 & 1/3 & 0 & 0 & 1/3 \\ 0 & 0 & 1/3 & 0 & 1/3 & 0 & 0 & 1/3 \\ 0 & 0 & 0 & 1/3 & 0 & 1/3 & 1/3 & 0 \end{pmatrix} \end{array}.$$

■

Example 2.9 (Birth-and-death chain) A birth-and-death Markov chain is a process with countably infinite state space and two types of transitions: *births* from i to $i+1$ and *deaths* from i to $i-1$. Define transition probabilities

$$P_{ij} = \begin{cases} q_i, & \text{if } j = i-1, \\ p_i, & \text{if } j = i+1, \\ 1 - p_i - q_i, & \text{if } j = i, \\ 0, & \text{otherwise,} \end{cases}$$

for $0 \le p_i, q_i$ and $p_i + q_i \le 1$. The infinite, tri-diagonal transition matrix is

$$P = \begin{array}{c} \\ 0 \\ 1 \\ 2 \\ 3 \\ \vdots \end{array} \begin{array}{c} \begin{array}{ccccc} 0 & 1 & 2 & 3 & \cdots \end{array} \\ \begin{pmatrix} 1-p_0 & p_0 & 0 & 0 & \cdots \\ q_1 & 1-q_1-p_1 & p_1 & 0 & \cdots \\ 0 & q_2 & 1-q_2-p_2 & p_2 & \cdots \\ 0 & 0 & q_3 & 1-q_3-p_3 & \cdots \\ \vdots & \vdots & \vdots & \vdots & \ddots \end{pmatrix} \end{array}.$$

Birth-and-death chains are used to model population size, spread of disease, and the number of customers in line at the supermarket. The case when $p_i = p$ and $q_i = q$ are constant gives random walk on the non-negative integers. ■

Example 2.10 (The lazy librarian and move-to-front) A library has k books and one very long bookshelf. Each book's popularity is measured by a probability. The chance that book i will be checked out is p_i, with $p_1 + \cdots + p_k = 1$. When patrons look for books they go to the bookshelf, start at the front (left end), and scan down the bookshelf, left to right, until they find the book they want.

The library wants to organize the bookshelf in such a way as to minimize patrons' search time, measured by how many books they need to scan before they find the one they want. If the probabilities p_i are known then the best way to organize the bookshelf is by order of popularity, with the most popular, high probability books at the front. However, the actual probabilities are not known.

A lazy librarian uses the following method for organizing the books. When a patron returns a book the librarian simply puts the book at the front of the shelf. All the other books move down the shelf to the right. For instance, assume that the library has six books labeled a to f. If the bookshelf is currently ordered (b, c, a, f, e, d) and book e is chosen, the new ordering of the shelf after the book is returned is (e, b, c, a, f, d). We assume that one book is chosen and returned at a time.

Such a scheme has the advantage that over time, as books get taken out and returned, the most popular books will gravitate to the front of the bookshelf and the least popular books will gravitate to the back. The process is known as the move-to-front self-organizing scheme. Move-to-front is studied in computer science as a dynamic data structure for maintaining a linked list.

Let X_n be the order of the books after n steps. Then, the move-to-front process $X_0, X_1, \ldots$ is a Markov chain whose state space is the set of all permutations (orderings) of k books. The transition matrix has dimension $k! \times k!$. Let $\sigma = (\sigma_1, \ldots, \sigma_k)$ and $\tau = (\tau_1, \ldots, \tau_k)$ denote permutations. Then,

$$P_{\sigma,\tau} = P(X_1 = \tau | X_0 = \sigma) = p_x,$$

if $\tau_1 = x$ and τ can be obtained from σ by moving item x to the front of σ. Here is the transition matrix for move-to-front for a $k = 3$ book library.

$$\boldsymbol{P} = \begin{array}{c} abc \\ acb \\ bac \\ bca \\ cab \\ cba \end{array} \overset{\begin{array}{cccccc} abc & acb & bac & bca & cab & cba \end{array}}{\begin{pmatrix} p_a & 0 & p_b & 0 & p_c & 0 \\ 0 & p_a & p_b & 0 & p_c & 0 \\ p_a & 0 & p_b & 0 & 0 & p_c \\ p_a & 0 & 0 & p_b & 0 & p_c \\ 0 & p_a & 0 & p_b & p_c & 0 \\ 0 & p_a & 0 & p_b & 0 & p_c \end{pmatrix}}.$$

Move-to-front is related to a card-shuffling scheme known as *random-to-top*. Given a standard deck of cards, pick a card uniformly at random and move it to the top of the deck. Random-to-top is obtained from move-to-front by letting $k = 52$

and $p_i = 1/52$, for all i. Of interest is how many such shuffles will mix up the deck of cards. ■

Example 2.11 (Weighted, directed graphs) A *weighted graph* associates a positive number (weight) with every edge. An example is shown in Figure 2.4. The graph contains *loops* at vertices b, c, and f, which are edges joining a vertex to itself.

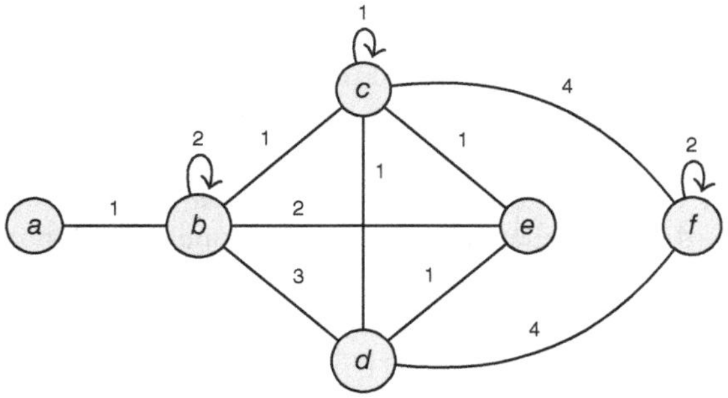

Figure 2.4 Weighted graph with loops.

For random walk on a weighted graph, transition probabilities are proportional to the sum of the weights. If vertices i and j are neighbors, let $w(i,j)$ denote the weight of the edge joining i and j. Let $w(i) = \sum_{i \sim k} w(i,k)$ be the sum of the weights on all edges joining i to its neighbors. The transition matrix is given by

$$P_{ij} = \begin{cases} \dfrac{w(i,j)}{w(i)}, & \text{if } i \sim j, \\ 0, & \text{otherwise.} \end{cases}$$

For the weighted graph in Figure 2.4, the transition matrix is

$$P = \begin{array}{c} \\ a \\ b \\ c \\ d \\ e \\ f \end{array} \begin{array}{c} \begin{array}{cccccc} a & b & c & d & e & f \end{array} \\ \begin{pmatrix} 0 & 1 & 0 & 0 & 0 & 0 \\ 1/9 & 2/9 & 1/9 & 3/9 & 2/9 & 0 \\ 0 & 1/8 & 1/8 & 1/8 & 1/8 & 4/8 \\ 0 & 3/9 & 1/9 & 0 & 1/9 & 4/9 \\ 0 & 2/4 & 1/4 & 1/4 & 0 & 0 \\ 0 & 0 & 4/10 & 4/10 & 0 & 2/10 \end{pmatrix} \end{array}.$$

A *directed graph* is a graph where edges have an associated direction. For every pair of vertices i and j one can have an edge from i to j and an edge from j to i. In a weighted, directed graph, there is a weight function $w(i,j)$ which gives the weight for the directed edge from i to j.

Every Markov chain can be described as a random walk on a weighted, directed graph whose vertex set is the state space of the chain. We call such a graph a *transition graph* for the Markov chain.

To create a transition graph from a transition matrix $\boldsymbol{P}$, for each pair of vertices i and j such that $P_{ij} > 0$, put a directed edge between i and j with edge weight P_{ij}.

Conversely, given a weighted, directed graph with non-negative weight function $w(i,j)$, to obtain the corresponding transition matrix let

$$P_{ij} = \frac{w(i,j)}{\sum_{i\sim k} w(i,k)} = \frac{w(i,j)}{w(i)}, \quad \text{for all } i,j.$$

Observe that matrix entries are non-negative and rows sum to 1, as for each i,

$$\sum_j P_{ij} = \sum_j \frac{w(i,j)}{w(i)} = \frac{1}{w(i)} \sum_{j\sim i} w(i,j) = \frac{w(i)}{w(i)} = 1.$$

For example, consider the transition matrix

$$\boldsymbol{P} = \begin{array}{c} \\ a \\ b \\ c \end{array} \begin{array}{c} \begin{array}{ccc} a & b & c \end{array} \\ \begin{pmatrix} 0 & 1 & 0 \\ 0.1 & 0.2 & 0.7 \\ 0.4 & 0 & 0.6 \end{pmatrix} \end{array}.$$

Two versions of the transition graph are shown in Figure 2.5. Note that multiplying the weights in the transition graph by a constant does not change the resulting transition matrix. ■

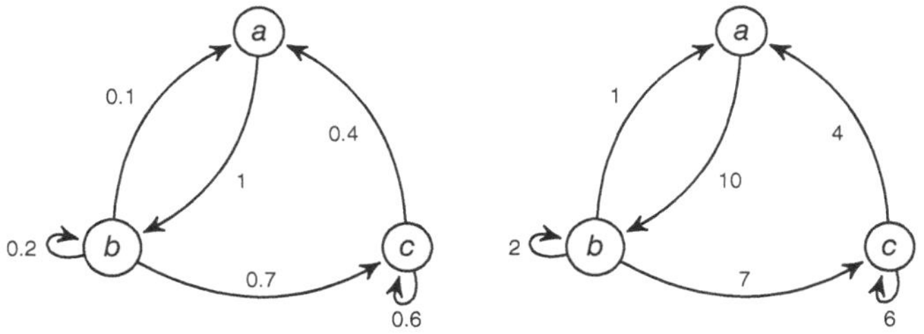

Figure 2.5 Markov transition graphs.

Example 2.12 The metastatic progression of lung cancer throughout the body is modeled in Newton et al. (2012). The 50 possible metastatic locations for cancer spread are the state space for a Markov chain model. Matrix entries were estimated from autopsy data extracted from 3,827 patients. The progress of the disease is observed as a random walk on the weighted, directed graph in Figure 2.6. Site 23 represents the lung.

An important quantity associated with this model is the *mean first-passage time*, the average number of steps it takes to move from the lung to each of the other locations in the body. ■

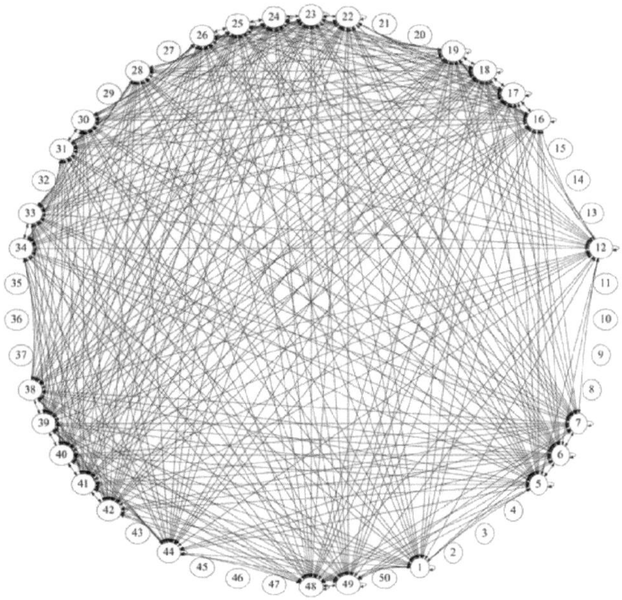

Figure 2.6 Lung cancer network as a weighted, directed graph (weights not shown). *Source:* Newton et al. (2012).

2.3 BASIC COMPUTATIONS

A powerful feature of Markov chains is the ability to use matrix algebra for computing probabilities. To use matrix methods, we consider probability distributions as vectors.

A *probability vector* is a row vector of non-negative numbers that sum to 1. Bold Greek letters, such as $\boldsymbol{\alpha}$, $\boldsymbol{\lambda}$, and $\boldsymbol{\pi}$, are used to denote such vectors.

Assume that X is a discrete random variable with $P(X = j) = \alpha_j$, for $j = 1, 2, \dots$. Then, $\boldsymbol{\alpha} = (\alpha_1, \alpha_2, \dots)$ is a probability vector. We say that *the distribution of* X *is* $\boldsymbol{\alpha}$. For matrix computations we will identify discrete probability distributions with row vectors.

For a Markov chain $X_0, X_1, \dots$, the distribution of X_0 is called the *initial distribution* of the Markov chain. If $\boldsymbol{\alpha}$ is the initial distribution, then $P(X_0 = j) = \alpha_j$, for all j.

n-Step Transition Probabilities

For states i and j, and $n \geq 1$, $P(X_n = j | X_0 = i)$ is the probability that the chain started in i hits j in n steps. The n-step transition probabilities can be arranged in a matrix. The matrix whose ijth entry is $P(X_n = j | X_0 = i)$ is the *n-step transition matrix* of the Markov chain. Of course, for $n = 1$, this is just the usual transition matrix $\boldsymbol{P}$.

For $n \geq 1$, one of the central computational results for Markov chains is that the n-step transition matrix is precisely $\boldsymbol{P}^n$, the nth matrix power of $\boldsymbol{P}$. To show that

$P(X_n = j|X_0 = i) = (P^n)_{ij}$, condition on X_{n-1}, which gives

$$\begin{aligned} P(X_n = j|X_0 = i) &= \sum_k P(X_n = j|X_{n-1} = k, X_0 = i)P(X_{n-1} = k|X_0 = i) \\ &= \sum_k P(X_n = j|X_{n-1} = k)P(X_{n-1} = k|X_0 = i) \\ &= \sum_k P_{kj}P(X_{n-1} = k|X_0 = i), \end{aligned}$$

where the second equality is by the Markov property, and the third equality is by time-homogeneity.

For $n = 2$, this gives

$$P(X_2 = j|X_0 = i) = \sum_k P_{kj}P(X_1 = k|X_0 = i) = \sum_k P_{kj}P_{ik} = \left(P^2\right)_{ij}.$$

Hence, the two-step transition matrix is $\boldsymbol{P}^2$. Similarly, for $n = 3$,

$$P(X_3 = j|X_0 = i) = \sum_k P_{kj}P(X_2 = k|X_0 = i) = \sum_k P_{kj}P^2_{ik} = \left(P^3\right)_{ij}.$$

The three-step transition matrix is $\boldsymbol{P}^3$. Induction establishes the general result. (See Section 2.6 for an introduction to mathematical induction.)

n-Step Transition Matrix

Let $X_0, X_1, \dots$ be a Markov chain with transition matrix $\boldsymbol{P}$. The matrix $\boldsymbol{P}^n$ is the n-step transition matrix of the chain. For $n \geq 0$,

$$P^n_{ij} = P(X_n = j|X_0 = i), \quad \text{for all } i, j.$$

Note that $P^n_{ij} = (P^n)_{ij}$. Do not confuse this with $(P_{ij})^n$, which is the number P_{ij} raised to the nth power. Also note that $\boldsymbol{P}^0$ is the identity matrix. That is,

$$P^0_{ij} = P(X_0 = j|X_0 = i) = \begin{cases} 1, & \text{if } i = j, \\ 0, & \text{if } i \neq j. \end{cases}$$

Example 2.13 Consider random walk on the cycle graph consisting of five vertices $\{0, 1, 2, 3, 4\}$. Describe the six-step transition probabilities of the chain.

Solution The transition matrix is

$$
P = \begin{array}{c} \\ 0 \\ 1 \\ 2 \\ 3 \\ 4 \end{array}
\begin{array}{c}
\begin{array}{ccccc} 0 & 1 & 2 & 3 & 4 \end{array} \\
\begin{pmatrix}
0 & 1/2 & 0 & 0 & 1/2 \\
1/2 & 0 & 1/2 & 0 & 0 \\
0 & 1/2 & 0 & 1/2 & 0 \\
0 & 0 & 1/2 & 0 & 1/2 \\
1/2 & 0 & 0 & 1/2 & 0
\end{pmatrix}
\end{array}.
$$

The six-step transition matrix is

$$
P^6 = \begin{array}{c} \\ 0 \\ 1 \\ 2 \\ 3 \\ 4 \end{array}
\begin{array}{c}
\begin{array}{ccccc} 0 & 1 & 2 & 3 & 4 \end{array} \\
\begin{pmatrix}
5/16 & 7/64 & 15/64 & 15/64 & 7/64 \\
7/64 & 5/16 & 7/64 & 15/64 & 15/64 \\
15/64 & 7/64 & 5/16 & 7/64 & 15/64 \\
15/64 & 15/64 & 7/64 & 5/16 & 7/64 \\
7/64 & 15/64 & 15/64 & 7/64 & 5/16
\end{pmatrix}
\end{array}.
$$

This matrix reveals that starting from any vertex on the cycle, in six steps the walk either (i) visits itself, with probability 5/16, (ii) visits a neighbor, with probability $7/64 + 7/64 = 7/32$, or (iii) visits a non-neighbor, with probability $15/64 + 15/64 = 15/32$. ■

Example 2.14 For gambler's ruin, assume that the gambler's initial stake is \$3 and the gambler plays until either gaining \$8 or going bust. At each play the gambler wins \$1, with probability 0.6, or loses \$1, with probability 0.4. Find the gambler's expected fortune after four plays.

Solution The transition matrix is

$$
P = \begin{array}{c} \\ 0 \\ 1 \\ 2 \\ 3 \\ 4 \\ 5 \\ 6 \\ 7 \\ 8 \end{array}
\begin{array}{c}
\begin{array}{ccccccccc} 0 & 1 & 2 & 3 & 4 & 5 & 6 & 7 & 8 \end{array} \\
\begin{pmatrix}
1 & 0 & 0 & 0 & 0 & 0 & 0 & 0 & 0 \\
0.4 & 0 & 0.6 & 0 & 0 & 0 & 0 & 0 & 0 \\
0 & 0.4 & 0 & 0.6 & 0 & 0 & 0 & 0 & 0 \\
0 & 0 & 0.4 & 0 & 0.6 & 0 & 0 & 0 & 0 \\
0 & 0 & 0 & 0.4 & 0 & 0.6 & 0 & 0 & 0 \\
0 & 0 & 0 & 0 & 0.4 & 0 & 0.6 & 0 & 0 \\
0 & 0 & 0 & 0 & 0 & 0.4 & 0 & 0.6 & 0 \\
0 & 0 & 0 & 0 & 0 & 0 & 0.4 & 0 & 0.6 \\
0 & 0 & 0 & 0 & 0 & 0 & 0 & 0 & 1
\end{pmatrix}
\end{array},
$$

with

$$
P^4 = \begin{array}{c} \\ 0 \\ 1 \\ 2 \\ 3 \\ 4 \\ 5 \\ 6 \\ 7 \\ 8 \end{array}
\begin{array}{c}
\begin{array}{ccccccccc} 0 & 1 & 2 & 3 & 4 & 5 & 6 & 7 & 8 \end{array} \\
\begin{pmatrix}
1 & 0 & 0 & 0 & 0 & 0 & 0 & 0 & 0 \\
0.496 & 0.115 & 0 & 0.259 & 0 & 0.130 & 0 & 0 & 0 \\
0.237 & 0 & 0.288 & 0 & 0.346 & 0 & 0.130 & 0 & 0 \\
0.064 & 0.115 & 0 & 0.346 & 0 & 0.346 & 0 & 0.130 & 0 \\
0.026 & 0 & 0.154 & 0 & 0.346 & 0 & 0.346 & 0 & 0.130 \\
0 & 0.026 & 0 & 0.154 & 0 & 0.346 & 0 & 0.259 & 0.216 \\
0 & 0 & 0.026 & 0 & 0.154 & 0 & 0.288 & 0 & 0.533 \\
0 & 0 & 0 & 0.026 & 0 & 0.115 & 0 & 0.115 & 0.744 \\
0 & 0 & 0 & 0 & 0 & 0 & 0 & 0 & 1
\end{pmatrix}
\end{array}.
$$

The gambler's expected fortune is

$$
\begin{aligned}
E(X_4|X_0 = 3) &= \sum_{j=0}^{8} jP(X_4 = j|X_0 = 3) = \sum_{j=0}^{8} jP^4_{3,j} \\
&= 0(0.064) + 1(0.115) + 3(0.346) + 5(0.346) + 7(0.130) \\
&= \$3.79.
\end{aligned}
$$

■

Chapman–Kolmogorov Relationship

For $m, n \geq 0$, the matrix identity $\boldsymbol{P}^{m+n} = \boldsymbol{P}^m \boldsymbol{P}^n$ gives

$$
P^{m+n}_{ij} = \sum_k P^m_{ik} P^n_{kj}, \quad \text{for all } i, j.
$$

By time-homogeneity, this gives

$$
\begin{aligned}
P(X_{n+m} = j|X_0 = i) &= \sum_k P(X_m = k|X_0 = i)P(X_n = j|X_0 = k) \\
&= \sum_k P(X_m = k|X_0 = i)P(X_{m+n} = j|X_m = k).
\end{aligned}
$$

The probabilistic interpretation is that transitioning from i to j in $m + n$ steps is equivalent to transitioning from i to some state k in m steps and then moving from that state to j in the remaining n steps. This is known as the *Chapman–Kolmogorov relationship*.

Distribution of X_n

In general, a Markov chain $X_0, X_1, \ldots$ is not a sequence of identically distributed random variables. For $n \geq 1$, the marginal distribution of X_n depends on the n-step transition matrix $\boldsymbol{P}^n$, as well as the initial distribution $\boldsymbol{\alpha}$. To obtain the probability mass function of X_n, condition on the initial state X_0. For all j,

$$P(X_n = j) = \sum_i P(X_n = j | X_0 = i) P(X_0 = i) = \sum_i P^n_{ij} \alpha_i. \tag{2.3}$$

The sum in the last expression of Equation (2.3) can be interpreted in terms of matrix operations on vectors. It is the dot product of the initial probability vector $\boldsymbol{\alpha}$ with the jth column of $\boldsymbol{P}^n$. That is, it is the jth component of the vector–matrix product $\boldsymbol{\alpha P}^n$. (Remember: $\boldsymbol{\alpha}$ is a *row* vector.)

Distribution of X_n

Let $X_0, X_1, \ldots$ be a Markov chain with transition matrix $\boldsymbol{P}$ and initial distribution $\boldsymbol{\alpha}$. For all $n \geq 0$, the distribution of X_n is $\boldsymbol{\alpha P}^n$. That is,

$$P(X_n = j) = (\alpha P^n)_j, \quad \text{for all } j.$$

Example 2.15 Consider the weather chain introduced in Example 2.3. For tomorrow, the meteorologist predicts a 50% chance of snow and a 50% chance of rain. Find the probability that it will snow 2 days later.

Solution As the ordered states of the chain are rain, snow, and clear, the initial distribution is $\boldsymbol{\alpha} = (0.5, 0.5, 0)$. We have

$$\boldsymbol{P} = \begin{matrix} & r & s & c \\ r & 0.2 & 0.6 & 0.2 \\ s & 0.1 & 0.8 & 0.1 \\ c & 0.1 & 0.6 & 0.3 \end{matrix} \quad \text{and} \quad \boldsymbol{P}^2 = \begin{matrix} & r & s & c \\ r & 0.12 & 0.72 & 0.16 \\ s & 0.11 & 0.76 & 0.13 \\ c & 0.11 & 0.72 & 0.17 \end{matrix}$$

This gives

$$\boldsymbol{\alpha P}^2 = (0.5, 0.5, 0) \begin{pmatrix} 0.12 & 0.72 & 0.16 \\ 0.11 & 0.76 & 0.13 \\ 0.11 & 0.72 & 0.17 \end{pmatrix} = (0.115, 0.74, 0.145).$$

The desired probability of snow is $P(X_2 = s) = (\alpha P^2)_s = 0.74$. ■

Present, Future, and Most Recent Past

The Markov property says that past and future are independent given the present. It is also true that past and future are independent, given *the most recent* past.

Markov Property

Let $X_0, X_1, \ldots$ be a Markov chain. Then, for all $m < n$,

$$
\begin{aligned}
P(X_{n+1} = j | X_0 = i_0, \ldots, X_{n-m-1} = i_{n-m-1}, X_{n-m} = i) \\
= P(X_{n+1} = j | X_{n-m} = i) \\
= P(X_{m+1} = j | X_0 = i) = P_{ij}^{m+1}, \qquad (2.4)
\end{aligned}
$$

for all $i, j, i_0, \ldots, i_{n-m-1}$, and $n \geq 0$.

Proof. With $m = 0$, Equation (2.4) reduces to the defining Markov relationship as stated in Equation (2.1).

Let $m = 1$. By conditioning on X_n,

$$
\begin{aligned}
P(X_{n+1} &= j | X_0 = i_0, \ldots, X_{n-1} = i) \\
&= \sum_k P(X_{n+1} = j | X_0 = i_0, \ldots, X_{n-1} = i, X_n = k) \\
&\quad \times P(X_n = k | X_0 = i_0, \ldots, X_{n-1} = i) \\
&= \sum_k P(X_{n+1} = j | X_n = k) P(X_n = k | X_{n-1} = i) \\
&= \sum_k P(X_1 = j | X_0 = k) P(X_1 = k | X_0 = i) \\
&= \sum_k P_{kj} P_{ik} = P_{ij}^2.
\end{aligned}
$$

The third equality is by time-homogeneity. For general m, induction gives the result by conditioning on X_{n-m+1}. ■

Joint Distribution

The marginal distributions of a Markov chain are determined by the initial distribution $\boldsymbol{\alpha}$ and the transition matrix $\boldsymbol{P}$. However, a much stronger result is true. In fact, $\boldsymbol{\alpha}$ and $\boldsymbol{P}$ determine all the joint distributions of a Markov chain, that is, the joint distribution of any finite subset of $X_0, X_1, X_2, \ldots$. In that sense, the initial distribution and transition matrix give a complete probabilistic description of a Markov chain.

To illustrate, consider an arbitrary joint probability, such as

$$P(X_5 = i, X_6 = j, X_9 = k, X_{17} = l), \text{ for some states } i, j, k, l.$$

For the underlying event, the chain moves to i in five steps, then to j in one step, then to k in three steps, and then to l in eight steps. With initial distribution $\boldsymbol{\alpha}$, intuition suggests that

$$P(X_5 = i, X_6 = j, X_9 = k, X_{17} = l) = (\alpha P^5)_i P_{ij} P^3_{jk} P^8_{kl}.$$

Indeed, conditional probability, the Markov property, and time-homogeneity give

$$\begin{aligned}
&P(X_5 = i, X_6 = j, X_9 = k, X_{17} = l)\\
&\quad = P(X_{17} = l | X_5 = i, X_6 = j, X_9 = k) P(X_9 = k | X_5 = i, X_6 = j)\\
&\qquad \times P(X_6 = j | X_5 = i) P(X_5 = i)\\
&\quad = P(X_{17} = l | X_9 = k) P(X_9 = k | X_6 = j) P(X_6 = j | X_5 = i) P(X_5 = i)\\
&\quad = P(X_8 = l | X_0 = k) P(X_3 = k | X_0 = j) P(X_1 = j | X_0 = i) P(X_5 = i)\\
&\quad = P^8_{kl} P^3_{jk} P_{ij} (\alpha P^5)_i.
\end{aligned}$$

The joint probability is obtained from just the initial distribution $\boldsymbol{\alpha}$ and the transition matrix $\boldsymbol{P}$. For completeness, here is the general formula.

Joint Distribution

Let $X_0, X_1, \ldots$ be a Markov chain with transition matrix $\boldsymbol{P}$ and initial distribution $\boldsymbol{\alpha}$. For all $0 \le n_1 < n_2 < \cdots < n_{k-1} < n_k$ and states $i_1, i_2, \ldots, i_{k-1}, i_k$,

$$\begin{aligned}
&P(X_{n_1} = i_1, X_{n_2} = i_2, \ldots, X_{n_{k-1}} = i_{k-1}, X_{n_k} = i_k)\\
&\qquad = (\alpha P^{n_1})_{i_1} (P^{n_2-n_1})_{i_1 i_2} \cdots (P^{n_k-n_{k-1}})_{i_{k-1} i_k}. \qquad (2.5)
\end{aligned}$$

Example 2.16 Danny's daily lunch choices are modeled by a Markov chain with transition matrix

$$\boldsymbol{P} = \begin{array}{c} \\ \text{Burrito} \\ \text{Falafel} \\ \text{Pizza} \\ \text{Sushi} \end{array} \begin{array}{c} \begin{array}{cccc} \text{Burrito} & \text{Falafel} & \text{Pizza} & \text{Sushi} \end{array} \\ \begin{pmatrix} 0.0 & 0.5 & 0.5 & 0.0 \\ 0.5 & 0.0 & 0.5 & 0.0 \\ 0.4 & 0.0 & 0.0 & 0.6 \\ 0.0 & 0.2 & 0.6 & 0.2 \end{pmatrix} \end{array}.$$

On Sunday, Danny chooses lunch uniformly at random. Find the probability that he chooses sushi on the following Wednesday and Friday, and pizza on Saturday.

Solution Let b, f, p, s denote Danny's lunch choices, respectively. Let X_0 denote Danny's lunch choice on Sunday. The desired probability is

$$P(X_3 = s, X_5 = s, X_6 = p) = (\alpha P^3)_s P^2_{ss} P_{sp},$$

where $\boldsymbol{\alpha} = (1/4, 1/4, 1/4, 1/4)$. We have

$$\boldsymbol{P}^2 = \begin{array}{c} \\ b \\ f \\ p \\ s \end{array}\begin{array}{c} \begin{array}{cccc} b & f & p & s \end{array} \\ \begin{pmatrix} 0.45 & 0.00 & 0.25 & 0.30 \\ 0.20 & 0.25 & 0.25 & 0.30 \\ 0.00 & 0.32 & 0.56 & 0.12 \\ 0.34 & 0.04 & 0.22 & 0.40 \end{pmatrix} \end{array}, \quad \boldsymbol{P}^3 = \begin{array}{c} \\ b \\ f \\ p \\ s \end{array}\begin{array}{c} \begin{array}{cccc} b & f & p & s \end{array} \\ \begin{pmatrix} 0.100 & 0.285 & 0.405 & 0.210 \\ 0.225 & 0.160 & 0.405 & 0.210 \\ 0.384 & 0.024 & 0.232 & 0.360 \\ 0.108 & 0.250 & 0.430 & 0.212 \end{pmatrix} \end{array}.$$

The desired probability is

$$(\alpha P^3)_s P^2_{ss} P_{sp} = (0.248)(0.40)(0.60) = 0.05952. \qquad \blacksquare$$

2.4 LONG-TERM BEHAVIOR—THE NUMERICAL EVIDENCE

In any stochastic—or deterministic—process, the long-term behavior of the system is often of interest.

The Canadian Forest Fire Weather Index is widely used as a means to estimate the risk of wildfire. The Ontario Ministry of Natural Resources uses the index to classify each day's risk of forest fire as either nil, low, moderate, high, or extreme.

Martell (1999) gathered daily fire risk data over 26 years at 15 weather stations across Ontario to construct a five-state Markov chain model for the daily changes in the index. The transition matrix from one location for the early summer subseason is

$$\boldsymbol{P} = \begin{array}{c} \\ \text{Nil} \\ \text{Low} \\ \text{Moderate} \\ \text{High} \\ \text{Extreme} \end{array}\begin{array}{c} \begin{array}{ccccc} \text{Nil} & \text{Low} & \text{Moderate} & \text{High} & \text{Extreme} \end{array} \\ \begin{pmatrix} 0.575 & 0.118 & 0.172 & 0.109 & 0.026 \\ 0.453 & 0.243 & 0.148 & 0.123 & 0.033 \\ 0.104 & 0.343 & 0.367 & 0.167 & 0.019 \\ 0.015 & 0.066 & 0.318 & 0.505 & 0.096 \\ 0.000 & 0.060 & 0.149 & 0.567 & 0.224 \end{pmatrix} \end{array}.$$

Of interest to forest managers is the long-term probability distribution of the daily index. Regardless of the risk on any particular day, what is the long-term likelihood of risk for a typical day in the early summer?

Consider the n-step transition matrix for several increasing values of n.

$$\boldsymbol{P}^2 = \begin{pmatrix} 0.404 & 0.164 & 0.218 & 0.176 & 0.038 \\ 0.388 & 0.173 & 0.212 & 0.185 & 0.042 \\ 0.256 & 0.234 & 0.259 & 0.210 & 0.041 \\ 0.079 & 0.166 & 0.304 & 0.372 & 0.079 \\ 0.051 & 0.117 & 0.277 & 0.446 & 0.109 \end{pmatrix},$$

$$\boldsymbol{P}^3 = \begin{pmatrix} 0.332 & 0.176 & 0.235 & 0.211 & 0.046 \\ 0.326 & 0.175 & 0.235 & 0.216 & 0.047 \\ 0.283 & 0.192 & 0.247 & 0.229 & 0.049 \\ 0.158 & 0.183 & 0.280 & 0.312 & 0.067 \\ 0.118 & 0.165 & 0.286 & 0.353 & 0.078 \end{pmatrix},$$

$$P^5 = \begin{pmatrix} 0.282 & 0.180 & 0.248 & 0.239 & 0.051 \\ 0.279 & 0.180 & 0.248 & 0.241 & 0.052 \\ 0.273 & 0.181 & 0.250 & 0.244 & 0.052 \\ 0.235 & 0.183 & 0.259 & 0.266 & 0.057 \\ 0.217 & 0.183 & 0.264 & 0.277 & 0.060 \end{pmatrix},$$

$$P^{10} = \begin{pmatrix} 0.264 & 0.181 & 0.252 & 0.249 & 0.053 \\ 0.264 & 0.181 & 0.252 & 0.249 & 0.054 \\ 0.264 & 0.181 & 0.252 & 0.249 & 0.054 \\ 0.263 & 0.181 & 0.252 & 0.25 & 0.054 \\ 0.262 & 0.181 & 0.252 & 0.251 & 0.054 \end{pmatrix},$$

$$P^{17} = \begin{pmatrix} 0.264 & 0.181 & 0.252 & 0.249 & 0.054 \\ 0.264 & 0.181 & 0.252 & 0.249 & 0.054 \\ 0.264 & 0.181 & 0.252 & 0.249 & 0.054 \\ 0.264 & 0.181 & 0.252 & 0.249 & 0.054 \\ 0.264 & 0.181 & 0.252 & 0.249 & 0.054 \end{pmatrix},$$

$$P^{18} = \begin{pmatrix} 0.264 & 0.181 & 0.252 & 0.249 & 0.054 \\ 0.264 & 0.181 & 0.252 & 0.249 & 0.054 \\ 0.264 & 0.181 & 0.252 & 0.249 & 0.054 \\ 0.264 & 0.181 & 0.252 & 0.249 & 0.054 \\ 0.264 & 0.181 & 0.252 & 0.249 & 0.054 \end{pmatrix}.$$

Numerical evidence suggests that matrix powers are converging to a limit. Furthermore, the rows of that limiting matrix are all the same. The fact that the rows of P^{17} are the same means that the probability of a particular fire index after 17 days does not depend on today's level of risk. After 17 days, the effect of the initial state has worn off, and no longer affects the distribution of the fire index.

Furthermore, $P^{18} = P^{17}$ (at least to three decimal places). In fact, $P^n = P^{17}$, for $n \geq 17$. The latter is intuitive since if the probability of hitting state j in 17 steps is independent of the initial state, then the probability of hitting j in 17 or more steps is also independent of the initial state. See also Exercise 2.16.

The long-term fire index distribution taken from the common row of P^{17} is

Nil	Low	Moderate	High	Extreme
0.264	0.181	0.252	0.249	0.054

R : Matrix Powers

The function `matrixpower(mat,n)` computes the nth power of a square matrix `mat`, for $n = 0, 1, 2, \ldots$ The function is found in the `R` script file **utilities.R**, which includes several useful utility functions for working with Markov chains.

R : Fire Weather Index

```
> P
              Nil      Low Moderate     High  Extreme
Nil         0.575    0.118    0.172    0.109    0.026
Low         0.453    0.243    0.148    0.123    0.033
Moderate 0.104    0.343    0.367    0.167    0.019
High        0.015    0.066    0.318    0.505    0.096
Extreme   0.000    0.060    0.149    0.567    0.224
> matrixpower(P,17)
                 Nil      Low Moderate     High  Extreme
Nil         0.263688 0.181273 0.251976 0.249484 0.0535768
Low         0.263688 0.181273 0.251976 0.249485 0.0535769
Moderate 0.263685 0.181273 0.251977 0.249486 0.0535772
High        0.263671 0.181273 0.251981 0.249495 0.0535789
Extreme   0.263663 0.181274 0.251982 0.249499 0.0535798
# round entries to three decimal places
> round(matrixpower(P,17),3)
              Nil      Low Moderate     High  Extreme
Nil         0.264    0.181    0.252    0.249    0.054
Low         0.264    0.181    0.252    0.249    0.054
Moderate 0.264    0.181    0.252    0.249    0.054
High        0.264    0.181    0.252    0.249    0.054
Extreme   0.264    0.181    0.252    0.249    0.054
```

Example 2.17 Changes in the distribution of wetlands in Yinchuan Plain, China are studied in Zhang et al. (2011). Wetlands are considered among the most important ecosystems on earth. A Markov model is developed to track yearly changes in wetland type. Based on imaging and satellite data from 1991, 1999, and 2006, researchers measured annual distributions of wetland type throughout the region and estimated the Markov transition matrix

$$P = \begin{array}{r} \\ \text{River} \\ \text{Lake} \\ \text{Pond} \\ \text{Paddy} \\ \text{Non} \end{array} \begin{array}{c} \begin{array}{ccccc} \text{River} & \text{Lake} & \text{Pond} & \text{Paddy} & \text{Non} \end{array} \\ \begin{pmatrix} 0.342 & 0.005 & 0.001 & 0.020 & 0.632 \\ 0.001 & 0.252 & 0.107 & 0.005 & 0.635 \\ 0.000 & 0.043 & 0.508 & 0.015 & 0.434 \\ 0.001 & 0.002 & 0.004 & 0.665 & 0.328 \\ 0.007 & 0.007 & 0.007 & 0.025 & 0.954 \end{pmatrix} \end{array}.$$

The state *Non* refers to nonwetland regions. Based on their model, the scientists predict that "The wetland distribution will essentially be in a steady state in Yinchuan Plain in approximately 100 years."

With technology one checks that $\boldsymbol{P}^{100} = \boldsymbol{P}^{101}$ has identical rows. The common row gives the predicted long-term, *steady-state* wetland distribution.

River	Lake	Pond	Paddy	Non
0.010	0.010	0.015	0.068	0.897

R : Limiting Distribution for Wetlands Type

```
> P
       River  Lake   Pond   Paddy    Non
River  0.342  0.005  0.001  0.020  0.632
Lake   0.001  0.252  0.107  0.005  0.635
Pond   0.003  0.043  0.507  0.014  0.433
Paddy  0.001  0.002  0.004  0.665  0.328
Non    0.007  0.007  0.007  0.025  0.954
> matrixpower(P,100)
       River  Lake   Pond   Paddy    Non
River   0.01  0.01   0.015  0.068  0.897
Lake    0.01  0.01   0.015  0.068  0.897
Pond    0.01  0.01   0.015  0.068  0.897
Paddy   0.01  0.01   0.015  0.068  0.897
Non     0.01  0.01   0.015  0.068  0.897
```

■

Random Walk on Cycle

Assume that the hopping frog of Example 2.8 now finds itself on a 25-lily pad cycle graph. If the frog starts hopping from vertex 1, where is it likely to be after many hops?

After a small number of hops the frog will tend to be close to its starting position at the top of the pond. But after a large number of hops the frog's position will tend to be randomly distributed about the cycle, that is, it will tend to be uniformly distributed on all the vertices. See Figure 2.7 and Table 2.1, where probabilities are shown for the frog's position after n steps, for several values of n.

Vertices 24, 25, 1, 2, and 3 are closest to the frog's starting position. We consider these vertices *near the top* of the cycle. Vertices 12, 13, 14, and 15 are the furthest away. We consider these *near the bottom* of the cycle. After just 12 hops, the frog is still relatively close to the starting vertex. The chance that the frog is near the top of the cycle is 0.61. After 25 steps, the probability of being near the top is eight times greater than the probability of being near the bottom—0.32 compared with 0.04. Even after 100 steps it is still almost twice as likely that the frog will be near the starting vertex as compared to being at the opposite side of the cycle.

After 400 steps, however, the frog's position is *mixed up* throughout the cycle and is very close to being uniformly distributed on all the vertices. The dependency on the frog's initial position has worn off and all vertices are essentially equally likely. Numerical evidence suggests that the long-term distribution of the frog's position is uniform on the vertices and independent of starting state.

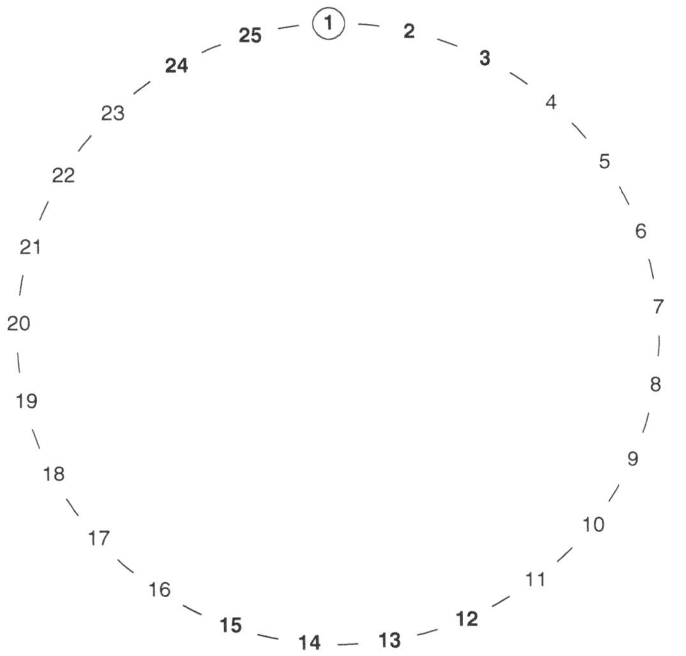

Figure 2.7 Frog starts hopping from vertex 1 of the 25-cycle graph.

TABLE 2.1 Probabilities, After *n* Steps, of the Frog's Position Near the Top and Bottom of the 25-Cycle Graph

n	24	25	1	2	3	12	13	14	15
1	0	0.5	0	0.5	0	0	0	0	0
2	0.25	0	0.5	0	0.25	0	0	0	0
6	0.23	0	0.31	0	0.23	0	0	0	0
12	0.19	0	0.23	0	0.19	0	0.00	0.00	0
25	0.00	0.16	0.00	0.16	0.0	0.01	0.01	0.01	0.01
60	0.10	0.00	0.10	0.00	0.10	0.02	0.03	0.03	0.02
100	0.08	0.01	0.08	0.01	0.08	0.03	0.04	0.04	0.03
400	0.04	0.04	0.04	0.04	0.04	0.04	0.04	0.04	0.04
401	0.04	0.04	0.04	0.04	0.04	0.04	0.04	0.04	0.04
402	0.04	0.04	0.04	0.04	0.04	0.04	0.04	0.04	0.04

Not all Markov chains exhibit the kind of long-term limiting behavior seen in random walk on the 25-cycle graph or in the fire index chain. Consider random walk on the cycle graph with six vertices. Here are several powers of the transition matrix.

$$
P = \begin{array}{c} \\ 0 \\ 1 \\ 2 \\ 3 \\ 4 \\ 5 \end{array}
\begin{array}{c} \begin{array}{cccccc} 0 & 1 & 2 & 3 & 4 & 5 \end{array} \\
\begin{pmatrix}
0 & 0.5 & 0 & 0 & 0 & 0.5 \\
0.5 & 0 & 0.5 & 0 & 0 & 0 \\
0 & 0.5 & 0 & 0.5 & 0 & 0 \\
0 & 0 & 0.5 & 0 & 0.5 & 0 \\
0 & 0 & 0 & 0.5 & 0 & 0.5 \\
0.5 & 0 & 0 & 0 & 0.5 & 0
\end{pmatrix} \end{array},
$$

$$
P^2 = \begin{pmatrix}
0.5 & 0 & 0.25 & 0 & 0.25 & 0 \\
0 & 0.5 & 0 & 0.25 & 0 & 0.25 \\
0.25 & 0 & 0.5 & 0 & 0.25 & 0 \\
0 & 0.25 & 0 & 0.5 & 0 & 0.25 \\
0.25 & 0 & 0.25 & 0 & 0.5 & 0 \\
0 & 0.25 & 0 & 0.25 & 0 & 0.5
\end{pmatrix},
$$

$$
P^3 = \begin{pmatrix}
0 & 0.375 & 0 & 0.25 & 0 & 0.375 \\
0.375 & 0 & 0.375 & 0 & 0.25 & 0 \\
0 & 0.375 & 0 & 0.375 & 0 & 0.25 \\
0.25 & 0 & 0.375 & 0 & 0.375 & 0 \\
0 & 0.25 & 0 & 0.375 & 0 & 0.375 \\
0.375 & 0 & 0.25 & 0 & 0.375 & 0
\end{pmatrix},
$$

$$
P^8 = \begin{pmatrix}
0.336 & 0 & 0.332 & 0 & 0.332 & 0 \\
0 & 0.336 & 0 & 0.332 & 0 & 0.332 \\
0.332 & 0 & 0.336 & 0 & 0.332 & 0 \\
0 & 0.332 & 0 & 0.336 & 0 & 0.332 \\
0.332 & 0 & 0.332 & 0 & 0.336 & 0 \\
0. & 0.332 & 0 & 0.332 & 0 & 0.336
\end{pmatrix},
$$

$$
P^{11} = \begin{pmatrix}
0 & 0.333 & 0 & 0.333 & 0 & 0.333 \\
0.336 & 0 & 0.336 & 0 & 0.333 & 0 \\
0 & 0.333 & 0 & 0.333 & 0 & 0.333 \\
0.333 & 0 & 0.333 & 0 & 0.333 & 0 \\
0 & 0.333 & 0 & 0.333 & 0 & 0.333 \\
0.333 & 0 & 0.333 & 0 & 0.333 & 0
\end{pmatrix},
$$

$$
P^{12} = \begin{pmatrix}
0.333 & 0 & 0.333 & 0 & 0.333 & 0 \\
0 & 0.333 & 0 & 0.333 & 0 & 0.333 \\
0.333 & 0 & 0.333 & 0 & 0.333 & 0 \\
0 & 0.333 & 0 & 0.333 & 0 & 0.333 \\
0.333 & 0 & 0.333 & 0 & 0.333 & 0 \\
0. & 0.333 & 0 & 0.333 & 0 & 0.333
\end{pmatrix}.
$$

The higher-order transition matrices flip-flop for odd and even powers. If the walk starts at an even vertex, it will always be on an even vertex after an even number of steps, and on an odd vertex after an odd number of steps. The parity of the position of the walk matches the parity of the starting vertex after an even number of steps. The parity switches after an odd number of steps.

High powers of the transition matrix do not converge to a limiting matrix. The long-term behavior of the walk depends on the starting state and depends on how many steps are taken.

2.5 SIMULATION

Simulation is a powerful tool for studying Markov chains. For many chains that arise in applications, state spaces are huge and matrix methods may not be practical, or even possible, to implement.

For instance, the card-shuffling chain introduced in Example 2.10 has a state space of $k!$ elements. The transition matrix for a standard deck of cards is $52! \times 52!$, which has about 6.5×10^{135} entries.

Even for a moderately sized 50×50 matrix, as in the cancer study of Example 2.12, numerical matrix computations can be difficult to obtain. The researchers of that study found it easier to derive their results by simulation.

A Markov chain can be simulated from an initial distribution and transition matrix. To simulate a Markov sequence $X_0, X_1, \ldots$, simulate each random variable sequentially conditional on the outcome of the previous variable. That is, first simulate X_0 according to the initial distribution. If $X_0 = i$, then simulate X_1 from the ith row of the transition matrix. If $X_1 = j$, then simulate X_2 from the jth row of the transition matrix, and so on.

Algorithm for Simulating a Markov Chain

Input: (i) initial distribution $\boldsymbol{\alpha}$, (ii) transition matrix $\boldsymbol{P}$, (iii) number of steps n.

Output: $X_0, X_1, \ldots, X_n$

Algorithm:

Generate X_0 according to $\boldsymbol{\alpha}$
FOR $i = 1, \ldots, n$
  Assume that $X_{i-1} = j$
  Set $\boldsymbol{p} = j$th row of $\boldsymbol{P}$
  Generate X_i according to $\boldsymbol{p}$
END FOR

To simulate a finite Markov chain, the algorithm is implemented in `R` by the function `markov(init,mat,n)`, which is contained in the file **utilities.R**. The arguments of the function are `init`, the initial distribution, `mat`, the transition matrix, and `n`, the number of steps to simulate. A call to `markov(init,mat,n)` generates the $(n + 1)$-element vector $(X_0, \ldots, X_n)$. The `markov` function allows for an optional fourth argument `states`, which is the state space given as a vector. If the state space has k elements, the function assigns the default value to `states` of $(1, \ldots, k)$.

Example 2.18 (Lung cancer study) Medical researchers can use simulation to study the progression of lung cancer in the body, as described in Example 2.12. The 50×50 transition matrix is stored in an Excel spreadsheet and can be downloaded into `R` from the file **cancerstudy.R**. The initial distribution is a vector of all 0s with a 1 at position 23, corresponding to the lung. See the documentation in the script file for the 50-site numbering system. Common sites are 24 and 25 (lymph nodes) and 22 (liver). Following are several simulations of the process for eight steps.

R : Simulating Lung Cancer Growth

```
> mat <- read.csv("lungcancer.csv",header=TRUE)
> init <- c(rep(0,22),1,rep(0,27))
    # all 0s with 1 at site 23 (lung)
> n <- 8
> markov(init,mat,n)
[1] 23 17 24 23 44  6  1 24 28
> markov(init,mat,n)
[1] 23 25 25 19 25  1 24 22 22
> markov(init,mat,n)
[1] 23 18 44  7 22 23 30 24 33
> markov(init,mat,n)
[1] 23 22 24 24 30 24 24 23 24
```

Newton et al. (2012) use simulation to estimate the *mean first passage time*—the number of steps, on average, it takes for cancer to pass from the lung to each other location in the body, "something a static autopsy data set cannot give us directly." The authors conclude that their study gives "important baseline quantitative insight into the structure of lung cancer progression networks." ■

Example 2.19 University administrators have developed a Markov model to simulate graduation rates at their school. Students might drop out, repeat a year, or move on to the next year. Students have a 3% chance of repeating the year. First-years and sophomores have a 6% chance of dropping out. For juniors and seniors, the drop-out rate is 4%. The transition matrix for the model is

$$
P = \begin{array}{c} \\ \text{Drop} \\ \text{Fr} \\ \text{So} \\ \text{Jr} \\ \text{Sr} \\ \text{Grad} \end{array}
\begin{array}{c} \begin{array}{cccccc} \text{Drop} & \text{Fr} & \text{So} & \text{Jr} & \text{Sr} & \text{Grad} \end{array} \\
\begin{pmatrix} 1 & 0 & 0 & 0 & 0 & 0 \\ 0.06 & 0.03 & 0.91 & 0 & 0 & 0 \\ 0.06 & 0 & 0.03 & 0.91 & 0 & 0 \\ 0.04 & 0 & 0 & 0.03 & 0.93 & 0 \\ 0.04 & 0 & 0 & 0 & 0.03 & 0.93 \\ 0 & 0 & 0 & 0 & 0 & 1 \end{pmatrix} \end{array}.
$$

Eventually, students will either drop out or graduate. See Figure 2.8 for the transition graph. To simulate the long-term probability that a new student graduates, the chain is run for 10 steps with initial distribution $\alpha = (0, 1, 0, 0, 0, 0)$, taking X_{10} as a *long-term* sample. (With high probability, a student will either drop out or graduate by 10 years.) The simulation is repeated 10,000 times, each time keeping track of whether a student graduates or drops out. The estimated long-term probability of graduating is 0.8037.

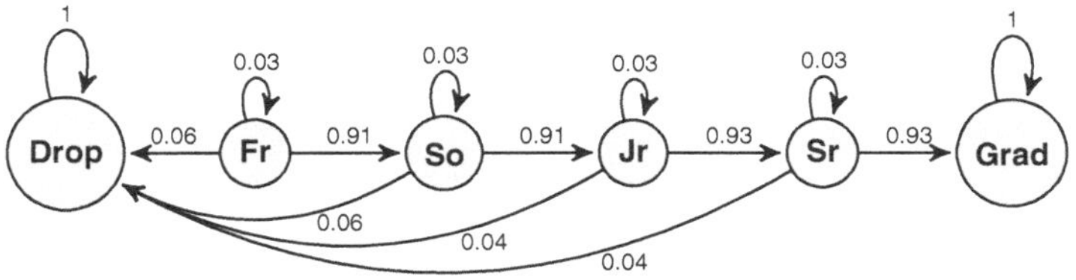

Figure 2.8 Transition graph of the graduation Markov chain.

R : Simulating Graduation, Drop-out Rate

```
# graduation.R
> init <- c(0,1,0,0,0,0) #Student starts as first-year
> P <- matrix(c(1,0,0,0,0,0,0.06,0.03,0.91,0,
  0,0,0.06,0,0.03,0.91,0,0,0.04,0,0,0.03,0.93,0,
  0.04,0,0,0,0.03,0.93,0,0,0,0,0,1),nrow=6,byrow=T)
> states <-  c("Drop","Fr","So","Jr","Se","Grad")
> rownames(P) <- states
> colnames(P) <- states
> P
     Drop   Fr   So   Jr   Se Grad
Drop 1.00 0.00 0.00 0.00 0.00 0.00
Fr   0.06 0.03 0.91 0.00 0.00 0.00
So   0.06 0.00 0.03 0.91 0.00 0.00
Jr   0.04 0.00 0.00 0.03 0.93 0.00
Se   0.04 0.00 0.00 0.00 0.03 0.93
Grad 0.00 0.00 0.00 0.00 0.00 1.00
> markov(init,P,10,states)
 [1] "Fr"   "So"   "Ju"   "Se"   "Grad" "Grad"
 [7] "Grad" "Grad" "Grad" "Grad"
> sim <- replicate(10000,markov(init,P,10,states)[11])
> table(sim)/10000
  Drop   Grad
0.1963 0.8037
```

The graduation transition matrix is small enough so that it is possible to use technology to take high matrix powers. We find that

$$P^{20} = P^{21} = \begin{array}{c} \\ \text{Drop} \\ \text{Fr} \\ \text{So} \\ \text{Jr} \\ \text{Sr} \\ \text{Grad} \end{array} \begin{array}{c} \begin{array}{cccccc} \text{Drop} & \text{Fr} & \text{So} & \text{Jr} & \text{Sr} & \text{Grad} \end{array} \\ \begin{pmatrix} 1 & 0 & 0 & 0 & 0 & 0 \\ 0.1910 & 0 & 0 & 0 & 0 & 0.8090 \\ 0.1376 & 0 & 0 & 0 & 0 & 0.8624 \\ 0.0808 & 0 & 0 & 0 & 0 & 0.9192 \\ 0.0412 & 0 & 0 & 0 & 0 & 0.9588 \\ 0 & 0 & 0 & 0 & 0 & 1 \end{pmatrix} \end{array}.$$

The interpretation of this limiting matrix as representing long-term probabilities shows that the probability that a first-year student eventually graduates is 0.809. The last column of the matrix gives the probability of eventually graduating for each class year.

Note that although matrix powers P^n converge to a limiting matrix, as $n \to \infty$, unlike the forest fire and frog-hopping examples, the rows of this matrix are not identical. In this case, the long-term probability of hitting a particular state depends on the initial state. ■

2.6 MATHEMATICAL INDUCTION*

Mathematical induction is a technique for proving theorems, or properties, which hold for the natural numbers $1, 2, 3, \ldots$ An example of such a theorem is that the sum of the first n positive integers is

$$1 + 2 + \cdots + n = \binom{n+1}{2} = \frac{n(n+1)}{2}.$$

An example for Markov chains is given in Section 2.3. Let $X_0, X_1, \ldots$ be a Markov chain with transition matrix P and initial distribution α. Then, for all $n \geq 1$, the distribution of X_n is αP^n. That is,

$$P(X_n = j) = (\alpha P^n)_j = \sum_i \alpha_i P^n_{ij} \text{for all states } j.$$

Both of these results can be proven using induction.

The *principle of mathematical induction* states that (i) if a statement is true for $n = 1$, and (ii) whenever the statement is true for a natural number $n = k$, it is also true for $n = k + 1$, then the statement will be true for all natural numbers.

Proving theorems by mathematical induction is a two-step process.

First, the *base case* is established by proving the result true for $n = 1$.

Second, one assumes the result true for a given n, and then shows that it is true for $n + 1$. Assuming that the result true for a given n is called the *induction hypothesis*.

To illustrate the proof technique, consider the claim that the sum of the first n integers is equal to $n(n + 1)/2$.

For the base case, when $n = 1$, the sum of the first n integers is trivially equal to 1. And

$$n(n + 1)/2 = 1(2)/2 = 1.$$

This establishes the base case.

Assume that the property true for a given n. We need to show that the sum of the first $n+1$ integers is $(n+1)(n+2)/2$. The sum of the first $n+1$ integers is

$$\sum_{k=1}^{n+1} k = \left(\sum_{k=1}^{n} k\right) + (n+1).$$

By the induction hypothesis, the latter is equal to

$$\frac{n(n+1)}{2} + (n+1) = \frac{n(n+1)+2(n+1)}{2} = \frac{n^2+3n+2}{2} = \frac{(n+2)(n+1)}{2}.$$

This establishes the sum formula.

For another example of an induction proof, take the Markov chain result that the distribution of X_n is $\boldsymbol{\alpha P}^n$. The base case $n = 1$ is shown in Section 2.3. The distribution of X_1 is $\boldsymbol{\alpha P}^1 = \boldsymbol{\alpha P}$. Assume the result true for a given n. For the distribution of X_{n+1}, condition on X_n. This gives

$$\begin{aligned} P(X_{n+1} = j) &= \sum_i P(X_{n+1} = j | X_n = i) P(X_n = i) \\ &= \sum_i P_{ij} P(X_n = i) \\ &= \sum_i P_{ij} (\alpha P^n)_i \\ &= (\alpha P^{n+1})_j, \end{aligned}$$

where the next-to-last equality is by the induction hypothesis.

If the reader would like more practice applying induction, see Exercise 2.22.

Figure 2.9 Andrei Andreyevich Markov (1856–1922). *Source:* Wikimedia Commons, https://commons.wikimedia.org/wiki/File:AAMarkov.jpg. Public domain.

EXERCISES

2.1 A Markov chain has transition matrix

$$P = \begin{array}{c} \\ 1 \\ 2 \\ 3 \end{array} \begin{array}{c} \begin{array}{ccc} 1 & 2 & 3 \end{array} \\ \begin{pmatrix} 0.1 & 0.3 & 0.6 \\ 0 & 0.4 & 0.6 \\ 0.3 & 0.2 & 0.5 \end{pmatrix} \end{array}$$

with initial distribution $\boldsymbol{\alpha} = (0.2, 0.3, 0.5)$. Find the following:

(a) $P(X_7 = 3|X_6 = 2)$

(b) $P(X_9 = 2|X_1 = 2, X_5 = 1, X_7 = 3)$

(c) $P(X_0 = 3|X_1 = 1)$

(d) $E(X_2)$

2.2 Let $X_0, X_1, \ldots$ be a Markov chain with transition matrix

$$\begin{array}{c} \\ 1 \\ 2 \\ 3 \end{array} \begin{array}{c} \begin{array}{ccc} 1 & 2 & 3 \end{array} \\ \begin{pmatrix} 0 & 1/2 & 1/2 \\ 1 & 0 & 0 \\ 1/3 & 1/3 & 1/3 \end{pmatrix} \end{array}$$

and initial distribution $\boldsymbol{\alpha} = (1/2, 0, 1/2)$. Find the following:

(a) $P(X_2 = 1|X_1 = 3)$

(b) $P(X_1 = 3, X_2 = 1)$

(c) $P(X_1 = 3|X_2 = 1)$

(d) $P(X_9 = 1|X_1 = 3, X_4 = 1, X_7 = 2)$

2.3 See Example 2.6. Consider the Wright–Fisher model with a population of $k = 3$ genes. If the population initially has one A allele, find the probability that there are no A alleles in three generations.

2.4 For the general two-state chain with transition matrix

$$P = \begin{array}{c} \\ a \\ b \end{array} \begin{array}{c} \begin{array}{cc} a & b \end{array} \\ \begin{pmatrix} 1-p & p \\ q & 1-q \end{pmatrix} \end{array}$$

and initial distribution $\boldsymbol{\alpha} = (\alpha_1, \alpha_2)$, find the following:

(a) the two-step transition matrix

(b) the distribution of X_1

2.5 Consider a random walk on $\{0, \dots, k\}$, which moves left and right with respective probabilities q and p. If the walk is at 0 it transitions to 1 on the next step. If the walk is at k it transitions to $k-1$ on the next step. This is called *random walk with reflecting boundaries*. Assume that $k = 3$, $q = 1/4$, $p = 3/4$, and the initial distribution is uniform. For the following, use technology if needed.

(a) Exhibit the transition matrix.

(b) Find $P(X_7 = 1 | X_0 = 3, X_2 = 2, X_4 = 2)$.

(c) Find $P(X_3 = 1, X_5 = 3)$.

2.6 A tetrahedron die has four faces labeled 1, 2, 3, and 4. In repeated independent rolls of the die $R_0, R_1, \dots,$ let $X_n = \max\{R_0, \dots, R_n\}$ be the maximum value after $n + 1$ rolls, for $n \geq 0$.

(a) Give an intuitive argument for why $X_0, X_1, \dots,$ is a Markov chain, and exhibit the transition matrix.

(b) Find $P(X_3 \geq 3)$.

2.7 Let $X_0, X_1, \dots$ be a Markov chain with transition matrix $\boldsymbol{P}$. Let $Y_n = X_{3n}$, for $n = 0, 1, 2, \dots$. Show that $Y_0, Y_1, \dots$ is a Markov chain and exhibit its transition matrix.

2.8 Give the Markov transition matrix for random walk on the weighted graph in Figure 2.10.

2.9 Give the transition matrix for the transition graph in Figure 2.11.

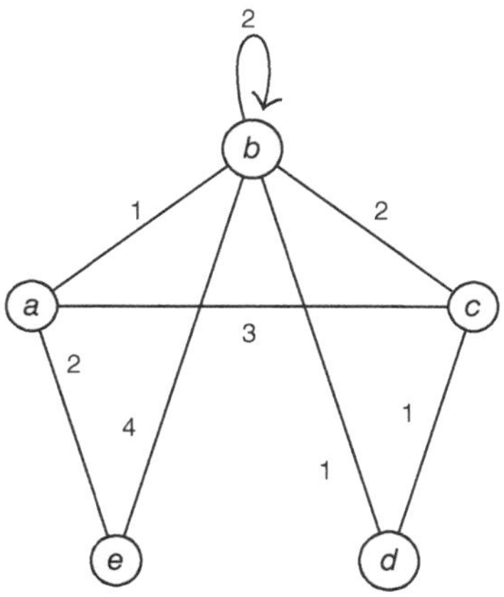

Figure 2.10

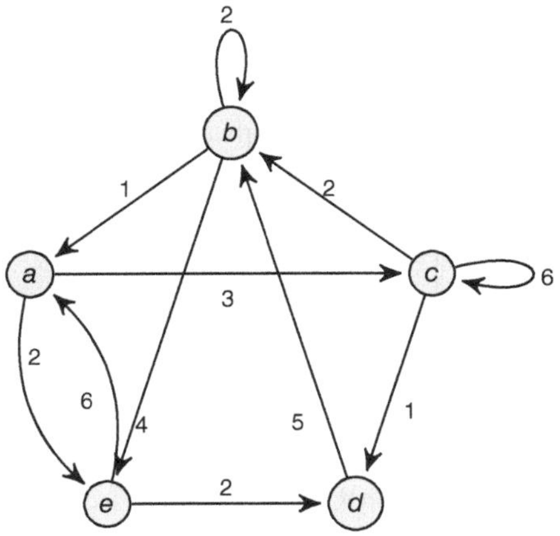

Figure 2.11

2.10 Consider a Markov chain with transition matrix

$$P = \begin{array}{c} \\ a \\ b \\ c \\ d \end{array} \begin{array}{c} \begin{array}{cccc} a & b & c & d \end{array} \\ \begin{pmatrix} 0 & 3/5 & 1/5 & 1/5 \\ 3/4 & 0 & 1/4 & 0 \\ 1/4 & 1/4 & 1/4 & 1/4 \\ 1/4 & 0 & 1/4 & 1/2 \end{pmatrix} \end{array}.$$

(a) Exhibit the directed, weighted transition graph for the chain.

(b) The transition graph for this chain can be given as a weighted graph without directed edges. Exhibit the graph.

2.11 You start with five dice. Roll all the dice and put aside those dice that come up 6. Then, roll the remaining dice, putting aside those dice that come up 6. And so on. Let X_n be the number of dice that are sixes after n rolls.

(a) Describe the transition matrix $\boldsymbol{P}$ for this Markov chain.

(b) Find the probability of getting all sixes by the third play.

(c) What do you expect $\boldsymbol{P}^{100}$ to look like? Use technology to confirm your answer.

2.12 Two urns contain k balls each. Initially, the balls in the left urn are all red and the balls in the right urn are all blue. At each step, pick a ball at random from each urn and exchange them. Let X_n be the number of blue balls in the left urn. (Note that necessarily $X_0 = 0$ and $X_1 = 1$.) Argue that the process is a Markov chain. Find the transition matrix. This model is called the Bernoulli–Laplace model of diffusion and was introduced by Daniel Bernoulli in 1769 as a model for the flow of two incompressible liquids between two containers.

2.13 See the move-to-front process in Example 2.10. Here is another way to organize the bookshelf. When a book is returned it is put back on the library shelf one position forward from where it was originally. If the book at the front of the shelf is returned it is put back at the front of the shelf. Thus, if the order of books is (a, b, c, d, e) and book d is picked, the new order is (a, b, d, c, e). This reorganization method is called the *transposition*, or *move-ahead-1*, scheme. Give the transition matrix for the transposition scheme for a shelf with three books.

2.14 There are k songs on Mary's music player. The player is set to *shuffle* mode, which plays songs uniformly at random, sampling with replacement. Thus, repeats are possible. Let X_n denote the number of *unique* songs that have been heard after the nth play.

(a) Show that $X_0, X_1, \ldots$ is a Markov chain and give the transition matrix.

(b) If Mary has four songs on her music player, find the probability that all songs are heard after six plays.

2.15 Assume that $X_0, X_1, \ldots$ is a two-state Markov chain on $\mathcal{S} = \{0, 1\}$ with transition matrix

$$\boldsymbol{P} = \begin{matrix} & \begin{matrix} 0 & 1 \end{matrix} \\ \begin{matrix} 0 \\ 1 \end{matrix} & \begin{pmatrix} 1-p & p \\ q & 1-q \end{pmatrix} \end{matrix}.$$

The present state of the chain only depends on the previous state. One can model a bivariate process that looks back two time periods by the following construction. Let $Z_n = (X_{n-1}, X_n)$, for $n \geq 1$. The sequence $Z_1, Z_2, \ldots$ is a Markov chain with state space $\mathcal{S} \times \mathcal{S} = \{(0, 0), (0, 1), (1, 0), (1, 1)\}$. Give the transition matrix of the new chain.

2.16 Assume that $\boldsymbol{P}$ is a stochastic matrix with equal rows. Show that $\boldsymbol{P}^n = \boldsymbol{P}$, for all $n \geq 1$.

2.17 Let $\boldsymbol{P}$ be a stochastic matrix. Show that $\lambda = 1$ is an eigenvalue of $\boldsymbol{P}$. What is the associated eigenvector?

2.18 A stochastic matrix is called *doubly stochastic* if its columns sum to 1. Let $X_0, X_1, \ldots$ be a Markov chain on $\{1, \ldots, k\}$ with doubly stochastic transition matrix and initial distribution that is uniform on $\{1, \ldots, k\}$. Show that the distribution of X_n is uniform on $\{1, \ldots, k\}$, for all $n \geq 0$.

2.19 Let $\boldsymbol{P}$ be the transition matrix of a Markov chain on k states. Let $\mathbf{I}$ denote the $k \times k$ identity matrix. Consider the matrix

$$\boldsymbol{Q} = (1 - p)\mathbf{I} + p\boldsymbol{P}, \text{ for } 0 < p < 1.$$

Show that $\boldsymbol{Q}$ is a stochastic matrix. Give a probabilistic interpretation for the dynamics of a Markov chain governed by the $\boldsymbol{Q}$ matrix in terms of the original Markov chain.

2.20 Let $X_0, X_1, \ldots$ be a Markov chain with transition matrix

$$P = \begin{array}{c} \\ 1 \\ 2 \\ 3 \end{array} \begin{array}{c} \begin{array}{ccc} 1 & 2 & 3 \end{array} \\ \begin{pmatrix} 0 & 1 & 0 \\ 0 & 0 & 1 \\ p & 1-p & 0 \end{pmatrix} \end{array},$$

for $0 < p < 1$. Let g be a function defined by

$$g(x) = \begin{cases} 0, & \text{if } x = 1, \\ 1, & \text{if } x = 2, 3. \end{cases}$$

Let $Y_n = g(X_n)$, for $n \geq 0$. Show that $Y_0, Y_1, \ldots$ is not a Markov chain.

2.21 Let P and Q be two transition matrices on the same state space. We define two processes, both started in some initial state i.
In process #1, a coin is flipped. If it lands heads, then the process unfolds according to the P matrix. If it lands tails, the process unfolds according to the Q matrix.
In process #2, at each step a coin is flipped. If it lands heads, the next state is chosen according to the P matrix. If it lands tails, the next state is chosen according to the Q matrix.
Thus, in #1, one coin is initially flipped, which governs the entire evolution of the process. And in #2, a coin is flipped at each step to decide the next step of the process.
Decide whether either of these processes is a Markov chain. If not, explain why, if yes, exhibit the transition matrix.

2.22 Prove the following using mathematical induction.

(a) $1 + 3 + 5 + \cdots + (2n - 1) = n^2$.

(b) $1^2 + 2^2 + \cdots + n^2 = n(n + 1)(2n + 1)/6$.

(c) For all real $x > -1$, $(1 + x)^n \geq 1 + nx$.

2.23 R : Simulate the first 20 letters (vowel/consonant) of the Pushkin poem Markov chain of Example 2.2.

2.24 R : Simulate 50 steps of the random walk on the graph in Figure 2.1. Repeat the simulation 10 times. How many of your simulations end at vertex c? Compare with the exact long-term probability the walk visits c.

2.25 R : The behavior of dolphins in the presence of tour boats in Patagonia, Argentina is studied in Dans et al. (2012). A Markov chain model is developed, with state space consisting of five primary dolphin activities (socializing, traveling, milling, feeding, and resting). The following transition matrix is obtained.

$$P = \begin{array}{c} \\ s \\ t \\ m \\ f \\ r \end{array} \begin{array}{c} \begin{array}{ccccc} s & t & m & f & r \end{array} \\ \begin{pmatrix} 0.84 & 0.11 & 0.01 & 0.04 & 0.00 \\ 0.03 & 0.80 & 0.04 & 0.10 & 0.03 \\ 0.01 & 0.15 & 0.70 & 0.07 & 0.07 \\ 0.03 & 0.19 & 0.02 & 0.75 & 0.01 \\ 0.03 & 0.09 & 0.05 & 0.00 & 0.83 \end{pmatrix} \end{array}.$$

Use technology to estimate the long-term distribution of dolphin activity.

2.26 R : In computer security applications, a *honeypot* is a trap set on a network to detect and counteract computer hackers. Honeypot data are studied in Kimou et al. (2010) using Markov chains. The authors obtain honeypot data from a central database and observe attacks against four computer ports—80, 135, 139, and 445—over 1 year. The ports are the states of a Markov chain along with a state corresponding to no port is attacked. Weekly data are monitored, and the port most often attacked during the week is recorded. The estimated Markov transition matrix for weekly attacks is

$$P = \begin{array}{c} \\ 80 \\ 135 \\ 139 \\ 445 \\ \text{No} \end{array} \begin{array}{c} \begin{array}{ccccc} 80 & 135 & 139 & 445 & \text{No attack} \end{array} \\ \begin{pmatrix} 0 & 0 & 0 & 0 & 1 \\ 0 & 8/13 & 3/13 & 1/13 & 1/13 \\ 1/16 & 3/16 & 3/8 & 1/4 & 1/8 \\ 0 & 1/11 & 4/11 & 5/11 & 1/11 \\ 0 & 1/8 & 1/2 & 1/8 & 1/4 \end{pmatrix} \end{array},$$

with initial distribution $\boldsymbol{\alpha} = (0, 0, 0, 0, 1)$.

(a) Which are the least and most likely attacked ports after 2 weeks?

(b) Find the long-term distribution of attacked ports.

2.27 R : See **gamblersruin.R.** Simulate gambler's ruin for a gambler with initial stake \$2, playing a fair game.

(a) Estimate the probability that the gambler is ruined before he wins \$5.

(b) Construct the transition matrix for the associated Markov chain. Estimate the desired probability in (a) by taking high matrix powers.

(c) Compare your results with the exact probability.

3

MARKOV CHAINS FOR THE LONG TERM

There exists everywhere a medium in things, determined by equilibrium.
—Dmitri Mendeleev

3.1 LIMITING DISTRIBUTION

In many cases, a Markov chain exhibits a long-term limiting behavior. The chain settles down to an equilibrium distribution, which is independent of its initial state.

Limiting Distribution

Let $X_0, X_1, \ldots$ be a Markov chain with transition matrix $\boldsymbol{P}$. A *limiting distribution* for the Markov chain is a probability distribution $\boldsymbol{\lambda}$ with the property that for all i and j,

$$\lim_{n\to\infty} P^n_{ij} = \lambda_j.$$

The definition of limiting distribution is equivalent to each of the following:

(i) For any initial distribution, and for all j,

$$\lim_{n\to\infty} P(X_n = j) = \lambda_j.$$

Introduction to Stochastic Processes with R, First Edition. Robert P. Dobrow.

(ii) For any initial distribution $\boldsymbol{\alpha}$,

$$\lim_{n\to\infty} \boldsymbol{\alpha P}^n = \boldsymbol{\lambda}.$$

(iii)

$$\lim_{n\to\infty} \boldsymbol{P}^n = \boldsymbol{\Lambda},$$

where $\boldsymbol{\Lambda}$ is a stochastic matrix all of whose rows are equal to $\boldsymbol{\lambda}$.

We interpret λ_j as the long-term probability that the chain hits state j. By the uniqueness of limits, if a Markov chain has a limiting distribution, then that distribution is unique.

If a limiting distribution exists, a *quick and dirty* numerical method to find it is to take high matrix powers of the transition matrix until one obtains an obvious limiting matrix with equal rows. The common row is the limiting distribution. Examples of this approach have been given in Section 2.5.

Numerical methods, however, have their limits (no pun intended), and the emphasis in this chapter is on exact solutions and theoretical results.

For the general two-state Markov chain, matrix powers can be found exactly in order to compute the limiting distribution.

Example 3.1 (Two-state Markov chain) The transition matrix for a general two-state chain is

$$\boldsymbol{P} = \begin{matrix} \\ 1 \\ 2 \end{matrix} \begin{matrix} \begin{matrix} 1 & \quad 2 \end{matrix} \\ \begin{pmatrix} 1-p & p \\ q & 1-q \end{pmatrix} \end{matrix},$$

for $0 \leq p, q \leq 1$. If $p + q = 1$,

$$\boldsymbol{P} = \begin{pmatrix} 1-p & p \\ 1-p & p \end{pmatrix},$$

and $\boldsymbol{P}^n = \boldsymbol{P}$ for all $n \geq 1$. Thus, $\boldsymbol{\lambda} = (1-p, p)$ is the limiting distribution.

Assume $p + q \neq 1$. To find $\boldsymbol{P}^n$, consider the entry P^n_{11}. As $\boldsymbol{P}^n = \boldsymbol{P}^{n-1}\boldsymbol{P}$,

$$\begin{aligned} P^n_{11} &= (P^{n-1}P)_{11} = P^{n-1}_{11}P_{11} + P^{n-1}_{12}P_{21} \\ &= P^{n-1}_{11}(1-p) + P^{n-1}_{12}q \\ &= P^{n-1}_{11}(1-p) + \left(1 - P^{n-1}_{11}\right)q \\ &= q + (1-p-q)P^{n-1}_{11} \quad \text{,for} \quad n \geq 1. \end{aligned}$$

The next-to-last equality uses the fact that $P^{n-1}_{11} + P^{n-1}_{12} = 1$, since $\boldsymbol{P}^{n-1}$ is a stochastic matrix. Unwinding the recurrence gives

$$\begin{aligned}
P^n_{11} &= q + (1-p-q)P^{n-1}_{11} \\
&= q + q(1-p-q) + (1-p-q)^2 P^{n-2}_{11} \\
&= q + q(1-p-q) + q(1-p-q)^2 + (1-p-q)^3 P^{n-3}_{11} \\
&= \cdots = q\sum_{k=0}^{n-1}(1-p-q)^k + (1-p-q)^n \\
&= (1-p-q)^n + q\frac{1-(1-p-q)^n}{1-(1-p-q)} \\
&= \frac{q}{p+q} + \frac{p}{p+q}(1-p-q)^n.
\end{aligned}$$

The matrix entry

$$P^n_{22} = \frac{p}{p+q} + \frac{q}{p+q}(1-p-q)^n$$

is found similarly. Since the rows of $\boldsymbol{P}^n$ sum to 1,

$$\boldsymbol{P}^n = \frac{1}{p+q}\begin{pmatrix} q + p(1-p-q)^n & p - p(1-p-q)^n \\ q - q(1-p-q)^n & p + q(1-p-q)^n \end{pmatrix}. \tag{3.1}$$

If p and q are not both 0, nor both 1, then $|1-p-q| < 1$ and

$$\lim_{n\to\infty} \boldsymbol{P}^n = \frac{1}{p+q}\begin{pmatrix} q & p \\ q & p \end{pmatrix}.$$

The limiting distribution is

$$\boldsymbol{\lambda} = \left(\frac{q}{p+q}, \frac{p}{p+q}\right).$$

Observe that in addition to giving the limiting distribution, Equation (3.1) reveals the rate of convergence to that limit. The convergence is exponential and governed by the quantity $(1-p-q)^n$. ∎

Proportion of Time in Each State

The limiting distribution gives the long-term probability that a Markov chain hits each state. It can also be interpreted as the long-term proportion of time that the chain visits each state. To make this precise, let $X_0, X_1, \ldots$ be a Markov chain with transition matrix $\boldsymbol{P}$ and limiting distribution $\boldsymbol{\lambda}$. For state j, define indicator random variables

$$I_k = \begin{cases} 1, & \text{if } X_k = j, \\ 0, & \text{otherwise,} \end{cases}$$

for $k = 0, 1, \ldots$ Then, $\sum_{k=0}^{n-1} I_k$ is the number of times the chain visits j in the first n steps (counting X_0 as the first step). From initial state i, the long-term expected proportion of time that the chain visits j is

$$\begin{aligned}
\lim_{n\to\infty} E\left(\frac{1}{n}\sum_{k=0}^{n-1} I_k | X_0 = i\right) &= \lim_{n\to\infty} \frac{1}{n}\sum_{k=0}^{n-1} E(I_k | X_0 = i) \\
&= \lim_{n\to\infty} \frac{1}{n}\sum_{k=0}^{n-1} P(X_k = j | X_0 = i) \\
&= \lim_{n\to\infty} \frac{1}{n}\sum_{k=0}^{n-1} P_{ij}^k \\
&= \lim_{n\to\infty} P_{ij}^n = \lambda_j.
\end{aligned}$$

The next-to-last equality applies a result in analysis known as Cesaro's lemma. The lemma states that if a sequence of numbers converges to a limit, then the sequence of partial averages also converges to that limit. That is, if $x_n \to x$, as $n \to \infty$, then $(x_1 + \cdots + x_n)/n \to x$, as $n \to \infty$.

Example 3.2 After work, Angel goes to the gym and either does aerobics, weights, yoga, or gets a massage. Each day, Angel decides her workout routine based on what she did the previous day according to the Markov transition matrix

$$\boldsymbol{P} = \begin{array}{l} \text{Aerobics} \\ \text{Massage} \\ \text{Weights} \\ \text{Yoga} \end{array} \overset{\begin{array}{cccc} \text{Aerobics} & \text{Massage} & \text{Weights} & \text{Yoga} \end{array}}{\begin{pmatrix} 0.1 & 0.2 & 0.4 & 0.3 \\ 0.4 & 0.0 & 0.4 & 0.2 \\ 0.3 & 0.3 & 0.0 & 0.4 \\ 0.2 & 0.1 & 0.4 & 0.3 \end{pmatrix}}.$$

Taking high matrix powers gives the limiting distribution

Aerobics	Massage	Weight	Yoga
0.238	0.164	0.286	0.312

See the following `R` code, where Angel's gym visits are simulated for 100 days. During that time, Angel did aerobics on 26 days, got a massage 14 times, did weights on 31 days, and did yoga 29 times. The proportion of visits to each state is

Aerobics	Massage	Weighs	Yoga
0.26	0.14	0.31	0.29

These proportions are relatively close to the actual limiting distribution of the chain notwithstanding the fact that the estimates are based on just 100 steps. Compare the results to the million step simulation, also given in the R code. ■

R: Angel at the Gym

```
# gym.R
> P
          Aerobics Massage Weights Yoga
Aerobics       0.1     0.2     0.4  0.3
Massage        0.4     0.0     0.4  0.2
Weights        0.3     0.3     0.0  0.4
Yoga           0.2     0.1     0.4  0.3
> init <- c(1/4,1/4,1/4,1/4) # initial distribution
> states <- c("a","m","w","y")
# simulate Markov chain for 100 steps
> simlist <- markov(init,P,100,states)
> mwyaaywyayawyayawymwywamwawyawywywyywywyawyamwaway
> amamawywmyawawmawywmwywmwmyaywaywywamwyyymwawamaay
> table(simlist)/100
Aerobics  Massage  Weights     Yoga
  0.26      0.14      0.31     0.29
> steps <- 1000000   # one million steps
> simlist <- markov(init,P,steps,states)
> table(simlist)/steps
Aerobics  Massage  Weights     Yoga
0.237425 0.164388 0.285548 0.312640
```

3.2 STATIONARY DISTRIBUTION

It is interesting to consider what happens if we assign the limiting distribution of a Markov chain to be the initial distribution of the chain.

For the two-state chain, as in Example 3.1, the limiting distribution is

$$\lambda = \left(\frac{q}{p+q}, \frac{p}{p+q}\right).$$

Let λ be the initial distribution for such a chain. Then, the distribution of X_1 is

$$\begin{aligned}\lambda \boldsymbol{P} &= \left(\frac{q}{p+q}, \frac{p}{p+q}\right) \begin{pmatrix} 1-p & p \\ q & 1-q \end{pmatrix} \\ &= \left(\frac{q(1-p)+pq}{p+q}, \frac{qp+p(1-q)}{p+q}\right)\end{aligned}$$

$$= \left(\frac{q}{p+q}, \frac{p}{p+q} \right) = \lambda.$$

That is, $\lambda P = \lambda$. A probability vector π that satisfies $\pi \mathbf{P} = \pi$ plays a special role for Markov chains.

Stationary Distribution

Let $X_0, X_1, \ldots$ be a Markov chain with transition matrix P. A *stationary distribution* is a probability distribution π, which satisfies

$$\pi = \pi P. \tag{3.2}$$

That is,

$$\pi_j = \sum_i \pi_i P_{ij}, \text{ for all } j.$$

If we assume that a stationary distribution π is the initial distribution, then Equation (3.2) says that the distribution of X_0 is the same as the distribution of X_1. Since the chain started at $n = 1$ is also a Markov chain with transition matrix P, it follows that X_2 has the same distribution as X_1. In fact, all of the X_n have the same distribution, as

$$\pi P^n = (\pi P)P^{n-1} = \pi P^{n-1} = (\pi P)P^{n-2} = \pi P^{n-2} = \cdots = \pi P = \pi.$$

If the initial distribution is a stationary distribution, then $X_0, X_1, X_2, \ldots$ is a sequence of identically distributed random variables.

The name *stationary* comes from the fact that if the chain starts in its stationary distribution, then it stays in that distribution. We refer to the *stationary Markov chain* or the Markov chain *in stationarity* for the chain started in its stationary distribution.

If $X_0, X_1, X_2, \ldots$ is a stationary Markov chain, then for any $n > 0$, the sequence $X_n, X_{n+1}, X_{n+2}, \ldots$ is also a stationary Markov chain with the same transition matrix and stationary distribution as the original chain.

(The fact that the stationary chain is a sequence of identically distributed random variables does not mean that the random variables are independent. On the contrary, the dependency structure between successive random variables in a Markov chain is governed by the transition matrix, regardless of the initial distribution.)

Other names for the stationary distribution are *invariant*, *steady-state*, and *equilibrium* distribution. The latter highlights the fact that there is an intimate connection between the stationary distribution and the limiting distribution. If a Markov chain has a limiting distribution then that distribution is a stationary distribution.

Limiting Distributions are Stationary Distributions

Lemma 3.1. *Assume that π is the limiting distribution of a Markov chain with transition matrix P. Then, π is a stationary distribution.*

Proof. Assume that $\boldsymbol{\pi}$ is the limiting distribution. We need to show that $\boldsymbol{\pi P} = \boldsymbol{\pi}$. For any initial distribution $\boldsymbol{\alpha}$,

$$\boldsymbol{\pi} = \lim_{n\to\infty} \boldsymbol{\alpha P}^n = \lim_{n\to\infty} \boldsymbol{\alpha}(\boldsymbol{P}^{n-1}\boldsymbol{P}) = \left(\lim_{n\to\infty} \boldsymbol{\alpha P}^{n-1}\right)\boldsymbol{P} = \boldsymbol{\pi P},$$

which uses the fact that if $\lim_{n\to\infty} x_n = x$, then $\lim_{n\to\infty} x_{n-1} = x$. ∎

Unfortunately, the converse of Lemma 3.1 is not true—stationary distributions are not necessarily limiting distributions. For a counterexample, take the Markov chain with transition matrix

$$\boldsymbol{P} = \begin{pmatrix} 0 & 1 \\ 1 & 0 \end{pmatrix}.$$

Solving $\boldsymbol{\pi P} = \boldsymbol{\pi}$, or

$$(\pi_1, \pi_2) \begin{pmatrix} 0 & 1 \\ 1 & 0 \end{pmatrix} = (\pi_1, \pi_2),$$

gives $\pi_1 = \pi_2$. Since the stationary distribution is a probability vector, the unique solution is $\boldsymbol{\pi} = (1/2, 1/2)$. The stationary distribution is uniform on each state. However, the chain has no limiting distribution. The process evolves by flip-flopping back and forth between states. As in the case of random walk on a cycle with an even number of vertices, the position of the walk after n steps depends on the starting vertex and the parity of n.

Another counterexample is the Markov chain with transition matrix

$$\boldsymbol{P} = \begin{pmatrix} 1 & 0 \\ 0 & 1 \end{pmatrix}.$$

This process is rather boring—the chain simply stays forever in its starting state. The chain has no limiting distribution, as the long-term state of the chain depends upon the starting state. However, *every* probability vector is a stationary distribution since $\boldsymbol{xP} = \boldsymbol{x}$, for all vectors $\boldsymbol{x}$.

Thus, there are Markov chains with more than one stationary distribution; there are Markov chains with unique stationary distributions that are not limiting distributions; and there are even Markov chains that do not have stationary distributions.

However, a large and important class of Markov chains has unique stationary distributions that are the limiting distribution of the chain. A goal of this chapter is to characterize such chains.

Regular Matrices

A matrix $\boldsymbol{M}$ is said to be *positive* if all the entries of $\boldsymbol{M}$ are positive. We write $\boldsymbol{M} > 0$. Similarly, write $\boldsymbol{x} > 0$ for a vector $\boldsymbol{x}$ with all positive entries.

Regular Transition Matrix

A transition matrix $\boldsymbol{P}$ is said to be *regular* if some power of $\boldsymbol{P}$ is positive. That is, $\boldsymbol{P}^n > 0$, for some $n \geq 1$.

For example,

$$\boldsymbol{P} = \begin{pmatrix} 0 & 1/2 & 1/2 \\ 1 & 0 & 0 \\ 1/2 & 1/2 & 0 \end{pmatrix}$$

is regular, since

$$\boldsymbol{P}^4 = \begin{pmatrix} 9/16 & 5/16 & 1/8 \\ 1/4 & 3/8 & 3/8 \\ 1/2 & 5/16 & 3/16 \end{pmatrix}$$

is positive. However,

$$\boldsymbol{P} = \begin{pmatrix} 0 & 1 & 0 \\ 0 & 0 & 1 \\ 1 & 0 & 0 \end{pmatrix}$$

is not regular, since the powers of $\boldsymbol{P}$ cycle through the matrices

$$\boldsymbol{P} = \begin{pmatrix} 0 & 1 & 0 \\ 0 & 0 & 1 \\ 1 & 0 & 0 \end{pmatrix}, \ \boldsymbol{P}^2 = \begin{pmatrix} 0 & 0 & 1 \\ 1 & 0 & 0 \\ 0 & 1 & 0 \end{pmatrix}, \text{ and } \boldsymbol{P}^3 = \begin{pmatrix} 1 & 0 & 0 \\ 0 & 1 & 0 \\ 0 & 0 & 1 \end{pmatrix}.$$

If the transition matrix of a Markov chain is regular, then the chain has a limiting distribution, which is the unique stationary distribution of the chain.

Limit Theorem for Regular Markov Chains

Theorem 3.2. *A Markov chain whose transition matrix* $\boldsymbol{P}$ *is regular has a limiting distribution, which is the unique, positive, stationary distribution of the chain. That is, there exists a unique probability vector* $\boldsymbol{\pi} > 0$*, such that*

$$\lim_{n\to\infty} P^n_{ij} = \pi_j,$$

for all i, j*, where*

$$\sum_i \pi_i P_{ij} = \pi_j.$$

Equivalently, there exists a positive stochastic matrix $\boldsymbol{\Pi}$ *such that*

$$\lim_{n\to\infty} \boldsymbol{P}^n = \boldsymbol{\Pi},$$

where $\boldsymbol{\Pi}$ has equal rows with common row $\boldsymbol{\pi}$, and $\boldsymbol{\pi}$ is the unique probability vector, which satisfies

$$\boldsymbol{\pi}\boldsymbol{P} = \boldsymbol{\pi}.$$

The proof of Theorem 3.2 is deferred to Section 3.10.

Example 3.3 Assume that a Markov chain has transition matrix

$$\boldsymbol{P} = \begin{pmatrix} 0 & 1-p & p \\ p & 0 & 1-p \\ 1-p & p & 0 \end{pmatrix},$$

for $0 < p < 1$. Find the limiting distribution.

Solution We find that

$$\boldsymbol{P}^2 = \begin{pmatrix} 2p(1-p) & p^2 & (1-p)^2 \\ (1-p)^2 & 2p(1-p) & p^2 \\ p^2 & (1-p)^2 & 2p(1-p) \end{pmatrix}.$$

Since $0 < p < 1$, the matrix $\boldsymbol{P}^2$ is positive. Thus, $\boldsymbol{P}$ is regular. By Theorem 3.2, the limiting distribution is the stationary distribution.

How to find the stationary distribution of a Markov chain is the topic of the next section. However, for now we give the reader a little help and urge them to try $\boldsymbol{\pi} = (1/3, 1/3, 1/3)$.

Indeed, $\boldsymbol{\pi}\boldsymbol{P} = \boldsymbol{\pi}$, as

$$\left(\frac{1}{3}, \frac{1}{3}, \frac{1}{3}\right) \begin{pmatrix} 0 & 1-p & p \\ p & 0 & 1-p \\ 1-p & p & 0 \end{pmatrix} = \left(\frac{1}{3}, \frac{1}{3}, \frac{1}{3}\right).$$

The uniform distribution $\boldsymbol{\pi}$ is the unique stationary distribution of the chain, and thus the desired limiting distribution.

The example is interesting because the limiting distribution is uniform for *all* choices of $0 < p < 1$. ■

Here is one way to tell if a stochastic matrix is not regular. If for some power n, all the 0s in $\boldsymbol{P}^n$ appear in the same locations as all the 0s in $\boldsymbol{P}^{n+1}$, then they will appear in the same locations for all higher powers, and the matrix is not regular.

Example 3.4 A Markov chain has transition matrix

$$P = \begin{pmatrix} 0 & 0 & 0 & 0 & 1 \\ \frac{1}{4} & 0 & 0 & \frac{1}{2} & \frac{1}{4} \\ 0 & 0 & 1 & 0 & 0 \\ 0 & 0 & \frac{1}{2} & \frac{1}{2} & 0 \\ \frac{1}{4} & \frac{1}{2} & 0 & 0 & \frac{1}{4} \end{pmatrix}.$$

Determine if the matrix is regular.

Solution We find that

$$P^4 = \begin{pmatrix} \frac{9}{64} & \frac{7}{32} & \frac{1}{8} & \frac{3}{16} & \frac{21}{64} \\ \frac{21}{256} & \frac{11}{128} & \frac{15}{32} & \frac{11}{64} & \frac{49}{256} \\ 0 & 0 & 1 & 0 & 0 \\ 0 & 0 & \frac{15}{16} & \frac{1}{16} & 0 \\ \frac{35}{256} & \frac{21}{128} & \frac{7}{32} & \frac{13}{64} & \frac{71}{256} \end{pmatrix} \quad \text{and} \quad P^5 = \begin{pmatrix} \frac{35}{256} & \frac{21}{128} & \frac{7}{32} & \frac{13}{64} & \frac{71}{256} \\ \frac{71}{1024} & \frac{49}{512} & \frac{71}{128} & \frac{33}{256} & \frac{155}{1024} \\ 0 & 0 & 1 & 0 & 0 \\ 0 & 0 & \frac{31}{32} & \frac{1}{32} & 0 \\ \frac{113}{1024} & \frac{71}{512} & \frac{41}{128} & \frac{47}{256} & \frac{253}{1024} \end{pmatrix}.$$

Since the 0s are in the same locations for both matrices, we conclude that $\boldsymbol{P}$ is not regular. ■

Finding the Stationary Distribution

Assume that $\boldsymbol{\pi}$ is a stationary distribution for a Markov chain with transition matrix $\boldsymbol{P}$. Then,

$$\sum_i \pi_i P_{ij} = \pi_j, \quad \text{for all states } j,$$

which gives a system of linear equations. If $\boldsymbol{P}$ is a $k \times k$ matrix, the system has k equations and k unknowns. Since the rows of $\boldsymbol{P}$ sum to 1, the $k \times k$ system will contain a redundant equation.

For the general two-state chain, with

$$P = \begin{pmatrix} 1-p & p \\ q & 1-q \end{pmatrix},$$

the equations are

$$(1-p)\pi_1 + q\pi_2 = \pi_1$$
$$p\pi_1 + (1-q)\pi_2 = \pi_2.$$

The equations are redundant and lead to $\pi_1 p = \pi_2 q$. If p and q are not both zero, then together with the condition $\pi_1 + \pi_2 = 1$, the unique solution is

$$\pi = \left(\frac{q}{p+q}, \frac{p}{p+q}\right).$$

Example 3.5 Find the stationary distribution of the weather Markov chain of Example 2.3, with transition matrix

$$P = \begin{array}{c} \\ \text{Rain} \\ \text{Snow} \\ \text{Clear} \end{array} \begin{array}{c} \begin{array}{ccc} \text{Rain} & \text{Snow} & \text{Clear} \end{array} \\ \begin{pmatrix} 1/5 & 3/5 & 1/5 \\ 1/10 & 4/5 & 1/10 \\ 1/10 & 3/5 & 3/10 \end{pmatrix} \end{array}.$$

Solution The linear system to solve is

$$\begin{aligned} (1/5)\pi_1 + (1/10)\pi_2 + (1/10)\pi_3 &= \pi_1 \\ (3/5)\pi_1 + (4/5)\pi_2 + (3/5)\pi_3 &= \pi_2 \\ (1/5)\pi_1 + (1/10)\pi_2 + (3/10)\pi_3 &= \pi_3 \\ \pi_1 + \pi_2 + \pi_3 &= 1. \end{aligned}$$

One of the first three equations is redundant. The unique solution is

$$\pi = \left(\frac{1}{9}, \frac{3}{4}, \frac{5}{36}\right).$$

■

R : Finding the Stationary Distribution

The R function `stationary(P)` finds the stationary distribution of a Markov chain with transition matrix P. The function is contained in the **utilities.R** file.

```
> P
       Rain Snow Clear
Rain    0.2  0.6   0.2
Snow    0.1  0.8   0.1
Clear   0.1  0.6   0.3
> stationary(P)
[1] 0.1111 0.7500 0.1389
# check that this is a stationary distribution
> stationary(P) %*% P
         Rain    Snow    Clear
[1]    0.1111    0.75   0.1389
```

Here is a useful technique for finding the stationary distribution, which reduces by one the number of equations to solve. It makes use of the fact that if $\boldsymbol{x}$ is a vector, not necessarily a probability vector, which satisfies $\boldsymbol{xP} = \boldsymbol{x}$, then $(c\boldsymbol{x})\boldsymbol{P} = c\boldsymbol{x}$, for all constants c. It follows that if one can find a non-negative $\boldsymbol{x}$, which satisfies $\boldsymbol{xP} = \boldsymbol{x}$, then a unique probability vector solution $\boldsymbol{\pi} = c\boldsymbol{x}$ can be gotten by an appropriate choice of c so that the rows of $c\boldsymbol{x}$ sum to 1. In particular, let $c = 1/\sum_j x_j$, the reciprocal of the sum of the components of $\boldsymbol{x}$.

The linear system $\sum_i \pi_i P_{ij} = \pi_j$, without the constraint $\sum_i \pi_i = 1$, has one redundant equation. Our solution method consists of (i) eliminating a redundant equation and (ii) solving the resulting system for $\boldsymbol{x} = (1, x_2, x_3, \ldots)$, where the first (or any) component of $\boldsymbol{x}$ is replaced by 1.

For a Markov chain with k states, this method reduces the problem to solving a $(k-1) \times (k-1)$ linear system. If the original chain has a unique stationary distribution, then the reduced linear system will have a unique solution, but one which is not necessarily a probability vector. To make it a probability vector whose components sum to 1, divide by the sum of the components. In other words, the unique stationary distribution is

$$\boldsymbol{\pi} = \frac{1}{1 + x_2 + \cdots + x_k}(1, x_2, \ldots, x_k).$$

To illustrate the method, consider the transition matrix

$$\boldsymbol{P} = \begin{pmatrix} 1/3 & 1/2 & 1/6 \\ 1/2 & 1/2 & 0 \\ 1/4 & 1/2 & 1/4 \end{pmatrix}.$$

To find the stationary distribution, first let $\boldsymbol{x} = (1, x_2, x_3)$. Then, $\boldsymbol{xP} = \boldsymbol{x}$ gives a 3×3 linear system. The first two equations are

$$\begin{aligned} (1/3)(1) + (1/2)x_2 + (1/4)x_3 &= 1, \\ (1/2)(1) + (1/2)x_2 + (1/2)x_3 &= x_2, \end{aligned}$$

or

$$\begin{aligned} (1/2)x_2 + (1/4)x_3 &= 2/3, \\ (-1/2)x_2 + (1/2)x_3 &= -1/2, \end{aligned}$$

with unique solution $\boldsymbol{x} = (1, 11/9, 2/9)$. The sum of the components is

$$1 + \frac{11}{9} + \frac{2}{9} = \frac{22}{9}.$$

The stationary distribution is

$$\boldsymbol{\pi} = \frac{9}{22}\left(1, \frac{11}{9}, \frac{2}{9}\right) = \left(\frac{9}{22}, \frac{11}{22}, \frac{2}{22}\right).$$

Example 3.6 A Markov chain on $\{1,2,3,4\}$ has transition matrix

$$P = \begin{pmatrix} p & 1-p & 0 & 0 \\ (1-p)/2 & p & (1-p)/2 & 0 \\ 0 & (1-p)/2 & p & (1-p)/2 \\ 0 & 0 & 1-p & p \end{pmatrix},$$

for $0 < p < 1$. Find the stationary distribution.

Solution Let $\boldsymbol{x} = (x_1, x_2, x_3, x_4)$, with $x_1 = 1$. Take

$$x_j = \sum_{i=1}^{4} x_i P_{ij},$$

for $j = 1, 2$, and 4. This gives

$$\begin{aligned} p + \left(\frac{1-p}{2}\right) x_2 &= 1, \\ 1 - p + p x_2 + \left(\frac{1-p}{2}\right) x_3 &= x_2, \\ \left(\frac{1-p}{2}\right) x_3 + p x_4 &= x_4, \end{aligned}$$

with solution $x_2 = x_3 = 2$ and $x_1 = x_4 = 1$. The stationary distribution is

$$\boldsymbol{\pi} = \frac{1}{1+2+2+1}(1,2,2,1) = \left(\frac{1}{6}, \frac{2}{6}, \frac{2}{6}, \frac{1}{6}\right).$$

■

Example 3.7 (The Ehrenfest dog–flea model) The Ehrenfest dog–flea model was originally proposed by physicists Tatyana and Paul Ehrenfest to describe the diffusion of gases. Mathematician Mark Kac called it "one of the most instructive models in the whole of physics."

Two dogs—Lisa and Cooper—share a population of N fleas. At each discrete unit of time, one of the fleas jumps from the dog it is on to the other dog. Let X_n denote the number of fleas on Lisa after n jumps. If there are i fleas on Lisa, then on the next jump the number of fleas on Lisa either goes up by one, if one of the $N - i$ fleas on Cooper jumps to Lisa, or goes down by one, if one of the i fleas on Lisa jumps to Cooper.

The process is a Markov chain on $\{0, 1, \ldots, N\}$, with transition matrix

$$P_{ij} = \begin{cases} i/N, & \text{if } j = i - 1, \\ (N-i)/N, & \text{if } j = i + 1, \\ 0, & \text{otherwise.} \end{cases}$$

Here is the Ehrenfest transition matrix for $N = 5$ fleas:

$$
\boldsymbol{P} = \begin{array}{c} \\ 0 \\ 1 \\ 2 \\ 3 \\ 4 \\ 5 \end{array}
\begin{array}{c} \begin{array}{cccccc} 0 & 1 & 2 & 3 & 4 & 5 \end{array} \\
\begin{pmatrix}
0 & 1 & 0 & 0 & 0 & 0 \\
1/5 & 0 & 4/5 & 0 & 0 & 0 \\
0 & 2/5 & 0 & 3/5 & 0 & 0 \\
0 & 0 & 3/5 & 0 & 2/5 & 0 \\
0 & 0 & 0 & 4/5 & 0 & 1/5 \\
0 & 0 & 0 & 0 & 1 & 0
\end{pmatrix} \end{array}.
$$

To find the stationary distribution for the general Ehrenfest chain, let $\boldsymbol{x} = (x_0, x_1, \ldots, x_N)$, with $x_0 = 1$. Set

$$
x_j = \sum_{i=0}^{N} x_i P_{ij} = x_{j-1} \frac{N - (j-1)}{N} + x_{j+1} \frac{j+1}{N}, \tag{3.3}
$$

for $j = 1, \ldots, N-1$. Also, $1 = (1/N)x_1$, so $x_1 = N$. Solving Equation (3.3) starting at $j = 1$ gives $x_2 = N(N-1)/2$, then $x_3 = N(N-1)(N-2)/6$. The general term is

$$
x_j = \frac{N(N-1)\cdots(N-j+1)}{j!} = \frac{N!}{j!(N-j)!} = \binom{N}{j}, \quad \text{for } j = 0, 1, \ldots, N,
$$

which can be derived by induction. The stationary distribution is

$$
\pi_j = \frac{1}{\sum_{i=0}^{N} \binom{N}{i}} \binom{N}{j} = \binom{N}{j} \frac{1}{2^N}, \quad \text{for } j = 0, \ldots, N.
$$

The distribution is a binomial distribution with parameters N and 1/2.

The Ehrenfest transition matrix is not regular, and the chain does not have a limiting distribution. However, we can interpret the stationary distribution as giving the long-term proportion of time spent in each state. See Example 3.19 for the description of a modified Ehrenfest scheme, which has a limiting distribution. ■

For the next example, rather than solve a linear system to find the stationary distribution, we take a *guess* at the distribution $\boldsymbol{\pi}$ based on our intuition for how the Markov chain evolves. The candidate distribution $\boldsymbol{\pi}$ is then checked to see if it satisfies $\boldsymbol{\pi} = \boldsymbol{\pi P}$.

> Guessing before proving! Need I remind you that it is so that all important discoveries have been made?
>
> —Henri Poincaré

Example 3.8 (Random walk on a graph) To find the stationary distribution for simple random walk on a graph, consider the interpretation of the distribution as the long-term fraction of time that the walk visits each vertex.

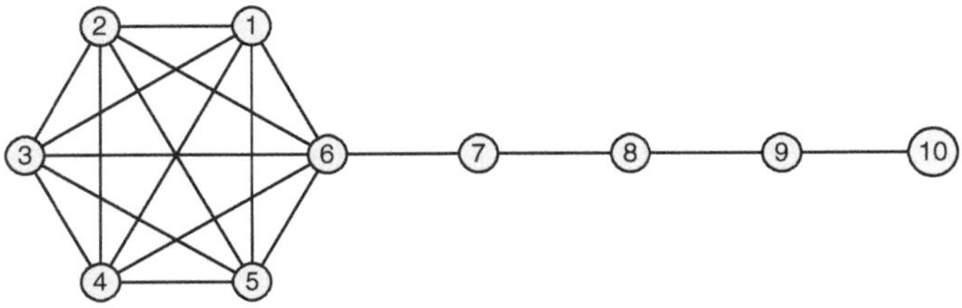

Figure 3.1 Lollipop graph.

For a concrete example, consider the *lollipop graph*, shown in Figure 3.1. One expects that in the long term a random walk on the graph is most likely to be on the *candy* at the leftmost end of the graph, and least likely to be at the right end of the stick. (The candy is a complete graph where all pairs of vertices are joined by edges.)

Intuition suggests that vertices that have more connections are more likely to be visited. That is, the time spent at vertex v is related to the degree of v.

This suggests considering a distribution on vertices that is related to the degree of the vertices. One possibility is a distribution that is proportional to the degree of the vertex. Let

$$\pi_v = \frac{\deg(v)}{\sum_w \deg(w)} = \frac{\deg(v)}{2e},$$

where e is the number of edges in the graph. The sum of the vertex degrees is equal to twice the number of edges since every edge contributes two vertices (its endpoints) to the sum of the vertex degrees.

It remains to check, for this choice of $\boldsymbol{\pi}$, whether in fact $\boldsymbol{\pi} = \boldsymbol{\pi P}$. For vertex v,

$$\begin{aligned}(\pi P)_v &= \sum_w \pi_w P_{wv} = \sum_{w \sim v} \left(\frac{\deg(w)}{2e}\right) \frac{1}{\deg(w)} \\ &= \frac{1}{2e} \sum_{w \sim v} 1 = \frac{\deg(v)}{2e} = \pi_v.\end{aligned}$$

Indeed, our candidate $\boldsymbol{\pi}$ is a stationary distribution.

For the lollipop graph in Figure 3.1, the sum of the vertex degrees is 38. Here is the stationary distribution.

Vertex	1	2	3	4	5	6	7	8	9	10
π_v	5/38	5/38	5/38	5/38	5/38	6/38	2/38	2/38	2/38	1/38

■

Simple random walk on a graph is a special case of random walk on a weighted graph, with edge weights all equal to 1. The stationary distribution for a weighted graph has a similar form as for an unweighted graph. If v is a vertex, say that an edge is *incident* to v if v is an endpoint of the edge.

Stationary Distribution for Random Walk on a Weighted Graph

Let G be a weighted graph with edge weight function $w(i,j)$. For random walk on G, the stationary distribution $\boldsymbol{\pi}$ is proportional to the sum of the edge weights incident to each vertex. That is,

$$\pi_v = \frac{w(v)}{\sum_z w(z)}, \quad \text{for all vertices } v, \tag{3.4}$$

where

$$w(v) = \sum_{z \sim v} w(v,z)$$

is the sum of the edge weights on all edges incident to v.

Stationary Distribution for Simple Random Walk on a Graph

For simple random walk on a nonweighted graph, set $w(i,j) = 1$, for all i,j. Then, $w(v) = \deg(v)$, which gives

$$\pi_v = \frac{\deg(v)}{\sum_z \deg(z)} = \frac{\deg(v)}{2e},$$

where e is the number of edges in the graph.

Example 3.9 The two-state Markov chain with transition matrix

$$P = \begin{matrix} & 1 & 2 \\ 1 & 1-p & p \\ 2 & q & 1-q \end{matrix}$$

can be expressed as a random walk on the weighted graph in Figure 3.2. Then, $w(1) = q(1-p) + pq = q$ and $w(2) = pq + p(1-q) = p$. By Equation (3.4), the stationary distribution is

$$\boldsymbol{\pi} = \left(\frac{q}{p+q}, \frac{p}{p+q}\right).$$

■

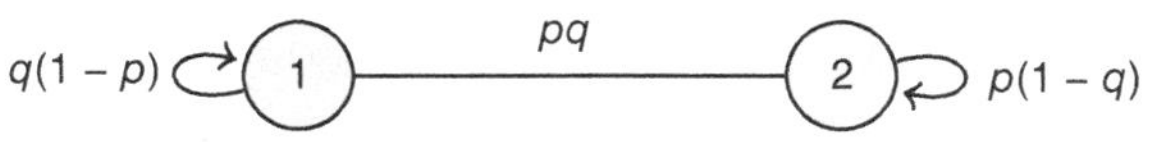

Figure 3.2 The general two-state Markov chain expressed as a weighted graph.

The reader may wonder how the weighted graph in Figure 3.2 is derived from the transition matrix $\boldsymbol{P}$. The algorithm will be explained in Section 3.7.

Example 3.10 Find the stationary distribution for random walk on the hypercube.

Solution The k-hypercube graph, as described in Example 2.8, has 2^k vertices. Each vertex has degree k. The sum of the vertex degrees is $k2^k$, and the stationary distribution $\boldsymbol{\pi}$ is given by

$$\pi_v = \frac{k}{k2^k} = \frac{1}{2^k}, \text{ for all } v.$$

That is, the stationary distribution is uniform on the set of vertices. ■

The hypercube is an example of a *regular* graph. A graph is regular if all the vertex degrees are the same. For simple random walk on a regular graph, the stationary distribution $\boldsymbol{\pi}$ is uniform on the set of vertices, since π_j is constant for all j. In addition to the hypercube, examples of regular graphs with uniform stationary distributions include the cycle graph ($\deg(v) = 2$) and the complete graph on k vertices ($\deg(v) = k - 1$).

The Eigenvalue Connection*

The stationary distribution of a Markov chain is related to the eigenstructure of the transition matrix.

First, a reminder on notation. For a matrix $\boldsymbol{M}$, the *transpose* of $\boldsymbol{M}$ is denoted $\boldsymbol{M}^T$. In this book, vectors are considered as row vectors. If $\boldsymbol{x}$ is a vector, $\boldsymbol{x}^T$ is a column vector.

Recall that an eigenvector of $\boldsymbol{M}$ is a column vector $\boldsymbol{x}^T$ such that $\boldsymbol{M}\boldsymbol{x}^T = \lambda\boldsymbol{x}^T$, for some scalar λ. We call such a vector a *right eigenvector* of $\boldsymbol{M}$. A *left eigenvector* of $\boldsymbol{M}$ is a row vector $\boldsymbol{y}$, which satisfies $\boldsymbol{y}\boldsymbol{M} = \mu\boldsymbol{y}$, for some scalar μ. A left eigenvector of $\boldsymbol{M}$ is simply a right eigenvector of $\boldsymbol{M}^T$.

If $\boldsymbol{\pi}$ is the stationary distribution of a Markov chain and satisfies $\boldsymbol{\pi}\boldsymbol{P} = \boldsymbol{\pi}$, then $\boldsymbol{\pi}$ is a left eigenvector of $\boldsymbol{P}$ corresponding to eigenvalue $\lambda = 1$.

Let $\mathbf{1}$ denote the column vector of all 1s. Since the rows of a stochastic matrix sum to 1, $\boldsymbol{P}\mathbf{1} = \mathbf{1} = (1)\mathbf{1}$. That is, $\mathbf{1}$ is a right eigenvector of $\boldsymbol{P}$ corresponding to eigenvalue $\lambda = 1$.

A matrix and its transpose have the same set of eigenvalues, with possibly different eigenvectors. It follows that $\lambda = 1$ is an eigenvalue of $\boldsymbol{P}^T$ with some corresponding right eigenvector $\boldsymbol{y}^T$. Equivalently, $\boldsymbol{y}$ is a left eigenvector of $\boldsymbol{P}$. That is, there exists a row vector $\boldsymbol{y}$ such that $\boldsymbol{y}\boldsymbol{P} = \boldsymbol{y}$. If a multiple of $\boldsymbol{y}$ can be normalized so that its components are non-negative and sum to 1, then this gives a stationary distribution. However, some of the entries of $\boldsymbol{y}$ might be negative, or complex-valued, and the vector might not be able to be normalized to give a probability distribution.

If a Markov chain has a unique stationary distribution, then the distribution is an eigenvector of $\boldsymbol{P}^T$ corresponding to $\lambda = 1$.

R: Stationary Distribution and Eigenvectors

The R command `eigen(P)` returns the eigenvalues and eigenvectors of a square matrix `P`. These are given in a list with two components: `values` contains the eigenvalues, and `vectors` contains the corresponding eigenvectors stored as a matrix. If `P` is a stochastic matrix, an eigenvector corresponding to eigenvalue $\lambda = 1$ will be stored in the first column of the `vectors` matrix. The R command `t(P)` gives the transpose of `P`.

In the following, an eigenvector is found for $\boldsymbol{P}^T$ corresponding to $\lambda = 1$. The vector is then normalized so that components sum to 1 in order to compute the stationary distribution. We illustrate on the weather matrix with stationary distribution $\boldsymbol{\pi} = (1/9, 3/4, 5/36) = (0.111,\ 0.750,\ 0.139)$.

```
> P
      Rain Snow Clear
Rain   0.2  0.6   0.2
Snow   0.1  0.8   0.1
Clear  0.1  0.6   0.3
> eigen(P)
$values
[1] 1.0 0.2 0.1
$vectors
          [,1]       [,2]       [,3]
[1,] 0.5773503  0.6882472 -0.9847319
[2,] 0.5773503 -0.2294157  0.1230915
[3,] 0.5773503  0.6882472  0.1230915
> eigen(P)$values  # eigenvalues
[1] 1.0 0.2 0.1
# eigenvalues of P and its transpose are the same
> eigen(t(P))$values
[1] 1.0 0.2 0.1
# eigenvectors of P-transpose
> eigen(t(P))$vectors
          [,1]          [,2]          [,3]
[1,] 0.1441500 -2.008469e-16  7.071068e-01
[2,] 0.9730125 -7.071068e-01  3.604182e-16
[3,] 0.1801875  7.071068e-01 -7.071068e-01

# first column gives eigenvector for eigenvalue 1
> x <- eigen(t(P))$vectors[,1]
> x
[1] 0.1441500 0.9730125 0.1801875
# normalize so rows sum to 1
> x/sum(x)
```

```
[1] 0.1111111 0.7500000 0.1388889

#  one-line command to find stationary distribution
> x <- eigen(t(P))$vectors[,1]; x/sum(x)
[1] 0.1111111 0.7500000 0.1388889
```

3.3 CAN YOU FIND THE WAY TO STATE *a*?

The long-term behavior of a Markov chain is related to how often states are visited. Here, we look more closely at the relationship between states and how reachable, or accessible, groups of states are from each other.

Say that state j is *accessible* from state i, if $P^n_{ij} > 0$, for some $n \geq 0$. That is, there is positive probability of reaching j from i in a finite number of steps. States i and j *communicate* if i is accessible from j and j is accessible from i.

Communication is an equivalence relation, which means that it satisfies the following three properties.

1. *(Reflexive)* Every state communicates with itself.
2. *(Symmetric)* If i communicates with j, then j communicates with i.
3. *(Transitive)* If i communicates with j, and j communicates with k, then i communicates with k.

Property 1 holds since $P^0_{ii} = P(X_0 = i|X_0 = i) = 1$. Property 2 follows since the definition of communication is symmetric. For Property 3, assume that i communicates with j, and j communicates with k. Then, there exists $n \geq 0$ and $m \geq 0$ such that $P^n_{ij} > 0$ and $P^m_{jk} > 0$. Therefore,

$$P^{n+m}_{ik} = \sum_t P^n_{it} P^m_{tk} \geq P^n_{ij} P^m_{jk} > 0.$$

Thus, k is accessible from i. Similarly, i is accessible from k.

Since communication is an equivalence relation the state space can be partitioned into equivalence classes, called *communication classes*. That is, the state space can be divided into disjoint subsets, each of whose states communicate with each other but do not communicate with any states outside their class.

A modified transition graph is a useful tool for finding the communication classes of a Markov chain. Vertices of the graph are the states of the chain. A directed edge is drawn between i and j if $P_{ij} > 0$. For purposes of studying the communication relationship between states, it is not necessary to label the edges with probabilities.

Example 3.11 Find the communication classes for the Markov chains with these transition matrices.

$$
P = \begin{array}{c} \\ a \\ b \\ c \\ d \\ e \end{array}
\begin{array}{c} \begin{array}{ccccc} a & b & c & d & e \end{array} \\
\begin{pmatrix} 0 & 0 & 0 & 2/3 & 1/3 \\ 1/2 & 0 & 1/6 & 0 & 1/3 \\ 0 & 1 & 0 & 0 & 0 \\ 0 & 0 & 3/4 & 1/4 & 0 \\ 0 & 0 & 0 & 0 & 1 \end{pmatrix} \end{array},
\quad
Q = \begin{array}{c} \\ a \\ b \\ c \\ d \\ e \\ f \end{array}
\begin{array}{c} \begin{array}{cccccc} a & b & c & d & e & f \end{array} \\
\begin{pmatrix} 1/6 & 1/3 & 0 & 0 & 1/2 & 0 \\ 0 & 1 & 0 & 0 & 0 & 0 \\ 0 & 0 & 0 & 0 & 3/4 & 1/4 \\ 1 & 0 & 0 & 0 & 0 & 0 \\ 4/5 & 0 & 0 & 1/5 & 0 & 0 \\ 0 & 0 & 1/2 & 0 & 1/2 & 0 \end{pmatrix} \end{array}.
$$

Solution The transition graphs are shown in Figure 3.3. For the $\boldsymbol{P}$-chain, the communication classes are $\{a, b, c, d\}$ and $\{e\}$. For the $\boldsymbol{Q}$-chain, the communication classes are $\{a, d, e\}$, $\{b\}$, and $\{c, f\}$.

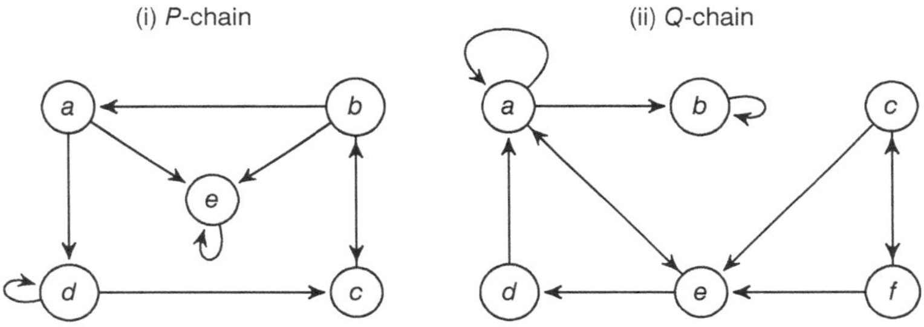

Figure 3.3 Transition graphs.

■

Most important is the case when the states of a Markov chain all communicate with each other.

Irreducibility

A Markov chain is called *irreducible* if it has exactly one communication class. That is, all states communicate with each other.

Example 3.12 The Markov chain with transition matrix

$$
P = \begin{array}{c} \\ a \\ b \\ c \\ d \\ e \\ f \end{array}
\begin{array}{c} \begin{array}{cccccc} a & b & c & d & e & f \end{array} \\
\begin{pmatrix} 0 & 1 & 0 & 0 & 0 & 0 \\ 1/2 & 0 & 0 & 0 & 1/2 & 0 \\ 0 & 0 & 0 & 0 & 1 & 0 \\ 0 & 0 & 0 & 0 & 1 & 0 \\ 1/4 & 1/4 & 0 & 1/4 & 0 & 1/4 \\ 0 & 0 & 1 & 0 & 0 & 0 \end{pmatrix} \end{array}
$$

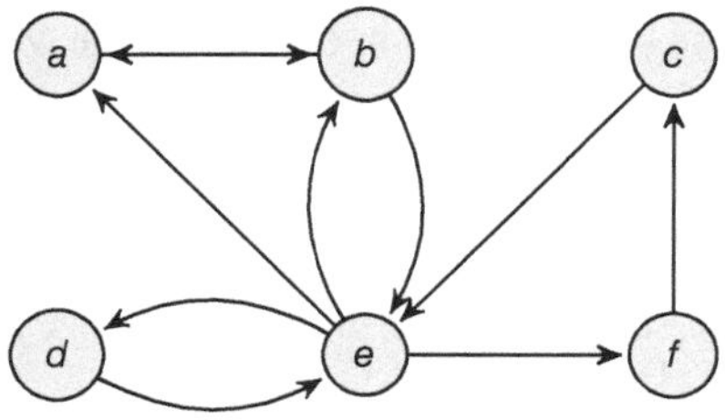

Figure 3.4 Transition graph for an irreducible Markov chain.

is irreducible, which can be seen by examining the transition graph in Figure 3.4. ■

Recurrence and Transience

Consider the transition graph in Figure 3.5. The communication classes are $\{a, b\}$ and $\{c\}$. From each state, consider the evolution of the chain started from that state and the probability that the chain eventually revisits that state.

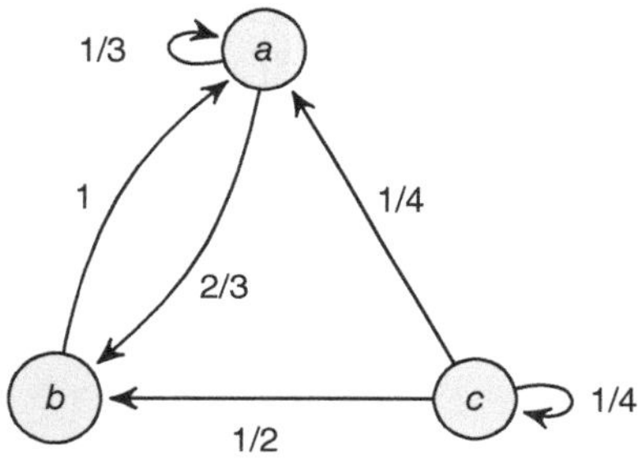

Figure 3.5

From a, the chain either returns to a in one step, or first moves to b and then returns to a on the second step. From a, the chain revisits a, with probability 1.

For the chain started in b, the chain first moves to a. It may continue to revisit a for many steps, but eventually it will return to b. This is because the probability that the chain stays at a forever is the probability that it continually transitions from a to a, which is equal to

$$\lim_{n\to\infty} (P_{aa})^n = \lim_{n\to\infty} \left(\frac{1}{3}\right)^n = 0.$$

Thus, from b, the chain revisits b, with probability 1.

For the chain started in c, the chain may revisit c for many successive steps. But with positive probability it will eventually move to either a or b. Once it does, it will never revisit c, as it is now *stuck* in the $\{a, b\}$ communication class. From c, there is positive probability that the chain started in c will never revisit c. In this case, that probability is $1 - 1/4 = 3/4$.

The states of a Markov chain, as this example illustrates, exhibit one of two contrasting behaviors. For the chain started in a given state, the chain either revisits that state, with probability 1, or there is positive probability that the chain will never revisit that state.

Given a Markov chain $X_0, X_1, \ldots$, let $T_j = \min\{n > 0 : X_n = j\}$ be the *first passage time* to state j. If $X_n \neq j$, for all $n > 0$, set $T_j = \infty$. Let

$$f_j = P(T_j < \infty | X_0 = j)$$

be the probability that the chain started in j eventually returns to j. For the three-state chain introduced in this section, $f_a = f_b = 1$, and $f_c = 1/4$. We classify the states j of a Markov chain according to whether or not $f_j = 1$.

Recurrent and Transient States

State j is said to be *recurrent* if the Markov chain started in j eventually revisits j. That is, $f_j = 1$.

State j is said to be *transient* if there is positive probability that the Markov chain started in j never returns to j. That is, $f_j < 1$.

Whether or not a state is eventually revisited is strongly related to how often that state is visited.

For the chain started in i, let

$$I_n = \begin{cases} 1, & \text{if } X_n = j, \\ 0, & \text{otherwise,} \end{cases}$$

for $n \geq 0$. Then, $\sum_{n=0}^{\infty} I_n$ is the number of visits to j. The expected number of visits to j is

$$E\left(\sum_{n=0}^{\infty} I_n\right) = \sum_{n=0}^{\infty} E(I_n) = \sum_{n=0}^{\infty} P(X_n = j | X_0 = i) = \sum_{n=0}^{\infty} P_{ij}^n,$$

where the infinite sum may possibly diverge to $+\infty$. For the chain started in i, the expected number of visits to j is the ijth entry of the matrix $\sum_{n=0}^{\infty} \boldsymbol{P}^n$.

Assume that j is recurrent. The chain started at j will eventually revisit j with probability 1. Once it hits j, the chain begins anew and behaves as if a new version of the chain started at j. We say the Markov chain *regenerates* itself. (This intuitive behavior is known as the *strong Markov property*. For a more formal treatment see Section 3.9.) From j, the chain will revisit j again, with probability 1, and so on. It follows that j will be visited infinitely many times, and

$$\sum_{n=0}^{\infty} P_{jj}^n = \infty.$$

On the other hand, assume that j is transient. Starting at j, the probability of eventually hitting j again is f_j, and the probability of never hitting j is $1 - f_j$. If the chain hits j, the event that it will eventually revisit j is independent of past history. It follows that the sequence of successive visits to j behaves like an i.i.d. sequence of coin tosses where heads occurs if j is eventually hit and tails occurs if j is never hit again. The number of times that j is hit is the number of coin tosses until tails occurs, which has a geometric distribution with parameter $1 - f_j$. Thus, the expected number of visits to j is $1/(1 - f_j)$, and

$$\sum_{n=0}^{\infty} P_{jj}^n = \frac{1}{1 - f_j} < \infty.$$

In particular, a transient state will only be visited a finite number of times.

This leads to another characterization of recurrence and transience.

Recurrence, Transience

(i) State j is recurrent if and only if

$$\sum_{n=0}^{\infty} P_{jj}^n = \infty.$$

(ii) State j is transient if and only if

$$\sum_{n=0}^{\infty} P_{jj}^n < \infty.$$

Assume that j is recurrent and accessible from i. For the chain started in i there is positive probability of hitting j. And from j, the expected number of visits to j is infinite. It follows that the expected number of visits to j for the chain started in i is also infinite, and thus

$$\sum_{n=0}^{\infty} P_{ij}^n = \infty.$$

Assume that j is transient and accessible from i. By a similar argument the expected number of visits to j for the chain started in i is finite, and thus

$$\sum_{n=0}^{\infty} P_{ij}^n < \infty,$$

from which it follows that

$$\lim_{n \to \infty} P_{ij}^n = 0. \tag{3.5}$$

The long-term probability that a Markov chain eventually hits a transient state is 0.

Recurrence and transience are *class* properties of a Markov chain as described by the following theorem.

Recurrence and Transience are Class Properties

Theorem 3.3. *The states of a communication class are either all recurrent or all transient.*

Proof. Let i and j be states in the same communication class. Assume that i is recurrent. Since i and j communicate, there exists $r \geq 0$ and $s \geq 0$ such that $P^r_{ji} > 0$ and $P^s_{ij} > 0$. For $n \geq 0$,

$$P^{r+n+s}_{jj} = \sum_k \sum_l P^r_{jk} P^n_{kl} P^s_{lj} \geq P^r_{ji} P^n_{ii} P^s_{ij}.$$

Summing over n gives

$$\sum_{n=0}^{\infty} P^{r+n+s}_{jj} \geq P^r_{ji} \left(\sum_{n=0}^{\infty} P^n_{ii} \right) P^s_{ji} = \infty.$$

Since

$$\sum_{n=0}^{\infty} P^n_{jj} \geq \sum_{n=r+s}^{\infty} P^n_{jj} = \sum_{n=0}^{\infty} P^{r+n+s}_{jj},$$

it follows that $\sum_{n=0}^{\infty} P^n_{jj}$ diverges to infinity, and thus j is recurrent. Hence, if one state of a communication class is recurrent, all states in that class are recurrent.

On the other hand, if one state is transient, the other states must be transient. By contradiction, if the communication class contains a recurrent state then by what was just proven all the states are recurrent. ■

Corollary 3.4. *For a finite irreducible Markov chain, all states are recurrent.*

Proof. The states of an irreducible chain are either all recurrent or all transient. Assume that they are all transient. Then, state 1 will be visited for a finite amount of time, after which it is never hit again, similarly with state 2, and with all states. Since there are finitely many states, it follows that *none* of the states will be visited after some finite amount of time, which is not possible. ■

By Corollary 3.4, a finite Markov chain cannot have all transient states. This is not true for infinite chains, as the following classic example illustrates.

Example 3.13 (Simple random walk) A random walk on the integer line starts at 0 and moves left, with probability p, or right, with probability $1 - p$. For $0 < p < 1$, the process is an irreducible Markov chain, as every state is accessible from every other state. Is the chain recurrent or transient?

Solution Since the chain is irreducible, it suffices to examine one state. Choose state 0 and consider $\sum_{n=0}^{\infty} P_{00}^n$. Observe that from 0 the walk can only revisit 0 in an even number of steps. So $P_{00}^n = 0$, if n is odd. To move from 0 to 0 in exactly $2n$ steps requires that the walk moves n steps to the left and n steps to the right, in some order. Such a path of length $2n$ can be identified with a sequence of n Ls and n Rs. There are $\binom{2n}{n}$ such sequences. Each left move occurs with probability p and each right move occurs with probability $1 - p$. This gives

$$P_{00}^{2n} = \binom{2n}{n} p^n (1-p)^n.$$

The binomial coefficient $\binom{2n}{n}$ is estimated using *Stirling's approximation*

$$n! \approx n^n e^{-n} \sqrt{2\pi n}, \text{ for large } n.$$

The more precise statement is

$$\lim_{n\to\infty} \frac{n!}{n^n e^{-n} \sqrt{2\pi n}} = 1.$$

By Stirling's approximation, for large n,

$$\binom{2n}{n} = \frac{(2n)!}{n!n!} \approx \frac{(2n)^{2n} e^{-2n} \sqrt{2\pi 2n}}{\left(n^n e^{-n} \sqrt{2\pi n}\right)^2} = \frac{4^n}{\sqrt{\pi n}}.$$

Thus,

$$\sum_{n=0}^{\infty} P_{00}^{2n} = \sum_{n=0}^{\infty} \binom{2n}{n} p^n (1-p)^n \approx \sum_{n=1}^{\infty} \frac{(4p(1-p))^n}{\sqrt{\pi n}}.$$

Convergence of the infinite series depends upon p. We have

$$\sum_{n=0}^{\infty} P_{00}^{2n} \approx \begin{cases} \sum_{n=1}^{\infty} \frac{1}{\sqrt{\pi n}} = \infty, & \text{if } p = 1/2, \\ \sum_{n=1}^{\infty} \frac{\epsilon^n}{\sqrt{\pi n}} < \infty, & \text{if } p \neq 1/2, \end{cases}$$

where $\epsilon = 4p(1-p)$. If $p \neq 1/2$, then $0 < \epsilon < 1$, and $\epsilon^n \to 0$, as $n \to \infty$.

For $p = 1/2$, the random walk is recurrent. Each integer, no matter how large, is visited infinitely often. For $p \neq 1/2$, the walk is transient. With positive probability the walk will never return to its starting point.

Surprises await for random walk in higher dimensions. Simple symmetric random walk on the integer points in the plane $\mathbb{Z}^2$ moves left, right, up, or down, with probability 1/4 each. The process has been called the *drunkard's walk*. As in one dimension, the walk is recurrent. The method of proof is similar. See Exercise 3.20. Letting 0 denote the origin in the plane, it can be shown that

$$\sum_{n=0}^{\infty} P_{00}^n \approx \sum_{n=1}^{\infty} \frac{1}{\pi n} = \infty.$$

Remarkably, in dimensions three and higher simple symmetric random walk is transient. This was first shown by George Pólya in 1921 and is known as *Pólya's Theorem*. It can be shown that in $\mathbb{Z}^3$,

$$\sum_{n=0}^{\infty} P_{00}^n \approx \sum_{n=1}^{\infty} \frac{1}{(\pi n)^{3/2}} < \infty.$$

The mathematician Shizuo Kakutani is quoted as explaining this result by saying, "A drunk man will find his way home, but a drunk bird may get lost forever." ■

Canonical Decomposition

A set of states C is said to be *closed* if no state outside of C is accessible from any state in C. If C is closed, then $P_{ij} = 0$ for all $i \in C$ and $j \notin C$.

Closed Communication Class

Lemma 3.5. *A communication class is closed if it consists of all recurrent states. A finite communication class is closed only if it consist of all recurrent states.*

Proof. Let C be a communication class made up of recurrent states. Assume that C is not closed. Then, there exists states $i \in C$ and $j \notin C$ such that $P_{ij} > 0$. Since j is accessible from i, i is not accessible from j, otherwise j would be contained in C. Start the chain in i. With positive probability, the chain will hit j and then never hit i again. But this contradicts the assumption that i is recurrent.

On the other hand, assume that C is closed and finite. By the same argument given in the proof of Corollary 3.4 the states cannot all be transient. Hence, they are all recurrent. ■

The state space S of a finite Markov chain can be partitioned into transient and recurrent states as $S = T \cup R_1 \cup \cdots \cup R_m$, where T is the set of all transient states and the R_i are closed communication classes of recurrent states. This is called the *canonical decomposition*. The computation of many quantities associated with Markov chains can be simplified by this decomposition.

Given a canonical decomposition, the state space can be reordered so that the Markov transition matrix has the block matrix form

$$\boldsymbol{P} = \begin{array}{c} \\ T \\ R_1 \\ \vdots \\ R_m \end{array} \begin{array}{c} \begin{array}{cccc} T & R_1 & \cdots & R_m \end{array} \\ \begin{pmatrix} * & * & \cdots & * \\ \mathbf{0} & \boldsymbol{P}_1 & \cdots & \mathbf{0} \\ \vdots & \vdots & \ddots & \vdots \\ \mathbf{0} & \mathbf{0} & \cdots & \boldsymbol{P}_m \end{pmatrix} \end{array}.$$

Each submatrix $\boldsymbol{P}_1, \ldots, \boldsymbol{P}_m$ is a square stochastic matrix corresponding to a closed recurrent communication class. By itself, each of these matrices is the matrix of an irreducible Markov chain with a restricted state space.

Example 3.14 Consider the Markov chain with transition matrix

$$\boldsymbol{P} = \begin{array}{c} \\ a \\ b \\ c \\ d \\ e \\ f \\ g \end{array} \begin{array}{c} \begin{array}{ccccccc} a & b & c & d & e & f & g \end{array} \\ \begin{pmatrix} 1/4 & 0 & 0 & 0 & 0 & 0 & 3/4 \\ 1/8 & 1/8 & 1/4 & 0 & 1/4 & 1/8 & 1/8 \\ 0 & 0 & 2/5 & 1/5 & 2/5 & 0 & 0 \\ 0 & 0 & 1/2 & 1/2 & 0 & 0 & 0 \\ 0 & 0 & 0 & 1/2 & 1/2 & 0 & 0 \\ 0 & 1/5 & 0 & 1/5 & 1/5 & 1/5 & 1/5 \\ 4/5 & 0 & 0 & 0 & 0 & 0 & 1/5 \end{pmatrix} \end{array}$$

described by the transition graph in Figure 3.6. Give the canonical decomposition.

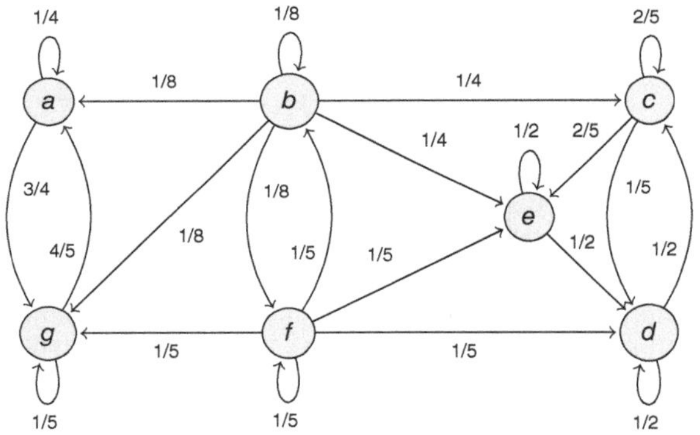

Figure 3.6

Solution States b and f are transient. The closed recurrent classes are $\{a, g\}$ and $\{c, d, e\}$. Reordering states gives the transition matrix in block matrix form

$$\boldsymbol{P} = \begin{array}{c} \\ b \\ f \\ a \\ g \\ c \\ d \\ e \end{array} \begin{array}{c} \begin{array}{ccccccc} b & f & a & g & c & d & e \end{array} \\ \begin{pmatrix} 1/8 & 1/8 & 1/8 & 1/8 & 1/4 & 0 & 1/4 \\ 1/5 & 1/5 & 0 & 1/5 & 0 & 1/5 & 1/5 \\ 0 & 0 & 1/4 & 3/4 & 0 & 0 & 0 \\ 0 & 0 & 4/5 & 1/5 & 0 & 0 & 0 \\ 0 & 0 & 0 & 0 & 2/5 & 1/5 & 2/5 \\ 0 & 0 & 0 & 0 & 1/2 & 1/2 & 0 \\ 0 & 0 & 0 & 0 & 0 & 1/2 & 1/2 \end{pmatrix} \end{array}.$$

■

The canonical decomposition is useful for describing the long-term behavior of a Markov chain. The block matrix form facilitates taking matrix powers. For $n \geq 1$,

$$\boldsymbol{P}^n = \begin{array}{c} \\ T \\ R_1 \\ \vdots \\ R_m \end{array} \begin{array}{c} \begin{array}{cccc} T & R_1 & \cdots & R_m \end{array} \\ \begin{pmatrix} * & * & \cdots & * \\ \mathbf{0} & \boldsymbol{P}_1^n & \cdots & \mathbf{0} \\ \vdots & \vdots & \ddots & \vdots \\ \mathbf{0} & \mathbf{0} & \cdots & \boldsymbol{P}_m^n \end{pmatrix} \end{array}.$$

Taking limits gives

$$\lim_{n\to\infty} \boldsymbol{P}^n = \begin{array}{c} \\ T \\ R_1 \\ \vdots \\ R_m \end{array} \begin{array}{c} \begin{array}{cccc} T & R_1 & \cdots & R_m \end{array} \\ \begin{pmatrix} \mathbf{0} & * & \cdots & * \\ \mathbf{0} & \lim_{n\to\infty} \boldsymbol{P}_1^n & \cdots & \mathbf{0} \\ \vdots & \vdots & \ddots & \vdots \\ \mathbf{0} & \mathbf{0} & \cdots & \lim_{n\to\infty} \boldsymbol{P}_m^n \end{pmatrix} \end{array}.$$

Note that the entries of the columns corresponding to transient states are all 0 as a consequence of Equation (3.5).

The recurrent, closed communication classes $R_1, \ldots, R_m$ behave like mini-irreducible Markov chains where all states communicate with each other. The asymptotic properties of the submatrices $\boldsymbol{P}_1, \ldots, \boldsymbol{P}_m$ lead us to consider the properties of irreducible Markov chains, which is where our path now goes.

3.4 IRREDUCIBLE MARKOV CHAINS

The next theorem characterizes the stationary distribution $\boldsymbol{\pi}$ for finite irreducible Markov chains. It relates the stationary probability π_j to the expected number of steps between visits to j. Recall that $T_j = \min\{n > 0 : X_n = j\}$ is the first passage time to state j.

Limit Theorem for Finite Irreducible Markov Chains

Theorem 3.6. *Assume that $X_0, X_1, \ldots$ is a finite irreducible Markov chain. For each state j, let $\mu_j = E(T_j|X_0 = j)$ be the expected return time to j. Then, μ_j is finite, and there exists a unique, positive stationary distribution $\boldsymbol{\pi}$ such that*

$$\pi_j = \frac{1}{\mu_j}, \quad \text{for all } j. \tag{3.6}$$

Furthermore, for all states i,

$$\pi_j = \lim_{n\to\infty} \frac{1}{n} \sum_{m=0}^{n-1} P_{ij}^m. \tag{3.7}$$

The theorem is proved in Section 3.10.

Remark:

1. The fact that $\pi_j = 1/\mu_j$ is intuitive. If there is one visit to j every μ_j steps, then the proportion of visits to j is $1/\mu_j$.
2. The theorem does not assert that $\boldsymbol{\pi}$ is a limiting distribution. The convergence in Equation (3.7) is a weaker form of convergence than $\pi_j = \lim_{n\to\infty} P^n_{ij}$. We discuss in the next section that an additional assumption is needed for $\boldsymbol{\pi}$ to be a limiting distribution.
3. For finite irreducible Markov chains, all states are recurrent, and the expected return time $E(T_j|X_0 = j)$ is finite, for all j. However, for an infinite Markov chain if j is a recurrent state, even though the chain will eventually revisit j with probability 1, the expected number of steps between such visits need not be finite. The theorem can be extended to infinite irreducible Markov chains for which the expected return time $E(T_j|X_0 = j)$ is finite, for all j.

   A recurrent state j is called *positive recurrent* if $E(T_j|X_0 = j) < \infty$, and *null recurrent* if $E(T_j|X_0 = j) = \infty$. Thus, the theorem holds for irreducible Markov chains for which all states are positive recurrent. See Exercise 3.27 for an example of a Markov chain in which all states are null recurrent.

In many applications, the expected time between visits to a given state is of particular importance.

Example 3.15 (Earthquake recurrences) The Chiayi–Tainan area of Taiwan was devastated by an earthquake on September 21, 1999. In Tsai (2002), Markov chains were used to study seismic activity in the region. A chain is constructed with states corresponding to Richter scale magnitudes of earthquake intensity. The state space is $\{M_2, M_3, M_4, M_5\}$, where M_k denotes an earthquake with a Richter level in the interval $[k, k+1)$. A Markov transition matrix is estimated from historical data for the period 1973–1975:

$$
\boldsymbol{P} = \begin{array}{c} \\ M_2 \\ M_3 \\ M_4 \\ M_5 \end{array}
\begin{array}{c} \begin{array}{cccc} M_2 & M_3 & M_4 & M_5 \end{array} \\
\begin{pmatrix} 0.785 & 0.194 & 0.018 & 0.003 \\ 0.615 & 0.334 & 0.048 & 0.003 \\ 0.578 & 0.353 & 0.069 & 0.000 \\ 0.909 & 0.000 & 0.091 & 0.000 \end{pmatrix} \end{array}.
$$

The stationary distribution $\boldsymbol{\pi} = (0.740, 0.230, 0.027, 0.003)$ is found with technology. Earthquakes of magnitude M_2 or greater tend to occur in this region about once every four months. If an M_5 earthquake occurs, investigators would like to know how long it will be before another M_5 earthquake.

The expected number of Markov chain transitions between M_5 earthquakes is $1/\pi_{M_5} = 1/0.003 = 333$. If earthquakes occur, on average, every four months, then according to the model it will take about $333 \times (4/12) = 111$ years before another M_5 earthquake. ■

Example 3.16 For the frog-jumping random walk on an n-cycle, how many hops does it take, on average, for the frog to return to its starting lily pad?

Solution In the cycle graph, all vertices have the same degree. Hence, for simple random walk on the cycle, the stationary distribution is uniform on the set of vertices. Since $\pi_v = 1/n$, for all vertices v, it takes the frog, on average, $1/\pi_v = n$ hops to return to its starting pad. ■

First-Step Analysis

The expected return time $E(T_j|X_0 = j)$ is found by taking the reciprocal of the stationary probability π_j. Another approach is to condition on the first step of the chain and use the law of total expectation. This is called *first-step analysis*.

Example 3.17 Consider a Markov chain with transition matrix

$$\boldsymbol{P} = \begin{matrix} & \begin{matrix} a & b & c \end{matrix} \\ \begin{matrix} a \\ b \\ c \end{matrix} & \begin{pmatrix} 0 & 1 & 0 \\ 1/2 & 0 & 1/2 \\ 1/3 & 1/3 & 1/3 \end{pmatrix} \end{matrix}.$$

From state a, find the expected return time $E(T_a|X_0 = a)$ using first-step analysis.

Solution Let $e_x = E(T_a|X_0 = x)$, for $x = a, b, c$. Thus, e_a is the desired expected return time, and e_b and e_c are the expected *first passage* times to a for the chain started in b and c, respectively.

For the chain started in a, the next state is b, with probability 1. From b, the further evolution of the chain behaves as if the original chain started at b. Thus,

$$e_a = 1 + e_b.$$

From b, the chain either hits a, with probability $1/2$, or moves to c, where the chain behaves as if the original chain started at c. It follows that

$$e_b = \frac{1}{2} + \frac{1}{2}(1 + e_c).$$

Similarly, from c, we have

$$e_c = \frac{1}{3} + \frac{1}{3}(1 + e_b) + \frac{1}{3}(1 + e_c).$$

Solving the three equations gives

$$e_c = \frac{8}{3}, \quad e_b = \frac{7}{3}, \quad \text{and} \quad e_a = \frac{10}{3}.$$

The desired expected return time is $10/3$.

We leave it to the reader to verify that the stationary distribution is

$$\pi = \left(\frac{3}{10}, \frac{2}{5}, \frac{3}{10}\right).$$

The expected return time is simulated in the following `R` code. The chain started at a is run for 25 steps (long enough to return to a with very high probability), and the return time to a is found. The mean return time is estimated based on 10,000 trials.

R : Simulating an Expected Return Time

```
#returntime.R
> P
        a       b       c
a 0.00000 1.00000 0.00000
b 0.50000 0.00000 0.50000
c 0.33333 0.33333 0.33333
> init
[1] 1 0 0
> states
[1] "a" "b" "c"
> markov(init,P,25,states)
 "a" "b" "c" "c" "c" "b" "a" "b" "a" "b" "a" "b" "c"
 "b" "c" "a" "b" "c" "b" "c" "b" "a" "b" "a" "b" "a"

> trials <- 10000
> simlist <- numeric(trials)
> for (i in 1:trials) {
   path <- markov(init,P,25,states)
     # find the index of the 2nd occurrence of "a"
     # subtract 1 to account for time 0
   returntime <- which(path == "a")[2] - 1
   simlist[i] <- returntime }
# expected return time to state a
> mean(simlist)
[1] 3.3346
```

■

3.5 PERIODICITY

Finite irreducible Markov chains have unique, positive stationary distributions. Although they may not have limiting distributions, they have almost limiting behavior in the sense that for all states i and j, the partial averages $(1/n)\sum_{m=0}^{n-1} P_{ij}^m$ converge.

An example of a finite irreducible Markov chain with no limiting distribution is random walk on the n-cycle, when n is even. The graph is regular (all vertex degrees are the same) and the unique stationary distribution is uniform. But there is no limiting distribution since the chain flip-flops back and forth between even and odd states. The chain's position after n steps depends on the parity of the initial state.

It is precisely the finite irreducible Markov chains that do not exhibit this type of *periodic* behavior, which have limiting distributions.

For a Markov chain started in state i, consider the set of times when the chain can return to i. For the chain described by the graph in Figure 3.7a, from any state the set of possible return times is $\{2, 4, 6, 8, \dots\}$. The same is true for the chain in Figure 3.7b. The chain started from any state returns to that state in multiples of two steps. For the chain in Figure 3.7c, from each state the set of return times is $\{3, 6, 9, 12, \dots\}$ The chain started in a returns to a in multiples of three steps.

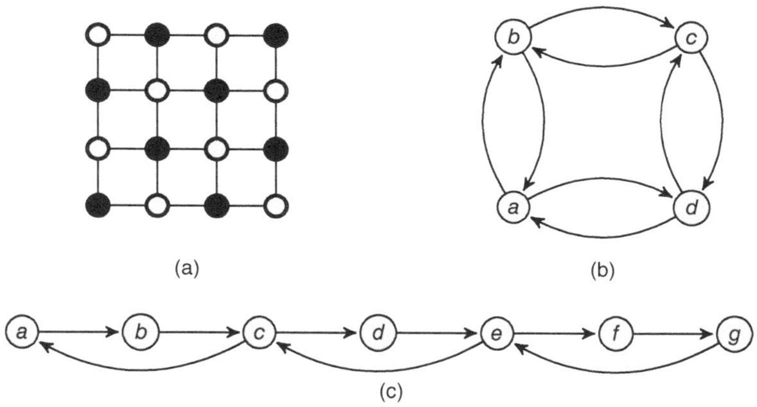

Figure 3.7 Periodic Markov chains.

The idea is formalized in the following definition. Recall that the *greatest common divisor (gcd)* of a set of positive integers is the largest integer that divides all the numbers of the set without a remainder.

Period

For a Markov chain with transition matrix $\boldsymbol{P}$, the *period* of state i, denoted $d(i)$, is the greatest common divisor of the set of possible return times to i. That is,

$$d(i) = \gcd\{n > 0 : P^n_{ii} > 0\}.$$

If $d(i) = 1$, state i is said to be *aperiodic*. If the set of return times is empty, set $d(i) = +\infty$.

The definition of period gives that from state i, returns to i can only occur in multiples of $d(i)$ steps. And the period $d(i)$ is the largest such number with this property.

Periodicity, similar to recurrence and transience, is a class property.

Periodicity is a Class Property

Lemma 3.7. *The states of a communication class all have the same period.*

Proof. Let i and j be states in the same communication class with respective periods $d(i)$ and $d(j)$. Since i and j communicate, there exist positive integers r and s, such that $P^r_{ij} > 0$ and $P^s_{ji} > 0$. Then,

$$P^{r+s}_{ii} = \sum_k P^r_{ik} P^s_{ki} \geq P^r_{ij} P^s_{ji} > 0.$$

Thus $r + s$ is a possible return time for i, and hence $d(i)$ is a divisor of $r + s$. Assume that $P^n_{jj} > 0$ for some positive integer n. Then,

$$P^{r+s+n}_{ii} \geq P^r_{ij} P^n_{jj} P^s_{ji} > 0,$$

and thus $d(i)$ is a divisor of $r + s + n$. Since $d(i)$ divides both $r + s$ and $r + s + n$, it must also divide n. Thus, $d(i)$ is a common divisor of the set $\{n > 0 : P^n_{jj} > 0\}$. Since $d(j)$ is the largest such divisor, it follows that $d(i) \leq d(j)$. By the same argument with i and j reversed, we have that $d(j) \leq d(i)$. Hence, $d(i) = d(j)$. ■

Example 3.18 Consider the transition graph in Figure 3.8. Identify the communication classes, their periods, and whether the class is recurrent or transient.

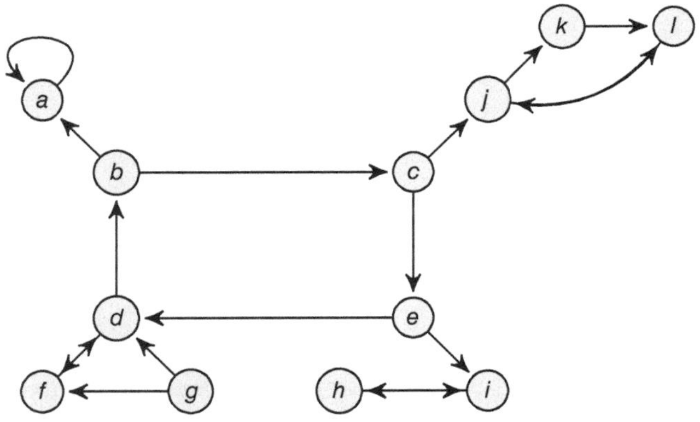

Figure 3.8

Solution Communication classes, with periods, are as follows:

(i) $\{a\}$ is recurrent with period 1,

(ii) $\{b, c, d, e, f\}$ is transient with period 2,

(iii) $\{g\}$ is transient with period $+\infty$,
(iv) $\{h, i\}$ is recurrent with period 2, and
(v) $\{j, k, l\}$ is recurrent with period 1.

Observe that return times for state j can occur in multiples of three (first k, then l, then j) or in multiples of two (first l then j). Since two and three are relatively prime, their great common divisor is one. Hence, $d(j) = d(k) = d(l) = 1$. ■

From Lemma 3.7, it follows that all states in an irreducible Markov chain have the same period.

Periodic, Aperiodic Markov Chain

A Markov chain is called *periodic* if it is irreducible and all states have period greater than 1.

A Markov chain is called *aperiodic* if it is irreducible and all states have period equal to 1.

Note that any state i with the property that $P_{ii} > 0$ is necessarily aperiodic. Thus, a sufficient condition for an irreducible Markov chain to be aperiodic is that $P_{ii} > 0$ for some i. That is, at least one diagonal entry of the transition matrix is nonzero.

3.6 ERGODIC MARKOV CHAINS

A Markov chain is called *ergodic* if it is irreducible, aperiodic, and all states have finite expected return times. The latter is always true for finite chains. Thus, a finite Markov chain is ergodic if it is irreducible and aperiodic. It is precisely the class of ergodic Markov chains that have positive limiting distributions.

Fundamental Limit Theorem for Ergodic Markov Chains

Theorem 3.8. *Let $X_0, X_1, \ldots$ be an ergodic Markov chain. There exists a unique, positive, stationary distribution $\boldsymbol{\pi}$, which is the limiting distribution of the chain. That is,*

$$\pi_j = \lim_{n\to\infty} P^n_{ij}, \text{ for all } i, j.$$

Recall Theorem 3.2, which asserts the same limit result for Markov chains with regular transition matrices. The proof of the fundamental limit theorem for ergodic

Markov chains is given in Section 3.10, where it is also shown that a Markov chain is ergodic if and only if its transition matrix is regular.

Example 3.19 (Modified Ehrenfest model) In the Ehrenfest scheme of Example 3.7, at each discrete step one of N fleas is picked uniformly at random. The flea jumps from one dog to the other. Our dogs are named Cooper and Lisa. Let X_n be the number of fleas on Lisa after n jumps.

The Ehrenfest chain $X_0, X_1, \ldots$ is irreducible and periodic with period 2. From state 0, the chain can only return to 0 in an even number of steps.

A modified Ehrenfest scheme picks a flea uniformly at random and then picks a dog uniformly at random for the flea to jump to. The transition probabilities are

$$P_{ij} = \begin{cases} i/(2N), & \text{if } j = i - 1, \\ (N - i)/(2N), & \text{if } j = i + 1, \\ 1/2, & \text{if } j = i, \\ 0, & \text{otherwise.} \end{cases}$$

Since $P_{ii} > 0$, the chain is aperiodic, and thus ergodic. The unique stationary distribution is binomial with parameters N and $1/2$, the same as in the regular Ehrenfest scheme. (We invite the reader to show this in Exercise 3.24.) By the fundamental limit theorem, the stationary distribution is the limiting distribution of the chain. The long-term stationary process can be simply described: for each of the N fleas, toss a fair coin. If heads, the flea jumps to Lisa, if tails it jumps to Cooper.

The modified Ehrenfest transition matrix for $N = 6$ fleas is

$$\boldsymbol{P} = \begin{array}{c} \\ 0 \\ 1 \\ 2 \\ 3 \\ 4 \\ 5 \\ 6 \end{array} \begin{array}{c} \begin{array}{ccccccc} 0 & 1 & 2 & 3 & 4 & 5 & 6 \end{array} \\ \begin{pmatrix} 1/2 & 1/2 & 0 & 0 & 0 & 0 & 0 \\ 5/12 & 1/2 & 1/12 & 0 & 0 & 0 & 0 \\ 0 & 1/3 & 1/2 & 1/6 & 0 & 0 & 0 \\ 0 & 0 & 1/4 & 1/2 & 1/4 & 0 & 0 \\ 0 & 0 & 0 & 1/6 & 1/2 & 1/3 & 0 \\ 0 & 0 & 0 & 0 & 1/12 & 1/2 & 5/12 \\ 0 & 0 & 0 & 0 & 0 & 1/2 & 1/2 \end{pmatrix} \end{array},$$

with limiting distribution

$$\boldsymbol{\pi} = \left(\frac{1}{64}, \frac{6}{64}, \frac{15}{64}, \frac{20}{64}, \frac{15}{64}, \frac{6}{64}, \frac{1}{64} \right).$$

■

Example 3.20 Consider a Markov chain with transition matrix

$$
\boldsymbol{P} = \begin{array}{c} \\ 1 \\ 2 \\ 3 \\ 4 \\ \vdots \\ \text{n-1} \\ \text{n} \end{array}
\begin{array}{c}
\begin{array}{ccccccc} 1 & 2 & 3 & 4 & \cdots & \text{n-1} & \text{n} \end{array} \\
\left(\begin{array}{ccccccc}
0 & 1 & 0 & 0 & \cdots & 0 & 0 \\
1/2 & 0 & 1/2 & 0 & \cdots & 0 & 0 \\
1/3 & 0 & 0 & 2/3 & \cdots & 0 & 0 \\
1/4 & 0 & 0 & 0 & \cdots & 0 & 0 \\
\vdots & \vdots & \vdots & \vdots & \ddots & \vdots & \vdots \\
1/(n-1) & 0 & 0 & 0 & \cdots & 0 & (n-2)/(n-1) \\
1 & 0 & 0 & 0 & \cdots & 0 & 0
\end{array}\right)
\end{array}.
$$

The transition graph is shown in Figure 3.9. Find the limiting distribution.

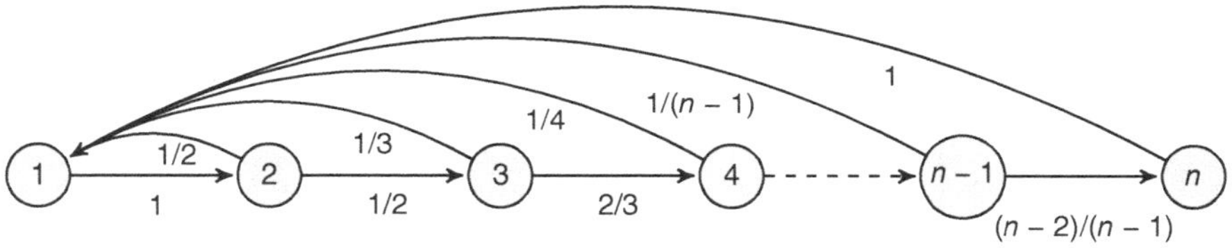

Figure 3.9

Solution State 1 is accessible from all states. Each state k is accessible from 1 by transitioning to 2, then 3, …, to k. Thus, the chain is irreducible. It is also aperiodic for $n \geq 3$. For instance, one can reach state 1 from 1 in either two steps or three steps. Thus, $d(1) = 1$. The chain is ergodic, and the limiting distribution is gotten by finding the stationary distribution.

Let $\boldsymbol{x} = (x_1, x_2, \ldots, x_n)$, with $x_1 = 1$. Solving

$$x_j = \sum_{i=1}^{n} x_i P_{ij} = x_{j-1} P_{j-1,j}, \quad \text{for } j = 2, \ldots, n,$$

gives $x_2 = x_1 = 1$ and

$$
\begin{aligned}
x_j &= x_{j-1}\left(\frac{j-2}{j-1}\right) = x_{j-2}\left(\frac{j-3}{j-2}\right)\left(\frac{j-2}{j-1}\right) \\
&= x_{j-2}\left(\frac{j-3}{j-1}\right) = \cdots = x_2\left(\frac{1}{j-1}\right) = \frac{1}{j-1}, \quad \text{for } j = 3, \ldots, n.
\end{aligned}
$$

Hence,

$$
\pi_j = \begin{cases} \dfrac{1}{c}, & \text{for } j = 1, \\ \dfrac{1}{c(j-1)}, & \text{for } j = 2, \ldots, n, \end{cases}
$$

where $c = 1 + \sum_{k=1}^{n-1} 1/k$.

An infinite version of this chain—see Exercise 3.27—does not have a stationary distribution. Although the chain is aperiodic and irreducible, it is not positive recurrent. That is, the expected return time between visits to the same state is infinite.

■

Example 3.21 (PageRank) Google's PageRank search algorithm is introduced in Chapter 1 in Example 1.1. The model is based on the *random surfer* model, which is a random walk on the *webgraph*. For this graph, each vertex represents an internet page. A directed edge connects i to j if there is a hypertext link from page i to page j. When the random surfer is at page i, they move to a new page by choosing from the available links on i uniformly at random.

Figure 3.10 shows a simplified network with seven pages. The network is described by the *network matrix*

$$N = \begin{array}{c} \\ a \\ b \\ c \\ d \\ e \\ f \\ g \end{array} \begin{array}{c} \begin{array}{ccccccc} a & b & c & d & e & f & g \end{array} \\ \begin{pmatrix} 0 & 0 & 0 & 0 & 1/2 & 1/2 & 0 \\ 1/3 & 0 & 1/3 & 0 & 0 & 1/3 & 0 \\ 0 & 0 & 0 & 1/2 & 0 & 1/2 & 0 \\ 0 & 0 & 0 & 0 & 0 & 1 & 0 \\ 1/4 & 0 & 0 & 1/4 & 0 & 1/4 & 1/4 \\ 1/2 & 1/2 & 0 & 0 & 0 & 0 & 0 \\ 0 & 0 & 0 & 0 & 0 & 0 & 0 \end{pmatrix} \end{array}.$$

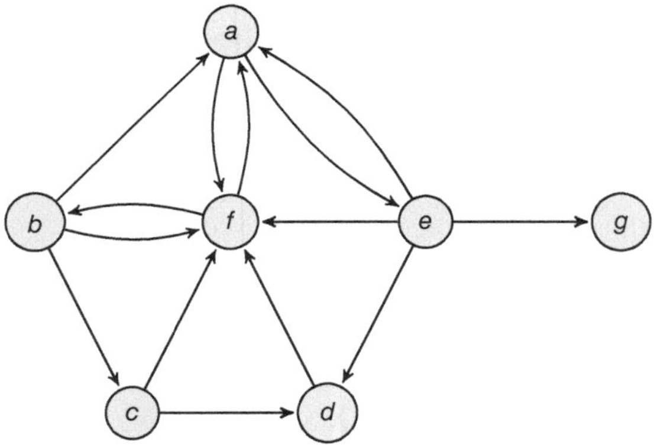

Figure 3.10

Note that N is not a stochastic matrix, as page g has no out-link. Row g consists of all 0s.

To insure that the walk reaches all pages in the network, the algorithm needs to account for (i) pages that have no out-links, called *dangling nodes*, and (ii) groups of pages that might result in the walk getting stuck in a subgraph. In the example network, node g is a dangling node.

Assume that the webgraph consists of k pages. In the PageRank algorithm, the fix for dangling nodes is to assume that when the random surfer reaches such a page they jump to a new page in the network uniformly at random. A new matrix $\boldsymbol{Q}$ is obtained where each row in the network matrix $\boldsymbol{N}$ corresponding to a dangling node is changed to one in which all entries are $1/k$. The matrix $\boldsymbol{Q}$ is a stochastic matrix.

For the problem of potentially getting stuck in small subgraphs of the webgraph, the solution proposed in the original paper by Brin and Page (1998) was to fix a *damping factor* $0 < p < 1$ for modifying the $\boldsymbol{Q}$ matrix. In their model, from a given page the random surfer, with probability $1 - p$, decides to not follow any links on the page and instead navigate to a new random page on the network. On the other hand, with probability p, they follow the links on the page as usual. This defines the PageRank transition matrix

$$\boldsymbol{P} = p\boldsymbol{Q} + (1 - p)\boldsymbol{A},$$

where $\boldsymbol{A}$ is a $k \times k$ matrix all of whose entries are $1/k$. The damping factor used by Google was originally set to $p = 0.85$.

With damping factor, the PageRank matrix $\boldsymbol{P}$ is stochastic, and the resulting random walk is aperiodic and irreducible. The PageRank of a page on the network is that page's stationary probability.

Consider the sample network in Figure 3.10. See the following `R` code for the relevant calculations. The PageRank stationary distribution is

a	b	c	d	e	f	g
0.222	0.153	0.071	0.084	0.122	0.294	0.054

Assume that you are searching this network with the query "stochastic process," and the search terms are found on pages a, c, e, and f. Ordered by their stationary probabilities, the PageRank algorithm would return the ordered pages f, a, e, c.

R: PageRank

```
# pagerank.R
> Q
       a      b      c      d      e      f      g
a 0.0000 0.0000 0.0000 0.0000 0.5000 0.5000 0.0000
b 0.3333 0.0000 0.3333 0.0000 0.0000 0.3333 0.0000
c 0.0000 0.0000 0.0000 0.5000 0.0000 0.5000 0.0000
d 0.0000 0.0000 0.0000 0.0000 0.0000 1.0000 0.0000
e 0.2500 0.0000 0.0000 0.2500 0.0000 0.2500 0.2500
f 0.5000 0.5000 0.0000 0.0000 0.0000 0.0000 0.0000
g 0.1429 0.1429 0.1429 0.1429 0.1429 0.1429 0.1429
> A <- matrix(rep(1/7,49),nrow=7)
```

```
> A
       a      b      c      d      e      f      g
a 0.1429 0.1429 0.1429 0.1429 0.1429 0.1429 0.1429
b 0.1429 0.1429 0.1429 0.1429 0.1429 0.1429 0.1429
c 0.1429 0.1429 0.1429 0.1429 0.1429 0.1429 0.1429
d 0.1429 0.1429 0.1429 0.1429 0.1429 0.1429 0.1429
e 0.1429 0.1429 0.1429 0.1429 0.1429 0.1429 0.1429
f 0.1429 0.1429 0.1429 0.1429 0.1429 0.1429 0.1429
g 0.1429 0.1429 0.1429 0.1429 0.1429 0.1429 0.1429
# Transition matrix with damping factor p=0.85
> P <- 0.85*Q + 0.15*A
> pr <- stationary(P)
> pr  # Stationary probabilities
     a      b      c      d      e      f      g
0.2220 0.1527 0.0713 0.0843 0.1223 0.2935 0.0540
```

■

3.7 TIME REVERSIBILITY

Some Markov chains exhibit a directional bias in their evolution. Take, for instance, simple random walk on the integers, which moves from i to $i+1$ with probability p, and from i to $i-1$ with probability $1-p$. If $p > 1/2$, the walk tends to move in the positive direction, mostly hitting ever larger integers. Similarly, if $p < 1/2$, over time the chain tends to hit ever smaller integers. However, if $p = 1/2$, the chain exhibits no directional bias.

The property of time reversibility can be explained intuitively as follows. If you were to take a movie of the Markov chain moving forward in time and then run the movie backwards, you could not tell the difference between the two.

Time Reversibility

An irreducible Markov chain with transition matrix $\boldsymbol{P}$ and stationary distribution $\boldsymbol{\pi}$ is *reversible*, or *time reversible*, if

$$\pi_i P_{ij} = \pi_j P_{ji}, \text{ for all } i,j. \tag{3.8}$$

Equations (3.8) are called the *detailed balance equations*. They say that for a chain in stationarity,

$$P(X_0 = i, X_1 = j) = P(X_0 = j, X_1 = i), \text{ for all } i,j.$$

That is, the frequency of transitions from i to j is equal to the frequency of transitions from j to i.

More generally (see Exercise 3.39), if a stationary Markov chain is reversible then

$$P(X_0 = i_0, X_1 = i_1, \dots, X_n = i_n) = P(X_0 = i_n, X_1 = i_{n-1}, \dots, X_n = i_0),$$

for all $i_0, i_1, \dots, i_n$.

Example 3.22 A Markov chain has transition matrix

$$P = \begin{array}{c} \\ 1 \\ 2 \\ 3 \end{array} \begin{array}{c} \begin{array}{ccc} 1 & 2 & 3 \end{array} \\ \begin{pmatrix} 0 & 2/5 & 3/5 \\ 1/2 & 1/4 & 1/4 \\ 1/2 & 1/6 & 1/3 \end{pmatrix} \end{array}.$$

Determine if the chain is reversible.

Solution The chain is irreducible and aperiodic. The stationary distribution is $\boldsymbol{\pi} = (1/3, 4/15, 2/5)$. We check the detailed balance equations

$$\pi_1 P_{12} = \left(\frac{1}{3}\right)\left(\frac{2}{5}\right) = \frac{2}{15} = \left(\frac{4}{15}\right)\left(\frac{1}{2}\right) = \pi_2 P_{21},$$

$$\pi_1 P_{13} = \left(\frac{1}{3}\right)\left(\frac{3}{5}\right) = \frac{1}{5} = \left(\frac{2}{5}\right)\left(\frac{1}{2}\right) = \pi_3 P_{31},$$

and

$$\pi_2 P_{23} = \left(\frac{4}{15}\right)\left(\frac{1}{4}\right) = \frac{1}{15} = \left(\frac{2}{5}\right)\left(\frac{1}{6}\right) = \pi_3 P_{32}.$$

Thus, the chain is time reversible.

■

If the stationary distribution of a Markov chain is uniform, it is apparent from Equation (3.8) that the chain is reversible if the transition matrix is symmetric.

Example 3.23 Consider random walk on the n-cycle with transition matrix

$$P = \begin{cases} p, & \text{if } j \equiv i+1 \mod n, \\ 1-p, & \text{if } j \equiv i-1 \mod n, \\ 0, & \text{otherwise.} \end{cases}$$

The stationary distribution is uniform. Hence, the random walk is time reversible for $p = 1/2$. For $p \neq 1/2$, directional bias of the walk will be apparent. For instance, if $p > 1/2$, the frequency of transitions from i to $i+1$ is greater than the frequency of transitions from $i+1$ to i. ■

Example 3.24 Simple random walk on a graph is time reversible. If i and j are neighbors, then

$$\pi_i P_{ij} = \left(\frac{\deg(i)}{2e}\right)\left(\frac{1}{\deg(i)}\right) = \frac{1}{2e} = \left(\frac{\deg(j)}{2e}\right)\left(\frac{1}{\deg(j)}\right) = \pi_j P_{ji}.$$

If i and j are not neighbors, $P_{ij} = P_{ji} = 0$. ■

Reversible Markov Chains and Random Walk on Weighted Graphs

Random walk on a weighted graph is time reversible. In fact, every reversible Markov chain can be considered as a random walk on a weighted graph.

Given a reversible Markov chain with transition matrix $\boldsymbol{P}$ and stationary distribution $\boldsymbol{\pi}$, construct a weighted graph on the state space by assigning edge weights $w(i,j) = \pi_i P_{ij}$. With these choice of weights, random walk on the weighted graph moves from i to j with probability

$$\frac{w(i,j)}{\sum_v w(i,v)} = \frac{\pi_i P_{ij}}{\sum_v \pi_i P_{iv}} = \frac{\pi_i P_{ij}}{\pi_i} = P_{ij}.$$

Conversely, given a weighted graph with edge weight function $w(i,j)$, the transition matrix of the corresponding Markov chain is obtained by letting

$$P_{ij} = \frac{w(i,j)}{\sum_v w(i,v)},$$

where the sum is over all neighbors of i. The stationary distribution is

$$\pi_i = \frac{\sum_y w(i,y)}{\sum_x \sum_y w(x,y)}.$$

One checks that

$$\begin{aligned}\pi_i P_{ij} &= \left(\frac{\sum_y w(i,y)}{\sum_x \sum_y w(x,y)}\right)\frac{w(i,j)}{\sum_y w(i,y)}\\ &= \frac{w(i,j)}{\sum_x \sum_y w(x,y)}\\ &= \left(\frac{\sum_y w(j,y)}{\sum_x \sum_y w(x,y)}\right)\frac{w(i,j)}{\sum_y w(j,y)} = \pi_j P_{ji}.\end{aligned}$$

Thus, the chain is time reversible.

Example 3.25 A reversible Markov chain has transition matrix

$$P = \begin{array}{c} \\ a \\ b \\ c \\ d \end{array} \begin{array}{c} \begin{array}{cccc} a & b & c & d \end{array} \\ \begin{pmatrix} 0 & 4/5 & 1/5 & 0 \\ 4/6 & 1/6 & 1/6 & 0 \\ 1/4 & 1/4 & 0 & 1/2 \\ 0 & 0 & 2/3 & 1/3 \end{pmatrix} \end{array}. \tag{3.9}$$

Find the associated weighted graph.

Solution The stationary distribution is found to be $\boldsymbol{\pi} = (5/18, 6/18, 4/18, 3/18)$. Arrange the quantities $\pi_i P_{ij}$ in a (nonstochastic) matrix

$$R = \begin{array}{c} \\ a \\ b \\ c \\ d \end{array} \begin{array}{c} \begin{array}{cccc} a & b & c & d \end{array} \\ \begin{pmatrix} 0 & 4/18 & 1/18 & 0 \\ 4/18 & 1/18 & 1/18 & 0 \\ 1/18 & 1/18 & 0 & 2/18 \\ 0 & 0 & 2/18 & 1/18 \end{pmatrix} \end{array},$$

where $R_{ij} = w(i,j) = \pi_i P_{ij}$. The matrix is symmetric. Multiplying the entries by 18 so that all weights are integers gives the weighted graph in Figure 3.11. ■

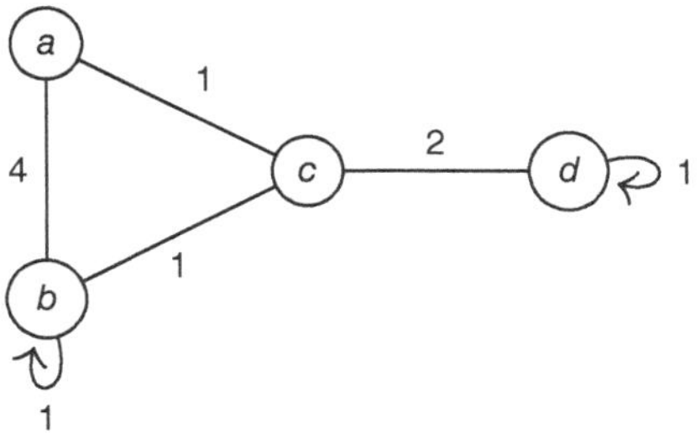

Figure 3.11 Weighted graph.

The next proposition highlights a key benefit of reversibility. It is often used to simplify computations for finding the stationary distribution of a Markov chain.

Proposition 3.9. *Let $\boldsymbol{P}$ be the transition matrix of a Markov chain. If $\boldsymbol{x}$ is a probability distribution which satisfies*

$$x_i P_{ij} = x_j P_{ji}, \text{for all } i,j, \tag{3.10}$$

then $\boldsymbol{x}$ is the stationary distribution, and the Markov chain is reversible.

Proof. Assume that $\boldsymbol{x}$ satisfies the detailed balance equations. We have to show that $\boldsymbol{x} = \boldsymbol{xP}$. For all j,

$$(xP)_j = \sum_i x_i P_{ij} = \sum_i x_j P_{ji} = x_j.$$

■

Example 3.26 (Birth-and-death chain) Birth-and-death chains were introduced in Example 2.9. From i, the process moves to either $i-1$, i, or $i+1$, with respective probabilities q_i, $1 - p_i - q_i$, and p_i. Show that a birth-and-death chain is time reversible and find the stationary distribution.

Solution To use Proposition 3.9, consider a vector $\boldsymbol{x}$ which satisfies the detailed balance equations. Then,

$$x_i P_{i,i+1} = x_{i+1} P_{i+1,i}, \text{ for } \quad i = 0, 1, \ldots$$

giving

$$x_i p_i = x_{i+1} q_{i+1}, \text{ for } \quad i = 0, 1, \ldots$$

Let $x_0 = 1$. Then, $x_1 = p_0/q_1$. Also, $x_2 = x_1 p_1/q_2 = (p_0 p_1)/(q_2 q_1)$. The general pattern is

$$x_k = \frac{p_0 p_1 \cdots p_{k-1}}{q_1 q_2 \cdots q_k} = \prod_{i=1}^{k} \frac{p_{i-1}}{q_i}, \text{ for } \quad k = 0, 1, \ldots$$

Normalizing gives the stationary distribution

$$\pi_k = \frac{x_k}{\sum_j x_j},$$

assuming that the infinite sum $\sum_j x_j$ converges. Thus, a necessary condition for a stationary distribution to exist, in which case the birth-and-death chain is time reversible, is that

$$\sum_j \prod_{i=1}^{j} \frac{p_{i-1}}{q_i} < \infty.$$

Birth-and-death chains encompass a large variety of special models. For instance, *random walk with a partially reflecting boundary* on $\{0, 1, \ldots, N\}$ is achieved as a birth-and-death chain by letting $p_i = p$ and $q_i = q$, for $0 \le i \le N$, with $p + q < 1$. See Figure 3.12.

Minor modification to this derivation gives the stationary probabilities

$$\pi_k = \frac{x_k}{\sum_{i=0}^{N} x_i}, \text{ for } \quad k = 0, 1, \ldots, N,$$

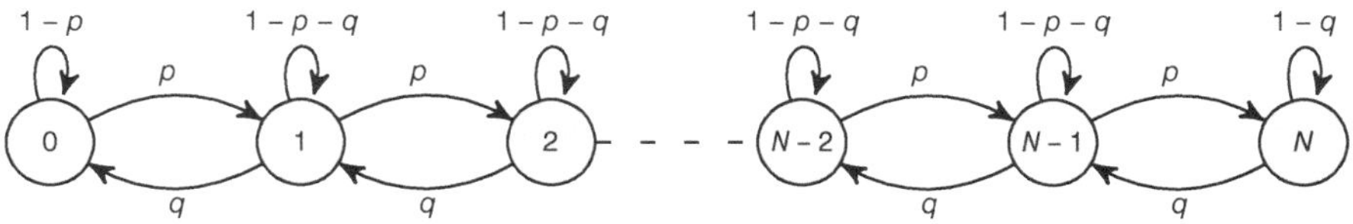

Figure 3.12 Random walk with partially reflecting boundaries.

where

$$x_k = \prod_{i=1}^{k} \left(\frac{p}{q}\right) = \left(\frac{p}{q}\right)^k, \text{ for } k = 0, 1, \ldots, N.$$

For $p \neq q$, this gives

$$\pi_k = \left(1 - \frac{p}{q}\right) \left(\frac{p}{q}\right)^k \Big/ \left(1 - \left(\frac{p}{q}\right)^{N+1}\right), \quad \text{for} \quad k = 0, 1, \ldots, N.$$

For $p = q$, the stationary distribution is uniform. ■

3.8 ABSORBING CHAINS

Many popular board games can be modeled as Markov chains. The children's game *Chutes and Ladders* is based on an ancient Indian game called *Snakes and Ladders*. It is played on a 100-square board, as in Figure 3.13. Players each have a token and take turns rolling a six-sided die and moving their token by the corresponding number of squares. If a player lands on a *ladder*, they immediately move up the ladder to a higher-numbered square. If they move on a *chute*, or *snake*, they drop down to a lower-numbered square. The finishing square 100 must be reached by an exact roll of the die (or by landing on square 80 whose ladder climbs to the finish). The first player to land on square 100 wins.

The game is a Markov chain since the player's position only depends on their previous position and the roll of the die. The chain has 101 states as the game starts with all players off the board (state 0). For the Markov model, once the chain hits state 100 it stays at 100. That is, if $\boldsymbol{P}$ is the transition matrix, then $P_{100,100} = 1$.

Of particular interest is the average number of plays needed to reach the finish. The R script file **snakes.R** contains code for building the 101×101 transition matrix $\boldsymbol{P}$. Here are the results from several simulations of the game.

R: Snakes and Ladders

```
> init <- c(1,rep(0,100)) # Start at square 0
> markov(init,P,150,0:100)
  [1]   0   6  11  12   6  12  15  20  42  43  44  45  46  47  26
 [16]  29  33  37  40  42  44  47  52  55  57  60  61  19  22  24
```

```
[31]  30  44  11  17  42  45  46  50  52  53  53  53  54  60  61
[46]  63  65  68  73  77  78  82  86  92  94  99 100
> markov(init,P,150,0:100)
 [1]   0  14  18  24  30  34  40  44  45  67  70  74 100
> markov(init,P,150,0:100)
 [1]   0   5   6   8  31  33  44  46  50  53  55  58  63  60  61
[16]  66  72  77  78 100
> markov(init,P,150,0:100)
 [1]   0   2   5   8  11  14  20  26  31  37  39  40  42  47  53
[16]  58  63  60  60  65  69  72  76  82  86  24  25  29  31  35
[31]  40  42  26  31  34  35  39  43  45  11  13  19  24  26  32
[46]  35  38  39  42  44  26  84  88  90  96 100
> markov(init,P,150,0:100)
 [1]   0   6  31  44  45  26  27  32  38  44  26  30  31  33  44
[16]  46  50  67  73  76  77  78  79  84  88  90  73  76 100
> markov(init,P,150,0:100)
 [1]   0   3  31  34  40  42  47  26  30  32  38  44  47  53  59
[16]  19  23  25  31  44  26  29  31  33  37  41  42  26  31  32
[31]  38  43  44  50  52  58  19  24  30  31  35  41  44  46  67
[46]  68  91  97  99  99  99  99  99  99  99  99  99 100
> markov(init,P,150,0:100)
 [1]   0   2   7  13   6  10  15  17  23  25  31  37  40  45  11
[16]  17  20  26  31  33  44  46  26  29  35  44  50  67  72  74
[31]  76 100
```

For these simulations, it took, respectively, 56, 12, 19, 55, 28, 57, and 31 steps to reach the winning square, for an average of 36.86 steps to win.

An exact analysis of the average time to win the game is given in this section, after establishing some theory.

In the Snakes and Ladders Markov chain every state, except 100, is transient. State 100 is recurrent. From 100, the chain stays at 100 forever. A state with this property is called *absorbing*.

Absorbing State, Absorbing Chain

State i is an *absorbing state* if $P_{ii} = 1$. A Markov chain is called an *absorbing chain* if it has at least one absorbing state.

Consider an absorbing Markov chain on k states for which t states are transient and $k - t$ states are absorbing. The states can be reordered, as in the canonical decomposition, with the transition matrix written in block matrix form

$$\boldsymbol{P} = \left(\begin{array}{c|c} \boldsymbol{Q} & \boldsymbol{R} \\ \hline \boldsymbol{0} & \boldsymbol{I} \end{array} \right) \tag{3.11}$$

where $\boldsymbol{Q}$ is a $t \times t$ matrix, $\boldsymbol{R}$ is a $t \times (k - t)$ matrix, $\boldsymbol{0}$ is a $(k - t) \times t$ matrix of 0s, and $\boldsymbol{I}$ is the $(k - t) \times (k - t)$ identity matrix.

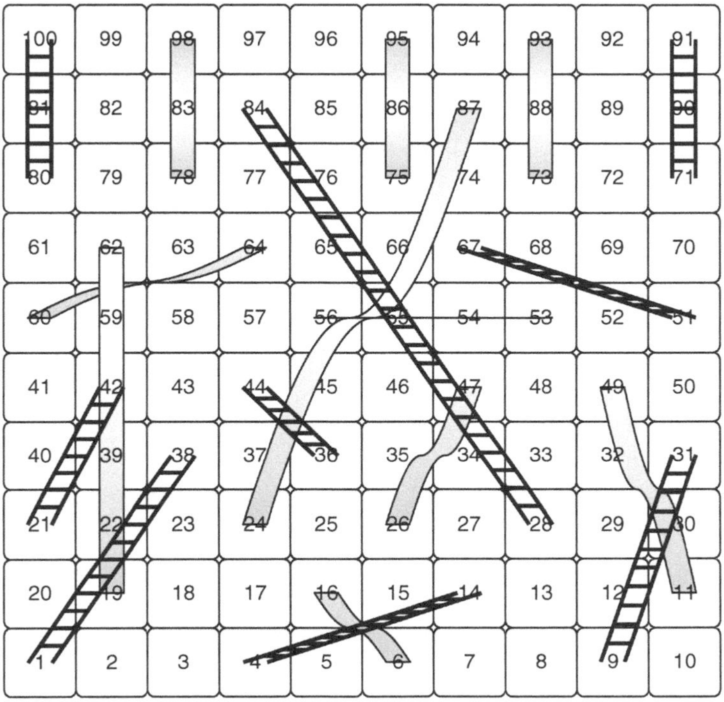

Figure 3.13 Children's Snakes and Ladders game. Board drawn using Ti*k*Z TeX package. *Source*: http://tex.stackexchange.com/questions/85411/chutes-and-ladders/. Reproduced with permission of Serge Ballif.

Computing powers of $\boldsymbol{P}$ is facilitated by the block matrix form. We have

$$\boldsymbol{P}^2 = \left(\begin{array}{c|c} \boldsymbol{Q} & \boldsymbol{R} \\ \hline \boldsymbol{0} & \boldsymbol{I} \end{array}\right)\left(\begin{array}{c|c} \boldsymbol{Q} & \boldsymbol{R} \\ \hline \boldsymbol{0} & \boldsymbol{I} \end{array}\right) = \left(\begin{array}{c|c} \boldsymbol{Q}^2 & (\boldsymbol{I}+\boldsymbol{Q})\boldsymbol{R} \\ \hline \boldsymbol{0} & \boldsymbol{I} \end{array}\right),$$

$$\boldsymbol{P}^3 = \left(\begin{array}{c|c} \boldsymbol{Q}^3 & (\boldsymbol{I}+\boldsymbol{Q}+\boldsymbol{Q}^2)\boldsymbol{R} \\ \hline \boldsymbol{0} & \boldsymbol{I} \end{array}\right),$$

and, in general,

$$\boldsymbol{P}^n = \left(\begin{array}{c|c} \boldsymbol{Q}^n & (\boldsymbol{I}+\boldsymbol{Q}+\cdots+\boldsymbol{Q}^{n-1})\boldsymbol{R} \\ \hline \boldsymbol{0} & \boldsymbol{I} \end{array}\right), \quad \text{for } n \geq 1. \tag{3.12}$$

To find the limiting matrix $\lim_{n\to\infty} \boldsymbol{P}^n$, we make use of the following lemma from linear algebra.

Lemma 3.10. *Let $\boldsymbol{A}$ be a square matrix with the property that $\boldsymbol{A}^n \to \boldsymbol{0}$, as $n \to \infty$. Then,*

$$\sum_{n=0}^{\infty} \boldsymbol{A}^n = (\boldsymbol{I} - \boldsymbol{A})^{-1}.$$

The lemma gives the matrix analog of the sum of a geometric series of real numbers. That is,

$$\sum_{n=0}^{\infty} r^n = (1 - r)^{-1},$$

if $r^n \to 0$, as $n \to \infty$.

Proof. For fixed n,

$$\begin{aligned}(\boldsymbol{I} - \boldsymbol{A})(\boldsymbol{I} + \boldsymbol{A} + \boldsymbol{A}^2 + \cdots + \boldsymbol{A}^n) &= \boldsymbol{I} + \boldsymbol{A} + \boldsymbol{A}^2 + \cdots + \boldsymbol{A}^n \\ &\quad - (\boldsymbol{A} + \boldsymbol{A}^2 + \cdots + \boldsymbol{A}^n + \boldsymbol{A}^{n+1}) \\ &= \boldsymbol{I} - \boldsymbol{A}^{n+1}.\end{aligned}$$

If $\boldsymbol{I} - \boldsymbol{A}$ is invertible, then

$$(\boldsymbol{I} + \boldsymbol{A} + \cdots + \boldsymbol{A}^n) = (\boldsymbol{I} - \boldsymbol{A})^{-1}(\boldsymbol{I} - \boldsymbol{A}^{n+1}).$$

Taking limits, as $n \to \infty$, gives

$$\sum_{n=0}^{\infty} \boldsymbol{A}^n = (\boldsymbol{I} - \boldsymbol{A})^{-1},$$

since $\boldsymbol{A}^{n+1} \to \boldsymbol{0}$, as $n \to \infty$.

To show that $\boldsymbol{I} - \boldsymbol{A}$ is invertible, consider the linear system $(\boldsymbol{I} - \boldsymbol{A})\boldsymbol{x} = \boldsymbol{0}$. Invertibility of $\boldsymbol{I} - \boldsymbol{A}$ is equivalent to the fact that the only solution to this system is $\boldsymbol{x} = \boldsymbol{0}$. We have that $\boldsymbol{0} = (\boldsymbol{I} - \boldsymbol{A})\boldsymbol{x} = \boldsymbol{x} - \boldsymbol{A}\boldsymbol{x}$. That is, $\boldsymbol{x} = \boldsymbol{A}\boldsymbol{x}$. Iterating gives

$$\boldsymbol{x} = \boldsymbol{A}\boldsymbol{x} = \boldsymbol{A}(\boldsymbol{A}\boldsymbol{x}) = \boldsymbol{A}^2\boldsymbol{x} = \cdots = \boldsymbol{A}^n\boldsymbol{x}, \text{ for all } n \geq 1.$$

Taking limits on both sides of the equation gives

$$\boldsymbol{x} = \lim_{n\to\infty} \boldsymbol{A}^n\boldsymbol{x} = \boldsymbol{0}.$$

■

To apply Lemma 3.10 to Equation (3.12), observe that $\boldsymbol{Q}^n \to \boldsymbol{0}$, as $n \to \infty$. The matrix $\boldsymbol{Q}$ is indexed by transient states. If i and j are transient,

$$\lim_{n\to\infty} Q_{ij}^n = \lim_{n\to\infty} P_{ij}^n = 0,$$

as the long-term probability that a Markov chain hits a transient state is 0. Taking limits in Equation (3.12) gives

$$\lim_{n\to\infty} \boldsymbol{P}^n = \lim_{n\to\infty} \left(\begin{array}{c|c} \boldsymbol{Q}^n & (\boldsymbol{I}+\boldsymbol{Q}+\cdots+\boldsymbol{Q}^{n-1})\boldsymbol{R} \\ \hline \boldsymbol{0} & \boldsymbol{I} \end{array}\right)$$
$$= \left(\begin{array}{c|c} \lim_{n\to\infty} \boldsymbol{Q}^n & \lim_{n\to\infty}(\boldsymbol{I}+\boldsymbol{Q}+\cdots+\boldsymbol{Q}^{n-1})\boldsymbol{R} \\ \hline \boldsymbol{0} & \boldsymbol{I} \end{array}\right)$$
$$= \left(\begin{array}{c|c} \boldsymbol{0} & (\boldsymbol{I}-\boldsymbol{Q})^{-1}\boldsymbol{R} \\ \hline \boldsymbol{0} & \boldsymbol{I} \end{array}\right).$$

Consider the interpretation of the limiting submatrix $(\boldsymbol{I}-\boldsymbol{Q})^{-1}\boldsymbol{R}$. The matrix is indexed by transient rows and absorbing columns. The ijth entry is the long-term probability that the chain started in transient state i is absorbed in state j. If the Markov chain has only one absorbing state, this submatrix will be a $(k-1)$-element column vector of 1s.

Example 3.27 (Gambler's ruin) A gambler starts with \$2 and plays a game where the chance of winning each round is 60%. The gambler either wins or loses \$1 on each round. The game stops when the gambler either gains \$5 or goes bust. Find the probability that the gambler is eventually ruined.

Solution The game is an absorbing Markov chain with absorbing states 0 and 5. The transition matrix in canonical form is

$$\boldsymbol{P} = \begin{array}{c} \\ 1 \\ 2 \\ 3 \\ 4 \\ 0 \\ 5 \end{array} \begin{array}{c} \begin{array}{cccccc} 1 & 2 & 3 & 4 & 0 & 5 \end{array} \\ \begin{pmatrix} 0 & 0.6 & 0 & 0 & 0.4 & 0 \\ 0.4 & 0 & 0.6 & 0 & 0 & 0 \\ 0 & 0.4 & 0 & 0.6 & 0 & 0 \\ 0 & 0 & 0.4 & 0 & 0 & 0.6 \\ 0 & 0 & 0 & 0 & 1 & 0 \\ 0 & 0 & 0 & 0 & 0 & 1 \end{pmatrix} \end{array},$$

with

$$\boldsymbol{Q} = \begin{array}{c} \\ 1 \\ 2 \\ 3 \\ 4 \end{array} \begin{array}{c} \begin{array}{cccc} 1 & 2 & 3 & 4 \end{array} \\ \begin{pmatrix} 0 & 0.6 & 0 & 0 \\ 0.4 & 0 & 0.6 & 0 \\ 0 & 0.4 & 0 & 0.6 \\ 0 & 0 & 0.4 & 0 \end{pmatrix} \end{array} \quad \text{and} \quad \boldsymbol{R} = \begin{array}{c} \\ 1 \\ 2 \\ 3 \\ 4 \end{array} \begin{array}{c} \begin{array}{cc} 0 & 5 \end{array} \\ \begin{pmatrix} 0.4 & 0 \\ 0 & 0 \\ 0 & 0 \\ 0 & 0.6 \end{pmatrix} \end{array}.$$

This gives

$$(I - Q)^{-1}R = \begin{pmatrix} 1 & -0.6 & 0 & 0 \\ -0.4 & 1 & -0.6 & 0 \\ 0 & -0.4 & 1 & -0.6 \\ 0 & 0 & -0.4 & 1 \end{pmatrix}^{-1} \begin{pmatrix} 0.4 & 0 \\ 0 & 0 \\ 0 & 0 \\ 0 & 0.6 \end{pmatrix}$$

$$= \begin{matrix} \\ 1 \\ 2 \\ 3 \\ 4 \end{matrix} \begin{matrix} \begin{matrix} 0 & \quad 5 \end{matrix} \\ \begin{pmatrix} 0.616 & 0.384 \\ 0.360 & 0.640 \\ 0.190 & 0.810 \\ 0.076 & 0.924 \end{pmatrix} \end{matrix}.$$

If the gambler starts with \$2, the probability of their eventual ruin is 0.36.

R: Gambler's Ruin

Commands for working with matrices in R are explained in Appendix E. The R command `solve(mat)` computes the matrix inverse of `mat`. The command `diag(k)` generates a $k \times k$ identity matrix.

```
> P
    1   2   3   4   0   5
1 0.0 0.6 0.0 0.0 0.4 0.0
2 0.4 0.0 0.6 0.0 0.0 0.0
3 0.0 0.4 0.0 0.6 0.0 0.0
4 0.0 0.0 0.4 0.0 0.0 0.6
0 0.0 0.0 0.0 0.0 1.0 0.0
5 0.0 0.0 0.0 0.0 0.0 1.0
> Q <- P[1:4,1:4]
> R <- P[1:4,5:6]
> Q
    1   2   3   4
1 0.0 0.6 0.0 0.0
2 0.4 0.0 0.6 0.0
3 0.0 0.4 0.0 0.6
4 0.0 0.0 0.4 0.0
> R
    0   5
1 0.4 0.0
2 0.0 0.0
3 0.0 0.0
4 0.0 0.6
```

```
> solve(diag(4)-Q) %*% R
         0       5
1 0.616114 0.38389
2 0.360190 0.63981
3 0.189573 0.81043
4 0.075829 0.92417
```

■

For absorbing Markov chains, the matrix $(\boldsymbol{I}-\boldsymbol{Q})^{-1}$ is called the *fundamental matrix*. Its importance is highlighted by the next theorem. Recall that if i is transient, then for the chain started in i, the expected number of visits to i is finite.

Expected Number of Visits to Transient States

Theorem 3.11. *Consider an absorbing Markov chain with t transient states. Let* $\boldsymbol{F}$ *be a* $t \times t$ *matrix indexed by transient states, where* F_{ij} *is the expected number of visits to j given that the chain starts in i. Then,*

$$\boldsymbol{F} = (\boldsymbol{I}-\boldsymbol{Q})^{-1}.$$

Two proofs are given. The first uses the method of first-step analysis.

Proof #1. Let T be the set of transient states. Assume that $i,j \in T$. Consider the chain started in i. On the first step, the chain moves to some state k. If k is an absorbing state, then the chain will never visit j, unless $j = i$, in which case it has visited j one time. If k is a transient state, then the expected number of visits to j is F_{kj}, if $j \neq i$, and $1 + F_{ki}$, if $j = i$. This gives

$$\begin{aligned} F_{ij} &= \begin{cases} \sum_{k\in T} P_{ik}(1+F_{ki}) + \sum_{k\notin T} P_{ik}, & \text{if } j = i, \\ \sum_{k\in T} P_{ik}F_{kj}, & \text{if } j \neq i, \end{cases} \\ &= \begin{cases} 1 + \sum\limits_{k\in T} Q_{ik}F_{ki}, & \text{if } j = i, \\ \sum_{k\in T} Q_{ik}F_{kj}, & \text{if } j \neq i, \end{cases} \\ &= \delta_{ij} + \sum_{k\in T} Q_{ik}F_{kj}, \end{aligned} \tag{3.13}$$

where $\delta_{ij} = 1$, if $i = j$, and 0, otherwise. The second equality is because if i and k are transient states, then $P_{ik} = Q_{ik}$.

In matrix terms, Equation (3.13) says that $\boldsymbol{F} = \boldsymbol{I} + \boldsymbol{QF}$, or $(\boldsymbol{I}-\boldsymbol{Q})\boldsymbol{F} = \boldsymbol{I}$. It follows that $\boldsymbol{I}-\boldsymbol{Q}$ is invertible and $(\boldsymbol{I}-\boldsymbol{Q})^{-1} = \boldsymbol{F}$. ■

Proof #2. For the chain started in i, define indicator variables

$$I_n = \begin{cases} 1, & \text{if } X_n = j, \\ 0, & \text{otherwise,} \end{cases}$$

for $n = 0, 1, \ldots$ Then, $\sum_{n=0}^{\infty} I_n$ is the number of visits to j. The expected number of visits is

$$\begin{aligned} F_{ij} &= E\left(\sum_{n=0}^{\infty} I_n \right) = \sum_{n=0}^{\infty} E(I_n) \\ &= \sum_{n=0}^{\infty} P(X_n = j | X_0 = i) \\ &= \sum_{n=0}^{\infty} P^n_{ij} = \sum_{n=0}^{\infty} Q^n_{ij} = \left(\sum_{n=0}^{\infty} Q^n \right)_{ij} = (I - Q)^{-1}_{ij}, \end{aligned}$$

as a consequence of Lemma 3.10. ∎

Expected Time to Absorption

For an absorbing Markov chain started in transient state i, let a_i be the expected absorption time, the expected number of steps to reach some absorbing state. The number of transitions from i to an absorbing state is simply the sum of the number of transitions from i to each of the transient states until eventual absorption. The expected number of steps from i to transient state j is F_{ij}. It follows that

$$a_i = \sum_{k \in T} F_{ik}.$$

In vector form, $\boldsymbol{a} = \boldsymbol{F}\mathbf{1}$, where $\mathbf{1}$ is the column vector of all 1s. That is, the expected absorption times are the row sums of the fundamental matrix.

We summarize results for absorbing Markov chains.

Absorbing Markov Chains

For an absorbing Markov chain with all states either transient or absorbing, let $\boldsymbol{F} = (\boldsymbol{I} - \boldsymbol{Q})^{-1}$.

1. *(Absorption probability)* The probability that from transient state i the chain is absorbed in state j is $(FR)_{ij}$.
2. *(Absorption time)* The expected number of steps from transient state i until the chain is absorbed in some absorbing state is $(F\mathbf{1})_i$.

Example 3.28 (Snakes and Ladders) We find the expected absorption time to square 100 in Snakes and Ladders. See the **snakes.R** script file. The transition matrix is a 101×101 matrix stored in the variable `P`.

R: Snakes and Ladders

```
> Q <- P[1:100,1:100]
> F <- solve(diag(100)-Q)
> a <- F %*% rep(1,100) # Expected absorption times
> round(t(a),2)
    0     1     2     3     4     5     6     7     8     9
39.60 40.25 40.07 39.57 39.84 39.67 39.46 39.23 38.98 39.05
   10    11    12    13    14    15    16    17    18    19
39.11 38.64 38.22 37.84 37.50 37.02 36.08 35.82 35.66 35.55
   20    21    22    23    24    25    26    27    28    29
35.50 35.50 33.82 34.31 34.66 34.94 35.15 35.30 37.52 37.25
   30    31    32    33    34    35    36    37    38    39
36.77 36.60 36.41 36.19 35.93 35.64 35.68 35.55 35.31 34.83
   40    41    42    43    44    45    46    47    48    49
34.37 33.93 34.10 34.75 33.87 31.95 31.61 31.27 30.20 28.47
   50    51    52    53    54    55    56    57    58    59
28.58 29.88 29.61 29.22 28.75 28.24 29.29 28.28 27.94 26.87
   60    61    62    63    64    65    66    67    68    69
25.95 25.16 22.57 22.19 21.16 20.42 20.43 20.44 20.03 19.85
   70    71    72    73    74    75    76    77    78    79
19.81 20.50 20.50 20.52 17.59 18.74 19.56 20.12 20.47 20.64
   80    81    82    83    84    85    86    87    88    89
24.12 25.62 24.48 23.53 22.56 21.62 20.90 18.54 17.63 17.79
   90    91    92    93    94    95    96    97    98    99
16.79 15.94 16.59 14.17 12.14 10.82 10.82 10.82  6.00  6.00
```

The desired expectation is a_0. At the start of the game, it takes, on average, 39.60 moves to reach the ending square. ■

Example 3.29 (Graduation) Recall the graduation Markov chain of Example 2.19. Students at a 4-year college either drop out, repeat a year, or move on to the next year. The chain is an absorbing chain with graduating and dropping out as absorbing states. Relevant matrices are

$$
\boldsymbol{P} = \begin{array}{c} \\ \text{Fr} \\ \text{So} \\ \text{Jr} \\ \text{Se} \\ \text{Drop} \\ \text{Grad} \end{array}
\begin{array}{c}
\begin{array}{cccccc} \text{Fr} & \text{So} & \text{Jr} & \text{Sr} & \text{Drop} & \text{Grad} \end{array} \\
\begin{pmatrix}
0.03 & 0.91 & 0 & 0 & 0.06 & 0 \\
0 & 0.03 & 0.91 & 0 & 0.06 & 0 \\
0 & 0 & 0.03 & 0.93 & 0.04 & 0 \\
0 & 0 & 0 & 0.03 & 0.04 & 0.93 \\
0 & 0 & 0 & 0 & 1 & 0 \\
0 & 0 & 0 & 0 & 0 & 1
\end{pmatrix}
\end{array},
$$

$$
\boldsymbol{Q} = \begin{matrix} & \text{Fr} & \text{So} & \text{Jr} & \text{Sr} \\ \text{Fr} & 0.03 & 0.91 & 0 & 0 \\ \text{So} & 0 & 0.03 & 0.91 & 0 \\ \text{Jr} & 0 & 0 & 0.03 & 0.93 \\ \text{Se} & 0 & 0 & 0 & 0.03 \end{matrix}, \quad \text{and } \boldsymbol{R} = \begin{matrix} & \text{Drop} & \text{Grad} \\ \text{Fr} & 0.06 & 0 \\ \text{So} & 0.06 & 0 \\ \text{Jr} & 0.04 & 0 \\ \text{Se} & 0.04 & 0.93 \end{matrix}.
$$

This gives

$$
(\boldsymbol{I} - \boldsymbol{Q})^{-1}\boldsymbol{R} = \begin{pmatrix} 0.97 & -0.91 & 0 & 0 \\ 0 & 0.97 & -0.91 & 0 \\ 0 & 0 & 0.97 & -0.93 \\ 0 & 0 & 0 & 0.97 \end{pmatrix}^{-1} \begin{pmatrix} 0.06 & 0 \\ 0.06 & 0 \\ 0.04 & 0 \\ 0.04 & 0.93 \end{pmatrix}
$$

$$
= \begin{matrix} & \text{Drop} & \text{Grad} \\ \text{Fr} & 0.191 & 0.809 \\ \text{So} & 0.138 & 0.862 \\ \text{Jr} & 0.081 & 0.919 \\ \text{Se} & 0.041 & 0.959 \end{matrix}.
$$

For a student who starts as a first-year, the probability of eventually graduating is 0.809. ■

R: Graduation

```
> P
       Fr   So   Jr   Se Drop Grad
Fr   0.03 0.91 0.00 0.00 0.06 0.00
So   0.00 0.03 0.91 0.00 0.06 0.00
Jr   0.00 0.00 0.03 0.93 0.04 0.00
Se   0.00 0.00 0.00 0.03 0.04 0.93
Drop 0.00 0.00 0.00 0.00 1.00 0.00
Grad 0.00 0.00 0.00 0.00 0.00 1.00
> Q <- P[1:4,1:4]
> R <- P[1:4,5:6]
> Absorb <- solve(diag(4)-Q)%*% R
> Absorb
       Drop    Grad
Fr 0.190975 0.80902
So 0.137633 0.86237
Jr 0.080774 0.91923
Se 0.041237 0.95876
```

Expected Hitting Times for Irreducible Chains

For an irreducible Markov chain, first hitting times can be analyzed as absorption times for a suitably modified chain. In particular, assume that $\boldsymbol{P}$ is the transition matrix of an irreducible Markov chain. To find the expected time until state i is first hit, consider a new chain in which i is an absorbing state. The transition matrix $\widetilde{\boldsymbol{P}}$ for the new chain is gotten by zeroing out the ith row of the $\boldsymbol{P}$ matrix and setting $\widetilde{P}_{ii} = 1$. The resulting $\boldsymbol{Q}$ matrix is obtained from $\widetilde{\boldsymbol{P}}$ by deleting the ith row and the ith column of $\boldsymbol{P}$. The time that the original $\boldsymbol{P}$-chain first hits i is equal to the time that the modified $\widetilde{\boldsymbol{P}}$-chain is absorbed in i.

Example 3.30 Consider random walk on the weighted graph in Figure 3.14. Starting from each vertex in the graph, find the expected number of steps until the walk first hits f.

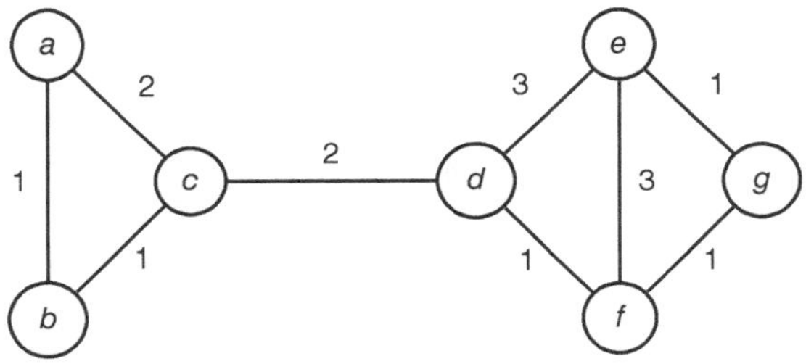

Figure 3.14

Solution The transition matrix is

$$
\boldsymbol{P} = \begin{array}{c} \\ a \\ b \\ c \\ d \\ e \\ f \\ g \end{array}
\begin{array}{c} \begin{matrix} a & b & c & d & e & f & g \end{matrix} \\
\begin{pmatrix}
0 & 1/3 & 2/3 & 0 & 0 & 0 & 0 \\
1/2 & 0 & 1/2 & 0 & 0 & 0 & 0 \\
2/5 & 1/5 & 0 & 2/5 & 0 & 0 & 0 \\
0 & 0 & 1/3 & 0 & 1/2 & 1/6 & 0 \\
0 & 0 & 0 & 3/7 & 0 & 3/7 & 1/7 \\
0 & 0 & 0 & 1/5 & 3/5 & 0 & 1/5 \\
0 & 0 & 0 & 0 & 1/2 & 1/2 & 0
\end{pmatrix} \end{array}.
$$

By making f an absorbing state, the resulting $\boldsymbol{Q}$ matrix is

$$
\boldsymbol{Q} = \begin{array}{c} \\ a \\ b \\ c \\ d \\ e \\ g \end{array}
\begin{array}{c} \begin{matrix} a & b & c & d & e & g \end{matrix} \\
\begin{pmatrix}
0 & 1/3 & 2/3 & 0 & 0 & 0 \\
1/2 & 0 & 1/2 & 0 & 0 & 0 \\
2/5 & 1/5 & 0 & 2/5 & 0 & 0 \\
0 & 0 & 1/3 & 0 & 1/2 & 0 \\
0 & 0 & 0 & 3/7 & 0 & 1/7 \\
0 & 0 & 0 & 0 & 1/2 & 0
\end{pmatrix} \end{array},
$$

which gives

$$F = (I - Q)^{-1} = \begin{array}{c} \\ a \\ b \\ c \\ d \\ e \\ g \end{array} \begin{array}{c} \begin{array}{cccccc} a & b & c & d & e & g \end{array} \\ \begin{pmatrix} 3.847 & 2.165 & 4.412 & 2.294 & 1.235 & 0.176 \\ 3.247 & 2.965 & 4.412 & 2.294 & 1.235 & 0.176 \\ 2.647 & 1.765 & 4.412 & 2.294 & 1.235 & 0.176 \\ 1.147 & 0.765 & 1.912 & 2.29 & 1.235 & 0.176 \\ 0.529 & 0.353 & 0.882 & 1.059 & 1.647 & 0.235 \\ 0.265 & 0.176 & 0.441 & 0.529 & 0.824 & 1.118 \end{pmatrix} \end{array}.$$

Row sums are

$$\begin{array}{c} a \\ b \\ c \\ d \\ e \\ g \end{array} \begin{pmatrix} 14.129 \\ 14.329 \\ 12.529 \\ 7.529 \\ 4.706 \\ 3.353 \end{pmatrix},$$

which gives the expected numbers of steps, starting from each vertex in the graph, to first hit f. ■

Example 3.31 A coin is flipped repeatedly until three heads in a row appear. What is the expected number of flips needed?

Solution One approach is to condition on the first coin flip and use the law of total expectation. Here is a Markov chain solution.

Let X_n be the most recent run of heads gotten after the nth coin flip. States are depicted as {∅, H, HH, HHH}, with HHH an absorbing state and ∅ representing the initial state. Note that if tails is ever flipped then the run of heads starts over again. Thus, ∅ also represents having gotten tails on the last flip. The transition matrix is

$$P = \begin{array}{c} \\ \emptyset \\ \text{H} \\ \text{HH} \\ \text{HHH} \end{array} \begin{array}{c} \begin{array}{cccc} \emptyset & \text{H} & \text{HH} & \text{HHH} \end{array} \\ \begin{pmatrix} 1/2 & 1/2 & 0 & 0 \\ 1/2 & 0 & 1/2 & 0 \\ 1/2 & 0 & 0 & 1/2 \\ 0 & 0 & 0 & 1 \end{pmatrix} \end{array}.$$

This gives

$$I - Q = \begin{pmatrix} 1 & 0 & 0 \\ 0 & 1 & 0 \\ 0 & 0 & 1 \end{pmatrix} - \begin{pmatrix} \frac{1}{2} & \frac{1}{2} & 0 \\ \frac{1}{2} & 0 & \frac{1}{2} \\ \frac{1}{2} & 0 & 0 \end{pmatrix} = \begin{pmatrix} \frac{1}{2} & -\frac{1}{2} & 0 \\ -\frac{1}{2} & 1 & -\frac{1}{2} \\ -\frac{1}{2} & 0 & 1 \end{pmatrix}$$

and

$$F = (I - Q)^{-1} = \begin{pmatrix} 8 & 4 & 2 \\ 6 & 4 & 2 \\ 4 & 2 & 2 \end{pmatrix},$$

with expected absorption times

$$
\boldsymbol{a} = \boldsymbol{F1} = \begin{array}{c} \\ \emptyset \\ \text{H} \\ \text{HH} \end{array} \begin{array}{c} \text{HHH} \\ \begin{pmatrix} 14 \\ 12 \\ 8 \end{pmatrix} \end{array}.
$$

It takes on average 14 flips to get three heads in a row. ■

Patterns in Sequences

The last example illustrates a general method for problems involving the occurrence of patterns in random trials. Assume that the elements of a set S are repeatedly sampled. A *pattern* is a sequence $(p_1, p_2, \ldots, p_n)$, such that each of the p_i is an element of S. In the last example, $S = \{\text{H, T}\}$, and the desired pattern is (H,H,H).

For $k = 1, \ldots, n$, let $s_k = (p_1, \ldots, p_k)$ be the subsequence consisting of the first k elements of the pattern. A Markov chain is constructed with state space $\{\emptyset, s_1, \ldots, s_n\}$, where s_n, the desired pattern, is an absorbing state. The absorption time for the Markov chain is equal to the time until the pattern first appears in repeated sampling from S.

Example 3.32 A biased coin comes up heads, with probability 2/3, and tails, with probability 1/3. The coin is repeatedly flipped. How many flips are needed, on average, until the pattern HTHTH first appears?

Solution An absorbing Markov chain is constructed with transition matrix

$$
\boldsymbol{P} = \begin{array}{l} \\ \emptyset \\ \text{H} \\ \text{HT} \\ \text{HTH} \\ \text{HTHT} \\ \text{HTHTH} \end{array} \begin{array}{c} \begin{array}{cccccc} \emptyset & \text{H} & \text{HT} & \text{HTH} & \text{HTHT} & \text{HTHTH} \end{array} \\ \begin{pmatrix} 1/3 & 2/3 & 0 & 0 & 0 & 0 \\ 0 & 2/3 & 1/3 & 0 & 0 & 0 \\ 1/3 & 0 & 0 & 2/3 & 0 & 0 \\ 0 & 2/3 & 0 & 0 & 1/3 & 0 \\ 1/3 & 0 & 0 & 0 & 0 & 2/3 \\ 0 & 0 & 0 & 0 & 0 & 1 \end{pmatrix} \end{array}.
$$

The fundamental matrix is

$$
\boldsymbol{F} = (\boldsymbol{I} - \boldsymbol{Q})^{-1} = \begin{pmatrix} 2/3 & -2/3 & 0 & 0 & 0 \\ 0 & 1/3 & -1/3 & 0 & 0 \\ -1/3 & 0 & 1 & -2/3 & 0 \\ 0 & -2/3 & 0 & 1 & -1/3 \\ -1/3 & 0 & 0 & 0 & 1 \end{pmatrix}^{-1}
$$

$$= \begin{pmatrix} 45/8 & 81/4 & 27/4 & 9/2 & 3/2 \\ 33/8 & 81/4 & 27/4 & 9/2 & 3/2 \\ 33/8 & 69/4 & 27/4 & 9/2 & 3/2 \\ 27/8 & 63/4 & 21/4 & 9/2 & 3/2 \\ 15/8 & 27/4 & 9/4 & 3/2 & 3/2 \end{pmatrix}.$$

The sum of the first row of the fundamental matrix is

$$\frac{45}{8} + \frac{81}{4} + \frac{27}{4} + \frac{9}{2} + \frac{3}{2} = \frac{309}{8} = 38.625.$$

It takes, on average, 38.625 flips before HTHTH first appears. ■

R : Simulation of Number of Flips Needed For HTHTH

```
# pattern.R
# P(Heads) = 2/3, P(Tails) = 1/3
> trials <- 100000
> simlist <- numeric(trials)
> for (i in 1:trials) {
+ pattern <- c(1,0,1,0,1) # 1:Heads, 0:Tails
+ state <- sample(c(0,1),5,prob=c(1/3,2/3),replace=T)
+ k <- 5
+ while (!prod(state==pattern))
+   { flip <- sample(c(0,1),1,prob=c(1/3,2/3))
+   state <- c(tail(state,4),flip)
+   k <- k + 1 }
+   simlist[i] <- k }
>   mean(simlist)
[1] 38.67718
# exact expectation is 38.625
```

In the last example, successive trials (e.g., coin flips) are independent and identically distributed. However, this is not necessary. In the next example, the trials themselves form a Markov chain.

Example 3.33 Assume that successive occurrences of DNA nucleotides on a chromosome are modeled by a Markov chain with transition matrix

$$\widetilde{\boldsymbol{P}} = \begin{array}{c} \\ a \\ c \\ g \\ t \end{array} \begin{array}{c} \begin{array}{cccc} a & c & g & t \end{array} \\ \begin{pmatrix} 0.3 & 0.2 & 0.3 & 0.2 \\ 0.4 & 0.3 & 0.1 & 0.2 \\ 0.25 & 0.15 & 0.4 & 0.2 \\ 0.2 & 0.2 & 0.3 & 0.3 \end{pmatrix} \end{array}.$$

Consider searching sequentially across the chromosome until the pattern *accgc* first appears. On average, how many steps (e.g., successive nucleotides), are needed?

Solution An absorbing Markov chain is built on $\{\emptyset, a, ac, acc, accg, accgc\}$, with transition matrix

$$P = \begin{array}{c} \emptyset \\ a \\ ac \\ acc \\ accg \\ accgc \end{array} \overset{\begin{array}{cccccc} \emptyset & a & ac & acc & accg & accgc \end{array}}{\begin{pmatrix} 0.7 & 0.3 & 0 & 0 & 0 & 0 \\ 0.5 & 0.3 & 0.2 & 0 & 0 & 0 \\ 0.3 & 0.4 & 0 & 0.3 & 0 & 0 \\ 0.5 & 0.4 & 0 & 0 & 0.1 & 0 \\ 0.6 & 0.25 & 0 & 0 & 0 & 0.15 \\ 0 & 0 & 0 & 0 & 0 & 1 \end{pmatrix}}.$$

To understand the matrix entries, observe, for instance, that from state acc, the chain moves to (i) $accg$, with probability $\widetilde{P}_{cg} = 0.1$, or to (ii) a, with probability $\widetilde{P}_{ca} = 0.4$, or to (iii) $\emptyset$, with the complementary probability 0.5. We have

$$F = (I - Q)^{-1} = \begin{pmatrix} 0.3 & -0.3 & 0 & 0 & 0 \\ -0.5 & 0.7 & -0.2 & 0 & 0 \\ -0.3 & -0.4 & 1 & -0.3 & 0 \\ -0.5 & -0.4 & 0 & 0.9 & 0 \\ -0.6 & -0.25 & 0 & 0 & 1 \end{pmatrix}^{-1}$$

$$= \begin{pmatrix} 2201.85 & 1111.11 & 222.22 & 66.67 & 6.67 \\ 2198.52 & 1111.11 & 222.22 & 66.67 & 6.67 \\ 2190.19 & 1106.11 & 222.22 & 66.67 & 6.67 \\ 2167.41 & 1094.44 & 218.89 & 66.67 & 6.67 \\ 1870.74 & 944.444 & 188.89 & 56.67 & 6.67 \end{pmatrix}.$$

The sum of the first row is

$$2201.85 + 1111.11 + 222.22 + 66.67 + 6.67 = 3608.52,$$

which gives the average number of nucleotides needed to reach the desired pattern. ■

3.9 REGENERATION AND THE STRONG MARKOV PROPERTY*

A Markov chain is sometimes observed from a fixed time $n > 0$ into the future. Assume that $X_0, X_1, \ldots$ is a Markov chain with transition matrix $\boldsymbol{P}$. Then, the process started at $n > 0$, $X_n, X_{n+1}, X_{n+2}, \ldots$ is also a Markov chain with transition matrix $\boldsymbol{P}$. This is a consequence of the Markov property, which says that given the present, past, and future are independent.

The *strong Markov property* asserts that for certain types of random times called *stopping times*, the Markov property holds. If S is a stopping time, then the sequence $X_S, X_{S+1}, X_{S+2}, \ldots$ is a Markov chain. Given that the present time is a stopping time, past and future are independent.

An integer-valued random variable S is a stopping time for a Markov chain if, for each s, the event $\{S = s\}$ can be determined from $X_0, \ldots, X_s$. In other words, if the outcomes $X_0, \ldots, X_s$ are known, then it can be determined whether or not $\{S = s\}$ occurs.

An important example of a stopping time is the *first hitting time* random variable

$$T_i = \min\{n \geq 0 : X_n = i\},$$

which is the first time that a Markov chain hits state i. For instance, consider the weather Markov chain. Let r denote rain. Then, T_r is the first day that it rains. For any day t, If we are given the succession of weather states up to time t, $X_0, \ldots, X_t$, then it can be determined whether or not the first day that it rained was on day t. This shows that T_r is a stopping time. By the strong Markov property, the sequence $X_{T_r}, X_{T_r+1}, \ldots$ is a Markov chain.

A closely related stopping time for a Markov chain started at i, is the *first return time*

$$T_i^+ = \min\{n \geq 1 : X_n = i\}.$$

The strong Markov property says that the chain started at T_i^+ looks the same as the chain started at i. We say that at time T_i^+ the Markov chain *regenerates itself* and probabilistically starts anew.

More generally, given a nonempty subset of states $A \subseteq \mathcal{S}$, the first time the chain hits a state in A

$$T_A = \min\{n \geq 0 : X_n \in A\}$$

is a stopping time.

A random time that is not a stopping time is the *last* visit to state i. Knowing whether or not the last visit to i occurs at time s cannot be determined from just $X_0, \ldots, X_s$. It requires knowledge of the entire Markov sequence $X_0, X_1, \ldots$

Strong Markov Property

Let $X_0, X_1, \ldots$ be a Markov chain with transition matrix $\boldsymbol{P}$. Let S be a stopping time. Then, $X_S, X_{S+1}, \ldots$ is a Markov chain with transition matrix $\boldsymbol{P}$.

Proof. For states $i, j, i_0, i_1, \ldots$, consider

$$\begin{aligned}
P(X_{S+1} &= j, X_S = i, X_u = i_u, 0 \leq u < S) \\
&= \sum_s P(S = s, X_{s+1} = j, X_s = i, X_u = i_u, 0 \leq u < s) \\
&= \sum_s P(S = s | X_{s+1} = j, X_s = i, X_u = i_u, 0 \leq u < s) \\
&\quad \times P(X_{s+1} = j | X_s = i, X_u = i_u, 0 \leq u < s) \\
&\quad \times P(X_s = i, X_u = i_u, 0 \leq u < s)
\end{aligned}$$

$$= \sum_s P(S = s | X_s = i, X_u = i_u, 0 \le u < s)$$
$$\times P_{ij} P(X_s = i, X_u = i_u, 0 \le u < s)$$
$$= P_{ij} P(X_S = i, X_u = i_u, 0 \le u < S).$$

The third equality is by conditional probability. The fourth equality is because (i) S is a stopping time and the event $\{S = s\}$ is determined by $X_0, \ldots, X_s$, and (ii) by the Markov property. It follows that

$$P(X_{S+1} = j | X_S = i, X_u = i_u, 0 \le u < S)$$
$$= \frac{P(X_{S+1} = j, X_S = i, X_u = i_u, 0 \le u < S)}{P(X_S = i, X_u = i_u, 0 \le u < S)}$$
$$= P_{ij}.$$

■

3.10 PROOFS OF LIMIT THEOREMS*

In this section, we prove the main limit theorems from this chapter. Each proof is given after restating the corresponding theorem.

Limit Theorem for Regular Markov Chains

Theorem 3.2. *A Markov chain whose transition matrix* $\boldsymbol{P}$ *is regular has a limiting distribution, which is the unique, positive, stationary distribution of the chain.*

Proof of Theorem 3.2. This is a direct consequence of two forthcoming results: Proposition 3.13, which gives that regular Markov chains are ergodic, and Theorem 3.8, the fundamental limit theorem for ergodic Markov chains. ■

Finite Irreducible Markov Chains

Theorem 3.6. *Assume that* $X_0, X_1, \ldots$ *is a finite irreducible Markov chain. For each state* j, *let* $\mu_j = E(T_j | X_0 = j)$ *be the expected return time to* j. *Then,* μ_j *is finite, and there exists a unique, positive stationary distribution* $\boldsymbol{\pi}$ *such that*

$$\pi_j = \frac{1}{\mu_j}, \text{ for all } j. \tag{3.14}$$

Furthermore, for all i *and* j,

$$\pi_j = \lim_{n \to \infty} \frac{1}{n} \sum_{m=0}^{n-1} P_{ij}^m. \tag{3.15}$$

Proof of Theorem 3.6. Let $X_0, X_1, \ldots$ be an irreducible Markov chain with state space $\{1, \ldots, k\}$. Given states i,j, consider the chain started in i. Since the chain is irreducible and all states are recurrent, the chain will visit j infinitely often.

Let Y_1 be the time the chain first hits j. Since j is recurrent, the chain will return to j infinitely often. For $n \geq 2$, let Y_n be the number of steps from the $(n-1)$st visit to j to the nth visit to j. By the strong Markov property, each time the chain visits j it probabilistically restarts itself independently of past history. Thus, $Y_1, Y_2, \ldots$ is an i.i.d. sequence with common mean $\mu_j = E(T_j|X_0 = j)$.

Assume that $\mu_j < \infty$. By the strong law of large numbers,

$$\lim_{n\to\infty} \frac{Y_1 + \cdots + Y_n}{n} = \mu_j, \text{ with probability } 1. \tag{3.16}$$

Define indicator variables

$$I_m = \begin{cases} 1, & \text{if } X_m = j, \\ 0, & \text{otherwise,} \end{cases}$$

for $m = 0, 1, \ldots$ Then, $\sum_{m=0}^{n-1} I_m$ is the number of visits to j in the first n steps of the chain. The long-term expected proportion of visits to j is

$$\lim_{n\to\infty} E\left(\frac{1}{n}\sum_{m=0}^{n-1} I_m\right) = \lim_{n\to\infty} \frac{1}{n}\sum_{m=0}^{n-1} E(I_m) = \lim_{n\to\infty} \frac{1}{n}\sum_{m=0}^{n-1} P_{ij}^m.$$

Since there are n visits to j by time $Y_1 + \cdots + Y_n$, for large n,

$$\frac{1}{n}\sum_{m=0}^{n-1} I_m \approx \frac{n}{Y_1 + \cdots + Y_n},$$

giving that

$$\lim_{n\to\infty} \frac{1}{n}\sum_{m=0}^{n-1} P_{ij}^m = \lim_{n\to\infty} \frac{n}{Y_1 + \cdots + Y_n} = \frac{1}{\mu_j}, \text{ with probability } 1. \tag{3.17}$$

Let $\pi_j = 1/\mu_j$, for all j. Then, $\boldsymbol{\pi} = (\pi_1, \ldots, \pi_k)$ is the desired stationary distribution by the following four properties.

1. Since the μ_j are positive and finite, $\boldsymbol{\pi}$ is positive.
2. Summing the entries of $\boldsymbol{\pi}$ gives

$$\sum_{j=1}^{k} \pi_j = \sum_{j=1}^{k} \lim_{n\to\infty} \frac{1}{n}\sum_{m=0}^{n-1} P_{ij}^m = \lim_{n\to\infty} \frac{1}{n}\sum_{m=0}^{n-1}\sum_{j=1}^{k} P_{ij}^m = \lim_{n\to\infty} \frac{1}{n}\sum_{m=0}^{n-1} 1 = 1.$$

Thus, $\boldsymbol{\pi}$ is a probability distribution.

3. To show $\boldsymbol{\pi}$ is a stationary distribution, we need to show that

$$\sum_{i=1}^{k} \frac{1}{\mu_i} P_{ij} = \frac{1}{\mu_j}.$$

We have that

$$\begin{aligned}
\sum_{i=1}^{k} \frac{1}{\mu_i} P_{ij} &= \sum_{i=1}^{k} \left(\lim_{n\to\infty} \frac{1}{n} \sum_{m=0}^{n-1} P_{ii}^m \right) P_{ij} \\
&= \lim_{n\to\infty} \frac{1}{n} \sum_{m=0}^{n-1} \sum_{i=1}^{k} P_{ii}^m P_{ij} \\
&= \lim_{n\to\infty} \frac{1}{n} \sum_{m=0}^{n-1} P_{ij}^{m+1} \\
&= \lim_{n\to\infty} \left(\frac{n+1}{n} \right) \left(\frac{1}{n+1} \right) \left(\sum_{m=0}^{n} P_{ij}^m - P_{ij}^0 \right) \\
&= \left(\lim_{n\to\infty} \frac{n+1}{n} \right) \left(\lim_{n\to\infty} \frac{1}{n+1} \sum_{m=0}^{n} P_{ij}^m \right) - \lim_{n\to\infty} \frac{P_{ij}^0}{n+1} \\
&= \frac{1}{\mu_j}.
\end{aligned}$$

4. For uniqueness, assume that $\boldsymbol{\pi} = \boldsymbol{\pi P}$ is a stationary distribution. Then, $\boldsymbol{\pi} = \boldsymbol{\pi P}^n$, for all n, and thus $\boldsymbol{\pi} = \lim_{n\to\infty} \boldsymbol{\pi P}^n$. Pointwise,

$$\begin{aligned}
\pi_j = \lim_{n\to\infty} \sum_{i=1}^{k} \pi_i P_{ij}^n &= \lim_{n\to\infty} \frac{1}{n} \sum_{m=0}^{n-1} \sum_{i=1}^{k} \pi_i P_{ij}^m \\
&= \sum_{i=1}^{k} \pi_i \lim_{n\to\infty} \frac{1}{n} \sum_{m=0}^{n-1} P_{ij}^m = \sum_{i=1}^{k} \pi_i \frac{1}{\mu_j} = \frac{1}{\mu_j}.
\end{aligned}$$

For the second equality, we use the fact that if a sequence converges to a limit, then the sequence of partial averages also converges to that limit. We have shown that if $\boldsymbol{\pi}$ is a stationary distribution, then necessarily $\pi_j = 1/\mu_j$.

To finish the proof of Theorem 3.6, it remains to show that for a finite Markov chain the expected return time μ_j is finite, for all j. A recurrent state j with finite expected return time is called positive recurrent. If the expected return time is infinite, the state is called null recurrent.

To show that the states of a finite irreducible Markov chain are all positive recurrent, we show that a finite irreducible Markov chain contains at least one positive recurrent state. The result will follow as a consequence of the following lemma.

Positive and Null Recurrence are Class Properties

Lemma 3.12. *All the states in a recurrent communication class are either positive recurrent or null recurrent.*

Proof of Lemma. Assume that i is a positive recurrent state. Let j be another state in the same communication class as i. Since both states communicate, there exist positive integers r and s such that $P^r_{ji} > 0$ and $P^s_{ij} > 0$. Thus,

$$\begin{aligned}
\frac{1}{\mu_j} &= \lim_{n\to\infty} \frac{1}{n} \sum_{m=0}^{n-1} P^m_{jj} \\
&\geq \lim_{n\to\infty} \frac{1}{n} \sum_{m=r+s}^{n-1} P^r_{ji} P^{m-r-s}_{ii} P^s_{ij} \\
&= \lim_{n\to\infty} \left(\frac{n-r-s}{n}\right) P^r_{ji} \left(\frac{1}{n-r-s} \sum_{m=r+s}^{n-1} P^{m-r-s}_{ii}\right) P^s_{ij} \\
&= P^r_{ji} \left(\frac{1}{\mu_i}\right) P^s_{ij} > 0.
\end{aligned}$$

Hence, $\mu_j < \infty$ and j is positive recurrent. Having shown that positive recurrence is a class property, it follows that null recurrence is a class property. For if the communication class of a null recurrent state contains a positive recurrent state, then all states in the class are positive recurrent, leading to a contradiction. ■

A finite irreducible Markov chain is recurrent. We show that at least one state must be positive recurrent. If not, then all states are null recurrent, and all expected return times are infinite. See Equation (3.16). Since the Y_i are non-negative, this equation still holds with $\mu_j = +\infty$. And by Equation (3.17),

$$\lim_{n\to\infty} \frac{1}{n} \sum_{m=1}^{n} P^m_{ij} = 0, \quad \text{for all } i,\ j.$$

Sum over j to obtain

$$\begin{aligned}
0 &= \sum_{j=1}^{k} \lim_{n\to\infty} \frac{1}{n} \sum_{m=1}^{n} P^m_{ij} = \lim_{n\to\infty} \frac{1}{n} \sum_{j=1}^{k} \sum_{m=1}^{n} P^m_{ij} \\
&= \lim_{n\to\infty} \frac{1}{n} \sum_{m=1}^{n} \sum_{j=1}^{k} P^m_{ij} = \lim_{n\to\infty} \frac{1}{n} \sum_{m=1}^{n} 1 = 1,
\end{aligned}$$

a contradiction. Thus, a finite irreducible Markov chain contains at least one positive recurrent state. By Lemma 3.12, all states are positive recurrent. And Theorem 3.6 is proved. ■

Remark: The theorem holds for infinite irreducible chains that are positive recurrent. For infinite irreducible chains that are null recurrent, no stationary distribution exists.

Fundamental Limit Theorem for Ergodic Markov Chains

Theorem 3.8. *Let $X_0, X_1, \ldots$ be an ergodic Markov chain. There exists a unique, positive, stationary distribution $\boldsymbol{\pi}$, which is the limiting distribution of the chain. That is,*

$$\pi_j = \lim_{n\to\infty} P^n_{ij}, \text{ for all } i,j.$$

Proof of Theorem 3.8. Two proofs will be given. One is probabilistic, based on an elegant technique known as *coupling*. The other relies on linear algebra and the eigenstructure of the transition matrix.

Both proofs are a consequence of the following proposition, which says that finite ergodic Markov chains are precisely those chains that have regular transition matrices. Recall that a square matrix is regular if some power of the matrix has all positive entries.

Ergodic Chains and Regular Matrices

Proposition 3.13. *Assume that $\boldsymbol{P}$ is the transition matrix of a finite Markov chain. The Markov chain is ergodic if and only if $\boldsymbol{P}$ is regular.*

The proof of this proposition relies on the following lemma.

Lemma 3.14. *If i is an aperiodic state, there exists a positive integer N such that $P^n_{ii} > 0$ for all $n \geq N$.*

Proof of Lemma. The following is based on Hoel et al. (1986) and uses results from number theory.

Assume that i is an aperiodic state. Let $T = \{n > 0 : P^n_{ii} > 0\}$. By definition, $\gcd(T) = 1$. The set T is closed under addition, since if $m, n \in T$, then

$$P^{m+n}_{ii} = \sum_k P^m_{ik} P^n_{ki} \geq P^m_{ii} P^m_{ii} > 0.$$

That is, $m + n \in T$.

We claim that T contains two consecutive integers. If not, then there exists $k, m \in T$ such that $k \geq 2$, $k + m \in T$, and any two integers in T differ by at least k. Furthermore, since $\gcd(T) = 1$, there is an $n \in T$ such that k is not a divisor of n. Write $n = qk + r$, where $q \geq 0$ and $0 < r < k$. Since T is closed under addition, $(q + 1)(m + k) \in T$ and $n + (q + 1)m \in T$. Their difference is

$$(q + 1)(m + k) - n - (q + 1)m = k + qk - n = k - r > 0.$$

Thus, we have found two elements of T whose difference is positive and smaller than k, giving a contradiction.

Hence, T contains consecutive integers, say m and $m+1$. Let $N = m^2$. We show that $n \in T$, for all $n \geq N$, which establishes the lemma. For $n \geq N$, write $n - N = qm + r$, for $q \geq 0$ and $0 \leq r < m$. Then,

$$n = m^2 + qm + r = (m - r + q)m + r(m+1) \in T. \qquad \blacksquare$$

Proof of Proposition 3.13. Assume that $\boldsymbol{P}$ is the transition matrix of a finite ergodic chain. Since the chain is irreducible, for states i and j there exists $m \geq 0$, such that $P^m_{ij} > 0$. The number $m = m(i,j)$ depends on i and j. Let $M^* = \max_{i,j} m(i,j)$. We can take the maximum since the chain is finite.

Since the chain is aperiodic, by Lemma 3.14, there exists $N > 0$ such that $P^n_{ii} > 0$, for all $n \geq N$. The number $N = N(i)$ depends on i. Let $N^* = \max_i N(i)$. Then, N^* does not depend on i, and for all $n \geq N^*$, $P^n_{ii} > 0$, for all i.

Let $X = M^* + N^*$. We claim that $\boldsymbol{P}^X$ is positive. For states i and j,

$$P^X_{ij} = P^{(X-m(i,j))+m(i,j)}_{ij} = \sum_{t=1}^{k} P^{X-m(i,j)}_{it} P^{m(i,j)}_{tj} \geq P^{X-m(i,j)}_{ii} P^{m(i,j)}_{ij} > 0.$$

The last inequality is because (i) $P^{m(i,j)}_{ij} > 0$, and (ii) $P^{X-m(i,j)}_{ii} > 0$, since

$$X - m(i,j) \geq X - M^* = N^*.$$

Thus, $\boldsymbol{P}$ is regular.

Conversely, assume that $\boldsymbol{P}$ is regular. Then, $\boldsymbol{P}^n > 0$ for some positive integer N. Thus, all states communicate and the chain is irreducible. It suffices to show that the chain is aperiodic. Stochastic matrices have the property that if $\boldsymbol{P}^n$ is positive then $\boldsymbol{P}^{N+m}$ is positive for all $m \geq 0$, a property we leave for the reader to prove. (See Exercise 3.33.) For any state i, the set of possible return times to i includes $N, N+1, \ldots$, and gcd $\{N, N+1, \ldots\} = 1$. That is, i is aperiodic. $\blacksquare$

Coupling Proof of Fundamental Limit Theorem

The method of proof is based on *coupling*, a probabilistic technique first introduced by the German mathematician Wolfgang Doeblin in the 1930s.

Here is a bird's-eye view. Let $X_0, X_1, \ldots$ be an ergodic Markov chain on S with transition matrix $\boldsymbol{P}$. Since the chain is irreducible, it has a unique stationary distribution $\boldsymbol{\pi}$. We need to show that for all i and j,

$$\lim_{n\to\infty} P(X_n = j | X_0 = i) = \pi_j.$$

Consider a second chain $Y_0, Y_1, \ldots$ with the same transition matrix $\boldsymbol{P}$, but with initial distribution $\boldsymbol{\pi}$. That is, the Y chain is a stationary chain. For all $n \geq 0$, the distribution of Y_n is $\boldsymbol{\pi}$.

The X and Y chains are run independently of each other. Eventually, at some finite time T, both chains will hit the same state. The chains are then *coupled* so that $X_n = Y_n$, for $n \geq T$. Since the Y chain is in stationarity, from time T onwards the X chain is also in stationarity, from which follows the result.

To elaborate, let $T = \min\{n \geq 0 : X_n = Y_n\}$ be the first time that the X and Y chains hit the same state. Define a new process by letting

$$Z_n = \begin{cases} X_n, & \text{if } n < T, \\ Y_n, & \text{if } n \geq T, \end{cases}$$

for $n \geq 0$. Then, $Z_0, Z_1, \ldots$ is a Markov chain with the same transition matrix and initial distribution as the X chain.

Consider the bivariate process $(Z_0, Y_0), (Z_1, Y_1), \ldots$ The bivariate process is a Markov chain on the state space $\mathcal{S} \times \mathcal{S}$ with transition matrix $\widetilde{\boldsymbol{P}}$ defined by

$$\widetilde{P}_{(i,k),(j,l)} = P_{ij}P_{kl}.$$

The bivariate process represents a *coupling* of the original X chain with the stationary Y chain. The chains are coupled in such a way so that once both chains hit the same state then from that time onward the chains march forward in lockstep. See Example 3.34 for an illustration of the construction.

We show that the bivariate Markov chain is ergodic. Since the X chain is ergodic, $\boldsymbol{P}$ is regular, and there exists some $N > 0$ such that $\boldsymbol{P}^N > 0$. For this choice of N, and for all i, j, k, l,

$$\widetilde{P}^N_{(i,k),(j,l)} = P^N_{ij}P^N_{kl} > 0.$$

Thus, $\widetilde{\boldsymbol{P}}$ is regular, and by Proposition 3.13 the bivariate chain is ergodic. From any state, the bivariate chain reaches any other state in finite time, with probability 1. In particular, it eventually hits a state of the form (j, j). The event that $(Z_n, Y_n) = (j, j)$ for some j implies that the two chains have coupled by time n. It follows that T, the first time the two chains meet, is finite with probability 1. Hence,

$$\lim_{n\to\infty} P(T > n) = \lim_{n\to\infty} 1 - P(T \leq n) = 1 - P(T < \infty) = 0.$$

Consider

$$P(Y_n = j) = P(Y_n = j, T \leq n) + P(Y_n = j, T > n).$$

Taking limits, as $n \to \infty$, the left-hand side converges to π_j, and the rightmost term converges to 0. Thus, $P(Y_n = j, T \leq n) \to \pi_j$, as $n \to \infty$.

For any initial distribution,

$$\begin{aligned}
P(X_n = j) &= P(Z_n = j) \\
&= P(Z_n = j, T \leq n) + P(Z_n = j, T > n) \\
&= P(Z_n = j | T \leq n)P(T \leq n) + P(Z_n = j, T > n) \\
&= P(Y_n = j | T \leq n)P(T \leq n) + P(Z_n = j, T > n) \\
&= P(Y_n = j, T \leq n) + P(Z_n = j, T > n).
\end{aligned}$$

Taking limits gives

$$\lim_{n\to\infty} P(X_n = j) = \lim_{n\to\infty} P(Y_n = j, T \le n) + \lim_{n\to\infty} P(Z_n = j, T > n) = \pi_j,$$

which completes the proof. ∎

Example 3.34 The coupling construction is illustrated on a two-state chain with state space $\mathcal{S} = \{a, b\}$ and transition matrix

$$\boldsymbol{P} = \begin{array}{c} \\ a \\ b \end{array} \begin{array}{c} \begin{array}{cc} a & b \end{array} \\ \begin{pmatrix} 9/10 & 1/10 \\ 1/5 & 4/5 \end{pmatrix} \end{array}.$$

The chain is ergodic with stationary distribution $\boldsymbol{\pi} = (2/3, 1/3)$.

For the following simulation, the X chain was started at a, the Y chain in $\boldsymbol{\pi}$. Both chains were run independently for 12 steps. The chains coupled at time $T = 8$.

Chain	0	1	2	3	4	5	6	7	8	9	10	11	12
X	a	a	a	a	a	b	b	b	a	a	b	a	a
Y	b	b	b	b	b	a	a	a	a	a	b	b	b
Z	a	a	a	a	a	b	b	b	a	a	b	b	b
(Z, Y)	ab	ab	ab	ab	ab	ba	ba	ba	aa	aa	bb	bb	bb

The transition matrix $\widetilde{\boldsymbol{P}}$ of the bivariate process is

$$\widetilde{\boldsymbol{P}} = \begin{array}{c} \\ aa \\ ab \\ ba \\ bb \end{array} \begin{array}{c} \begin{array}{cccc} aa & ab & ba & bb \end{array} \\ \begin{pmatrix} (9/10)^2 & (9/10)(1/10) & (1/10)(9/10) & (1/10)^2 \\ (9/10)(1/5) & (9/10)(4/5) & (1/10)(1/5) & (1/10)(4/5) \\ (1/5)(9/10) & (1/5)(1/10) & (4/5)(9/10) & (4/5)(1/10) \\ (1/5)^2 & (1/5)(4/5) & (4/5)(1/5) & (4/5)^2 \end{pmatrix} \end{array}$$

$$= \begin{array}{c} \\ aa \\ ab \\ ba \\ bb \end{array} \begin{array}{c} \begin{array}{cccc} aa & ab & ba & bb \end{array} \\ \begin{pmatrix} 81/100 & 9/100 & 9/100 & 1/100 \\ 9/50 & 36/50 & 1/50 & 4/50 \\ 9/50 & 1/50 & 36/50 & 4/50 \\ 1/25 & 4/25 & 4/25 & 16/25 \end{pmatrix} \end{array}.$$

∎

Linear Algebra Proof of Fundamental Limit Theorem

Asymptotic properties of Markov chains are related to the eigenstructure of the transition matrix.

Eigenvalues of a Stochastic Matrix

Lemma 3.15. *A stochastic matrix* $\boldsymbol{P}$ *has an eigenvalue* $\lambda^* = 1$. *All other eigenvalues* λ *of* $\boldsymbol{P}$ *are such that* $|\lambda| \leq 1$.

If $\boldsymbol{P}$ *is a regular matrix, then the inequality is strict. That is,* $|\lambda| < 1$ *for all* $\lambda \neq \lambda^*$.

Proof. Let $\boldsymbol{P}$ be a $k \times k$ stochastic matrix. Since the rows of $\boldsymbol{P}$ sum to 1, we have that $\boldsymbol{P1} = \boldsymbol{1}$, where $\boldsymbol{1}$ is a column vector of all 1s. Thus, $\lambda^* = 1$ is an eigenvalue of $\boldsymbol{P}$. Let λ be any other eigenvalue of $\boldsymbol{P}$ with corresponding eigenvector z. Let $|z_m| = \max_{1\leq i\leq k}|z_i|$. That is, z_m is the component of z of maximum absolute value. Then,

$$|\lambda||z_m| = |\lambda z_m| = |(Pz)_m| = \left|\sum_{i=1}^{k} P_{mi}z_i\right| \leq \sum_{i=1}^{k} P_{mi}|z_i| \leq |z_m|\sum_{i=1}^{k} P_{mi} = |z_m|.$$

Thus, $|\lambda| \leq 1$.

Assume that $\boldsymbol{P}$ is regular. Then, $\boldsymbol{P}^N > 0$, for some $N > 0$. Since $\boldsymbol{P}^N$ is a stochastic matrix, the first part of the lemma holds for $\boldsymbol{P}^N$. If λ is an eigenvalue of $\boldsymbol{P}$, then λ^N is an eigenvalue of $\boldsymbol{P}^N$. Let $\boldsymbol{x}$ be the corresponding eigenvector, with $|x_m| = \max_{1\leq i\leq k}|x_i|$. Then,

$$|\lambda|^N|x_m| = |(P^N x)_m| = \left|\sum_{i=1}^{k} P^N_{mi}x_i\right| \leq \sum_{i=1}^{k} P^N_{mi}|x_i| \leq |x_m|\sum_{i=1}^{k} P^N_{mi} = |x_m|.$$

Since the entries of $\boldsymbol{P}^N$ are all positive, the last inequality is an equality only if $|x_1| = |x_2| = \cdots = |x_k|$. And the first inequality is an equality only if $x_1 = \cdots = x_k$. But the constant vector whose components are all the same is an eigenvector associated with the eigenvalue 1. Hence, if $\lambda \neq 1$, one of the inequalities is strict. Thus, $|\lambda|^N < 1$, and the result follows. ■

The fundamental limit theorem for ergodic Markov chains is a consequence of the Perron–Frobenius theorem for positive matrices.

Perron–Frobenius Theorem

Theorem 3.16. *Let* $\boldsymbol{M}$ *be a* $k \times k$ *positive matrix. Then, the following statements hold.*

1. *There is a positive real number* λ^* *which is an eigenvalue of* $\boldsymbol{M}$. *For all other eigenvalues* λ *of* $\boldsymbol{M}$, $|\lambda| < \lambda^*$. *The eigenvalue* λ^* *is called the Perron–Frobenius eigenvalue.*
2. *The eigenspace of eigenvectors associated with* λ^* *is one-dimensional.*

3. There exists a positive right eigenvector $\boldsymbol{v}$ *associated with* λ^**, and a positive left eigenvector* $\boldsymbol{w}$ *associated with* λ^**. Furthermore,*

$$\lim_{n\to\infty} \frac{1}{(\lambda^*)^n} \boldsymbol{M}^n = \boldsymbol{v}\boldsymbol{w}^T,$$

where the eigenvectors are normalized so that $\boldsymbol{w}^T\boldsymbol{v} = 1$.

The proof of the Perron–Frobenius theorem can be found in many advanced linear algebra textbooks, including Horn and Johnson (1990).

For an ergodic Markov chain, the transition matrix $\boldsymbol{P}$ is regular and $\boldsymbol{P}^N$ is a positive matrix for some integer N. The Perron–Frobenius theorem applies. The Perron–Frobenius eigenvalue of $\boldsymbol{P}^n$ is $\lambda^* = 1$, with associated right eigenvector $\boldsymbol{v} = \mathbf{1}$, and associated left eigenvector $\boldsymbol{w}$.

If $\lambda^* = 1$ is an eigenvalue of $\boldsymbol{P}^N$, then $(\lambda^*)^{1/N} = 1$ is an eigenvalue of $\boldsymbol{P}$, with associated right and left eigenvectors $\boldsymbol{v}$ and $\boldsymbol{w}$, respectively. Normalizing $\boldsymbol{w}$ so that its components sum to 1 gives the unique, positive stationary distribution $\boldsymbol{\pi}$, which is the limiting distribution of the chain. The limiting matrix $\boldsymbol{v}\boldsymbol{w}^T$ is a stochastic matrix all of whose rows are equal to $\boldsymbol{w}^T$. ∎

EXERCISES

3.1 Consider a Markov chain with transition matrix

$$\boldsymbol{P} = \begin{pmatrix} 1/2 & 1/4 & 0 & 1/4 \\ 0 & 1/2 & 1/2 & 0 \\ 1/4 & 1/4 & 1/2 & 0 \\ 0 & 1/4 & 1/2 & 1/4 \end{pmatrix}.$$

Find the stationary distribution. Do not use technology.

3.2 A stochastic matrix is called *doubly stochastic* if its rows and columns sum to 1. Show that a Markov chain whose transition matrix is doubly stochastic has a stationary distribution, which is uniform on the state space.

3.3 Determine which of the following matrices are regular.

$$\boldsymbol{P} = \begin{pmatrix} 0.4 & 0.6 & 0 & 0 \\ 0 & 0 & 1 & 0 \\ 0 & 0 & 0 & 1 \\ 1 & 0 & 0 & 0 \end{pmatrix}, \quad \boldsymbol{Q} = \begin{pmatrix} 0 & 1 \\ p & 1-p \end{pmatrix}, \quad \boldsymbol{R} = \begin{pmatrix} 0 & 1 & 0 \\ 0.25 & 0.5 & 0.25 \\ 1 & 0 & 0 \end{pmatrix}.$$

3.4 Consider a Markov chain with transition matrix

$$\boldsymbol{P} = \begin{pmatrix} 1-a & a & 0 \\ 0 & 1-b & b \\ c & 0 & 1-c \end{pmatrix},$$

where $0 < a, b, c < 1$. Find the stationary distribution.

3.5 A Markov chain has transition matrix

$$P = \begin{pmatrix} 0 & 1/4 & 0 & 0 & 3/4 \\ 3/4 & 0 & 0 & 0 & 1/4 \\ 0 & 0 & 1 & 0 & 0 \\ 0 & 0 & 0 & 1 & 0 \\ 1/4 & 3/4 & 0 & 0 & 0 \end{pmatrix}.$$

(a) Describe the set of stationary distributions for the chain.

(b) Use technology to find $\lim_{n\to\infty} P^n$. Explain the long-term behavior of the chain.

(c) Explain why the chain does not have a limiting distribution, and why this does not contradict the existence of a limiting matrix as shown in (b).

3.6 Consider a Markov chain with transition matrix

$$P = \begin{matrix} & \begin{matrix} 1 & 2 & 3 & 4 & 5 & \cdots \end{matrix} \\ \begin{matrix} 1 \\ 2 \\ 3 \\ 4 \\ 5 \\ \vdots \end{matrix} & \begin{pmatrix} 1/2 & 1/2 & 0 & 0 & 0 & \cdots \\ 2/3 & 0 & 1/3 & 0 & 0 & \cdots \\ 3/4 & 0 & 0 & 1/4 & 0 & \cdots \\ 4/5 & 0 & 0 & 0 & 1/5 & \cdots \\ 5/6 & 0 & 0 & 0 & 0 & \cdots \\ \vdots & \vdots & \vdots & \vdots & \vdots & \ddots \end{pmatrix} \end{matrix},$$

defined by

$$P_{ij} = \begin{cases} i/(i+1), & \text{if } j = 1, \\ 1/(i+1), & \text{if } j = i+1, \\ 0, & \text{otherwise.} \end{cases}$$

(a) Does the chain have a stationary distribution? If yes, exhibit the distribution. If no, explain why.

(b) Classify the states of the chain.

(c) Repeat part (a) with the row entries of $\boldsymbol{P}$ switched. That is, let

$$P_{ij} = \begin{cases} 1/(i+1), & \text{if } j = 1, \\ i/(i+1), & \text{if } j = i+1, \\ 0, & \text{otherwise.} \end{cases}$$

3.7 A Markov chain has n states. If the chain is at state k, a coin is flipped, whose heads probability is p. If the coin lands heads, the chain stays at k. If the coin lands tails, the chain moves to a different state uniformly at random. Exhibit the transition matrix and find the stationary distribution.

3.8 Let

$$P_1 = \begin{pmatrix} 1/4 & 3/4 \\ 1/2 & 1/2 \end{pmatrix} \quad \text{and} \quad P_2 = \begin{pmatrix} 1/5 & 4/5 \\ 4/5 & 1/5 \end{pmatrix}.$$

Consider a Markov chain on four states whose transition matrix is given by the block matrix

$$P = \begin{pmatrix} P_1 & \mathbf{0} \\ \mathbf{0} & P_2 \end{pmatrix}.$$

(a) Does the Markov chain have a unique stationary distribution? If so, find it.

(b) Does $\lim_{n\to\infty} P^n$ exist? If so, find it.

(c) Does the Markov chain have a limiting distribution? If so, find it.

3.9 Let P be a stochastic matrix.

(a) If P is regular, is P^2 regular?

(b) If P is the transition matrix of an irreducible Markov chain, is P^2 the transition matrix of an irreducible Markov chain?

3.10 A Markov chain has transition matrix P and limiting distribution π. Further assume that π is the initial distribution of the chain. That is, the chain is in stationarity. Find the following:

(a) $\lim_{n\to\infty} P(X_n = j | X_{n-1} = i)$

(b) $\lim_{n\to\infty} P(X_n = j | X_0 = i)$

(c) $\lim_{n\to\infty} P(X_{n+1} = k, X_n = j | X_0 = i)$

(d) $\lim_{n\to\infty} P(X_0 = j | X_n = i)$

3.11 Consider a simple symmetric random walk on $\{0, 1, \dots, k\}$ with reflecting boundaries. If the walk is at state 0, it moves to 1 on the next step. If the walk is at k, it moves to $k - 1$ on the next step. Otherwise, the walk moves left or right, with probability 1/2.

(a) Find the stationary distribution.

(b) For $k = 1{,}000$, if the walk starts at 0, how many steps will it take, on average, for the walk to return to 0?

3.12 A Markov chain has transition matrix

$$P = \begin{pmatrix} 0 & 0 & 1 & 0 \\ 0 & 1/4 & 0 & 3/4 \\ 1/2 & 0 & 1/2 & 0 \\ 0 & 3/4 & 0 & 1/4 \end{pmatrix}.$$

Find the set of all stationary distributions.

3.13 Find the communication classes of a Markov chain with transition matrix

$$P = \begin{array}{c} \\ 1 \\ 2 \\ 3 \\ 4 \\ 5 \end{array} \begin{array}{c} \begin{array}{ccccc} 1 & 2 & 3 & 4 & 5 \end{array} \\ \begin{pmatrix} 1/2 & 0 & 0 & 0 & 1/2 \\ 1/3 & 1/2 & 1/6 & 0 & 0 \\ 0 & 1/4 & 0 & 1/2 & 1/4 \\ 0 & 0 & 0 & 1 & 0 \\ 1 & 0 & 0 & 0 & 0 \end{pmatrix} \end{array}.$$

Rewrite the transition matrix in canonical form.

3.14 The California Air Resources Board warns the public when smog levels are above certain thresholds. Days when the board issues warnings are called *episode* days. Lin (1981) models the daily sequence of episode and nonepisode days as a Markov chain with transition matrix

$$P = \begin{array}{c} \text{Nonepisode} \\ \text{Episode} \end{array} \begin{array}{c} \begin{array}{cc} \text{Nonepisode} & \text{Episode} \end{array} \\ \begin{pmatrix} 0.77 & 0.23 \\ 0.24 & 0.76 \end{pmatrix} \end{array}.$$

(a) What is the long-term probability that a given day will be an episode day?

(b) Over a year's time about how many days are expected to be episode days?

(c) In the long-term, what is the average number of days that will transpire between episode days?

3.15 On a chessboard a single random knight performs a simple random walk. From any square, the knight chooses from among its permissible moves with equal probability. If the knight starts on a corner, how long, on average, will it take to return to that corner?

3.16 As in the previous exercise, find the expected return time from a corner square for the following chess pieces: (i) queen, (ii) rook, (iii) king, (iv) bishop. Order the pieces by which pieces return quickest.

3.17 Consider a Markov chain with transition matrix

$$P = \begin{pmatrix} 1/2 & 1/2 \\ 0 & 1 \end{pmatrix}.$$

Obtain a closed form expression for P^n. Exhibit the matrix $\sum_{n=0}^{\infty} P^n$ (some entries may be $+\infty$). Explain what this shows about the recurrence and transience of the states.

3.18 Use first-step analysis to find the expected return time to state b for the Markov chain with transition matrix

$$\boldsymbol{P} = \begin{array}{c} \\ a \\ b \\ c \end{array} \begin{array}{c} \begin{array}{ccc} a & b & c \end{array} \\ \begin{pmatrix} 1/2 & 1/2 & 0 \\ 1/4 & 0 & 3/4 \\ 1/2 & 1/2 & 0 \end{pmatrix} \end{array}.$$

3.19 Consider random walk on the graph in Figure 3.15. Use first-step analysis to find the expected time to hit d for the walk started in a. (*Hint*: By exploiting symmetries in the graph, the solution can be found by solving a 3×3 linear system.)

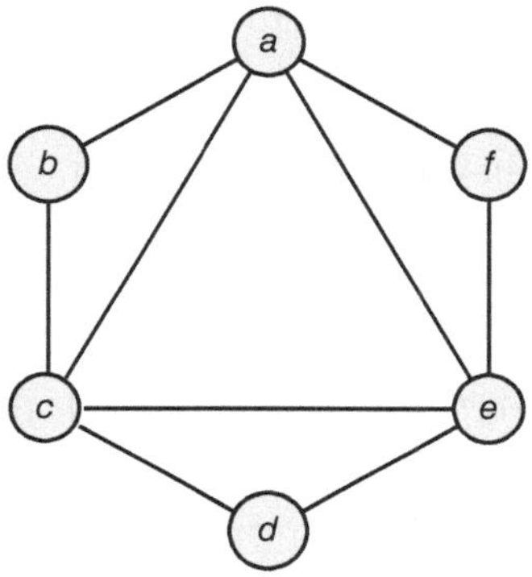

Figure 3.15

3.20 Show that simple symmetric random walk on $\mathbb{Z}^2$, that is, on the integer points in the plane, is recurrent. As in the one-dimensional case, consider the origin.

3.21 Show that simple symmetric random walk on $\mathbb{Z}^3$ is transient. As in the one-dimensional case, consider the origin and show

$$\begin{aligned} P_{00}^{2n} &= \frac{1}{6^{2n}} \sum_{0 \leq j+k \leq n} \frac{(2n)!}{j!j!k!k!(n-j-k)!(n-j-k)!}, \\ &\leq \frac{1}{2^{2n}} \binom{2n}{n} \left(\frac{1}{3^n} \frac{n!}{(n/3)!(n/3)!(n/3)!} \right). \end{aligned}$$

Then, use Stirling's approximation.

3.22 Consider the general two-state chain

$$\boldsymbol{P} = \begin{array}{c} \\ 1 \\ 2 \end{array} \begin{array}{c} \begin{array}{cc} 1 & 2 \end{array} \\ \begin{pmatrix} 1-p & p \\ q & 1-q \end{pmatrix} \end{array},$$

where p and q are not both 0. Let T be the first return time to state 1, for the chain started in 1.

(a) Show that $P(T \geq n) = p(1-q)^{n-2}$, for $n \geq 2$.

(b) Find $E(T)$ and verify that $E(T) = 1/\pi_1$, where $\boldsymbol{\pi}$ is the stationary distribution of the chain.

3.23 Consider a k-state Markov chain with transition matrix

$$\boldsymbol{P} = \begin{array}{c} \\ 1 \\ 2 \\ 3 \\ \vdots \\ k-2 \\ k-1 \\ k \end{array} \begin{array}{c} \begin{array}{ccccccc} 1 & 2 & 3 & \cdots & k-2 & k-1 & k \end{array} \\ \begin{pmatrix} 1/k & 1/k & 1/k & \cdots & 1/k & 1/k & 1/k \\ 1 & 0 & 0 & \cdots & 0 & 0 & 0 \\ 0 & 1 & 0 & \cdots & 0 & 0 & 0 \\ \vdots & \vdots & \vdots & \vdots & \vdots & \vdots & \vdots \\ 0 & 0 & 0 & \cdots & 0 & 0 & 0 \\ 0 & 0 & 0 & \cdots & 1 & 0 & 0 \\ 0 & 0 & 0 & \cdots & 0 & 1 & 0 \end{pmatrix} \end{array}.$$

Show that the chain is ergodic and find the limiting distribution.

3.24 Show that the stationary distribution for the modified Ehrenfest chain of Example 3.19 is binomial with parameters N and 1/2.

3.25 Read about the Bernoulli–Laplace model of diffusion in Exercise 2.12.

(a) Find the stationary distribution for the cases $k = 2$ and $k = 3$.

(b) For general k, show that $\pi_j = \binom{k}{j}^2 / \binom{2k}{k}$, for $j = 0, 1, \ldots, k$, satisfies the equations for the stationary distribution and is thus the unique limiting distribution of the chain.

3.26 Assume that $(p_1, \ldots, p_k)$ is a probability vector. Let $\boldsymbol{P}$ be a $k \times k$ transition matrix defined by

$$P_{ij} = \begin{cases} p_j, & \text{if } i = 1, \ldots, k-1, \\ 0, & \text{if } i = k, j < k, \\ 1, & \text{if } i = k, j = k. \end{cases}$$

Describe all the stationary distributions for $\boldsymbol{P}$.

3.27 Sinclair (2005). Consider the infinite Markov chain on the non-negative integers described by Figure 3.16.

(a) Show that the chain is irreducible and aperiodic.

(b) Show that the chain is recurrent by computing the first return time to 0 for the chain started at 0.

(c) Show that the chain is null recurrent.

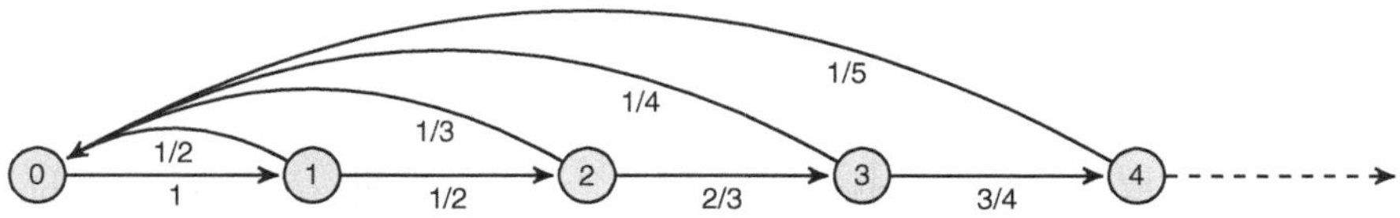

Figure 3.16

3.28 Consider a Markov chain with transition matrix

$$
P = \begin{array}{c} \\ 1 \\ 2 \\ 3 \\ 4 \\ 5 \\ 6 \\ 7 \end{array}
\begin{array}{c} \begin{array}{ccccccc} 1 & 2 & 3 & 4 & 5 & 6 & 7 \end{array} \\
\begin{pmatrix}
1/3 & 1/3 & 1/3 & 0 & 0 & 0 & 0 \\
2/3 & 0 & 1/3 & 0 & 0 & 0 & 0 \\
0 & 2/3 & 1/3 & 0 & 0 & 0 & 0 \\
0 & 1/4 & 0 & 1/2 & 0 & 1/4 & 0 \\
0 & 0 & 1/4 & 0 & 1/4 & 0 & 1/2 \\
0 & 0 & 0 & 0 & 0 & 0 & 1 \\
0 & 1/8 & 1/8 & 1/8 & 1/8 & 1/4 & 1/4
\end{pmatrix} \end{array}.
$$

Identify the communication classes. Classify the states as recurrent or transient. For all i and j, determine $\lim_{n\to\infty} P_{ij}^n$ without using technology.

3.29 Consider a Markov chain with transition matrix

$$
P = \begin{array}{c} \\ a \\ b \\ c \\ d \\ e \\ f \\ g \end{array}
\begin{array}{c} \begin{array}{ccccccc} a & b & c & d & e & f & g \end{array} \\
\begin{pmatrix}
0.6 & 0.2 & 0 & 0 & 0.2 & 0 & 0 \\
0 & 0 & 0 & 0 & 0 & 0.5 & 0.5 \\
0 & 0 & 0.3 & 0.7 & 0 & 0 & 0 \\
0 & 0.3 & 0.4 & 0.3 & 0 & 0 & 0 \\
0 & 0 & 0 & 0 & 1 & 0 & 0 \\
0 & 0.1 & 0 & 0 & 0 & 0 & 0.9 \\
0 & 0.2 & 0 & 0 & 0 & 0.8 & 0
\end{pmatrix} \end{array}.
$$

Identify the communication classes. Classify the states as recurrent or transient, and determine the period of each state.

3.30 A graph is *bipartite* if the vertex set can be colored with two colors black and white such that every edge in the graph joins a black vertex and a white vertex. See Figure 3.7(a) for an example of a bipartite graph. Show that for simple random walk on a connected graph, the walk is periodic if and only if the graph is bipartite.

3.31 For the network graph in Figure 3.17, find the PageRank for the nodes of the network using a damping factor of $p = 0.90$. See Example 3.21.

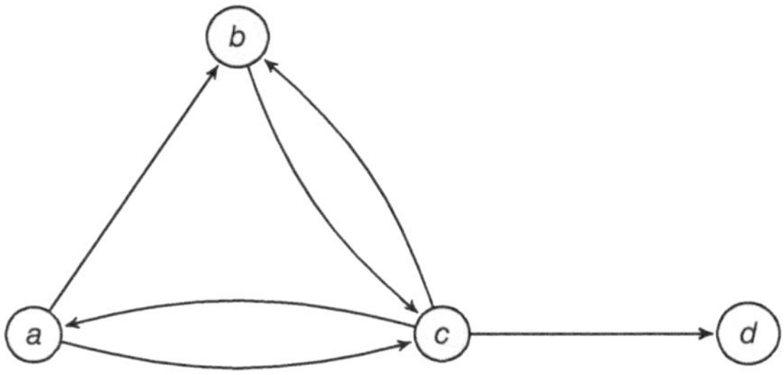

Figure 3.17

3.32 Let $X_0, X_1, \ldots$ be an ergodic Markov chain with transition matrix $\boldsymbol{P}$ and stationary distribution $\boldsymbol{\pi}$. Define the bivariate process $Z_n = (X_n, X_{n-1})$, for $n \geq 1$, with $Z_0 = (X_0, X_0)$.

(a) Give an intuitive explanation for why $Z_0, Z_1, \ldots$ is a Markov chain.

(b) Determine the transition probabilities in terms of $\boldsymbol{P}$. That is, find

$$P(Z_n = (i,j) | Z_{n-1} = (s,t)).$$

(c) Find the limiting distribution.

3.33 Assume that $\boldsymbol{P}$ is a stochastic matrix. Show that if $\boldsymbol{P}^N$ is positive, then $\boldsymbol{P}^{N+m}$ is positive for all $m \geq 0$.

3.34 Let $\boldsymbol{P}$ be the transition matrix of an irreducible, but not necessarily ergodic, Markov chain. For $0 < p < 1$, let

$$\widetilde{\boldsymbol{P}} = p\boldsymbol{P} + (1-p)\boldsymbol{I},$$

where $\boldsymbol{I}$ is the identity matrix. Show that $\widetilde{\boldsymbol{P}}$ is a stochastic matrix for an ergodic Markov chain with the same stationary distribution as $\boldsymbol{P}$. Give an intuitive description for how the $\widetilde{\boldsymbol{P}}$ chain evolves compared to the $\boldsymbol{P}$-chain.

3.35 Let $\boldsymbol{Q}$ be a $k \times k$ stochastic matrix. Let $\boldsymbol{A}$ be a $k \times k$ matrix each of whose entries is $1/k$. For $0 < p < 1$, let

$$\boldsymbol{P} = p\boldsymbol{Q} + (1-p)\boldsymbol{A}.$$

Show that $\boldsymbol{P}$ is the transition matrix for an ergodic Markov chain.

3.36 Let $X_0, X_1, \ldots$ be an ergodic Markov chain on $\{1, \ldots, k\}$ with stationary distribution $\boldsymbol{\pi}$. Assume that the chain is in stationarity.

(a) Find $\text{Cov}(X_m, X_{m+n})$.

(b) Find $\lim_{n\to\infty} \text{Cov}(X_m, X_{m+n})$.

3.37 Show that all two-state Markov chains, except for the trivial chain whose transition matrix is the identity matrix, are time reversible.

3.38 You throw five dice and set aside those dice that are sixes. Throw the remaining dice and again set aside the sixes. Continue until you get all sixes.

(a) Exhibit the transition matrix for the associated Markov chain, where X_n is the number of sixes after n throws. See also Exercise 2.11.

(b) How many turns does it take, on average, before you get all sixes?

3.39 Show that if $X_0, X_1, \dots$ is reversible, then for the chain in stationarity

$$P(X_0 = i_0, X_1 = i_1, \dots, X_n = i_n) = P(X_n = i_0, X_{n-1} = i_1, \dots, X_0 = i_n),$$

for all $i_0, i_1, \dots, i_n$.

3.40 Consider a *biased random walk* on the n-cycle, which moves one direction with probability p and the other direction with probability $1 - p$. Determine whether the walk is time reversible.

3.41 Show that the Markov chain with transition matrix

$$P = \begin{array}{c} \\ a \\ b \\ c \\ d \end{array} \begin{array}{c} \begin{array}{cccc} a & b & c & d \end{array} \\ \begin{pmatrix} 1/6 & 1/6 & 0 & 2/3 \\ 1/5 & 2/5 & 2/5 & 0 \\ 0 & 1/3 & 1/6 & 1/2 \\ 4/9 & 0 & 1/3 & 2/9 \end{pmatrix} \end{array}$$

is reversible. The chain can be described by a random walk on a weighted graph. Exhibit the graph such that all the weights are integers.

3.42 Consider random walk on $\{0, 1, 2, \dots\}$ with one reflecting boundary. If the walk is at 0, it moves to 1 on the next step. Otherwise, it moves left, with probability p, or right, with probability $1 - p$. For what values of p is the chain reversible? For such p, find the stationary distribution.

3.43 A Markov chain has transition matrix

$$P = \begin{pmatrix} 1/2 & 1/4 & 0 & 1/4 \\ p & 0 & 1-p & 0 \\ 0 & 1/4 & 1/2 & 1/4 \\ q & 0 & 1-q & 0 \end{pmatrix}.$$

(a) For what values of p and q is the chain ergodic?

(b) For what values of p and q is the chain reversible?

3.44 Markov chains are used to model nucleotide substitutions and mutations in DNA sequences. Kimura gives the following transition matrix for such a model.

$$P = \begin{array}{c} \\ a \\ g \\ c \\ t \end{array} \begin{array}{c} \begin{array}{cccc} a & g & c & t \end{array} \\ \begin{pmatrix} 1-p-2r & p & r & r \\ p & 1-p-2r & r & r \\ q & q & 1-p-2q & p \\ q & q & p & 1-p-2q \end{pmatrix} \end{array}.$$

Find a vector $\boldsymbol{x}$ that satisfies the detailed-balance equations. Show that the chain is reversible and find the stationary distribution. Confirm your result for the case $p = 0.1$, $q = 0.2$, and $r = 0.3$.

3.45 If $\boldsymbol{P}$ is the transition matrix of a reversible Markov chain, show that $\boldsymbol{P}^2$ is, too. Conclude that $\boldsymbol{P}^n$ is the transition matrix of a reversible Markov chain for all $n \geq 1$.

3.46 Given a Markov chain with transition matrix $\boldsymbol{P}$ and stationary distribution $\boldsymbol{\pi}$, the *time reversal* is a Markov chain with transition matrix $\widetilde{\boldsymbol{P}}$ defined by

$$\widetilde{P}_{ij} = \frac{\pi_j P_{ji}}{\pi_i}, \quad \text{for all } i, j.$$

(a) Show that a Markov chain with transition matrix $\boldsymbol{P}$ is reversible if and only if $\boldsymbol{P} = \widetilde{\boldsymbol{P}}$.

(b) Show that the time reversal Markov chain has the same stationary distribution as the original chain.

3.47 Consider a Markov chain with transition matrix

$$\boldsymbol{P} = \begin{matrix} & \begin{matrix} 1 & 2 & 3 \end{matrix} \\ \begin{matrix} 1 \\ 2 \\ 3 \end{matrix} & \begin{pmatrix} 1/3 & 0 & 2/3 \\ 1/2 & 1/2 & 0 \\ 1/6 & 1/3 & 1/2 \end{pmatrix} \end{matrix}.$$

Find the transition matrix of the time reversal chain (see Exercise 3.46).

3.48 Consider a Markov chain with transition matrix

$$\boldsymbol{P} = \begin{matrix} & \begin{matrix} a & b & c \end{matrix} \\ \begin{matrix} a \\ b \\ c \end{matrix} & \begin{pmatrix} 1-\alpha & \alpha & 0 \\ 0 & 1-\beta & \beta \\ \gamma & 0 & 1-\gamma \end{pmatrix} \end{matrix},$$

where $0 < \alpha, \beta, \gamma < 1$. Find the transition matrix of the time reversal chain (see Exercise 3.46).

3.49 Consider an absorbing chain with t transient and $k - t$ absorbing states. For transient state i and absorbing state j, let B_{ij} denote the probability starting at i that the chain is absorbed in j. Let $\boldsymbol{B}$ be the resulting $t \times (k - t)$ matrix. By first-step analysis show that $\boldsymbol{B} = (\boldsymbol{I} - \boldsymbol{Q})^{-1}\boldsymbol{R}$.

3.50 Consider the following method for shuffling a deck of cards. Pick two cards from the deck uniformly at random and then switch their positions. If the same two cards are chosen, the deck does not change. This is called the *random transpositions* shuffle.

(a) Argue that the chain is ergodic and the stationary distribution is uniform.

(b) Exhibit the 6×6 transition matrix for a three-card deck.

(c) How many shuffles does it take on average to reverse the original order of the deck of cards?

3.51 A deck of k cards is shuffled by the *top-to-random* method: the top card is placed in a uniformly random position in the deck. (After one shuffle, the top card stays where it is with probability $1/k$.) Assume that the top card of the deck is the ace of hearts. Consider a Markov chain where X_n is the position of the ace of hearts after n top-to-random shuffles, with $X_0 = 1$. The state space is $\{1, \dots, k\}$. Assume that $k = 6$.

(a) Exhibit the transition matrix and find the expected number of shuffles for the ace of hearts to return to the top of the deck.

(b) Find the expected number of shuffles for the bottom card to reach the top of the deck.

3.52 The board for a modified Snakes and Ladder game is shown in Figure 3.18. The game is played with a tetrahedron (four-faced) die.

(a) Find the expected length of the game.

(b) Assume that the player is on square 6. Find the probability that they will find themselves on square 3 before finishing the game.

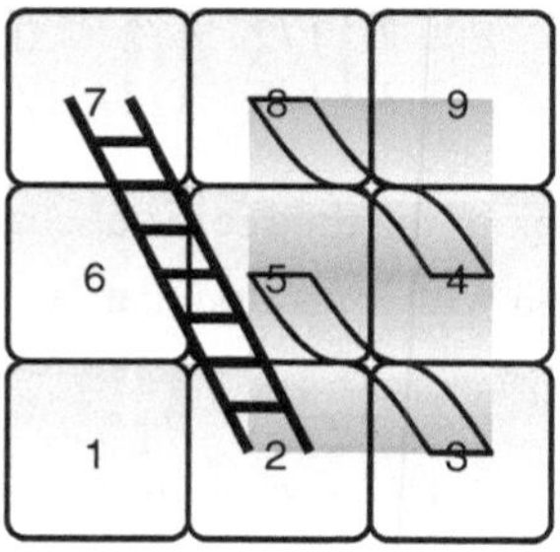

Figure 3.18

3.53 When an NFL football game ends in a tie, under *sudden-death* overtime the two teams play at most 15 extra minutes and the team that scores first wins the game. A Markov chain analysis of sudden-death is given in Jones (2004). Assuming two teams A and B are evenly matched, a four-state absorbing Markov chain is given with states *PA*: team A gains possession, *PB*: team B gains possession, *A*: A wins, and *B*: B wins. The transition matrix is

$$
\boldsymbol{P} = \begin{array}{c} \\ PA \\ PB \\ A \\ B \end{array} \begin{array}{c} \begin{array}{cccc} PA & PB & A & B \end{array} \\ \begin{pmatrix} 0 & 1-p & p & 0 \\ 1-p & 0 & 0 & p \\ 0 & 0 & 1 & 0 \\ 0 & 0 & 0 & 1 \end{pmatrix} \end{array},
$$

where p is the probability that a team scores when it has the ball. Which team first receives the ball in overtime is decided by a coin flip.

(a) If team A receives the ball in overtime, find the probability that A wins.

(b) An alternate overtime procedure is the *first-to-six rule*, where the first time to score six points in overtime wins the game. Consider two evenly matched teams. Let α be the probability that a team scores a touchdown (six points). Let β be the probability that a team scores a field goal (three points). Assume for simplicity that touchdowns and field goals are the only way points can be scored. Develop a 10-state Markov chain model for overtime play.

(c) For the 2002 regular NFL season, there were 6,049 possessions, 1,270 touchdowns, and 737 field goals. Using these data compare the probability that A wins the game for each of the two overtime procedures.

3.54 A mouse is placed in the maze in Figure 3.19 starting in box A. A piece of cheese is put in box I. From each room the mouse moves to an adjacent room through an open door, choosing from the available doors with equal probability.

(a) How many rooms, on average, will the mouse visit before it finds the cheese?

(b) How many times, on average, will the mouse visit room A before it finds the cheese?

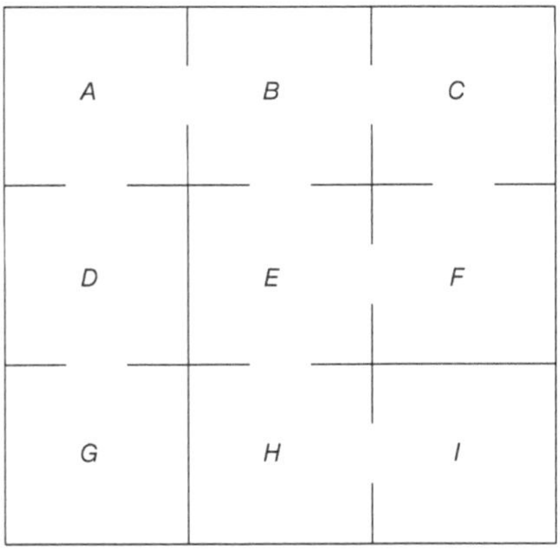

Figure 3.19 Mouse in a maze.

3.55 In a sequence of fair coin flips, how many flips, on average, are required to first see the pattern H-H-T-H?

3.56 A biased coin has heads probability 1/3 and tails probability 2/3. If the coin is tossed repeatedly, find the expected number of flips required until the pattern H-T-T-H-H appears.

3.57 In repeated coin flips, consider the set of all three-element patterns:

$$\{\text{HHH, HHT, HTH, HTT, THH, THT, TTH, TTT}\}.$$

Which patterns take the longest time, on average, to appear in repeated sampling? Which take the shortest?

3.58 A sequence of 0s and 1s is generated by a Markov chain with transition matrix

$$P = \begin{array}{c} \\ 0 \\ 1 \end{array}\begin{array}{c} \begin{array}{cc} 0 & 1 \end{array} \\ \begin{pmatrix} 1/4 & 3/4 \\ 3/4 & 1/4 \end{pmatrix} \end{array}.$$

The first element of the sequence is decided by a fair coin flip. On average, how many steps are required for the pattern 0-0-1-1 to first appear?

3.59 Consider random walk on the weighted graph in Figure 3.20.

(a) If the walk starts in a, find the expected number of steps to return to a.

(b) If the walk starts in a, find the expected number of steps to first hit b.

(c) If the walk starts in a, find the probability that the walk hits b before c.

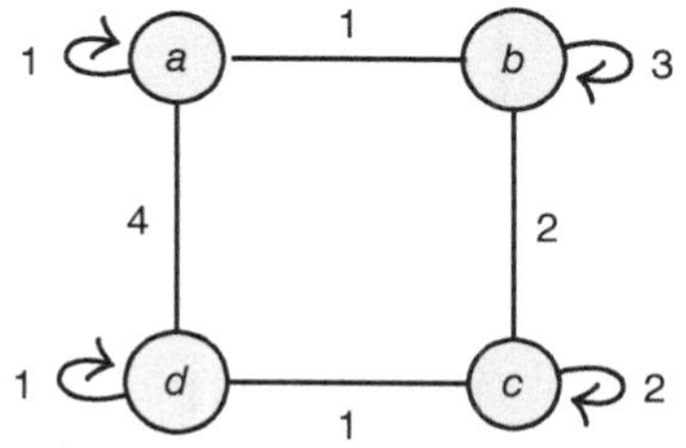

Figure 3.20

3.60 For a Markov chain started in state i, let T denote the *fifth time* the chain visits state i. Is T a stopping time? Explain.

3.61 Consider the weather Markov chain $X_0, X_1, \ldots$ of Example 2.3. Let T be the first time that it rains for 40 days in a row. Is $X_T, X_{T+1}, X_{T+2}, \ldots$ a Markov chain? Explain.

3.62 Let S be a random variable that is constant, with probability 1, where that constant is some positive integer. Show that S is a stopping time. Conclude that the Markov property follows from the strong Markov property.

3.63 `R` : Hourly wind speeds in a northwestern region of Turkey are modeled by a Markov chain in Sahin and Sen (2001). Seven wind speed levels are the states of the chain. The transition matrix is

$$P = \begin{array}{c} \\ 1 \\ 2 \\ 3 \\ 4 \\ 5 \\ 6 \\ 7 \end{array} \begin{array}{c} \begin{array}{ccccccc} 1 & 2 & 3 & 4 & 5 & 6 & 7 \end{array} \\ \begin{pmatrix} 0.756 & 0.113 & 0.129 & 0.002 & 0 & 0 & 0 \\ 0.174 & 0.821 & 0.004 & 0.001 & 0 & 0 & 0 \\ 0.141 & 0.001 & 0.776 & 0.082 & 0 & 0 & 0 \\ 0.003 & 0 & 0.192 & 0.753 & 0.052 & 0 & 0 \\ 0 & 0 & 0.002 & 0.227 & 0.735 & 0.036 & 0 \\ 0 & 0 & 0 & 0.007 & 0.367 & 0.604 & 0.022 \\ 0 & 0 & 0 & 0 & 0.053 & 0.158 & 0.789 \end{pmatrix} \end{array}.$$

(a) Find the limiting distribution by (i) taking high matrix powers, and (ii) using the `stationary` command in the **utilities.R** file. How often does the highest wind speed occur? How often does the lowest speed occur?

(b) Simulate the chain for 100,000 steps and estimate the proportion of times that the chain visits each state.

3.64 R: The evolution of forest ecosystems in the United States and Canada is studied in Strigul et al. (2012) using Markov chains. Five-year changes in the state of the forest soil are modeled with a 12-state Markov chain. The transition matrix can be found in the R script file **forest.R**. About how many years does it take for the ecosystem to move from state 1 to state 12?

3.65 R: Simulate the gambler's ruin problem for a gambler who starts with \$15 and quits when he reaches \$50 or goes bust. Use your code to simulate the probability of eventual ruin and compare to the exact probability.

3.66 R: Simulate the expected hitting time for the random walk on the hexagon in Exercise 3.19.

3.67 R: Simulate the dice game of Exercise 3.38. Verify numerically the theoretical expectation for the number of throws needed to get all sixes.

3.68 R : Write a function `reversal(mat)`, whose input is the transition matrix of an irreducible Markov chain and whose output is the transition matrix of the reversal chain.

3.69 R : Make up your own board game which can be modeled as a Markov chain. Ask interesting questions and answer them by simulation and/or an exact analysis.

4

BRANCHING PROCESSES

> Every moment dies a man, every moment one is born.
>
> —Alfred Tennyson, *The Vision of Sin*
>
> Every moment dies a man, every moment one and one-sixteenth is born.
>
> —Mathematician Charles Babbage in a letter to Alfred Tennyson suggesting a change "in your otherwise beautiful poem."

4.1 INTRODUCTION

Branching processes are a class of stochastic processes that model the growth of populations. They are widely used in biology and epidemiology to study the spread of infectious diseases and epidemics. Applications include nuclear chain reactions and the spread of computer software viruses. Their original motivation was to study the extinction of family surnames, an issue of concern to the Victorian aristocracy in 19th century Britain.

In 1873, the British statistician Sir Francois Galton posed the following question in the *Educational Times*.

> Problem 4001: A large nation, of whom we will only concern ourselves with adult males, N in number, and who each bear separate surnames colonize a district. Their law of population is such that, in each generation, a_0 percent of the adult males have no male children who reach adult life; a_1 have one such male child; a_2 have two; and so on up to

Introduction to Stochastic Processes with R, First Edition. Robert P. Dobrow.

> a_5 who have five. Find (1) what proportion of their surnames will have become extinct after r generations; and (2) how many instances there will be of the surname being held by m persons.

The Reverend Henry William Watson replied with a solution. The study of branching processes grew out of Watson and Galton's collaboration. Their results were independently discovered by the French statistician Irénée-Jules Bienaymé. The basic branching process model is sometimes called a Bienaymé–Galton–Watson process.

We use the imagery of populations, generations, children, and offspring. Assume that we have a population of individuals, each of which independently produces a random number of children according to a probability distribution $\boldsymbol{a} = (a_0, a_1, a_2, \dots)$. That is, an individual gives birth to k children with probability a_k, for $k \geq 0$, independent of other individuals. Call $\boldsymbol{a}$ the *offspring distribution*.

The population grows or declines from generation to generation. Let Z_n be the size (e.g., number of individuals) of the nth generation, for $n \geq 0$. Assume $Z_0 = 1$. That is, the population starts with one individual. The sequence $Z_0, Z_1, \dots$ is a *branching process*. See Figure 4.1 for a realization of such a process through three generations.

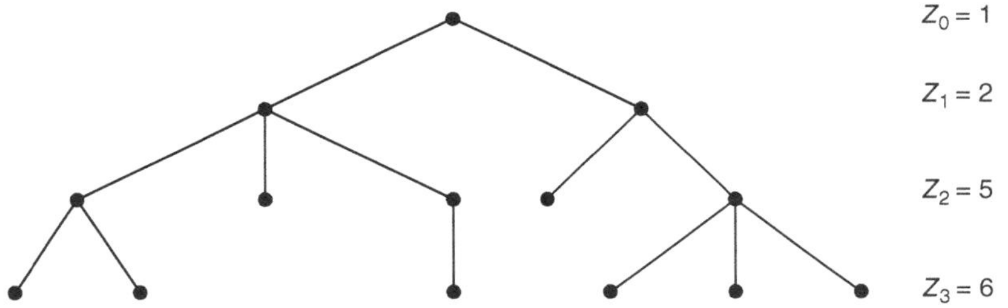

Figure 4.1 Branching process.

A branching process is a Markov chain since the size of a generation only depends on the size of the previous generation and the number of their offspring. If Z_n is given, then the size of the next generation Z_{n+1} is independent of $Z_0, \dots, Z_{n-1}$.

Assume $0 < a_0 < 1$. If $a_0 = 0$, then the population only grows and 0 is not in the state space. If $a_0 = 1$, then $Z_n = 0$, for all $n \geq 1$. We also assume that there is positive probability that an individual gives birth to more than one offspring, that is, $a_0 + a_1 < 1$.

Galton's first question, "What proportion of their surnames will have become extinct after r generations?" leads one to examine the recurrence and transience properties of the Markov chain.

It should be clear that 0 is an absorbing state. If a generation has no individuals, there will be no offspring. Under the initial assumptions, all other states of a branching process are transient.

Lemma 4.1. *In a branching process, all nonzero states are transient.*

Proof. If $Z_n = 0$, say the process has become *extinct* by generation n. Consider the probability that a population of size $i > 0$ goes extinct in one generation, that is, $P(Z_{n+1} = 0|Z_n = i)$. If a generation has i individuals and the next generation has none, then each individual produced zero offspring, which occurs with probability $(a_0)^i$, by independence.

To show i is transient, we need to show that f_i, the probability of eventually hitting i for the chain started in i, is less than one. If the chain starts with i individuals, then the event that the chain eventually hits i is $\{Z_n = i \text{ for some } n \geq 1\} \subseteq \{Z_1 > 0\}$. Hence,

$$\begin{aligned} f_i &= P(Z_n = i \text{ for some } n \geq 1 | Z_0 = i) \\ &\leq P(Z_1 > 0|Z_0 = i) \\ &= 1 - P(Z_1 = 0|Z_0 = i) \\ &= 1 - (a_0)^i < 1, \end{aligned}$$

since $a_0 > 0$. ■

Since all nonzero states are transient and the chain has infinite state space, there are two possibilities for the long-term evolution of the process: either it gets absorbed in state 0, that is, the population eventually goes extinct, or the population grows without bound.

4.2 MEAN GENERATION SIZE

In a branching process, the size of the nth generation is the sum of the total offspring of the individuals of the previous generation. That is,

$$Z_n = \sum_{i=1}^{Z_{n-1}} X_i, \tag{4.1}$$

where X_i denotes the number of children born to the ith person in the $(n-1)$th generation. Because of the independence assumption, $X_1, X_2, \ldots$ is an i.i.d. sequence with common distribution $\boldsymbol{a}$. Furthermore, Z_{n-1} is independent of the X_i.

Equation (4.1) represents Z_n as a random sum of i.i.d. random variables. Results for such random sums can be applied to find the moments of Z_n.

Let $\mu = \sum_{k=0}^{\infty} k a_k$ be the mean of the offspring distribution. To find the mean of the size of the nth generation $E(Z_n)$, condition on Z_{n-1}. By the law of total expectation,

$$\begin{aligned} E(Z_n) &= \sum_{k=0}^{\infty} E(Z_n | Z_{n-1} = k) P(Z_{n-1} = k) \\ &= \sum_{k=0}^{\infty} E\left(\sum_{i=1}^{Z_{n-1}} X_i \middle| Z_{n-1} = k \right) P(Z_{n-1} = k) \end{aligned}$$

$$= \sum_{k=0}^{\infty} E\left(\sum_{i=1}^{k} X_i \middle| Z_{n-1} = k \right) P(Z_{n-1} = k)$$

$$= \sum_{k=0}^{\infty} E\left(\sum_{i=1}^{k} X_i \right) P(Z_{n-1} = k)$$

$$= \sum_{k=0}^{\infty} k\mu P(Z_{n-1} = k) = \mu E(Z_{n-1}),$$

where the fourth equality is because the X_i are independent of Z_{n-1}. Iterating the resulting recurrence relation gives

$$E(Z_n) = \mu E(Z_{n-1}) = \mu^2 E(Z_{n-2}) = \cdots = \mu^n E(Z_0) = \mu^n, \text{ for } n \geq 0,$$

since $Z_0 = 1$.

Three Cases

For the long-term expected generation size,

$$\lim_{n\to\infty} E(Z_n) = \lim_{n\to\infty} \mu^n = \begin{cases} 0, & \text{if } \mu < 1, \\ 1, & \text{if } \mu = 1, \\ \infty, & \text{if } \mu > 1. \end{cases}$$

A branching process is said to be *subcritical* if $\mu < 1$, *critical* if $\mu = 1$, and *supercritical* if $\mu > 1$. For a subcritical branching process, mean generation size declines exponentially to zero. For a supercritical process, mean generation size exhibits long-term exponential growth. The limits suggest three possible regimes depending on μ: long-term extinction, stability, and boundless growth. However, behavior of the *mean* generation size does not tell the whole story.

Insight into the evolution of a branching process is gained by simulation. We simulated 10 generations $Z_0, \ldots, Z_{10}$ of a branching process with Poisson offspring distribution, choosing three values for the Poisson mean parameter corresponding to three types of branching process: $\mu = 0.75$ (subcritical), $\mu = 1$ (critical), and $\mu = 1.5$ (supercritical). Each process was simulated five times. See the R script file **branching**.R. Results are shown in Table 4.1.

For $\mu = 0.75$, all simulated paths result in eventual extinction. Furthermore, the extinction occurs fairly rapidly.

When $\mu = 1$, all but one of the simulations in Table 4.1 become extinct by the 10th generation.

TABLE 4.1 Simulations of a Branching Process for Three Choices of μ

μ	Z_0	Z_1	Z_2	Z_3	Z_4	Z_5	Z_6	Z_7	Z_8	Z_9	Z_{10}
0.75	1	2	3	3	2	1	1	0	0	0	0
0.75	1	0	0	0	0	0	0	0	0	0	0
0.75	1	2	0	0	0	0	0	0	0	0	0
0.75	1	0	0	0	0	0	0	0	0	0	0
0.75	1	3	3	1	3	1	0	0	0	0	0
1	1	1	1	0	0	0	0	0	0	0	0
1	1	2	0	0	0	0	0	0	0	0	0
1	3	6	6	5	6	7	8	8	8	6	5
1	1	3	4	1	2	1	0	0	0	0	0
1	1	2	1	1	2	1	2	1	0	0	0
1.5	1	2	3	10	22	41	93	173	375	763	1,597
1.5	1	1	1	1	2	4	7	9	11	19	29
1.5	1	4	5	18	34	68	127	246	521	1,011	2,065
1.5	1	1	2	0	0	0	0	0	0	0	0
1.5	1	2	5	3	2	6	9	17	18	13	19

Indeed, for a general branching process, in the subcritical and critical cases ($\mu \leq 1$), the population becomes extinct with probability 1.

In the supercritical case $\mu = 1.5$, most simulations in Table 4.1 seem to grow without bound. However, one realization goes extinct. We will see that in the general supercritical case, the probability that the population eventually dies out is less than one, but typically greater than zero.

Extinction in the Subcritical Case

Assume that $Z_0, Z_1, \ldots$ is a subcritical branching process. Let $E_n = \{Z_n = 0\}$ be the event that the population is extinct by generation n, for $n \geq 1$. Let E be the event that the population is ultimately extinct. Then,

$$E = \{Z_n = 0, \quad \text{for some } n \geq 1\} = \bigcup_{n=1}^{\infty} E_n,$$

and $E_1 \subseteq E_2 \subseteq \cdots$. It follows that the probability that the population eventually goes extinct is

$$P(E) = P\left(\bigcup_{n=1}^{\infty} E_n\right) = \lim_{n\to\infty} P(E_n) = \lim_{n\to\infty} P(Z_n = 0). \tag{4.2}$$

The probability that the population is extinct by generation n is

$$\begin{aligned}P(Z_n = 0) &= 1 - P(Z_n \geq 1)\\ &= 1 - \sum_{k=1}^{\infty} P(Z_n = k)\\ &\geq 1 - \sum_{k=1}^{\infty} kP(Z_n = k)\\ &= 1 - E(Z_n) = 1 - \mu^n.\end{aligned}$$

Taking limits gives

$$P(E) = \lim_{n\to\infty} P(Z_n = 0) \geq \lim_{n\to\infty} 1 - \mu^n = 1,$$

since $\mu < 1$. Thus, $P(E) = 1$. With probability 1, a subcritical branching process eventually goes extinct.

Example 4.1 Subcritical branching processes have been used to model the spread of infections and disease in highly vaccinated populations. Farrington and Grant (1999) cite several examples, including the spread of measles and mumps, the outbreak of typhoidal salmonellae reported in Scotland in 1967–1990, and outbreaks of human monkeypox virus in past decades. Becker (1974) finds evidence of subcriticality in European smallpox data from 1950 to 1970.

Often the goal of these studies is to use data on observed outbreaks to estimate the unknown mean offspring parameter μ as well as the number of generations of spread until extinction. ■

Variance of Generation Size

To explore the process of extinction in the critical and supercritical cases ($\mu \geq 1$), we first consider the variance of the size of the nth generation $Var(Z_n)$. Let σ^2 denote the variance of the offspring distribution. By the law of total variance,

$$Var(Z_n) = Var(E(Z_n|Z_{n-1})) + E(Var(Z_n|Z_{n-1})).$$

We have shown that

$$E(Z_n|Z_{n-1} = k) = E\left(\sum_{i=1}^{k} X_i\right) = \sum_{i=1}^{k} E(X_i) = \mu k,$$

which gives $E(Z_n|Z_{n-1}) = \mu Z_{n-1}$. Similarly, $Var(Z_n|Z_{n-1}) = \sigma^2 Z_{n-1}$, since

$$Var(Z_n|Z_{n-1} = k) = Var\left(\sum_{i=1}^{k} X_i\right) = \sum_{i=1}^{k} Var(X_i) = \sigma^2 k,$$

using the independence of the X_i. Applying the law of total variance,

$$\begin{aligned} Var(Z_n) &= Var(\mu Z_{n-1}) + E(\sigma^2 Z_{n-1}) \\ &= \mu^2 Var(Z_{n-1}) + \sigma^2 \mu^{n-1}, \quad \text{for } n \geq 1. \end{aligned} \tag{4.3}$$

With $Var(Z_0) = 0$, Equation (4.3) yields

$$\begin{aligned} Var(Z_1) &= \mu^2 Var(Z_0) + \sigma^2 = \sigma^2, \\ Var(Z_2) &= \mu^2 Var(Z_1) + \sigma^2 \mu = \sigma^2 \mu(1 + \mu), \text{ and} \\ Var(Z_3) &= \mu^2 Var(Z_2) + \sigma^2 \mu^2 = \sigma^2 \mu^2(1 + \mu + \mu^2). \end{aligned}$$

The general pattern, proved by induction on n, gives

$$Var(Z_n) = \sigma^2 \mu^{n-1} \sum_{k=0}^{n-1} \mu^k = \begin{cases} n\sigma^2, & \text{if } \mu = 1, \\ \sigma^2 \mu^{n-1}(\mu^n - 1)/(\mu - 1), & \text{if } \mu \neq 1. \end{cases}$$

In the subcritical case, both the mean and variance of generation size tend to 0.

In the critical case, the mean size of every generation is one, but the variance is a linearly growing function of n.

In the supercritical case, the variance grows exponentially large. The potentially large difference between the mean μ^n and variance suggests that in some cases both extinction and boundless growth are possible outcomes.

To explore the issue more carefully, we will find the probability of ultimate extinction when $\mu \geq 1$. First, however, new tools are needed.

4.3 PROBABILITY GENERATING FUNCTIONS

For a discrete random variable X taking values in $\{0, 1, \ldots\}$, the *probability generating function of* X is the function

$$\begin{aligned} G(s) = E\left(s^X\right) &= \sum_{k=0}^{\infty} s^k P(X = k) \\ &= P(X = 0) + sP(X = 1) + s^2 P(X = 2) + \cdots \end{aligned}$$

The function is a power series whose coefficients are probabilities. Observe that $G(1) = 1$. The series converges absolutely for $|s| \leq 1$. To emphasize the underlying random variable X, we may write $G(s) = G_X(s)$.

The generating function represents the distribution of a discrete random variable as a power series. If two power series are equal, then they have the same coefficients. Hence, if two discrete random variables X and Y have the same probability generating function, that is, $G_X(s) = G_Y(s)$ for all s, then X and Y have the same distribution.

Example 4.2 Let X be uniformly distributed on $\{0, 1, 2\}$. Find the probability generating function of X.

Solution

$$G(s) = E\left(s^X\right) = \frac{1}{3} + s\left(\frac{1}{3}\right) + s^2\left(\frac{1}{3}\right) = \frac{1}{3}(1 + s + s^2).$$

■

Example 4.3 Assume that X has a geometric distribution with parameter p. Find the probability generating function of X.

Solution

$$G(s) = E\left(s^X\right) = \sum_{k=1}^{\infty} s^k p(1-p)^{k-1} = sp\sum_{k=1}^{\infty}(s(1-p))^{k-1} = \frac{sp}{1 - s(1-p)},$$

for $|s| < 1$. ■

Probabilities for X can be obtained from the generating function by successive differentiation. We have that

$$G(0) = P(X = 0),$$

$$G'(0) = \sum_{k=1}^{\infty} ks^{k-1}P(X = k)\bigg|_{s=0} = P(X = 1),$$

$$G''(0) = \sum_{k=2}^{\infty} k(k-1)s^{k-2}P(X = k)\bigg|_{s=0} = 2P(X = 2),$$

and so on. In general,

$$G^{(j)}(0) = \sum_{k=j}^{\infty} k(k-1)\cdots(k-j+1)s^{k-j}P(X = j)\bigg|_{s=0} = j!P(X = j),$$

and thus

$$P(X = j) = \frac{G^{(j)}(0)}{j!}, \quad \text{for } j = 0, 1, \ldots,$$

where $G^{(j)}$ denotes the jth derivative of G.

Example 4.4 A random variable X has probability generating function

$$G(s) = (1 - p + sp)^n.$$

Find the distribution of X.

Solution We have $P(X = 0) = G(0) = (1 - p)^n$. For $1 \leq j \leq n$, the jth derivative of G is

$$G^{(j)}(s) = n(n-1)\cdots(n-j+1)p^j(1-p+sp)^{n-j},$$

which gives

$$P(X = j) = \frac{G^{(j)}(0)}{j!}$$

$$= \frac{n(n-1)\cdots(n-j+1)}{j!}p^j(1-p)^{n-j}$$

$$= \binom{n}{j} p^j(1-p)^{n-j}.$$

For $j > n$, $G^{(j)}(0) = 0$, and thus $P(X = j) = 0$. We see that X has a binomial distribution with parameters n and p. ■

Sums of Independent Random Variables

Generating functions are useful tools for working with sums of independent random variables. Assume that $X_1, \ldots, X_n$ are independent. Let $Z = X_1 + \cdots + X_n$. The probability generating function of Z is

$$G_Z(s) = E\left(s^Z\right) = E\left(s^{X_1+\cdots+X_n}\right)$$

$$= E\left(\prod_{k=1}^{n} s^{X_k}\right) = \prod_{k=1}^{n} E\left(s^{X_k}\right)$$

$$= G_{X_1}(s)\cdots G_{X_n}(s),$$

where the fourth equality is by independence. The generating function of an independent sum is the product of the individual generating functions. If the X_i are also identically distributed, then

$$G_Z(s) = G_{X_1}(s)\cdots G_{X_n}(s) = [G_X(s)]^n,$$

where X is a random variable with the same distribution as the X_i.

Example 4.5 In Example 4.4, it is shown that the generating function of a binomial random variable with parameters n and p is $G(s) = (1 - p + ps)^n$. Here is a derivation using the fact that a sum of i.i.d. Bernoulli random variables has a binomial distribution.

Solution Let $X_1, \ldots, X_n$ be an i.i.d. sequence of Bernoulli random variables with parameter p. The common generating function of the X_i is

$$G(s) = E\left(s^{X_i}\right) = s^0 P(X_i = 0) + s^1 P(X_i = 1) = (1-p) + sp.$$

The sum $Z = X_1 + \cdots + X_n$ has a binomial distribution with parameters n and p. The probability generating function of Z is thus

$$G_Z(s) = [G(s)]^n = (1 - p + ps)^n.$$

■

Moments

The probability generating function of X can be used to find the mean, variance, and higher moments of X. Observe that

$$G'(1) = E\left(Xs^{X-1}\right)\Big|_{s=1} = E(X).$$

Also,

$$G''(1) = E\left(X(X-1)s^{X-2}\right)\Big|_{s=1} = E(X(X-1)) = E\left(X^2\right) - E(X),$$

which gives

$$\begin{aligned} Var(X) &= E\left(X^2\right) - E(X)^2 = \left(E\left(X^2\right) - E(X)\right) + E(X) - E(X)^2 \\ &= G''(1) + G'(1) - G'(1)^2. \end{aligned}$$

Example 4.6 For a geometric random variable with parameter p, the generating function is

$$G(s) = \frac{sp}{1 - s(1-p)},$$

as shown in Example 4.3. Use the generating function to find the mean and variance of the geometric distribution.

Solution For the mean,

$$G'(s) = \frac{p}{(1 - s(1-p))^2},$$

which gives $E(X) = G'(1) = 1/p$. For the variance,

$$G''(s) = \frac{2p(1-p)}{(1-s(1-p))^3}, \quad \text{and} \quad G''(1) = \frac{2(1-p)p}{p^2}.$$

This gives

$$Var(X) = G''(1) + G'(1) - G'(1)^2 = \frac{2(1-p)p}{p^2} + \frac{1}{p} - \frac{1}{p^2} = \frac{1-p}{p^2}.$$

We summarize some key properties of probability generating functions. ■

Properties of Probability Generating Function

1. Let $G(s) = E\left(s^X\right)$ be the probability generating function of a discrete random variable X. Then,
   (a) $G(1) = 1$,
   (b) $P(X = k) = G^{(k)}(0)/k!$, for $k \geq 0$,
   (c) $E(X) = G'(1)$,
   (d) $Var(X) = G''(1) + G'(1) - G'(1)^2$.
2. If X and Y are random variables such that $G_X(s) = G_Y(s)$ for all s, then X and Y have the same distribution.
3. If X and Y are independent, then $G_{X+Y}(s) = G_X(s)G_Y(s)$.

4.4 EXTINCTION IS FOREVER

Probability generating functions are especially useful for analyzing branching processes. We use them to find the probability that a branching process eventually goes extinct.

For $n \geq 0$, let

$$G_n(s) = \sum_{k=0}^{\infty} s^k P(Z_n = k)$$

be the generating function of the nth generation size Z_n. Let

$$G(s) = \sum_{k=0}^{\infty} s^k a_k$$

be the generating function of the offspring distribution. We have

$$G_n(s) = E\left(s^{Z_n}\right) = E\left(s^{\sum_{k=1}^{Z_{n-1}} X_k}\right) = E\left(E\left(s^{\sum_{k=1}^{Z_{n-1}} X_k} | Z_{n-1}\right)\right),$$

where the last equality is by the law of total expectation. From the independence of Z_{n-1} and the X_k,

$$\begin{aligned} E\left(s^{\sum_{k=1}^{Z_{n-1}} X_k} \middle| Z_{n-1} = z\right) &= E\left(s^{\sum_{k=1}^{z} X_k} \middle| Z_{n-1} = z\right) \\ &= E\left(s^{\sum_{k=1}^{z} X_k}\right) = E\left(\prod_{k=1}^{z} s^{X_k}\right) \\ &= \prod_{k=1}^{z} E\left(s^{X_k}\right) = [G(s)]^z, \end{aligned}$$

for all z. This gives

$$E\left(s^{\sum_{k=1}^{Z_{n-1}} X_k} | Z_{n-1}\right) = [G(s)]^{Z_{n-1}}.$$

Taking expectations,

$$G_n(s) = E\left(G(s)^{Z_{n-1}}\right) = G_{n-1}(G(s)), \quad \text{for } n \geq 1.$$

The probability generating function of Z_n is the composition of the generating function of Z_{n-1} and the generating function of the offspring distribution.

Observe that $G_0(s) = s$, and $G_1(s) = G_0(G(s)) = G(s)$. From the latter we see that the distribution of Z_1 is the offspring distribution $\boldsymbol{a}$.

Continuing,

$$G_2(s) = G_1(G(s)) = G(G(s)) = G(G_1(s)),$$

and

$$G_3(s) = G_2(G(s)) = G(G(G(s))) = G(G_2(s)).$$

In general,

$$G_n(s) = G_{n-1}(G(s)) = \underbrace{G(\cdots G(G(s))\cdots)}_{n\text{-fold}} = G(G_{n-1}(s)). \tag{4.4}$$

The generating function of Z_n is the n-fold composition of the offspring distribution generating function.

Equation (4.4) is typically not useful for computing the actual distribution of Z_n. For an arbitrary offspring distribution, the distribution of Z_n will be complicated with no tractable closed-form expression. However, the equation is central to the proof of the following theorem, which characterizes the extinction probability for a branching process.

Extinction Probability

Theorem 4.2. *Given a branching process, let G be the probability generating function of the offspring distribution. Then, the probability of eventual extinction is the smallest positive root of the equation $s = G(s)$.*

If $\mu \leq 1$, that is, in the subcritical and critical cases, the extinction probability is equal to 1.

Remark: We have already shown that in the subcritical $\mu < 1$ case, the population goes extinct with probability 1. The theorem gives that this is also true for $\mu = 1$, even though for each generation the expected generation size is $E(Z_n) = \mu^n = 1$. For the supercritical case $\mu > 1$, the expected generation size Z_n grows without bound.

However, the theorem gives that even in this case there is positive probability of eventual extinction.

Before proving Theorem 4.2, we offer some examples of its use. Let e denote the probability of eventual extinction.

Example 4.7 Find the extinction probability for a branching process with offspring distribution $\boldsymbol{a} = (1/6, 1/2, 1/3)$.

Solution The mean of the offspring distribution is

$$\mu = 0(1/6) + 1(1/2) + 2(1/3) = 7/6 > 1,$$

so this is the supercritical case. The offspring generating function is

$$G(s) = \frac{1}{6} + \frac{s}{2} + \frac{s^2}{3}.$$

Solving

$$s = G(s) = \frac{1}{6} + \frac{s}{2} + \frac{s^2}{3}$$

gives the quadratic equation $s^2/3 - s/2 + 1/6 = 0$, with roots $s = 1$ and $s = 1/2$. The smallest positive root is the probability of eventual extinction $e = 1/2$. ■

Example 4.8 A branching process has offspring distribution

$$a_k = (1-p)^k p, \quad \text{for } k = 0, 1, \ldots$$

Find the extinction probability in the supercritical case.

Solution The offspring distribution is a variant of the geometric distribution. The generating function is

$$G(s) = \sum_{k=0}^{\infty} s^k (1-p)^k p = p \sum_{k=0}^{\infty} (s(1-p))^k = \frac{p}{1 - s(1-p)}, \quad \text{for } |s(1-p)| < 1.$$

The mean of the offspring distribution is

$$\mu = G'(1) = \left. \frac{p(1-p)}{(1-s(1-p))^2} \right|_{s=1} = \frac{1-p}{p}.$$

The supercritical case $\mu > 1$ corresponds to $p < 1/2$.

To find the extinction probability, solve

$$s = G(s) = \frac{p}{1 - s(1-p)},$$

which gives the quadratic equation

$$(1-p)s^2 - s + p = 0,$$

with roots

$$s = \frac{1 \pm \sqrt{1-(4(1-p)p}}{2(1-p)} = \frac{1 \pm (1-2p)}{2(1-p)}.$$

The roots are 1 and $p/(1-p)$. For $0 < p < 1/2$, the smaller root is $p/(1-p) = 1/\mu$, that is, $e = 1/\mu$. ■

We explore the extinction probability result in Example 4.8 with the use of simulation. Let $p = 1/4$. We simulated a branching process with offspring distribution $a_k = (3/4)^k(1/4)$, for $k \geq 0$. The process is supercritical with $\mu = 3$. Results are collected in Table 4.2. Four of the 12 runs went extinct by time $n = 10$. The exact extinction probability is $e = 1/3$. (The fact that 4 out of 12 is exactly $1/3$ is, we assure the reader, pure coincidence.)

R: Simulating the Extinction Probability

The branching process is simulated 10,000 times, keeping track of the number of times the process goes extinct by the 10th generation. See the file **branching.R**.

```
> branch(10,1/4)
 1   4   6 16   71   205   569   1559 4588   13726 40800

> trials <- 10000
> simlist <- replicate(trials,branch(10,1/4)[11])
# Estimate of extinction probability
> sum(simlist==0)/trials
[1] 0.332
```

The function `branch(n,p)` simulates n steps of a branching process whose offspring distribution is geometric with parameter p. The `replicate` command repeats the simulation 10,000 times, storing the outcome of Z_{10} for each trial in the vector `simlist`. The proportion of 0s in `simlist` estimates the extinction probability e.

Our conclusions are slightly biased since we assume that if extinction takes place it will occur by time $n = 10$. Of course, extinction could occur later. However, it appears from the simulations that if extinction occurs it happens very rapidly.

TABLE 4.2 Simulation of a Supercritical Branching Process, with $\mu = 3$. Four of the 12 runs go extinct by the 10th generation.

Z_0	Z_1	Z_2	Z_3	Z_4	Z_5	Z_6	Z_7	Z_8	Z_9	Z_{10}
1	4	25	97	394	1160	3475	10685	31885	95757	287130
1	1	1	3	7	11	47	165	515	1525	4689
1	10	37	115	350	1124	3455	10073	29896	88863	267386
1	1	1	2	2	3	0	0	0	0	0
1	0	0	0	0	0	0	0	0	0	0
1	8	31	71	248	779	2282	6864	19895	59196	178171
1	0	0	0	0	0	0	0	0	0	0
1	4	7	13	34	106	380	1123	3385	10200	30090
1	16	49	163	447	1284	3794	11592	34626	104390	312704
1	1	1	5	16	51	155	559	1730	5378	15647
1	1	3	31	79	267	883	2637	8043	23970	71841
1	0	0	0	0	0	0	0	0	0	0

Example 4.9 (Lotka's estimate of the extinction probability) One of the earliest applications of branching processes is contained in the work of Alfred Lotka, considered the father of demographic analysis, who estimated the probability that a male line of descent would ultimately become extinct. Based on the 1920 census data, Lotka fitted the distribution of male offspring to a *zero-adjusted* geometric distribution of the form

$$a_0 = 0.48235 \quad \text{and} \quad a_k = (0.2126)(0.5893)^{k-1}, \quad \text{for } k \geq 1.$$

The generating function of the offspring distribution is

$$G(s) = 0.48235 + 0.2126 \sum_{k=1}^{\infty} (0.5893)^{k-1} s^k = 0.48235 + \frac{(0.2126)s}{1 - (0.5893)s}.$$

Lotka found the extinction probability as the numerical solution to $G(s) = s$, giving the value $e = 0.819$.

The mean of the male offspring distribution is $\mu = 1.26$. It is interesting that despite a mean number of children (sons and daughters) per individual of about 2.5, the probability of extinction of family surnames is over 80% See Lotka (1931) and Hull (2001). ■

Example 4.10 A *worm* is a self-replicating computer virus, which exploits computer network security vulnerabilities to spread itself. The *Love Letter* was a famous worm, which attacked tens of millions of Windows computers in 2000. It was spread as an attachment to an email message with the subject line ILOVEYOU. When users clicked on the attachment the worm automatically downloaded onto their machines.

Sellke et al. (2008) modeled the spread of computer worms as a branching process. The worms they analyzed are spread by randomly scanning from the 2^{32} current IP addresses to find a vulnerable host. Let V denote the total number of vulnerable hosts. Then, $p = V/2^{32}$ is the probability of finding a vulnerable host in one scan. If a worm scans at most M hosts, then infected hosts represent the individuals of a branching process whose offspring distribution is binomial with parameters M and p. Since the mean of the offspring distribution is Mp, it follows that the spread of the worm will eventually die out, with probability 1, if $Mp \leq 1$, or $M \leq 1/p$.

Typically, M is large and p is small and thus the offspring binomial distribution is well approximated by a Poisson distribution with parameter $\lambda = Mp$. The total number of infected hosts T before the worm eventually dies out is the *total progeny* of the branching process. See Exercises 4.24 and 4.25, where the mean and variance of the total progeny of a branching process are derived. If the worm starts out with I infected hosts then the mean and variance of the total number of infected hosts before the virus dies out is

$$E(T) = \frac{I}{1-\lambda} \quad \text{and} \quad Var(T) = \frac{I}{(1-\lambda)^3}.$$

■

Proof of Extinction Probability Theorem 4.2

Proof. Let $e_n = P(Z_n = 0)$ denote the probability that the population goes extinct by generation n. We have

$$\begin{aligned} e_n &= P(Z_n = 0) = G_n(0) = G(G_{n-1}(0)) \\ &= G(P(Z_{n-1} = 0)) = G(e_{n-1}), \end{aligned} \tag{4.5}$$

for $n \geq 1$. From Equation (4.2), $e_n \to e$, as $n \to \infty$. Taking limits on both sides of Equation (4.5), as $n \to \infty$, and using the fact that the probability generating function is continuous, gives $e = G(e)$. Thus, e is a root of the equation $s = G(s)$.

Let x be a positive solution of $s = G(s)$. We need to show that $e \leq x$. Since $G(s) = \sum_k s^k P(X = k)$ is an increasing function on $(0, 1]$, and $0 < x$,

$$e_1 = P(Z_1 = 0) = G_1(0) = G(0) \leq G(x) = x.$$

By induction, assuming $e_k \leq x$, for all $k < n$,

$$e_n = P(Z_n = 0) = G_n(0) = G(G_{n-1}(0)) = G(e_{n-1}) \leq G(x) = x.$$

Taking limits as $n \to \infty$, gives $e \leq x$. This proves the first part of the theorem.

The remainder of the theorem is essentially revealed by Figure 4.2. Consider the intersection of the graph of $y = G(s)$ with the line $y = s$ on the interval $[0, 1]$. Observe

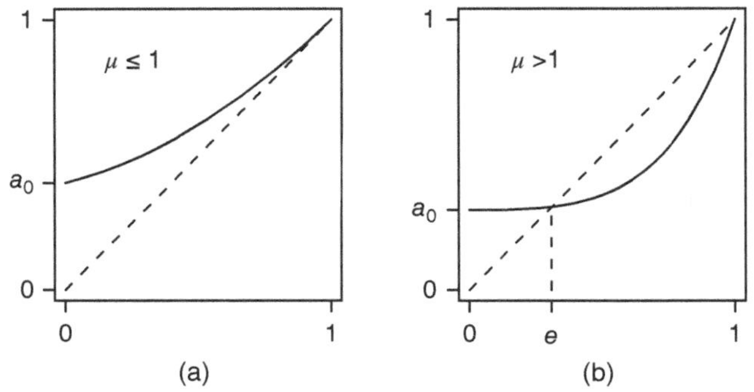

Figure 4.2 Graph of $G(s)$.

that $G(0) = a_0$ and $G(1) = 1$. Furthermore, the continuous and differentiable function G is convex (concave up) as

$$G''(s) = \sum_{k=2}^{\infty} k(k-1)s^{k-2}P(X = k) > 0.$$

It follows that the graph of $y = G(s)$ can intersect the line $y = s$ at either one or two points.

What distinguishes the two cases is the derivative of $G(s)$ at $s = 1$. Recall that $G'(1) = \mu$.

(i) If $\mu = G'(1) \le 1$, we are in the setting of Figure 4.2(b). Since

$$G'(s) = \sum_{k=1}^{\infty} ks^{k-1}P(X = k) > 0$$

is a strictly increasing function of s, we have $G'(s) < G'(1) = 1$, for $0 < s < 1$. Let $h(s) = s - G(s)$. Then, $h'(s) = 1 - G'(s) > 0$, for $0 < s < 1$. Since h is increasing and $h(1) = 0$, it follows that $h(s) < 0$, for $0 < s < 1$. That is, $s < G(s)$. Hence, the graph of $G(s)$ lies above the line $y = s$, for $0 < s < 1$, and $s = 1$ is the only point of intersection. Hence, the extinction probability is $e = 1$.

(ii) If $\mu = G'(1) > 1$, we are in the setting of Figure 4.2(b). Here,

$$h(0) = 0 - G(0) = -a_0 < 0.$$

Also, $h'(1) = 1 - G'(1) = 1 - \mu < 0$, thus $h(s)$ is decreasing at $s = 1$. Since $h(1) = 0$, there is some $0 < t < 1$ such that $h(t) > 0$. It follows that there is a number e between 0 and 1 such that $h(e) = 0$. That is, $e = G(e)$. This is the desired extinction probability. ■

EXERCISES

4.1 Consider a branching process with offspring distribution $\boldsymbol{a} = (a, b, c)$, where $a + b + c = 1$. Let $\boldsymbol{P}$ be the Markov transition matrix. Exhibit the first three rows of $\boldsymbol{P}$. That is, find P_{ij} for $i = 0, 1, 2$ and $j = 0, 1, \ldots$

4.2 Find the probability generating function of a Poisson random variable with parameter λ. Use the pgf to find the mean and variance of the Poisson distribution.

4.3 Let $X \sim \text{Poisson}(\lambda)$ and $Y \sim \text{Poisson}(\mu)$. Assume that X and Y are independent. Use probability generating functions to find the distribution of $X + Y$.

4.4 If X is a negative binomial distribution with parameters r and p, then X can be written as the sum of r i.i.d. geometric random variables with parameter p. Use this fact to find the pgf of X. Then, use the pgf to find the mean and variance of the negative binomial distribution.

4.5 The *kth factorial moment* of a random variable X is

$$E(X(X-1)\cdots(X-k+1)) = E\left(\frac{X!}{(X-k)!}\right), \quad \text{for } k \geq 0.$$

(a) Given the probability generating function G of X, show how to find the kth factorial moment of X.

(b) Find the kth factorial moment of a binomial random variable with parameters n and p.

4.6 Let $X_1, X_2, \ldots$ be a sequence of i.i.d. Bernoulli random variables with parameter p. Let N be a Poisson random variable with parameter λ, which is independent of the X_i.

(a) Find the probability generating function of $Z = \sum_{i=1}^{N} X_i$.

(b) Use (a) to identify the probability distribution of Z.

4.7 Give the probability generating function for an offspring distribution in which an individual either dies, with probability $1 - p$, or gives birth to three children, with probability p, Also find the mean and variance of the number of children in the fourth generation.

4.8 If X is a discrete random variable with generating function G. Show that

$$P(X \text{ is even}) = \frac{1 + G(-1)}{2}.$$

4.9 Let $Z_0, Z_1, \ldots$ be a branching process whose offspring distribution mean is μ. Let $Y_n = Z_n/\mu^n$, for $n \neq 0$. Show that $E(Y_{n+1}|Y_n) = Y_n$.

4.10 Show by induction that for $\mu \neq 1$,

$$Var(Z_n) = \sigma^2 \mu^{n-1} \frac{\mu^n - 1}{\mu - 1}.$$

4.11 Use the generating function representation of Z_n in Equation (4.4) to find $E(Z_n)$.

4.12 A branching process has offspring distribution $\boldsymbol{a} = (1/4, 1/4, 1/2)$. Find the following:

(a) μ.

(b) $G(s)$.

(c) The extinction probability.

(d) $G_2(s)$.

(e) $P(Z_2 = 0)$.

4.13 Use numerical methods to find the extinction probability for a branching process with Poisson offspring distribution with parameter $\lambda = 1.5$.

4.14 A branching process has offspring distribution with $a_0 = p, a_1 = 1 - p - q$, and $a_2 = q$. For what values of p and q is the process supercritical? In the supercritical case, find the extinction probability.

4.15 Assume that the offspring distribution is uniform on $\{0, 1, 2, 3, 4\}$. Find the extinction probability.

4.16 Consider a branching process where $Z_0 = k$. That is, the process starts with k individuals. Let $G(s)$ be the probability generating function of the offspring distribution. Let $G_n(s)$ be the probability generating function of Z_n for $n = 0, 1, \ldots$

(a) Find the probability generating function $G_1(s)$ in terms of $G(s)$.

(b) True or False: $G_{n+1}(s) = G_n(G(s))$, for $n = 1, 2, \ldots$

(c) True or False: $G_{n+1}(s) = G(G_n(s))$, for $n = 1, 2, \ldots$

4.17 For $0 < p < 1$, let $\boldsymbol{a} = (1 - p, 0, p)$ be the offspring distribution of a branching process. Each individual in the population can have either two or no offspring. Assume that the process starts with two individuals.

(a) Find the extinction probability.

(b) Write down the general term P_{ij} for the Markov transition matrix of the branching process.

4.18 Consider a branching process with offspring distribution

$$\boldsymbol{a} = \left(p^2, 2p(1 - p), (1 - p)^2\right), \quad \text{for } 0 < p < 1.$$

The offspring distribution is binomial with parameters 2 and $1 - p$. Find the extinction probability.

4.19 Let $T = \min\{n : Z_n = 0\}$ be the time of extinction for a branching process. Show that $P(T = n) = G_n(0) - G_{n-1}(0)$, for $n \geq 1$.

4.20 Consider the offspring distribution defined by $a_k = (1/2)^{k+1}$, for $k \geq 0$.

(a) Find the extinction probability.

(b) Show by induction that

$$G_n(s) = \frac{n - (n - 1)s}{n + 1 - ns}.$$

(c) See Exercise 4.19. Find the distribution of the time of extinction.

4.21 The *linear fractional case* is one of the few branching process examples in which the generating function $G_n(s)$ can be explicitly computed. For $0 < p < 1$, let

$$a_0 = \frac{1-c-p}{1-p}, \quad a_k = cp^{k-1}, \quad \text{for } k = 1, 2, \ldots,$$

where $0 < c < 1 - p$ is a parameter. The offspring distribution is a geometric distribution rescaled at 0.

(a) Find μ, the mean of the offspring distribution.

(b) Assume that $\mu = 1$. Show, by induction, that

$$G_n(s) = \frac{np - (np + p - 1)s}{1 - p + np - nps}.$$

(c) For $\mu > 1$, Athreya and Ney (1972) show

$$G_n(s) = \frac{(\mu^n e - 1)s + e(1 - \mu^n)}{(\mu^n - 1)s + e - \mu^n},$$

where $e = (1 - c - p)/(p(1 - p))$ is the extinction probability. See Example 4.9. Observe that Lotka's model falls in the linear fractional case. For Lotka's data, find the probability that a male line of descent goes extinct by the third generation.

4.22 *Linear fractional case, continued.* A rumor-spreading process evolves as follows. At time 0, one person has heard a rumor. At each discrete unit of time every person who has heard the rumor decides how many people to tell according to the following mechanism. Each person flips a fair coin. If heads, they tell no one. If tails, they proceed to roll a fair die until 5 appears. The number of rolls needed determines how many people they will tell the rumor.

(a) After four *generations*, how many people, on average, have heard the rumor?

(b) Find the probability that the rumor-spreading process will stop after four generations.

(c) Find the probability that the rumor-spreading process will eventually stop.

4.23 Let $\boldsymbol{a}$ be an offspring distribution with generating function G. Let X be a random variable with distribution $\boldsymbol{a}$. Let Z be a random variable whose distribution is that of X conditional on $X > 0$. That is, $P(Z = k) = P(X = k | X > 0)$. Find the generating function of Z in terms of G.

4.24 Let $T_n = Z_0 + Z_1 + \cdots + Z_n$ be the total number of individuals up through generation n. Let $T = \lim_{n\to\infty} T_n$ be the *total progeny* of the branching process. Find $E(T)$ for the subcritical, critical, and supercritical cases.

4.25 *Total progeny, continued.* Let $\phi_n(s) = E\left(s^{T_n}\right)$ be the probability generating function of T_n, as defined in Exercise 4.24.

(a) Show that ϕ_n satisfies the recurrence relation

$$\phi_n(s) = sG(\phi_{n-1}(s)), \quad \text{for } n = 1, 2, \dots ,$$

where $G(s)$ is the pgf of the offspring distribution. Hint: Condition on Z_1 and use Exercise 4.16(a).

(b) From (a), argue that

$$\phi(s) = sG(\phi(s)),$$

where $\phi(s)$ is the pgf of the total progeny T.

(c) Use (b) to find the mean of T in the subcritical case.

4.26 In a lottery game, three winning numbers are chosen uniformly at random from $\{1, \dots , 100\}$, sampling without replacement. Lottery tickets cost \$1 and allow a player to pick three numbers. If a player matches the three winning numbers they win the jackpot prize of \$1,000. For matching exactly two numbers, they win \$15. For matching exactly one number they win \$3.

(a) Find the distribution of net winnings for a random lottery ticket. Show that the expected value of the game is −70.8 cents.

(b) *Parlaying* bets in a lottery game occurs when the winnings on a lottery ticket are used to buy tickets for future games. Hoppe (2007) analyzes the effect of parlaying bets on several lottery games. Assume that if a player matches either one or two numbers they parlay their bets, buying respectively 3 or 15 tickets for the next game. The number of tickets obtained by parlaying can be considered a branching process. Find the mean of the offspring distribution and show that the process is subcritical.

(c) See Exercise 4.19. Let T denote the duration of the process, that is, the length of the parlay. Find $P(T = k)$, for $k = 1, \dots , 4$.

(d) Hoppe shows that the probability that a single parlayed ticket will ultimately win the jackpot is approximately $p/(1 - m)$, where p is the probability that a single ticket wins the jackpot, and m is the mean of the offspring distribution of the associated branching process. Find this probability and show that the parlaying strategy increases the probability that a ticket will ultimately win the jackpot by slightly over 40%.

4.27 Consider a branching process whose offspring distribution is Bernoulli with parameter p.

(a) Find the probability generating function for the nth generation size Z_n. Describe the distribution of Z_n.

(b) For $p = 0.9$, find the extinction probability and the expectation of total progeny.

4.28 In a *branching process with immigration*, a random number of immigrants W_n is independently added to the population at the nth generation.

(a) Let H_n be the probability generating function of W_n. If G_n is the generating function of the size of the nth generation, show that

$$G_n(s) = G_{n-1}(G(s))H_n(s).$$

(b) Assume that the offspring distribution is Bernoulli with parameter p, and the immigration distribution is Poisson with parameter λ. Find the generating function $G_n(s)$, and show that

$$\lim_{n\to\infty} G_n(s) = e^{-\lambda(1-s)/(1-p)}.$$

What can you conclude about the limiting distribution of generation size?

4.29 R: Examine the proof of Theorem 4.2 and observe that

$$e_n = G\left(e_{n-1}\right), \quad \text{for } n \geq 1, \tag{4.6}$$

where $e_n = P(Z_n = 0)$ is the probability that the population goes extinct by generation n. Since $e_n \to e$, as $n \to \infty$, Equation (4.6) is the basis for a numerical, recursive method to approximate the extinction probability in the supercritical case. To find e:

1. Initialize with $e_0 \in (0, 1)$.
2. Successively compute $e_n = G(e_{n-1})$, for $n \geq 1$.
3. Set $e = e_n$, for large n.

Convergence can be shown to be exponentially fast, so that n can often be taken to be relatively small (e.g., $n \approx$ 10-20). Use this numerical method to find the extinction probability for the following cases.

(a) $a_0 = 0.8, a_4 = 0.1, a_9 = 0.1$.

(b) Offspring distribution is uniform on $\{0, 1, \ldots, 10\}$.

(c) $a_0 = 0.6, a_3 = 0.2, a_6 = 0.1, a_{12} = 0.1$.

4.30 R: Simulate the branching process in Exercise 4.12. Use your simulation to estimate the extinction probability e.

4.31 R: Simulating a branching process whose offspring distribution is uniformly distributed on $\{0, 1, 2, 3, 4\}$.

(a) Use your simulation to estimate the probability that the process goes extinct by the third generation. Compare with the exact result obtained by numerical methods.

(b) See Exercise 4.15. Use your simulation to estimate the extinction probability e. Assume that if the process goes extinct it will do so by the 10th generation with high probability.

4.32 R: Simulate the branching process with immigration in Exercise 4.28(b), with $p = 3/4$ and $\lambda = 1.2$. Illustrate the limit result in Exercise 4.28(c) with $n = 100$.

4.33 R: Simulate the total progeny for a branching process whose offspring distribution is Poisson with parameter $\lambda = 0.60$. Estimate the mean and variance of the total progeny distribution.

4.34 R: Based on the numerical algorithm in Exercise 4.29, write an R function `extinct(offspring)` to find the extinction probability for any branching process with a finite offspring distribution.

5

MARKOV CHAIN MONTE CARLO

An algorithm must be seen to be believed.

—Donald Knuth

5.1 INTRODUCTION

How to simulate from complex and high-dimensional probability distributions is a fundamental problem in science, statistics, and numerous applied fields. Markov chain Monte Carlo (MCMC) is a remarkable methodology, which utilizes Markov sequences to effectively simulate from what would otherwise be intractable distributions. MCMC has been described as a "revolution" in applied mathematics, a "paradigm shift" for the field of statistics, and one of the "ten most important algorithms" of the 20th century. See Diaconis (2009), Robert and Casella (2011), and Dongarra and Sullivan (2000).

Given a probability distribution $\boldsymbol{\pi}$, the goal of MCMC is to simulate a random variable X whose distribution is $\boldsymbol{\pi}$. The distribution may be continuous or discrete, although we assume at the beginning that $\boldsymbol{\pi}$ is discrete. Often, one wants to estimate an expectation or other function of a joint distribution from a high-dimensional space.

The MCMC algorithm constructs an ergodic Markov chain whose limiting distribution is the desired $\boldsymbol{\pi}$. One then runs the chain long enough for the chain to converge,

Introduction to Stochastic Processes with R, First Edition. Robert P. Dobrow.

or nearly converge, to its limiting distribution, and outputs the final element or elements of the Markov sequence as a sample from $\boldsymbol{\pi}$.

MCMC relies on the fact that the limiting properties of ergodic Markov chains have some similarities to independent and identically distributed sequences. In particular, the strong law of large numbers holds.

Law of Large Numbers

The law of large numbers is one of the fundamental limit theorems of probability. If $Y_1, Y_2, \ldots$ is an i.i.d. sequence with common mean $\mu < \infty$, then the strong law of large numbers says that, with probability 1,

$$\lim_{n\to\infty} \frac{Y_1 + \cdots + Y_n}{n} = \mu.$$

Equivalently, let Y be a random variable with the same distribution as the Y_i and assume that r is a bounded, real-valued function. Then, $r(Y_1), r(Y_2), \ldots$ is also an i.i.d. sequence with finite mean, and, with probability 1,

$$\lim_{n\to\infty} \frac{r(Y_1) + \cdots + r(Y_n)}{n} = E(r(Y)).$$

Remarkably, the i.i.d. assumption of the strong law can be significantly weakened.

Strong Law of Large Numbers for Markov Chains

Theorem 5.1. *Assume that $X_0, X_1, \ldots$ is an ergodic Markov chain with stationary distribution $\boldsymbol{\pi}$. Let r be a bounded, real-valued function. Let X be a random variable with distribution $\boldsymbol{\pi}$. Then, with probability 1,*

$$\lim_{n\to\infty} \frac{r(X_1) + \cdots + r(X_n)}{n} = E(r(X)),$$

where $E(r(X)) = \sum_j r(j)\pi_j$.

Although Markov chains are not independent sequences, the theorem is a consequence of the fact that, for ergodic chains, successive excursions between visits to the same state are independent. A proof of the strong law of large numbers for Markov chains may be found in Norris (1998).

Given an ergodic Markov chain with stationary distribution $\boldsymbol{\pi}$, let A be a nonempty subset of the state space. Write $\pi_A = \sum_{j\in A}\pi_j$. From Theorem 5.1, we can interpret π_A

as the long-term average number of visits of an ergodic Markov chain to A. Define the indicator variable

$$I_A(x) = \begin{cases} 1, & \text{if } x \in A, \\ 0, & \text{if } x \notin A. \end{cases}$$

Then, $\sum_{k=0}^{n-1} I_A(X_k)$ is the number of visits to A in the first n steps of the chain. Let X be a random variable with distribution $\boldsymbol{\pi}$. With probability 1,

$$\lim_{n\to\infty} \frac{1}{n} \sum_{k=0}^{n-1} I_A(X_k) = E\big(I_A(X)\big) = P(X \in A) = \pi_A.$$

One setting where the strong law for Markov chains arises is when there is a reward, or cost, function associated with the states of the chain.

Example 5.1 Bob's daily lunch choices at the cafeteria are described by a Markov chain with transition matrix

$$\boldsymbol{P} = \begin{array}{l} \\ \text{Yogurt} \\ \text{Salad} \\ \text{Hamburger} \\ \text{Pizza} \end{array} \begin{array}{c} \begin{array}{cccc} \text{Yogurt} & \text{Salad} & \text{Hamburger} & \text{Pizza} \end{array} \\ \begin{pmatrix} 0 & 0 & 1/2 & 1/2 \\ 1/4 & 1/4 & 1/4 & 1/4 \\ 1/4 & 0 & 1/4 & 1/2 \\ 1/4 & 0 & 1/4 & 1/2 \end{pmatrix} \end{array}.$$

Yogurt costs \$3.00, hamburgers cost \$7.00, and salad and pizza cost \$4.00 each. Over the long term, how much, on average, does Bob spend for lunch?

Solution Let

$$r(x) = \begin{cases} 3, & \text{if } x = \text{yogurt}, \\ 4, & \text{if } x = \text{salad or pizza}, \\ 7, & \text{if } x = \text{hamburger}. \end{cases}$$

The lunch chain is ergodic with stationary distribution

Yogurt	Salad	Hamburger	Pizza
7/65	2/13	18/65	6/13

With probability 1, Bob's average lunch cost converges to

$$\sum_x r(x)\pi_x = 3\left(\frac{7}{65}\right) + 4\left(\frac{2}{13} + \frac{6}{13}\right) + 7\left(\frac{18}{65}\right) = \$4.72 \text{ per day}.$$

■

Armed with the strong law of large numbers for Markov chains, we now describe the details of MCMC. As is often the case, the knowing is in the doing, so we start with an expanded example.

Binary Sequences with No Adjacent 1s

The following is a toy problem for which an exact analysis is possible. Consider sequences of length m consisting of 0s and 1s. Call a sequence *good* if it has no adjacent 1s. What is the expected number of 1s in a good sequence if all good sequences are equal likely?

For $m = 4$, there are $2^4 = 16$ binary sequences of 0s and 1s. The eight good sequences are

$$(0000), (1000), (0100), (0010), (0001), (1010), (1001), (0101),$$

and the desired expectation is

$$\frac{1}{8}(0 + 1 + 1 + 1 + 1 + 2 + 2 + 2) = \frac{10}{8} = 1.25.$$

For general m, the expected number of 1s is $\mu = \sum_k k\pi_k$, where π_k is the probability that a good sequence of length m has exactly k 1s. While π_k can be derived by a combinatorial argument (see Exercise 5.4), there is no simple closed form expression for μ. So we approach the problem using simulation.

If x is a sequence of 0s and 1s of length m, let $r(x)$ be the number of 1s in the sequence. Assume that we are able to generate a uniformly random sequence of length m with no adjacent 1s. Doing this repeatedly and independently, generating good sequences $Y_1, Y_2, \ldots, Y_n$, a Monte Carlo estimate of the desired expectation is

$$\mu \approx \frac{r(Y_1) + r(Y_2) + \cdots + r(Y_n)}{n},$$

for large n. This is by the (regular) law of large numbers.

Thus, the problem of estimating μ reduces itself to the problem of simulating a good sequence.

Here is one way to generate such a sequence, which does not use Markov chains, based on the *rejection method*. Generate a sequence of 0s and 1s with no constraints. This is easy to do by just flipping m coins, where heads represents 1 and tails represents 0. If the resulting sequence is good, then keep it. If the sequence has adjacent 1s then reject it, and try again.

The approach works in principle, but not in practice. Even for moderate m, most binary sequences will have some adjacent 1s, and thus be rejected. For instance, for $m = 100$, there are $2^{100} \approx 10^{30}$ binary sequences, of which *only* 10^{21} are good. The probability that a binary sequence is good is about $10^{21}/10^{30} = 10^{-9}$. Thus, the rejection algorithm will typically see about one billion sequences before one good sequence appears. This method is prohibitively slow.

Here is an alternate approach using Markov chains. The idea is to construct an ergodic Markov chain $X_0, X_1, \ldots$ whose state space is the set of good sequences and whose limiting distribution is uniform on the set of good sequences. The Markov chain is then generated and, as in the i.i.d. case, we take

$$\mu \approx \frac{r(X_1) + r(X_2) + \cdots + r(X_n)}{n},$$

for large n, as an MCMC estimate for the desired expectation.

The Markov chain is constructed as a random walk on a graph whose vertices are good sequences. The random walk proceeds as follows. From a given good sequence, pick one of its m components uniformly at random.

If the component is 1, then switch it to 0. Since the number of 1s is reduced, the resulting sequence is good. The walk moves to the new sequence.

If the component is 0, then switch it to 1 only if the resulting sequence is good. In that case, the walk moves to the new sequence. However, if the switch to 1 would result in a bad sequence, then stay put. The walk stays at the current state.

The walk can be described as a random walk on a weighted graph with weights assigned as follows. Two good sequences that differ in only one component are joined by an edge of weight 1. Also, each sequence possibly has a loop. The weight of the loop is such that the sum of the weights of all the edges incident to the sequence is equal to m.

Since all vertices have the same total weight, the stationary distribution is uniform. See Figure 5.1 for the weighted graph construction for $m = 4$.

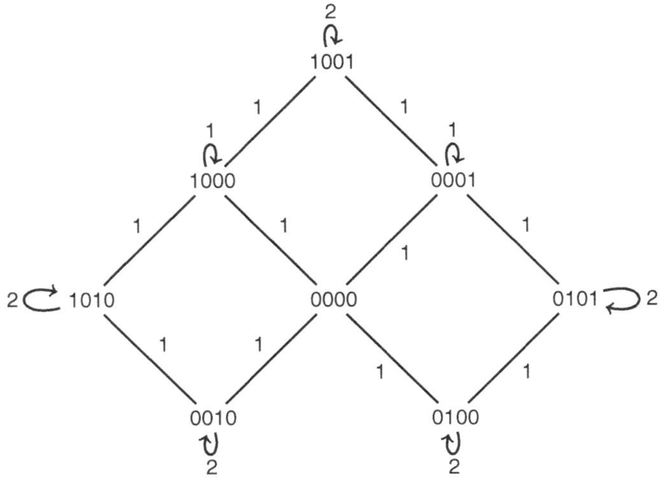

Figure 5.1 Weighted graph on sequences with no adjacent 1s.

The MCMC random walk algorithm was implemented for sequences of length $m = 100$. The chain was run for 100,000 steps (taking 2 seconds on a laptop computer), starting with the sequence of all 0s as the initial state of the chain. The MCMC estimate for the expected number of 1s is $\mu \approx 27.833$. An exact analysis gives $\mu = 27.7921$.

R : Expected Number of 1s in Good Sequences

```
> # adjacent.R
> # init: initial sequence
> # n: number of steps to run the chain
> adjacent <- function(init, n)
+ { k <- length(init)
+  tot <- 0  # total number of 1s
+  new <-c(2, init,2)  # pad sequence at the ends
+  for (i in 1:n) {
+  index <- 1 +sample(1:k,1)
+  newbit <- 0 + !new[index]   # flip the bit
+  if (newbit==0) {
+    new[index] <- 0
+    tot <- tot+sum(new)
+    next} else {
+         if (new[index-1]==1 | new[index+1] ==1) {
+         tot <-tot + sum(new)
+         next}
+         else {new[index] <- 1}
+         tot <- tot + sum(new)}
+ }
+ tot/n- 4  }  # subtract both endpoints
> m <- 100
> init <- rep(0,m)  # Start at sequence of all 0s
> adjacent(init,100000)
[1] 27.83333
```

Remarks:

1. The number of vertices of the random walk graph is huge—about 10^{21} good sequences for $m = 100$. However, the number of steps required for the walk to get sufficiently close to the limiting distribution is a small fraction of that—in this implementation 100,000 steps.
2. The uniform distribution on the set of good sequences assigns probability $1/c$ to each sequence, where c is the number of good sequences. The actual value of c was never needed in the implementation, and for all practical purposes could have been unknown.
3. A benefit of the algorithm that cannot be overstated is that it is easily and efficiently coded and is intuitive based on the structure of the problem.

5.2 METROPOLIS–HASTINGS ALGORITHM

The most common Markov chain Monte Carlo method is the Metropolis–Hastings algorithm. The algorithm is named after Nicholas Metropolis, a physicist who led the research group at Los Alamos National Laboratory, which first proposed the method in the early 1950s, and W.K. Hastings, a Canadian statistician, who extended the scope of the algorithm in 1970.

Let $\boldsymbol{\pi} = (\pi_1, \pi_2, \dots)$ be a discrete probability distribution. The algorithm constructs a reversible Markov chain $X_0, X_1, \dots$ whose stationary distribution is $\boldsymbol{\pi}$.

Let $\boldsymbol{T}$ be a transition matrix for *any* irreducible Markov chain with the same state space as $\boldsymbol{\pi}$. It is assumed that the user knows how to sample from $\boldsymbol{T}$. The $\boldsymbol{T}$ chain will be used as a *proposal* chain, generating elements of a sequence that the algorithm decides whether or not to *accept*.

To describe the transition mechanism for $X_0, X_1, \dots$, assume that at time n the chain is at state i. The next step of the chain X_{n+1} is determined by a two-step procedure.

1. Choose a new state according to $\boldsymbol{T}$. That is, choose j with probability T_{ij}. State j is called the *proposal state*.
2. Decide whether to accept j or not. Let

$$a(i,j) = \frac{\pi_j T_{ji}}{\pi_i T_{ij}}.$$

The function a is called the *acceptance function*. If $a(i,j) \geq 1$, then j is accepted as the next state of the chain. If $a(i,j) < 1$, then j is accepted with probability $a(i,j)$. If j is not accepted, then i is kept as the next step of the chain.

In other words, assume that $X_n = i$. Let U be uniformly distributed on (0, 1). Set

$$X_{n+1} = \begin{cases} j, & \text{if } U \leq a(i,j), \\ i, & \text{if } U > a(i,j). \end{cases}$$

Metropolis–Hastings Algorithm

The sequence $X_0, X_1, \dots$ constructed by the Metropolis–Hastings algorithm is a reversible Markov chain whose stationary distribution is $\boldsymbol{\pi}$.

Proof. It should be clear that $X_0, X_1, \dots$ is in fact a Markov chain, as each X_{n+1}, given the past history $X_0, \dots, X_n$, only depends on X_n. Let $\boldsymbol{P}$ be the transition matrix of the chain. The theorem will be proven by showing that the detailed balance equations $\pi_i P_{ij} = \pi_j P_{ji}$ are satisfied.

For $i \neq j$, consider $P_{ij} = P(X_1 = j | X_0 = i)$. Given $X_0 = i$, then $X_1 = j$ if and only if (i) j is proposed, and (ii) j is accepted. The first occurs with probability T_{ij}.

The second occurs if $U \le a(i,j)$, where U is uniform on (0, 1). We have that

$$P(U \le a(i,j)) = \begin{cases} a(i,j), & \text{if } a(i,j) \le 1, \\ 1, & \text{if } a(i,j) > 1, \end{cases}$$

$$= \begin{cases} a(i,j), & \text{if } \pi_j T_{ji} \le \pi_i T_{ij}, \\ 1, & \text{if } \pi_j T_{ji} > \pi_i T_{ij}. \end{cases}$$

Thus, for $i \ne j$,

$$P_{ij} = \begin{cases} T_{ij} a(i,j), & \text{if } \pi_j T_{ji} \le \pi_i T_{ij}, \\ T_{ij}, & \text{if } \pi_j T_{ji} > \pi_i T_{ij}. \end{cases}$$

The diagonal entries of $\boldsymbol{P}$ are not needed for the proof. They are determined by the fact that the rows of $\boldsymbol{P}$ sum to 1.

To show the detailed balance equations are satisfied, assume that $\pi_j T_{ji} \le \pi_i T_{ij}$. Then,

$$\pi_i P_{ij} = \pi_i T_{ij} a(i,j) = \pi_i T_{ij} \frac{\pi_j T_{ji}}{\pi_i T_{ij}} = \pi_j T_{ji} = \pi_j P_{ji}.$$

Similarly, if $\pi_j T_{ji} > \pi_i T_{ij}$,

$$\pi_i P_{ij} = \pi_i T_{ij} = \pi_j T_{ji} \frac{\pi_i T_{ij}}{\pi_j T_{ji}} = \pi_j T_{ji} a(j,i) = \pi_j P_{ji}.$$

■

Remarks:

1. The exact form of $\boldsymbol{\pi}$ is not necessary to implement Metropolis–Hastings. The algorithm only uses ratios of the form π_j/π_i. Thus, $\boldsymbol{\pi}$ needs only to be specified up to proportionality. For instance, if $\boldsymbol{\pi}$ is uniform on a set of size c, then $\pi_j/\pi_i = 1$, and the acceptance function reduces to $a(i,j) = T_{ji}/T_{ij}$.
2. If the proposal transition matrix $\boldsymbol{T}$ is symmetric, then $a(i,j) = \pi_j/\pi_i$.
3. The algorithm works for *any* irreducible proposal chain. Thus, the user has wide latitude to find a proposal chain that is efficient in the context of their problem.
4. If the proposal chain is ergodic (irreducible and aperiodic in the finite case) then the resulting Metropolis–Hastings chain is also ergodic with limiting distribution $\boldsymbol{\pi}$.
5. The generated sequence $X_0, X_1, \ldots, X_n$ gives an approximate sample from $\boldsymbol{\pi}$. However, if the chain requires many steps to get close to stationarity, there may be initial bias. *Burn-in* refers to the practice of discarding the initial iterations and retaining $X_m, X_{m+1}, \ldots, X_n$, for some m. In that case, the strong law of large numbers for Markov chains gives

$$\lim_{n\to\infty} \frac{r(X_m) + \cdots + r(X_n)}{n - m + 1} = \sum_x r(x)\pi_x.$$

Example 5.2 Power-law distributions are positive probability distributions of the form $\pi_i \propto i^s$, for some constant S. Unlike distributions with exponentially decaying tails (e.g., Poisson, geometric, exponential, normal), power-law distributions have *fat tails*, and thus are often used to model skewed data. Let

$$\pi_i = \frac{i^{-3/2}}{\sum_{k=1}^{\infty} k^{-3/2}}, \text{ for } i = 1, 2, \ldots$$

Implement a Metropolis–Hastings algorithm to simulate from $\boldsymbol{\pi}$.

Solution For a proposal distribution, we use simple symmetric random walk on the positive integers with reflecting boundary at 1. From 1, the walk always moves to 2. Otherwise, the walk moves left or right with probability 1/2. The proposal chain transition matrix is

$$T_{ij} = \begin{cases} 1/2, & \text{if } j = i \pm 1 \text{ for } i > 1, \\ 1, & \text{if } i = 1 \text{ and } j = 2, \\ 0 & \text{otherwise.} \end{cases}$$

The acceptance function is

$$a(i, i+1) = \left(\frac{i}{i+1}\right)^{3/2}, \text{ and } \quad a(i+1, i) = \left(\frac{i+1}{i}\right)^{3/2}, \text{ for } i \geq 2,$$

with

$$a(1,2) = \frac{\pi_2 T_{21}}{\pi_1 T_{12}} = \left(\frac{1}{2}\right)^{3/2} \frac{1}{2} = \left(\frac{1}{2}\right)^{5/2} \quad \text{and} \quad a(2,1) = 2^{5/2}.$$

Observe that $a(i+1, i) \geq 1$, for all i.

The Metropolis–Hastings algorithm can be described as follows. From state $i \geq 2$, flip a fair coin. If heads, go to $i-1$. If tails, set $p = (i/(i+1))^{3/2}$. Flip another coin whose heads probability is p. If heads, go to $i+1$. If tails, stay at i. If the chain is at 1, then move to 2 with probability $(1/2)^{5/2} \approx 0.177$. Otherwise, stay at 1.

The chain was run for one million steps. See Table 5.1 to compare the simulated and exact probabilities.

TABLE 5.1 Comparison of Markov chain Monte Carlo Estimates with Exact Probabilities for Power-Law Distribution

i	1	2	3	4	5	6	7	8	≥ 9
Simulation	0.389	0.137	0.075	0.048	0.034	0.026	0.021	0.017	0.252
Exact	0.383	0.135	0.074	0.048	0.034	0.026	0.021	0.017	0.262

R : Power Law Distribution

```
# powerlaw.R
> trials <- 1000000
> simlist <- numeric(trials)
> simlist[1] <- 2
> for (i in 2:trials) {
+    if (simlist[i-1] ==1) {
+    p <- (1/2)^(5/2)
+    new <- sample(c(1,2),1,prob=c(1-p,p))
+    simlist[i] <- new
+    } else { leftright <- sample(c(-1,1),1)
+    if  (leftright == -1)  {
+        simlist[i] <- simlist[i-1] - 1} else {
+    p <- (simlist[i-1]/(simlist[i-1]+1))^(3/2)
+    simlist[i] <- sample(c(simlist[i-1],
        1+simlist[i-1]), 1,prob=c(1-p,p))
+     } } }
> tab <- table(simlist)/trials
> tab[1:8]
    1     2     3     4     5     6     7     8
0.389 0.137 0.075 0.048 0.034 0.026 0.021 0.017
```

■

Example 5.3 (Cryptography)

ahicainqcaqx ic zqcqwbl bwq zwqbj xjustlicz tlhamx ic jyq kbr ho jybj albxx ho jyicmqwx kyh ybgq tqqc qnuabjqn jh mchk chjyicz ho jyq jyqhwr ho dwhtbtilijiqx jybj jyqhwr jh kyiay jyq shxj zlhwihux htpqajx ho yusbc wqxqbway bwq icnqtjqn ohw jyq shxj zlhwihux ho illuxjwbjihcx

—qnzbw bllqc dhq jyq suwnqwx ic jyq wuq shwzuq

The coded message was formed by a simple substitution cipher—each letter standing for one letter in the alphabet. The hidden message can be thought of as a coding function from the letters of the coded message to the regular alphabet. For example, given an encrypted message *xoaaoaaoggo*, the function that maps x to m, o to i, a to s, and g to p decodes the message to *mississippi*.

If one keeps tracks of all 26 letters and spaces, there are $27! \approx 10^{28}$ possible coding functions. The goal is to find the one that decodes the message.

This example is based on Diaconis (2009), who gives a compelling demonstration of the power and breadth of MCMC applied to cryptography.

My colleague Jack Goldfeather assisted in downloading the complete works of Jane Austen (about four million characters) from Project Gutenberg's online site and recording the number of transitions of consecutive text symbols. For simplicity, we ignored case and only kept track of spaces and the letters a to z. The counts are kept in a 27×27 matrix $\boldsymbol{M}$ of transitions indexed by $(a, b, \ldots, z, [\text{space}])$. For example, there are 6,669 places in Austen's work where b follows a and thus $M_{ab} = 6,669$.

The encoded message has 320 characters, denoted as $(c_1, \ldots, c_{320})$. For each coding function f, associate a score

$$\text{score}(f) = \prod_{i=1}^{319} M_{f(c_i),f(c_{i+1})}.$$

The score is a product over all successive pairs of letters in the decrypted text $(f(c_i), f(c_{i+1}))$ of the number of occurrences of that pair in the reference Austen library. The score is higher when successive pair frequencies in the decrypted message match those of the reference text. Coding functions with high scores are good candidates for decryption. The goal is to find the coding function of maximum score.

A probability distribution proportional to the scores is obtained by letting

$$\pi_f = \frac{\text{score}(f)}{\sum_g \text{score}(g)}. \tag{5.1}$$

From a Monte Carlo perspective, we want to sample from $\boldsymbol{\pi}$, with the idea that a sample is most likely to return a value of high probability. The denominator in Equation (5.1) is intractable, being the sum of 27! terms. But the beauty of Metropolis–Hastings is that the denominator is not needed since the algorithm relies on ratios of the form π_f / π_g.

The MCMC implementation runs a random walk on the set of coding functions. Given a coding function f, the transition to a proposal function f^* is made by picking two letters at random and switching the values that f assigns to these two letters. This method of *random transpositions* gives a symmetric proposal matrix T, simplifying the acceptance function

$$a(f, f^*) = \frac{\pi_{f^*} T_{f^*,f}}{\pi_f T_{f,f^*}} = \frac{\pi_{f^*}}{\pi_f} = \frac{\text{score}(f^*)}{\text{score}(f)}.$$

The algorithm is as follows:

1. Start with any coding function f. For convenience, we use the identity function.
2. Pick two letters uniformly at random and switch the values that f assigns to these two letters. Call the new proposal function f^*.
3. Compute the acceptance function $a(f, f^*) = \text{score}(f^*)/\text{score}(f)$.
4. Let U be uniformly distributed on (0, 1). If $U \leq a(f, f^*)$, accept f^*. Otherwise, stay with f.

We ran the algorithm on the coded message at the start of this example. See the script file **decode.R**. At iteration 2658, the message was decoded.

R : Decoding the Message

[100] xlitxisatxau it hatawnc nwa hwand udepocith oclxfu it dra mng ly drnd xcnuu ly dritfawu mrl rnza oaat asexndas dl ftlm tldrith ly dra dralwg ly bwlonoi-cidiau drnd dralwg dl mrixr dra plud hclwileu lojaxdu ly repnt wauanwxr nwa itsaodas ylw dra plud hclwileu ly icceudwndiltu – ashnw nccat bla dra pewsawu it dra wea plwhea

[500] gsadgaredgen ad wedeoil ioe woeit ntubcladw clsgyn ad the kif sm thit glinn sm thadyeon khs hive ceed erugiter ts ydsk dsthadw sm the thesof sm posci-calataen thit thesof ts khagh the bsnt wlsoasun sczegtn sm hubid oeneiogh ioe adrecter mso the bsnt wlsoasun sm alluntoitasdn – erwio illed pse the buoreon ad the oue bsowue

[1000] goidgimedges id wedenal ane wneat strpclidw clogus id the fak oy that glass oy thiduens fho have ceed emrgatem to udof dothidw oy the theonk oy bno-cacilities that theonk to fhigh the post wloniors ocxegts oy hrpad neseangh ane idmectem yon the post wloniors oy illrstnatiods – emwan alled boe the prnmens id the nre ponwre

[1400] goingidenges in beneral are breat stumplinb plogks in the way of that glass of thinkers who have peen edugated to know nothinb of the theory of cropapilities that theory to whigh the most blorious opxegts of human researgh are indepted for the most blorious of illustrations – edbar allen coe the murders in the rue morbue

[1600] goingidenges in ceneral are creat stumplinc plogks in the way of that glass of thinkers who have peen edugated to know nothinc of the theory of bropapilities that theory to whigh the most clorious opxegts of human researgh are indepted for the most clorious of illustrations – edcar allen boe the murders in the rue morcue

[1800] coincidences in general are great stumpling plocks in the way of that class of thinkers who have peen educated to know nothing of the theory of bropapilities that theory to which the most glorious opzects of human research are indepted for the most glorious of illustrations – edgar allen boe the murders in the rue morgue

[2658] coincidences in general are great stumbling blocks in the way of that class of thinkers who have been educated to know nothing of the theory of probabilities that theory to which the most glorious objects of human research are indebted for the most glorious of illustrations – edgar allen poe the murders in the rue morgue

■

Example 5.4 (Darwin's finches) The following example is based on Cobb and Chen (2003).

Charles Darwin, on his visit to the Galapagos Islands in 1835, discovered several species of finches, which varied from island to island. It is said that the variation and diversity of the birds he observed was a significant factor which led Darwin to develop his theory of natural selection.

Darwin chronicled the presence of 13 species of finches on 17 islands. The data are presented in the *co-occurrence matrix* in Figure 5.2. Ecologists use such species-presence matrices to study levels of competition and cooperation among species.

	Islands																	
Species	1	2	3	4	5	6	7	8	9	10	11	12	13	14	15	16	17	Total
A	1	1	1	1	1	1	1	1	1	1	1	1	1	1	1	1	1	17
B	1	1	1	1	1	1	1	1	1	1	1	0	0	1	0	1	1	14
C	1	1	1	1	1	1	1	1	1	1	1	1	1	0	1	0	0	14
D	1	1	1	1	1	1	1	1	1	1	1	1	1	0	0	0	0	13
E	1	0	1	1	1	1	1	1	1	1	1	1	1	0	0	0	0	12
F	1	1	1	1	1	1	1	1	1	1	1	0	0	0	0	0	0	11
G	1	1	1	1	1	0	1	1	0	0	0	0	0	1	0	1	1	10
H	1	1	1	1	1	1	0	1	1	1	1	0	0	0	0	0	0	10
I	1	1	1	1	1	1	1	1	1	1	0	0	0	0	0	0	0	10
J	1	1	1	1	0	1	1	0	0	0	0	0	0	0	0	0	0	6
K	0	0	0	0	0	0	0	0	0	0	0	0	0	1	1	0	0	2
L	1	1	0	0	0	0	0	0	0	0	0	0	0	0	0	0	0	2
M	0	0	0	0	1	0	0	0	0	0	0	0	0	0	0	0	0	1
Total	11	10	10	10	10	9	9	9	8	8	7	4	4	4	3	3	3	122

Figure 5.2 Co-occurrence matrix for Darwin's finches.

A species pair and an island pair constitute a *checkerboard* if each species appears on different islands. A checkerboard is a 2×2 submatrix of the form

$$\begin{pmatrix} 1 & 0 \\ 0 & 1 \end{pmatrix} \text{ or } \begin{pmatrix} 0 & 1 \\ 1 & 0 \end{pmatrix}.$$

For instance, in Figure 5.2, species E and G on islands 10 and 14, respectively, form a checkerboard.

The number of checkerboards in a co-occurrence matrix is a measure of competition. A large number of checkerboards would indicate a high degree of competition. For the finches data, there are 333 checkerboards. Is this number large or small?

To explore the question, consider the expected number of checkerboards for a typical 0–1 co-occurrence matrix, which is constrained by the row and column totals of the finches matrix.

From a theoretical point of view, the expectation is obtained by listing all such 0–1 matrices, counting the number of checkerboards in each, and taking an overall average. From a practical point of view, however, things are not so neat.

Counting 0–1 matrices with fixed marginal totals is a well-studied problem in applied mathematics. The number of matrices with the same row and column sums as the finches data are not known. Liu (2001) reports that Susan Holmes at Stanford University estimated the number as about 6.715×10^{16}. It is not computationally feasible to list all such matrices and count the checkerboards in each.

We implement a Metropolis–Hastings algorithm to generate a uniformly random co-occurrence matrix and to estimate the probability that such a matrix has at least 333 checkerboards.

Consider the following *swap* operation. Assume that $\boldsymbol{M}$ is a co-occurrence matrix with at least one checkerboard. Pick a checkerboard in the matrix and swap 0s and 1s. That is, if the chosen checkerboard submatrix is $\begin{pmatrix} 1 & 0 \\ 0 & 1 \end{pmatrix}$, then change it to $\begin{pmatrix} 0 & 1 \\ 1 & 0 \end{pmatrix}$. If the submatrix is $\begin{pmatrix} 0 & 1 \\ 1 & 0 \end{pmatrix}$, change it to $\begin{pmatrix} 1 & 0 \\ 0 & 1 \end{pmatrix}$. Let $\boldsymbol{N}$ be the matrix that results from $\boldsymbol{M}$ after the swap operation. The swap operation preserves row and columns sums. That is, $\boldsymbol{N}$ is a co-occurrence matrix with the same row and column totals as $\boldsymbol{M}$.

The swap operation is the basis of an MCMC random walk algorithm on a graph whose vertex set is the set of all co-occurrence matrices with fixed row and column sums. Matrices $\boldsymbol{M}$ and $\boldsymbol{N}$ are joined by an edge if $\boldsymbol{N}$ can be obtained from $\boldsymbol{M}$ by one checkerboard swap.

For example, there are five 3×3 co-occurrence matrices with successive row sums 2, 2, and 1, and column sums 2, 2, and 1. The corresponding graph is shown in Figure 5.3. Simple random walk on this graph has transition matrix

$$\boldsymbol{T} = \begin{array}{c} \\ A \\ B \\ C \\ D \\ E \end{array} \begin{array}{c} \begin{array}{ccccc} A & B & C & D & E \end{array} \\ \begin{pmatrix} 0 & 1/3 & 1/3 & 1/3 & 0 \\ 1/3 & 0 & 1/3 & 0 & 1/3 \\ 1/4 & 1/4 & 0 & 1/4 & 1/4 \\ 1/3 & 0 & 1/3 & 0 & 1/3 \\ 0 & 1/3 & 1/3 & 1/3 & 0. \end{pmatrix} \end{array}$$

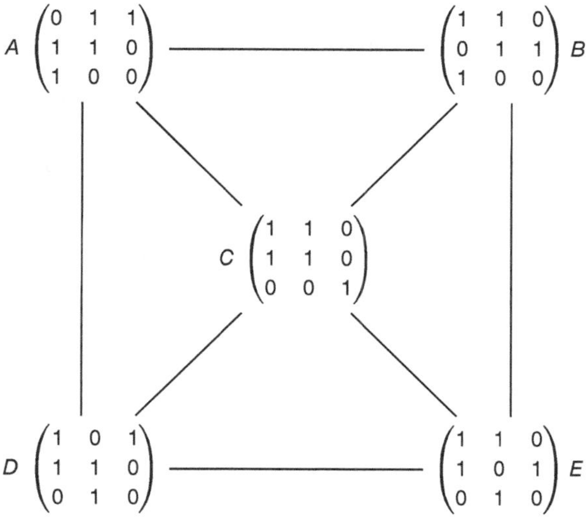

Figure 5.3 Transition graph for co-occurrence matrices.

Consider simple random walk on a general co-occurrence graph. Although it is not completely obvious, such a Markov chain is irreducible (see Ryser 1957) and has a unique stationary distribution. Unfortunately, the stationary distribution is not the uniform distribution, which is the one we want. The stationary distribution for simple random walk is proportional to the vertex degrees. For the graph in Figure 5.3, the stationary distribution $\boldsymbol{s}$ is

$$\boldsymbol{s} = (s_A, s_B, s_C, s_D, s_E) = \left(\frac{3}{16}, \frac{3}{16}, \frac{4}{16}, \frac{3}{16}, \frac{3}{16}\right).$$

Let $\boldsymbol{\pi}$ denote the uniform distribution on the set of co-occurrence matrices with fixed row and column totals. That is, $\pi_M = 1/c$, for all co-occurrence matrices $\boldsymbol{M}$, where c is the number of such matrices. For MCMC, simple random walk on the co-occurrence graph is used as the proposal distribution. Since $\pi_i/\pi_j = 1$ for all i and j, the acceptance function, for adjacent vertices i and j, is

$$a(i,j) = \frac{\pi_j T_{ji}}{\pi_i T_{ij}} = \frac{\deg(j)}{\deg(i)} = \frac{\text{Number of checkerboards in } j}{\text{Number of checkerboards in } i}.$$

The algorithm is described as follows. From a co-occurrence matrix i, pick two rows and two columns uniformly at random, until the resulting submatrix is swappable. Do the swap and let j be the resulting co-occurrence matrix. Count the number of checkerboards in i and j, and accept the new matrix according to the acceptance function $a(i,j)$.

The algorithm is implemented in **darwin.R** for 5,000 steps, obtaining a distribution of the number of checkerboards in an approximately uniform co-occurrence matrix. See Figure 5.4. Estimates for the mean and standard deviation are 246.9 and 15.4, respectively. Of 5,000 matrices, only two had 333 or more checkerboards. Thus, a *typical* co-occurrence matrix contains about 247 checkerboards give or take about 30, and the estimated probability that a random co-occurrence matrix has at least 333 checkerboards is $2/5,000 = 0.0004$. The large number of co-occurrence matrices in the data is unusual if the distribution of finches on the islands was random. An ecologist might conclude that the distribution of finch species in the Galapagos Islands shows evidence of competition.

■

Continuous State Space

Although the target distributions in the previous examples are discrete, MCMC can also be used in the continuous case, when $\boldsymbol{\pi}$ is a probability density function.

We have not yet studied continuous state space stochastic processes. Intuitively, for a continuous state space *Markov process* a transition *function* replaces the transition matrix, where P_{ij} is the value of a conditional density function given $X_0 = i$.

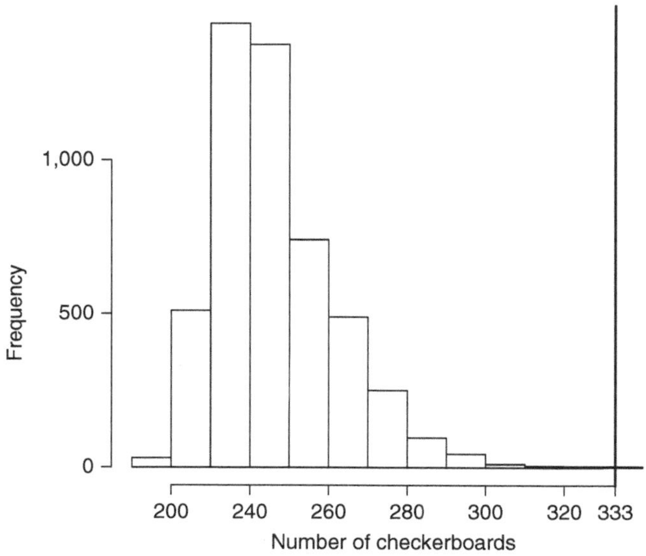

Figure 5.4 Simulated distribution of the number of checkerboards in a uniformly random co-occurrence matrix. The probability of at least 333 checkerboards is 0.0004.

The Metropolis–Hastings algorithm is essentially the same as in the discrete case, with densities replacing discrete distributions. Without delving into any new theory, we present the continuous case by example.

Example 5.5 Using only a uniform random number generator, simulate a standard normal random variable using MCMC.

Solution Write the standard normal density function as $\pi_t = e^{-t^2/2}/\sqrt{2\pi}$. For the Metropolis–Hastings proposal distribution, we choose the uniform distribution on an interval of length two centered at the current state. From state s, the proposal chain moves to t, where t is uniformly distributed on $(s-1, s+1)$. Hence, the conditional density given s is constant, with

$$T_{st} = \frac{1}{2}, \text{ for } s-1 \le t \le s+1.$$

The acceptance function is

$$a(s,t) = \frac{\pi_t T_{ts}}{\pi_s T_{st}} = \left(\frac{e^{-t^2/2}}{\sqrt{2\pi}}\right)(1/2)\Bigg/\left(\frac{e^{-s^2/2}}{\sqrt{2\pi}}\right)(1/2) = e^{-(t^2-s^2)/2}.$$

Notice that the length of the interval for the uniform proposal distribution does not affect the acceptance function.

See Figure 5.5 for the results of one million iterations of the MCMC sequence. The overlaid curve is the standard normal density function.

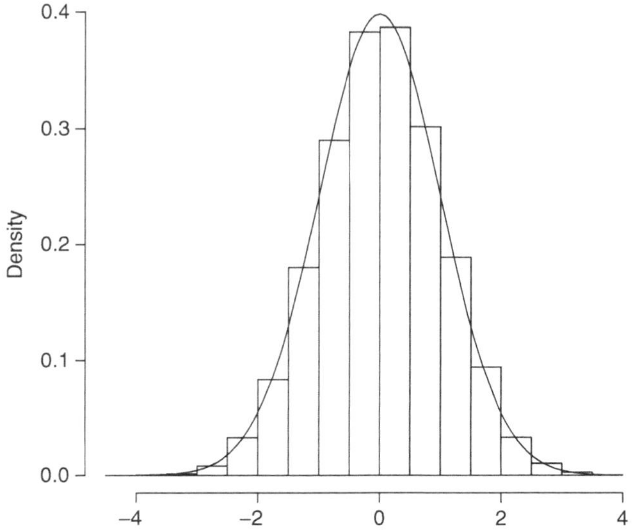

Figure 5.5 Simulation of standard normal distribution with MCMC.

R : Standard Normal Simulation

```
# standardnormal.R
> trials <- 1000000
> simlist <- numeric(trials)
> state <- 0
> for (i in 2:trials) {
+   prop <- runif(1,state-1,state+1)
+   acc <-  exp(-(prop^2 -state^2)/2)
+   if (runif(1) < acc) state <- prop
+   simlist[i] <- state  }
> hist(simlist,xlab="",main="",prob=T)
> curve(dnorm(x),-4,4,add=T)
```

■

5.3 GIBBS SAMPLER

The original Metropolis–Hastings algorithm was discovered in 1953 and motivated by problems in physics. In 1984, a landmark paper by brothers Donald and Stuart Geman showed how the algorithm could be adapted to the high-dimensional problems that arise in Bayesian statistics. Their new algorithm was named after the physicist Josiah Gibbs with reference to connections to statistical physics.

In the Gibbs sampler, the target distribution $\boldsymbol{\pi}$ is an m-dimensional joint density

$$\boldsymbol{\pi}(\boldsymbol{x}) = \pi(x_1, \ldots, x_m).$$

A multivariate Markov chain is constructed whose limiting distribution is $\boldsymbol{\pi}$, and which takes values in an m-dimensional space. The algorithm generates elements of the form

$$\begin{aligned}&\boldsymbol{X}^{(0)}, \boldsymbol{X}^{(1)}, \boldsymbol{X}^{(2)}, \ldots \\ &\quad = (X_1^{(0)}, \ldots, X_m^{(0)}), (X_1^{(1)}, \ldots, X_m^{(1)}), (X_1^{(2)}, \ldots, X_m^{(2)}), \ldots\end{aligned}$$

by iteratively updating each component of an m-dimensional vector conditional on the other $m - 1$ components. We show that the Gibbs sampler is a special case of the Metropolis–Hastings algorithm, with a particular choice of proposal distribution.

The algorithm is sometimes hard to understand because of an abundance of notation. We start off with a relatively simple two-dimensional example.

Example 5.6 Consider a bivariate standard normal distribution with correlation ρ. For background, see Appendix B, Section B.4. The bivariate normal distribution has the property that conditional distributions are normal. If (X, Y) has a bivariate standard normal distribution, then the conditional distribution of X given $Y = y$ is normal with mean ρy and variance $1 - \rho^2$. Similarly, the conditional distribution of Y given $X = x$ is normal with mean ρx and variance $1 - \rho^2$.

The Gibbs sampler is implemented to simulate (X, Y) from a bivariate standard normal distribution with correlation ρ. At each step of the algorithm, one component of a two-element vector is updated by sampling from its conditional distribution given the other component. Updates switch back and forth. The resulting sequence of bivariate samples converges to the target distribution.

1. Initialize: $(x_0, y_0) \leftarrow (0, 0)$
   $m \leftarrow 1$.
2. Generate x_m from the conditional distribution of X given $Y = y_{m-1}$. That is, simulate from a normal distribution with mean ρy_{m-1} and variance $1 - \rho^2$.
3. Generate y_m from the conditional distribution of Y given $X = x_m$. That is, simulate from a normal distribution with mean ρx_m and variance $1 - \rho^2$.
4. $m \leftarrow m + 1$.
5. Return to Step 2.

We simulated a bivariate standard normal distribution with $\rho = 0.60$ using the Gibbs sampler. The chain was run for 2,000 steps. In R, the output is a 2000×2 matrix. A scatterplot of the output is seen in Figure 5.6.

■

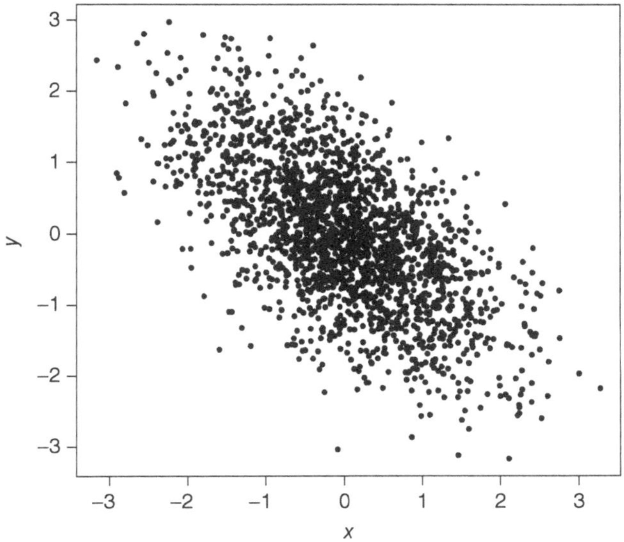

Figure 5.6 Bivariate standard normal simulation.

R : Bivariate Standard Normal

```
# bivariatenormal.R
> rho <- -0.60
> trials <- 2000
> sdev <- sqrt(1-rho^2)
> simlist <- matrix(rep(0,2*trials),ncol=2)
> for (i in 2:trials) {
+ simlist[i,1] <- rnorm(1,rho*simlist[i-1,2],sdev)
+ simlist[i,2] <- rnorm(1,rho*simlist[i,1],sdev)   }
> plot(simlist,pch=20,xlab="x",ylab="y",main="")
```

Example 5.7 The following implementation of the Gibbs sampler for a three-dimensional joint distribution is a classic example based on Casella and George (1992). Random variables X, P, and N have joint density

$$\pi(x, p, n) \propto \binom{n}{x} p^x (1-p)^{n-x} \frac{4^n}{n!},$$

for $x = 0, 1, \dots, n$, $0 < p < 1$, $n = 0, 1, \dots$ The p variable is continuous; x and n are discrete.

The Gibbs sampler requires being able to simulate from the conditional distributions of each component given the remaining variables. The trick to identifying these conditional distributions is to treat the two conditioning variables in the joint density function as fixed constants.

The conditional distribution of X given $N = n$ and $P = p$ is proportional to $\binom{n}{x} p^x(1-p)^{n-x}$, for $x = 0, 1, \ldots, n$, which is binomial with parameters n and p.

The conditional distribution of P given $X = x$ and $N = n$ is proportional to $p^x(1-p)^{n-x}$, for $0 < p < 1$, which gives a beta distribution with parameters $x + 1$ and $n - x + 1$.

The conditional distribution of N given $X = x$ and $P = p$ is proportional to $(1-p)^{n-x}4^n/(n-x)!$, for $n = x, x+1, \ldots$ This is a shifted Poisson distribution with parameter $4(1-p)$. That is, the conditional distribution is equal to the distribution of $Z + x$, where Z has a Poisson distribution with parameter $4(1-p)$.

The Gibbs sampler, with arbitrary initial value, is implemented as follows:

1. Initialize: $(x_0, p_0, n_0) \leftarrow (1, 0.5, 2)$
   $m \leftarrow 1$
2. Generate x_m from a binomial distribution with parameters n_{m-1} and p_{m-1}.
3. Generate p_m from a beta distribution with parameters $x_m + 1$ and $n_{m-1} - x_m + 1$.
4. Let $n_m = z + x_m$, where z is simulated from a Poisson distribution with parameter $4(1 - p_m)$.
5. $m \leftarrow m + 1$
6. Return to Step 2.

The output of the Gibbs sampler is a sequence of samples

$$(X_0, P_0, N_0), (X_1, P_1, N_1), (X_2, P_2, N_2), \ldots$$

In R, the output is stored in a matrix with three columns. Each column gives a sample from the marginal distribution. Each pair of columns gives a sample from a bivariate joint distribution. See Figure 5.7 for graphs of joint and marginal distributions. ■

R : Trivariate Distribution

```
# trivariate.R
> trials <- 5000
> sim <- matrix(rep(0,3*trials),ncol=3)
> sim[1,] <- c(1,0.5,2)
> for (i in 2:trials) {
+ sim[i,1] <- rbinom(1,sim[i-1,3],simlist[i-1,2])
+ sim[i,2] <- rbeta(1,sim[i,1]+1,sim[i-1,3]-sim[i,1]+1)
+ sim[i,3] <- rpois(1,4*(1-sim[i,2])+sim[i,1]   }
```

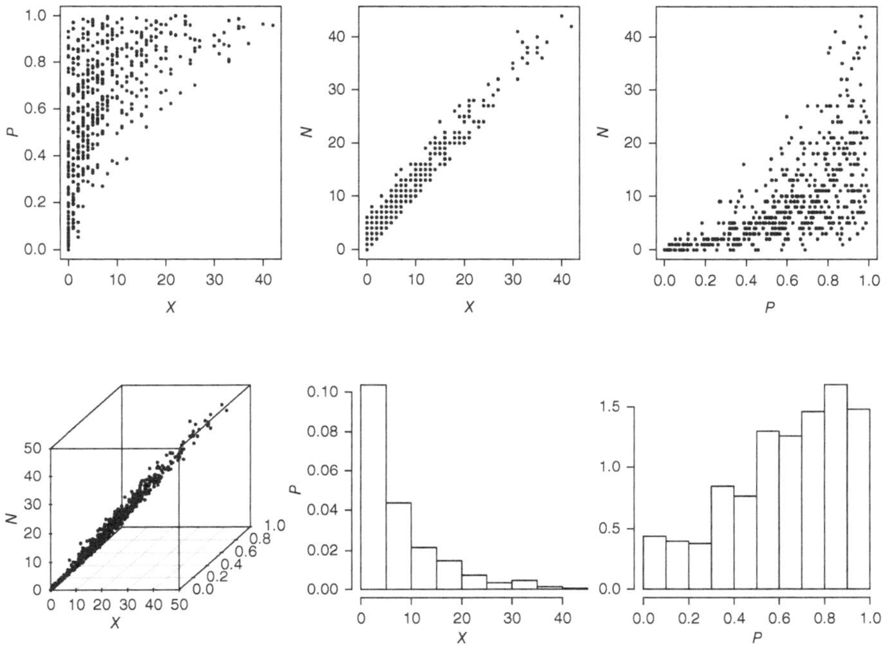

Figure 5.7 Joint and marginal distributions for trivariate distribution.

The Gibbs sampler is a special case of the Metropolis–Hastings algorithm. To see this, assume that $\boldsymbol{\pi}$ is an m-dimensional joint distribution. To avoid excessive notation and simplify the presentation, we just consider one step of the Gibbs sampler when the first component is being updated.

Assume that $\boldsymbol{i} = (x_1, x_2, \dots, x_m)$ is the current state, and $\boldsymbol{j} = (x'_1, x_2, \dots, x_m)$ is the proposed state. The proposal distribution T is the conditional distribution of X_1 given $X_2, \dots, X_m$. The acceptance function is $a(\boldsymbol{i},\boldsymbol{j}) = \pi_j T_{ji} / \pi_i T_{ij}$.

We have that

$$\begin{aligned}
\pi_j T_{ji} &= \pi(x'_1, x_2, \dots, x_m) f_{X_1|X_2,\dots,X_m}(x_1|x_2, \dots, x_m) \\
&= \pi(x'_1, x_2, \dots, x_m) \left(\frac{\pi(x_1, x_2, \dots, x_m)}{\int \pi(x, x_2, \dots, x_m)\, dx} \right) \\
&= \pi(x_1, x_2, \dots, x_m) \left(\frac{\pi(x'_1, x_2, \dots, x_m)}{\int \pi(x, x_2, \dots, x_m)\, dx} \right) \\
&= \pi(x_1, x_2, \dots, x_m) f_{X_1|X_2,\dots,X_m}(x'_1|x_2, \dots, x_m) \\
&= \pi_i T_{ij}.
\end{aligned}$$

Thus, $a(\boldsymbol{i},\boldsymbol{j}) = 1$. The proposal state is always accepted. The same is true for all of the m components. The algorithm can be implemented by either successively updating each component of the m-dimensional distribution, or by selecting components to update uniformly at random.

The Gibbs sampler is remarkably versatile, and can be applied to a large variety of complex multidimensional problems.

Example 5.8 (Ising model) The Ising model was originally proposed in physics as a model for magnetism. It also arises in image processing.

Consider a graph consisting of *sites* (vertices), in which each site v is assigned a *spin* of $+1$ or -1. A *configuration* $\boldsymbol{\sigma}$ is an assignment of spins to each site. That is, $\sigma_v = \pm 1$, for all v. We will assume a square $n \times n$ grid of sites, where each site is connected to four neighbors (up, down, left, and right), except at the boundary. Thus, there are n^2 sites and 2^{n^2} possible configurations.

Associated with each configuration $\boldsymbol{\sigma}$ is its *energy*, defined as

$$E(\boldsymbol{\sigma}) = -\sum_{v \sim w} \sigma_v \sigma_w,$$

where the sum is over all pairs of sites v and w, which are neighbors. Note that if most neighbors have similar spins, the energy is negative; if most neighbors have different spins, the energy is positive; and for a uniformly random assignment of spins, the energy is about 0. One checks that for the configuration $\boldsymbol{\sigma}$ in Figure 5.8, $E(\boldsymbol{\sigma}) = 4$.

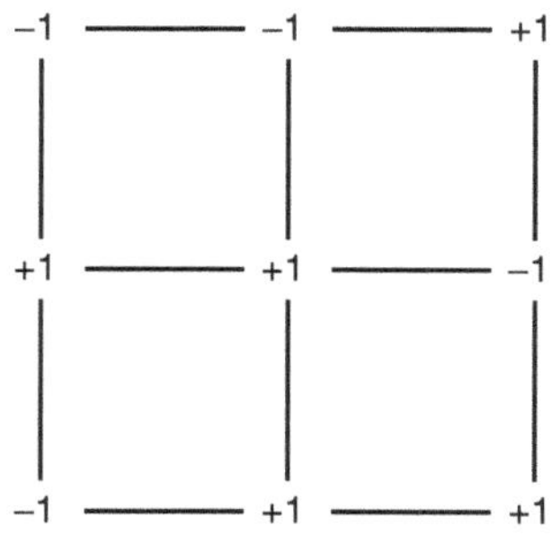

Figure 5.8 Sample configuration $\boldsymbol{\sigma}$ on 3×3 grid.

The *Gibbs distribution* is a probability distribution on the set of configurations, defined by

$$\pi_{\boldsymbol{\sigma}} = \frac{e^{-\beta E(\boldsymbol{\sigma})}}{\sum_{\boldsymbol{\tau}} e^{-\beta E(\boldsymbol{\tau})}},$$

where the sum in the denominator is over all configurations. The parameter β has a physical interpretation as the reciprocal of temperature. If $\beta = 0$ (infinite temperature), the distribution is uniform on the set of configurations. For $\beta > 0$, the Gibbs distribution puts more mass on low-energy configurations, which favors neighbors of similar spin. For $\beta < 0$, the distribution puts more mass on high-energy configurations.

The Ising model is studied, in part, because of its *phase transition* properties. In two dimensions, the system undergoes a radical change of behavior at the critical temperature $1/\beta = 2/\ln(1+\sqrt{2}) \approx 2.269$ ($\beta = 0.441$). Above that temperature, the system appears disorganized and chaotic. Below the critical temperature, the system is *magnetized,* and a phase transition occurs.

The Gibbs sampler is ideally suited for simulating the Ising model. Given a configuration $\boldsymbol{\sigma}$, a site v is chosen uniformly at random. The spin at that site is then updated from the conditional distribution of that site given the other sites of $\boldsymbol{\sigma}$.

Denote the sites as $1, \ldots, m$. Let σ_k be the spin at site k. Let

$$\boldsymbol{\sigma}_{-k} = (\sigma_1, \ldots, \sigma_{k-1}, \sigma_{k+1}, \ldots, \sigma_m)$$

denote the other $m-1$ sites of the configuration. For fixed k, write

$$\boldsymbol{\sigma}^+ = (\sigma_1, \ldots, \sigma_{k-1}, +1, \sigma_{k+1}, \ldots, \sigma_m)$$

and

$$\boldsymbol{\sigma}^- = (\sigma_1, \ldots, \sigma_{k-1}, -1, \sigma_{k+1}, \ldots, \sigma_m).$$

Then,

$$\begin{aligned} P(\sigma_k = +1 | \boldsymbol{\sigma}_{-k}) &= \frac{P(\boldsymbol{\sigma}^+)}{P(\boldsymbol{\sigma}_{-k})} = \frac{P(\boldsymbol{\sigma}^+)}{P(\boldsymbol{\sigma}^+) + P(\boldsymbol{\sigma}^-)} \\ &= \frac{e^{-\beta E(\boldsymbol{\sigma}^+)}}{e^{-\beta E(\boldsymbol{\sigma}^+)} + e^{-\beta E(\boldsymbol{\sigma}^-)}} \\ &= \frac{1}{1 + e^{\beta[E(\boldsymbol{\sigma}^+) - E(\boldsymbol{\sigma}^-)]}}. \end{aligned}$$

Observe that

$$E(\boldsymbol{\sigma}^+) = -\left(\sum_{\substack{i \sim j \\ i,j \nsim k}} \sigma_i \sigma_j + \sum_{i \sim k} \sigma_i \right),$$

where the first sum is over all sites that are not neighbors of k. Also,

$$E(\boldsymbol{\sigma}^-) = -\left(\sum_{\substack{i \sim j \\ i,j \nsim k}} \sigma_i \sigma_j - \sum_{i \sim k} \sigma_i \right).$$

This gives

$$E(\boldsymbol{\sigma}^+) - E(\boldsymbol{\sigma}^-) = -2 \sum_{i \sim k} \sigma_i,$$

and thus

$$P(\sigma_k = +1|\sigma_{-k}) = \frac{1}{1 + e^{\beta[E(\sigma^+)-E(\sigma^-)]}} = \frac{1}{1 + e^{-2\beta \sum_{i \sim k} \sigma_i}}.$$

Also, $P(\sigma_k = -1|\sigma_{-k}) = 1 - P(\sigma_k = +1|\sigma_{-k})$.

The probabilities are easily computed as they only depend on the neighbors of k. The Gibbs sampler is implemented by randomly picking sites on the grid and updating each site according to these conditional distributions.

See Figure 5.9 for simulations of the Ising model on a 60×60 grid with varying values of β. The Gibbs sampler was run for 100,000 steps for each realization. The state space has 2^{3600} elements. The simulations took less than 10 seconds on a laptop computer. The model in graph B was run at the critical inverse temperature $\beta = \ln\,(1 + \sqrt{2})/2$. Simulations B and C are for attractive systems, with $\beta > 0$. In D, $\beta < 0$ and the system is repelling. ■

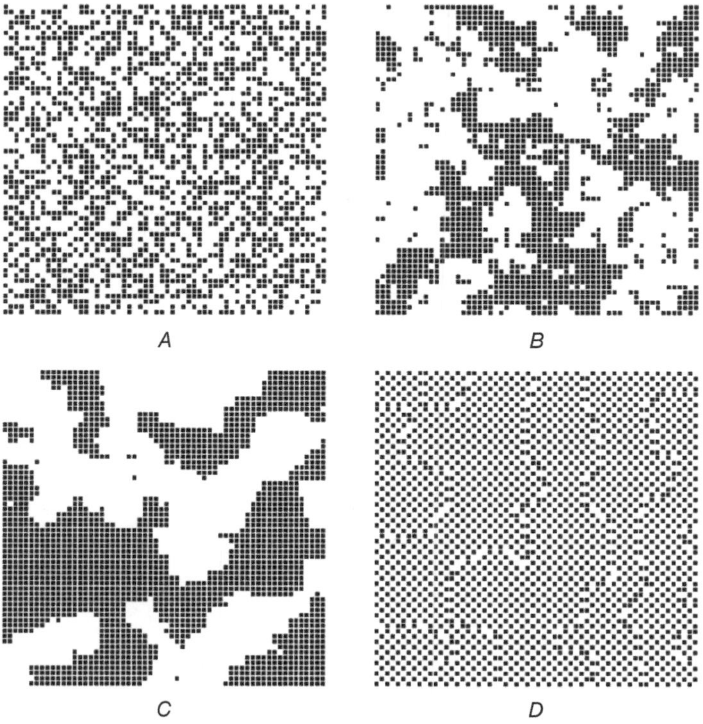

Figure 5.9 Ising simulation. Parameter values are A: $\beta = 0$, B: $\beta = 0.441$, C: $\beta = 0.75$, D: $\beta = -1.5$.

R : Ising Model

```
# ising.R
> par(mfrow=c(2,2))
> betalist <- c(0,0.441,0.75,-1.5)
> for (z in 1:4) {
+ g <- 100
+ beta <- betalist[z]
+ trials <- 100000
+ grid <- matrix(sample(c(-1,1),(g+2)^{2},rep=T), nrow=g+2)
+ grid[c(1,g+2),]<- 0
+ grid[,c(1,g+2)] <-0
+ for (m in 1:trials) {
+   i <- sample(2:(g+1), 1 )
+   j <- sample(2:(g+1),1 )
+ deg <- grid[i,j+1]+grid[i,j-1]+grid[i-1,j]+grid[i+1,j]
+  p <- 1/(1 + exp(-beta*2*deg))
+  if (runif(1) < p) grid[i,j] <- 1 else grid[i,j] <- -1
+  }
+  final <- grid[2:(g+1),2:(g+1)]
+ image(final, yaxt="n", xaxt="n", col=c(0,1))  }
```

5.4 PERFECT SAMPLING*

A central question when implementing Markov chain Monte Carlo is how long to run the chain to get close to stationarity. In the mid-1990s, an algorithm for implementing MCMC was introduced by Jim Propp and David Wilson, called *coupling from the past*. This *perfect sampling* algorithm "determines on its own when to stop, and then outputs samples in exact accordance with the desired [stationary] distribution." (From Propp and Wilson (1996).)

The idea is disarmingly simple. Assume that an ergodic Markov chain with limiting distribution $\boldsymbol{\pi}$ is started in the infinite past and run forward in time. Then, at $t = 0$, after having run for an infinitely long time, the chain will have reached its stationary distribution. Thus, if one could simulate $\ldots, X_{-2}, X_{-1}, X_0$, then X_0 would give a sample from $\boldsymbol{\pi}$.

If a chain is in stationarity then the distribution of the current state is independent of the initial state. For a k-state chain, the algorithm proceeds by running k copies of the chain, each started from a different state, but with each chain using the same source of randomness to determine transitions. Say the chain has *coalesced* at time t if all k copies of the chain are at the same state at time t. The algorithm moves backwards from $t = -1$ until there is a time $t = s < 0$ such that each of the k copies of the chain $X_s, \ldots, X_{-2}, X_{-1}, X_0$ have coalesced at $t = 0$. Then, X_0 is taken as a sample—an exact sample—from $\boldsymbol{\pi}$.

To explain the perfect sampling algorithm more carefully, we represent Markov transitions with an *update function*, based on the following basic method for simulating discrete random variables.

Given a probability distribution $\boldsymbol{p} = (p_1, \ldots, p_k)$ on $(s_1, \ldots, s_k)$, the *inverse transform method* is a simple and intuitive way to simulate an outcome X from $\boldsymbol{p}$, using a uniform random number generator. Let U be uniformly distributed on $(0, 1)$. If $U = u$, then set $X = s_j$, where j is the smallest index such that $u < p_1 + \cdots + p_j$. The method works since

$$\begin{aligned}
P(X = s_j) &= P(p_1 + \cdots + p_{j-1} < U < p_1 + \cdots + p_j) \\
&= (p_1 + \cdots + p_j) - (p_1 + \cdots + p_{j-1}) \\
&= p_j.
\end{aligned}$$

For example, consider the distribution $\boldsymbol{p} = (0.1, 0.2, 0.3, 0.4)$ on $\{a, b, c, d\}$. To simulate X from $\boldsymbol{p}$, let U be uniformly distributed on $(0, 1)$. If $U = u$, set

$$X = \begin{cases} a, & \text{if } u < 0.1, \\ b, & \text{if } 0.1 \le u < 0.3, \\ c, & \text{if } 0.3 \le u < 0.6, \\ d, & \text{if } 0.6 \le u < 1. \end{cases}$$

For a Markov chain $X_0, X_1, \ldots$, the update function value $g(x, u)$ is the outcome obtained from the conditional distribution of X_1 given $X_0 = x$ based on using the inverse transform method with $U = u$. For example, for a Markov chain with transition matrix

$$\boldsymbol{P} = \begin{array}{c} \\ a \\ b \\ c \\ d \end{array} \begin{array}{c} \begin{array}{cccc} a & b & c & d \end{array} \\ \begin{pmatrix} 1/4 & 1/4 & 1/4 & 1/4 \\ 1/3 & 1/3 & 1/3 & 0 \\ 0 & 1/3 & 1/3 & 1/3 \\ 0 & 1 & 0 & 0 \end{pmatrix} \end{array}, \tag{5.2}$$

the update function is

$$g(a, u) = \begin{cases} a, & \text{for } 0 \le u < 0.25, \\ b, & \text{for } 0.25 \le u < 0.50, \\ c, & \text{for } 0.50 \le u < 0.75, \\ d, & \text{for } 0.75 \le u < 1, \end{cases} \qquad g(b, u) = \begin{cases} a, & \text{for } 0 \le u < 0.33, \\ b, & \text{for } 0.33 \le u < 0.67, \\ c, & \text{for } 0.67 \le u < 1, \end{cases}$$

$$g(c,u) = \begin{cases} b, & \text{for } 0 \le u < 0.33, \\ c, & \text{for } 0.33 \le u < 0.67, \\ d, & \text{for } 0.67 \le u < 1, \end{cases} \quad \text{and} \quad g(d,u) = b, \text{ for } 0 \le u < 1.$$

For a general Markov chain, we can represent the chain recursively as $X_{n+1} = g(X_n, U_n)$, for $n = 0, 1, \ldots$, where g is an update function, and $U_0, U_1, \ldots$, is an i.i.d. sequence of uniform (0,1) random variables.

For an ergodic Markov chain on k states, coupling from the past works as follows:

1. Let $U_{-1}, U_{-2}, \ldots$ be an i.i.d. sequence of uniform (0, 1) random variables.
2. Let $t = -1$. Run k copies of the chain from time t, each started from a different state. For the chain started in x, set $X_0 = g(x, U_{-1})$. Thus, k copies of X_0 will be generated, one from each state. The key point of the algorithm is that even though k different chains are generated, they all use the *same* source of randomness U_{-1}. If all the X_0 agree, the chains have coalesced, the algorithm stops, and the common X_0 is output as a sample from $\boldsymbol{\pi}$.
3. If the chains have not coalesced, move back in time to $t = -2$. Again, run k copies of the chain each from different starting states. If $X_{-2} = x$, generate $X_{-1} = g(x, U_{-2})$, and $X_0 = g(X_{-1}, U_{-1})$. Thus,

$$X_0 = g(g(x, U_{-2}), U_{-1}).$$

   A new random variable U_{-2} is used in the simulation, but the old value of U_{-1} from the previous step is retained. Again, if the chains coalesce at $t = 0$ return the common value of X_0 as a sample from $\boldsymbol{\pi}$.
4. The algorithm proceeds by iteratively moving back in time. Propp and Wilson suggest, for efficiency, to double the number of steps at each iteration. Thus, for the next iteration, start at $t = -4$, then $t = -8$, and so on. At $t = -4$, uniform variables U_{-4} and U_{-3} are generated, while retaining U_{-2} and U_{-1} from the previous steps.

See Figure 5.10 for an illustration of the algorithm for the transition matrix $\boldsymbol{P}$ of Equation (5.2). The stationary distribution is $\boldsymbol{\pi} = (2/11, 9/22, 3/11, 3/22)$.

At the first iteration, transitions from each state are based on $U_{-1} = 0.394$. This gives $g(a, u_{-1}) = b$, $g(b, u_{-1}) = b$, $g(c, u_{-1}) = c$, and $g(d, u_{-1}) = b$. No coalescence occurs.

At the second iteration, the chain is started from $t = -2$, with $U_{-2} = 0.741$. Keeping the previous value of U_{-1}, the chain moves forward as shown. Again, coalescence has not occurred by $t = 0$.

At the third iteration, the chain is run from $t = -4$. Two new uniform variables are generated $U_{-3} = 0.0407$ and $U_{-4} = 0.123$. Even though coalescence occurs

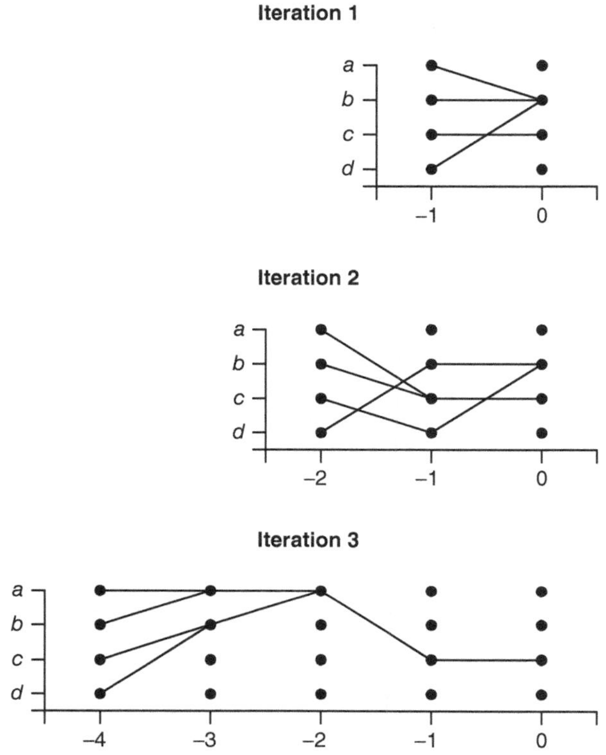

Figure 5.10 Perfect sampling on a 4-state chain. Coalescence occurs for the chain started at time $t = -4$. The algorithm outputs state c as a sample from the stationary distribution.

at $t = -2$, the algorithm outputs state c at $t = 0$ as a sample from $\boldsymbol{\pi}$. See the R file **perfect.R** for the implementation.

Monotonicity and the Ising Model

For Markov chains with large state spaces, running copies of the chain from each state is not practical. However, for chains where the state space exhibits a certain *monotonicity* property, the efficiency of the algorithm can be dramatically improved.

Say a Markov chain with update fuction g is *monotone* if the state space can be ordered in such a way that if $x \leq y$, then $g(x, u) \leq g(y, u)$, for $0 < u < 1$. For example, the chain on $\{1, 2, 3\}$ with transition matrix

$$\boldsymbol{P} = \begin{array}{c} \\ 1 \\ 2 \\ 3 \end{array} \begin{array}{c} \begin{array}{ccc} 1 & 2 & 3 \end{array} \\ \begin{pmatrix} 1/2 & 1/3 & 1/6 \\ 3/8 & 3/8 & 1/4 \\ 1/4 & 1/2 & 1/4 \end{pmatrix} \end{array}$$

is monotone. We invite the reader to check this claim with the update function

$$g(1,u) = \begin{cases} 1, & \text{for } 0 \le u < 0.50, \\ 2, & \text{for } 0.50 \le u < 0.83, \\ 3, & \text{for } 0.83 \le u < 1, \end{cases} \quad g(2,u) = \begin{cases} 1, & \text{for } 0 \le u < 0.375, \\ 2, & \text{for } 0.375 \le u < 0.75, \\ 3, & \text{for } 0.75 \le u < 1, \end{cases}$$

and

$$g(3,u) = \begin{cases} 1, & \text{for } 0 \le u < 0.25, \\ 2, & \text{for } 0.25 \le u < 0.75, \\ 3, & \text{for } 0.75 \le u < 1. \end{cases}$$

Assume that there exists a minimal state m and a maximal state M such that for all x in the state space, $m \le x \le M$. Then, a monotone chain has the property that the path of every chain is sandwiched between the paths of the two chains started in m and M, respectively. An example is shown in Figure 5.11.

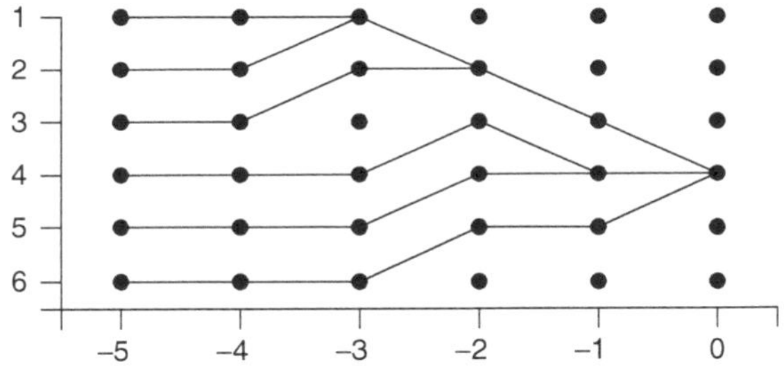

Figure 5.11 Monotone chain. All paths are sandwiched between the maximal and minimal paths.

When implementing coupling from the past for a monotone Markov chain one need only keep track of the chains started from m and M, since if those two chains coalesce at a state x, then all other paths of the chain are sandwiched between the maximal and minimal states and thus also coalesce at x.

The Ising model Gibbs sampler is a chain with an exponentially large state space. For an $n \times n$ grid there are 2^{n^2} configurations. Running chains from each state is practically impossible. However, the Ising model is monotone for $\beta > 0$ (the attractive case), and thus admits a remarkably efficient implementation of coupling from the past.

Given configurations $\boldsymbol{\sigma}$ and $\boldsymbol{\tau}$, say that $\boldsymbol{\sigma} \le \boldsymbol{\tau}$, if $\sigma_v \le \tau_v$ for all sites v. This defines a *partial ordering* on the set of all configurations. The configuration $\boldsymbol{\sigma}^m$, in which all sites are -1, is the minimal configuration in this ordering, and the configuration $\boldsymbol{\sigma}^M$, in which all sites are $+1$, is the maximal configuration.

To implement the Propp–Wilson algorithm on the Ising model, it suffices at each step to run the chain from $\boldsymbol{\sigma}^m$ and $\boldsymbol{\sigma}^M$, checking whether or not coalescence occurs.

To show the Ising chain is monotone, assume that $\boldsymbol{\sigma} \leq \boldsymbol{\tau}$. Then, for any site v,

$$\sum_{i \sim v} \sigma_i \leq \sum_{i \sim v} \tau_i.$$

Updates occur one site at a time. For site v, let X denote the spin at v. For $\beta > 0$,

$$P(X = +1|\boldsymbol{\sigma}_{-v}) = \frac{1}{1 + e^{-2\beta \sum_{i \sim v} \sigma_i}} \leq \frac{1}{1 + e^{-2\beta \sum_{i \sim v} \tau_i}} = P(X = +1|\boldsymbol{\tau}_{-v}),$$

which gives the result. See Figure 5.12 for a perfect sample of the Ising model at the critical temperature.

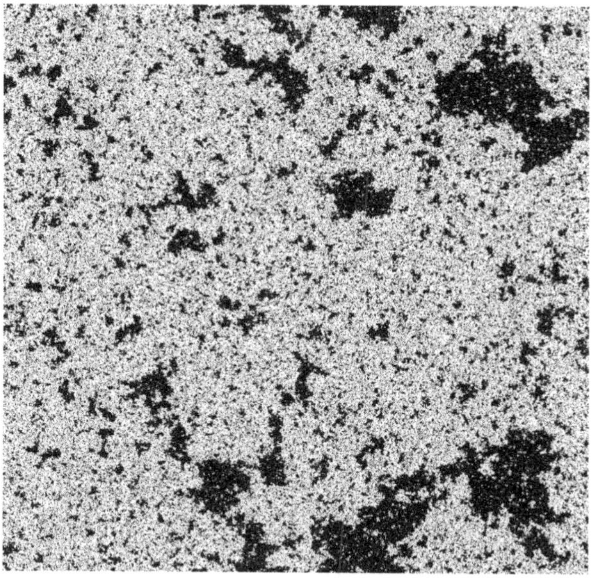

Figure 5.12 An exact sample of the Ising model on a 4200×4200 grid at the critical temperature $1/\beta = 2.269$. *Source:* Propp and Wilson (1996). Reproduced with permission of John Wiley and Sons, Inc.

5.5 RATE OF CONVERGENCE: THE EIGENVALUE CONNECTION*

Perfect sampling is an impressive algorithm, but limited in its use. For all but the simplest chains, it is impractical to run simultaneous copies of the chain from each starting state. And most Markov chains are not monotone. Thus, for all practical purposes users of MCMC must confront the issue of how long to run the chain in order to reach convergence, or near convergence, to the targeted stationary distribution.

In this section, we show that the second largest eigenvalue of the transition matrix is a lead actor in the story.

Assume a finite, reversible ergodic Markov chain, with transition matrix $\boldsymbol{P}$ and limiting distribution $\boldsymbol{\pi} = (\pi_1, \ldots, \pi_k)$. Let

$$\boldsymbol{Q} = \begin{pmatrix} \sqrt{\pi_1} & 0 & \cdots & 0 \\ 0 & \sqrt{\pi_2} & \cdots & 0 \\ \vdots & \vdots & \ddots & \vdots \\ 0 & 0 & \cdots & \sqrt{\pi_k} \end{pmatrix}.$$

Since $\boldsymbol{\pi}$ is positive, the matrix is invertible, and

$$\boldsymbol{Q}^{-1} = \begin{pmatrix} 1/\sqrt{\pi_1} & 0 & \cdots & 0 \\ 0 & 1/\sqrt{\pi_2} & \cdots & 0 \\ \vdots & \vdots & \ddots & \vdots \\ 0 & 0 & \cdots & 1/\sqrt{\pi_k} \end{pmatrix}.$$

Let $\boldsymbol{A} = \boldsymbol{QPQ}^{-1}$. Then,

$$A_{ij} = \sum_{r=1}^{k} \sum_{s=1}^{k} Q_{ir} P_{rs} Q_{sj}^{-1} = Q_{ii} P_{ij} Q_{jj}^{-1} = \sqrt{\pi_i} P_{ij} \frac{1}{\sqrt{\pi_j}}.$$

Since the chain is reversible,

$$A_{ij} = \sqrt{\pi_i} P_{ij} \frac{1}{\sqrt{\pi_j}} = \frac{\pi_i P_{ij}}{\sqrt{\pi_i}\sqrt{\pi_j}} = \frac{\pi_j P_{ji}}{\sqrt{\pi_i}\sqrt{\pi_j}} = \sqrt{\pi_j} P_{ji} \frac{1}{\sqrt{\pi_i}} = A_{ji}.$$

Thus, $\boldsymbol{A}$ is symmetric. That is, $\boldsymbol{A} = \boldsymbol{A}^T$. The eigenvalues of a symmetric matrix are real. Furthermore, a symmetric matrix is orthogonally diagonalizable. That is, we can write $\boldsymbol{A} = \boldsymbol{SDS}^T$, where $\boldsymbol{D}$ is a diagonal matrix whose diagonal entries are the eigenvalues of $\boldsymbol{A}$, and $\boldsymbol{S}$ is a matrix whose columns are an orthonormal set of eigenvectors corresponding to those eigenvalues.

The eigenvalues of a Markov transition matrix are described by Lemma 3.15 in Section 3.10. Since the chain is ergodic, the transition matrix is regular, and there is a single largest eigenvalue in absolute value $\lambda_1 = 1$. As the eigenvalues are real, they can be written in decreasing order

$$1 = \lambda_1 > \lambda_2 \geq \lambda_3 \geq \cdots \geq \lambda_k > -1.$$

Since $\boldsymbol{A} = \boldsymbol{QPQ}^{-1}$, the matrices $\boldsymbol{P}$ and $\boldsymbol{A}$ are similar and thus have the same eigenvalues. This gives

$$\boldsymbol{P} = \boldsymbol{Q}^{-1}\boldsymbol{A}\boldsymbol{Q} = \boldsymbol{Q}^{-1}(\boldsymbol{SDS}^T)\boldsymbol{Q} = (\boldsymbol{Q}^{-1}\boldsymbol{S})\boldsymbol{D}(\boldsymbol{S}^T\boldsymbol{Q}),$$

where

$$D = \begin{pmatrix} 1 & 0 & \cdots & 0 \\ 0 & \lambda_2 & \cdots & 0 \\ \vdots & \vdots & \ddots & \vdots \\ 0 & 0 & \cdots & \lambda_k \end{pmatrix}.$$

For $n \geq 0$, $\boldsymbol{P}^n = \left(\boldsymbol{Q}^{-1}\boldsymbol{S}\right)\boldsymbol{D}^n\left(\boldsymbol{S}^T\boldsymbol{Q}\right)$. Taking the ijth entry,

$$\begin{aligned} P^n_{ij} &= \sum_{t=1}^{k}\sum_{u=1}^{k} \left(Q^{-1}S\right)_{it} D^n_{tu} \left(S^T Q\right)_{uj} \\ &= \sum_{t=1}^{k} \left(Q^{-1}S\right)_{it} \lambda^n_t \left(S^T Q\right)_{tj} \\ &= \frac{\sqrt{\pi_j}}{\sqrt{\pi_i}} \sum_{t=1}^{k} \lambda^n_t S_{it} S_{jt} \\ &= \frac{\sqrt{\pi_j}}{\sqrt{\pi_i}} S_{i1} S_{j1} + \frac{\sqrt{\pi_j}}{\sqrt{\pi_i}} \sum_{t=2}^{k} \lambda^n_t S_{it} S_{jt}. \end{aligned} \tag{5.3}$$

Equation (5.3) is known as the *spectral representation formula* for the n-step transition probabilities. Since $P^n_{ij} \to \pi_j$, as $n \to \infty$, and $\lambda^n_t \to 0$, as $n \to \infty$, for $2 \leq t \leq k$, we have that

$$\pi_j = \frac{\sqrt{\pi_j}}{\sqrt{\pi_i}} S_{i1} S_{j1},$$

and

$$P^n_{ij} - \pi_j = \frac{\sqrt{\pi_j}}{\sqrt{\pi_i}} \sum_{t=2}^{k} \lambda^n_t S_{it} S_{jt}.$$

To bound the difference between the n-step and stationary probabilities,

$$|P^n_{ij} - \pi_j| \leq \frac{\sqrt{\pi_j}}{\sqrt{\pi_i}} \sum_{t=2}^{k} |\lambda^n_t S_{it} S_{jt}| \leq \frac{\sqrt{\pi_j}}{\sqrt{\pi_i}} |\lambda_*|^n \sum_{t=2}^{k} |S_{it} S_{jt}| = |\lambda_*|^n c_{ij}, \tag{5.4}$$

where $|\lambda_*| = \max_{2 \leq t \leq k} |\lambda_t|$, and c_{ij} is a constant that does not depend on n. The quantity λ_* is the second largest eigenvalue in absolute value. The bounds in Equation (5.4) show that the convergence to stationarity is exponentially fast and the rate of convergence is dominated by the second largest eigenvalue.

5.6 CARD SHUFFLING AND TOTAL VARIATION DISTANCE*

The asymptotic rate of convergence only tells part of the story of a Markov chain's path to stationarity. In many applications where a Markov chain is run until it is close to its limiting distribution, such as MCMC, what a user is interested in is not an asymptotic result about what happens *at infinity*, but rather the more practical issue of how many steps of the chain are *close enough*. The issue is illustrated by card shuffling.

Many card shuffling schemes can be modeled as Markov chains. For a deck of k cards, the state space is the set of $k!$ permutations (orderings) of the deck. A *shuffle* is one step of the chain. If repeated shuffles eventually mix up the deck, then the card shuffling scheme is modeled as an ergodic Markov chain whose limiting distribution is uniform on the set of permutations. If σ is a permutation, then $\pi_\sigma = 1/k!$.

Theoretically, the uniform distribution is achieved as the number of shuffles tends to infinity. This is not much help to a card player, however, who wants to know how many shuffles are needed to bring the deck sufficiently close to random.

To quantify the idea of *how far* a Markov chain is from its limiting distribution, a measure of distance is needed between the distribution of the chain at a fixed time and the limiting distribution. A common measure is *total variation distance*.

Total Variation Distance

Let $\boldsymbol{P}$ be the transition matrix of an ergodic Markov chain with limiting distribution $\boldsymbol{\pi}$. The *total variation distance at time n* is

$$v(n) = \max_{i \in S} \max_{A \subseteq S} |P(X_n \in A | X_0 = i) - \pi_A|.$$

In words, total variation distance at time n is the maximum absolute difference over all events A and all starting states between the probabilities $P(X_n \in A)$ and π_A. The distance measure takes values between 0 and 1. If $v(n) = 0$, then the chain is in stationarity at time n. Over time, $v(n) \to 0$, as $n \to \infty$. It can be shown that the definition is equivalent to

$$v(n) = \max_i \frac{1}{2} \sum_j |P^n_{ij} - \pi_j|.$$

Example 5.9 Compute total variation distance for the two-state Markov chain with transition matrix

$$\boldsymbol{P} = \begin{array}{c} \\ 1 \\ 2 \end{array} \begin{array}{c} \begin{array}{cc} 1 & 2 \end{array} \\ \begin{pmatrix} 1-p & p \\ q & 1-q \end{pmatrix} \end{array}.$$

Solution Recall that

$$\boldsymbol{P}^n = \frac{1}{p+q} \begin{pmatrix} q + p(1-p-q)^n & p - p(1-p-q)^n \\ q - q(1-p-q)^n & p + q(1-p-q)^n \end{pmatrix}$$

and

$$\boldsymbol{\pi} = \left(\frac{q}{p+q} \quad \frac{p}{p+q} \right).$$

For $i = 1$,

$$\sum_{j=1}^{2} |P^n_{1j} - \pi_j| = \frac{2p}{p+q}(1-p-q)^n.$$

For $i = 2$,

$$\sum_{j=1}^{2} |P^n_{2j} = \pi_j| = \frac{2q}{p+q}(1-p-q)^n.$$

Thus,

$$v(n) = \frac{\max(p,q)}{p+q}(1-p-q)^n.$$

Convergence to stationarity, as measured by total variation distance, occurs exponentially fast, with the rate of convergence governed by $1 - p - q$. Note that the second largest eigenvalue of $\boldsymbol{P}$ is $\lambda_2 = 1 - p - q$.

■

Top-to-Random Shuffle

For the *top-to-random* shuffling scheme, a shuffle consists of placing the card at the top of the deck into a uniformly random position in the deck. How many such shuffles does it take to mix up a deck of cards?

Admittedly this is not a very efficient shuffling method for mixing up cards. But it is a good example to illustrate our analysis. We show, in a fairly precise sense, that for a k-card deck it takes about $k \ln k$ such shuffles to mix up the deck. For a 52-card deck, that is $52 \ln 52$, which is about 200 shuffles.

Central to the analysis is the notion of a strong stationary time. Recall the discussion of stopping times and the strong Markov property in Section 3.9. A strong stationary time is a stopping time. When a Markov chain stops at a strong stationary time it is in its stationary distribution.

Strong Stationary Time

A *strong stationary time* for an ergodic Markov chain $X_0, X_1, \ldots$ is a stopping time T such that

$$P(X_n = j, T = n) = \pi_j P(T = n), \text{ for all states } j \text{ and } n \geq 0.$$

The definition says that if T is a strong stationary time and $T = n$, then the Markov chain is in stationarity at time n. Furthermore, the state of the chain at time T is independent of T.

In Aldous and Diaconis (1986), the top-to-random shuffle is analyzed by constructing a suitable strong stationary time.

Starting with the deck in a fixed order, let b denote the card initially at the bottom of the deck. As the deck is shuffled, cards will eventually be inserted below b, and the position of b in the deck will rise. We show that as soon as b reaches the top of the deck and then is inserted in a random position, the deck will be mixed up in the sense that every ordering is equally likely.

Initially, the probability that the top card is inserted at the bottom of the deck below b is $1/k$, since there are k available positions. Since successive shuffles are independent, the number of shuffles needed for a card to be inserted below b has a geometric distribution with parameter $1/k$.

After b has moved up one position from the bottom, consider the time until a second card moves below b. There are now two available slots, and the number of shuffles needed has a geometric distribution with parameter $2/k$. Furthermore, once two cards are below b, each of the two orderings of those two cards is equally likely.

More generally, assume that there are $i - 1$ cards below b. The probability that the top card is inserted below b is i/k. The number of shuffles needed before the top card is inserted below b has a geometric distribution with parameter i/k. Once i cards are below b, each of the $i!$ orderings of the i cards is equally likely.

Finally, after the original bottom card has risen to the top of the deck it takes one more shuffle, which sends b to a uniformly random position, before all $k!$ orderings of the deck are equally likely. At that (random) time, the deck is mixed up.

Let T be the number of top-to-random shuffles needed for the original bottom card to reach the top of the deck, plus 1. Then, T is a strong stationary time.

The analysis also shows that T can be expressed as the sum of k independent geometric random variables, with successive parameters $1/k, 2/k, \ldots, (k-1)/k, 1$. The expectation of a geometric random variable is the reciprocal of the parameter. Hence,

$$E(T) = \frac{k}{1} + \frac{k}{2} + \cdots + \frac{k}{k-1} + \frac{k}{k} = k\left(1 + \frac{1}{2} + \cdots + \frac{1}{k-1} + \frac{1}{k}\right) \approx k \ln k.$$

Strong stationary times are used to bound total variation distance by the following lemma.

Strong Stationary Times and Total Variation Distance

Lemma 5.2. *Let T be a strong stationary time for an ergodic Markov chain. For $n > 0$,*

$$v(n) \leq P(T > n).$$

The lemma is proved at the end of the chapter. To apply the lemma to the top-to-random shuffle, we estimate $P(T > n)$ by relating the strong stationary time T to the classic coupon collector's problem.

Assume that a set of k distinct items (coupons) is sampled repeatedly with replacement. Let C be the number of samples needed to obtain a full set of coupons, that is, for each distinct item to be sampled at least once. We show that the distributions of C and T are the same.

Let C_i be the number of trials needed before the ith new coupon has been selected, for $i = 1, \ldots, k$, so that $C = C_k$. Necessarily, $C_1 = 1$. For $i = 2, \ldots, k$, $C_i - C_{i-1}$, the number of samples needed after the $(i-1)$th coupon is picked until the ith new coupon is picked has a geometric distribution with parameter $(k - i + 1)/k$. This is because after the $(i-1)$th coupon is selected, there are $k - i + 1$ remaining coupons, and each one can be chosen with probability $(k - i + 1)/k$. Furthermore, the random variables $C_i - C_{i-1}$ are independent. We have that

$$C = C_1 + (C_2 - C_1) + \cdots + (C_k - C_{k-1})$$

is a sum of independent geometric random variables with respective parameters $1, (k-1)/k, \ldots, 2/k, 1/k$. It follows that C and T have the same distribution, and $P(T > n) = P(C > n)$, for all n.

The event $\{C > n\}$ is the event that after n trials, some coupon has not been sampled. Let A_i denote the event that coupon i has not been picked after n trials. Then,

$$\begin{aligned} P(C > n) &= P\left(\bigcup_{i=1}^{n} A_i\right) \leq \sum_{i=1}^{k} P(A_i) \\ &= \sum_{i=1}^{k} \left(1 - \frac{1}{k}\right)^n \leq \sum_{i=1}^{k} e^{-n/k} \\ &= ke^{-n/k}, \end{aligned}$$

using that $1 - x \leq e^{-x}$, for all x.

For the top-to-random shuffle, with Lemma 5.2, this gives

$$v(n) \leq P(T > n) = P(C > n) \leq ke^{-n/k}. \tag{5.5}$$

To make total variation distance small, find a value of n that makes the righthand side of Equation (5.5) small. For $c > 0$, let $n = k \ln k + ck$. Then,

$$v(n) = v(k \ln k + ck) \leq ke^{-(\ln k + c)} = e^{-c} \approx 0.$$

The bound is interpreted to mean that for a *little more* than $k \ln k$ shuffles the distance to stationarity is small.

Aldous and Diaconis (1986) also show, by a different analysis, that for a *little less* than $k \ln k$ shuffles, that is, for $n = k \ln k - ck$, with $c > 0$, total variation distance $v(n)$ stays close to 1.

Thus, the claim that for large k it takes about $k \ln k$ top-to-random shuffles to mix up the deck of cards. Fewer than $k \ln k$ steps are insufficient. More than $k \ln k$ steps are not necessary.

The top-to-random shuffle exhibits a remarkable property, shared by many card shuffling Markov chains, known as the *cutoff phenomenon*. The approach to stationarity, as measured by total variation distance, behaves like a phase transition in physics. As the number of shuffles increases, the chain stays far from uniform and then, in a relatively short time window, the chain rapidly gets close to uniform. See Figure 5.13.

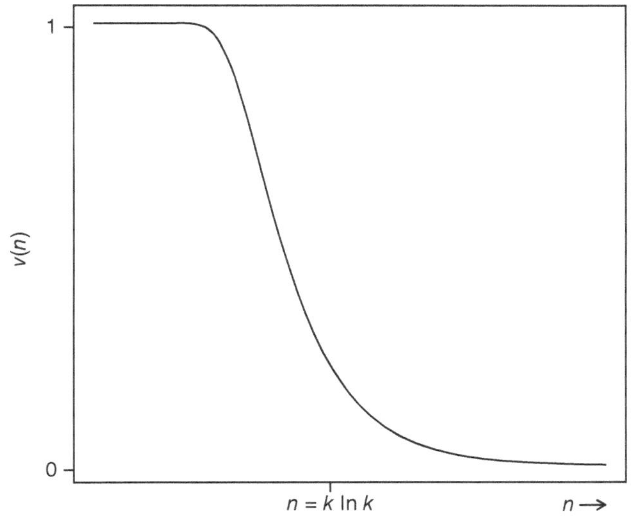

Figure 5.13 Cutoff phenomenon for top-to-random shuffle.

Seven Shuffles is Enough

In 1990, the *New York Times* reported the "fascinating result," discovered by mathematicians David Bayer and Persi Diaconis, that, "It takes just seven ordinary, imperfect [riffle] shuffles to mix a deck of cards thoroughly" (Kolata, 1990). The analysis by Bayer and Diaconis (1992) was based on calculating the total variation distance for a Markov model for how most people shuffle cards.

The riffle-shuffling scheme is known as the Gilbert–Shannon–Reeds model. A deck of k cards is cut into two piles according to a binomial distribution. That is, the probability that one pile has i cards, and the other pile has $k - i$ cards, is $\binom{k}{i} 2^{-k}$, for $i = 0, 1, \ldots, k$. The two piles are then riffled together by successively dropping cards from either pile with probability proportional to the size of the pile. Thus, if there are a cards in the first pile, and b cards in the second pile, the probability that the next card dropped comes from the first pile is $a/(a + b)$.

Bayer and Diaconis give a thorough analysis of the total variation distance for the resulting Markov chain. They show that about $(3/2)\log_2(k)$ shuffles are necessary and sufficient to make total variation distance small. They further demonstrate the existence of the cutoff phenomenon, as shown in Figure 5.14 and Table 5.2, which contains values for total variation distance after n riffle shuffles of 52 cards.

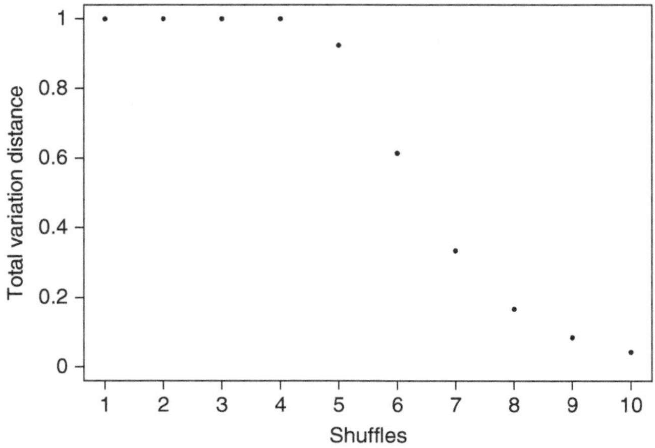

Figure 5.14 Total variation distance for riffle shuffling.

TABLE 5.2 Total Variation Distance for n Shuffles of 52 Cards

n	1	2	3	4	5	6	7	8	9	10
$v(n)$	1.000	1.000	1.000	1.000	0.924	0.614	0.334	0.167	0.085	0.043

Source: Data from Bayer and Diaconis (1992).

For an introduction to the mathematics of riffle shuffling, see Mann (1994).

Proof of Lemma 5.2

Let T be a strong stationary time for a Markov chain $X_0, X_1, \ldots$ We show that the definition of strong stationary time implies that

$$P(X_n = j, T \leq n) = \pi_j P(T \leq n).$$

We have that

$$\begin{aligned}
P(X_n = j, T \leq n) &= \sum_{m \leq n} P(T = m | X_n = j) P(X_n = j) \\
&= \sum_{m \leq n} \sum_{x} P(T = m | X_n = j, X_m = x) P(X_m = x | X_n = j) P(X_n = j) \\
&= \sum_{m \leq n} \sum_{x} P(T = m | X_m = x) P(X_m = x, X_n = j) \\
&= \sum_{m \leq n} P(T = m) \sum_{x} \pi_x \frac{P(X_m = x, X_n = j)}{P(X_m = x)}
\end{aligned}$$

$$= P(T \leq m) \sum_x \pi_x P(X_n = j | X_m = x)$$

$$= P(T \leq m) \sum_x \pi_x P^{n-m}_{xj} = P(T \leq m)\pi_j.$$

The third equality is because T is a stopping time and the event $\{T = m\}$ depends on $X_0, \ldots, X_m$, and not on X_n for $n > m$. The fourth equality is by the definition of strong stationary time. The last equality is because $\boldsymbol{\pi}$ is a stationary distribution and $\boldsymbol{\pi} \boldsymbol{P}^{n-m} = \boldsymbol{\pi}$.

For the chain started in i,

$$\begin{aligned} P(X_n \in A) &= P(X_n \in A, T \leq n) + P(X_n \in A, T > n) \\ &= \pi_A P(T \leq n) + P(X_n \in A | T > n) P(T > n) \\ &= \pi_A + [P(X_n \in A | T > n) - \pi_A] P(T > n), \end{aligned}$$

which gives

$$|P(X_n \in A) - \pi_A| = |P(X_n \in A | T > n) - \pi_A| P(T > n) \leq P(T > n),$$

where the last inequality is because the absolute difference of two probabilities is at most 1. Maximizing over initial states i gives the result.

■

EXERCISES

5.1 Four out of every five trucks on the highway are followed by a car, while only one out of every four cars is followed by a truck. At a toll booth, cars pay \$1.50 and trucks pay \$5.00. If 1,000 vehicles pass through the tollbooth in one day, how much toll is collected?

5.2 Consider simple random walk on $\{0, 1, \ldots, k\}$ with reflecting boundaries at 0 and k, that is, random walk on the path from 0 to k. A random walker earns \$k every time the walk reaches 0 or k, but loses \$1 at each internal vertex (from 1 to $k - 1$). In 10,000 steps of the walk, how much, on average, will be gained?

5.3 Commuters in an urban center either drive by car, take the bus, or bike to work. Based on recent changes to local transportation systems, urban planners predict yearly annual behavior changes of commuters based on a Markov model with transition matrix

$$\boldsymbol{P} = \begin{array}{c} \\ \text{Car} \\ \text{Bus} \\ \text{Bike} \end{array} \begin{array}{c} \begin{array}{ccc} \text{Car} & \text{Bus} & \text{Bike} \end{array} \\ \begin{pmatrix} 0.7 & 0.2 & 0.1 \\ 0.1 & 0.8 & 0.1 \\ 0.05 & 0.15 & 0.8 \end{pmatrix} \end{array}.$$

(a) In a city of 10,000 commuters, 6,000 currently drive cars, 3,000 take the bus, and 1,000 bike to work. Two years after transportation changes are made, how many commuters will use each type of transportation? Over the long term, how many commuters will use each type of transportation?

(b) The European Cyclists Federation reports the following levels of CO_2 emissions (in grams) per passenger per kilometer traveled: car: 271, bus: 101, bike: 21. What is the current average amount of CO_2 emissions per kilometer traveled? How does the average change over the long-term?

5.4 The Fibonacci sequence $1, 1, 2, 3, 5, 8, 13, \ldots$ is described by the recurrence $f_n = f_{n-1} + f_{n-2}$, for $n \geq 3$, with $f_1 = f_2 = 1$.

(a) See the MCMC example for binary sequences with no adjacent 1s. Show that the number of binary sequences of length m with no adjacent 1s is f_{m+2}.

(b) Let $p_{k,m}$ be the number of good sequences of length m with exactly k 1s. Show that

$$p_{k,m} = \binom{m-k+1}{k}, \text{ for } k = 0, 1, \ldots, \lceil m/2 \rceil.$$

(c) Let μ_m be the expected number of 1s in a good sequence of length m under the uniform distribution. Find μ_m, for $m = 10, 100, 1000$.

(d) If you have access to a mathematical symbolic software system, such as *Mathematica*, use it to find

$$\lim_{m\to\infty} \frac{\mu_m}{m}.$$

5.5 Exhibit a Metropolis–Hastings algorithm to sample from the distribution

1	2	3	4	5	6
0.01	0.39	0.11	0.18	0.26	0.05

with proposal distribution based on one fair die roll.

5.6 Show how to generate a Poisson random variable with parameter λ using Metropolis–Hastings. Use simple symmetric random walk as the proposal distribution.

5.7 Exhibit a Metropolis–Hastings algorithm to sample from a binomial distribution with parameters n and p. Use a proposal distribution that is uniform on $\{0, 1, \ldots, n\}$.

5.8 The Metropolis–Hastings algorithm is used to simulate a binomial random variable with parameters $n = 4$ and $p = 1/4$. The proposal distribution is simple symmetric random walk on $\{0, 1, 2, 3, 4\}$ with reflecting boundaries.

(a) Exhibit the T matrix.

(b) Exhibit the transition matrix for the Markov chain created by the Metropolis–Hastings algorithm.

(c) Show explicitly that this transition matrix gives a reversible Markov chain whose stationary distribution is the desired binomial distribution.

5.9 Show how to use the Metropolis–Hastings algorithm to simulate from the *double exponential distribution*, with density

$$f(x) = \frac{\lambda}{2}e^{-\lambda|x|}, \text{ for } -\infty < x < \infty.$$

Use the normal distribution as a proposal distribution.

5.10 A Markov chain has transition matrix

$$P = \begin{array}{c} \\ 1 \\ 2 \\ 3 \end{array}\begin{array}{c} \begin{array}{ccc} 1 & 2 & 3 \end{array} \\ \begin{pmatrix} 1/2 & 1/2 & 0 \\ 1/2 & 0 & 1/2 \\ 0 & 1/2 & 1/2. \end{pmatrix} \end{array}$$

Show that the chain is monotone.

5.11 Consider the following random walk on $\{1, \ldots, n\}$. From state 1, the walk moves to 1 or 2 with probability 1/2 each. From state n, the walk moves to n or $n-1$ with probability 1/2 each. From all other states i, the walk moves to $i-1$ or $i+1$ with probability 1/2 each.

(a) Show that the Markov chain is monotone.

(b) R : Implement perfect sampling, making use of monotonicity, to simulate the random walk on $\{1, \ldots, 100\}$.

5.12 Consider random walk on the k-hypercube graph. At each step, one of k coordinates is selected uniformly at random. A coin is then tossed. If heads, the coordinate is set to 0. If tails, it is set to 1. Let T be the first time that all k coordinates are selected. Argue that T is a strong stationary time. What can be said on how many steps it takes for the walk to get uniformly distributed? (*Hint:* consider the coupon collector's problem.)

5.13 Consider the lazy librarian and move-to-front process as described in Example 2.10. It can be shown that the first time that all the books have been selected is a strong stationary time. Assume that p_j is the probability that book j is selected, for $j = 1, \ldots, k$. Show that

$$v(n) \leq \sum_{j=1}^{k} e^{-p_j n}.$$

5.14 Show that the two definitions of total variation distance given in Section 5.6 are equivalent.

5.15 See the description of the Gilbert–Shannon–Reeds model for riffle shuffles in Section 5.6. Give the Markov transition matrix for a three-card deck.

5.16 See the discussion on riffle shuffling and total variation distance. Bayer and Diaconis (1992) prove the following. If k cards are riffle shuffled n times with $n = (3/2)\log_2(k) + \theta$, then for large k,

$$v(n) = 1 - 2\Phi\left(\frac{-2^{-\theta}}{4\sqrt{3}}\right) + g(k), \tag{5.6}$$

where Φ is the cumulative distribution function of the standard normal distribution, and $g(k)$ is a slow-growing function of order $k^{-1/4}$.

(a) Show that total variation distance tends to 1 with θ small, and to 0 with θ large.

(b) Verify that Equation (5.6) gives values that are close to the entries in Table 5.2.

5.17 R : Random variables X and N have joint distribution, defined up to a constant of proportionality,

$$f(x, n) \propto \frac{e^{-3x}x^n}{n!}, \text{ for } n = 0, 1, 2, \ldots \text{ and } x > 0.$$

Note that X is continuous and N is discrete.

(a) Implement a Gibbs sampler to sample from this distribution.

(b) Use your simulation to estimate (i) $P(X^2 < N)$ and (ii) $E(XN)$.

5.18 R : A random variable X has density function, defined up to proportionality,

$$f(x) \propto e^{-(x-1)^2/2} + e^{-(x-4)^2/2}, \text{ for } 0 < x < 5.$$

Implement a Metropolis–Hastings algorithm for simulating from the distribution. Use your algorithm to approximate the mean and variance of X.

5.19 R : Implement the algorithm in Exercise 5.7 for $n = 50$ and $p = 1/4$. Use your simulation to estimate $P(10 \leq X \leq 15)$, where X has the given binomial distribution. Compare your estimate to the exact probability.

5.20 R : Consider a Poisson distribution with parameter $\lambda = 3$ conditioned to be nonzero. Implement an MCMC algorithm to simulate from this distribution, using a proposal distribution that is geometric with parameter $p = 1/3$. Use your simulation to estimate the mean and variance.

6

POISSON PROCESS

The count has arrived ...

–Thomas Carlyle, *Count Cagliostro*

6.1 INTRODUCTION

Text messages arrive on your cell phone at irregular times throughout the day. Accidents occur on the highway in a seemingly random distribution of time and place. Babies are born at chance moments on a maternity ward.

All of these phenomena are well modeled by the Poisson process, a stochastic process used to model the occurrence, or arrival, of events over a continuous interval. Typically, the interval represents time.

A Poisson process is a special type of *counting process*. Given a stream of events that arrive at random times starting at $t = 0$, let N_t denote the number of arrivals that occur by time t, that is, the number of events in $[0, t]$. For instance, N_t might be the number of text messages received up to time t.

For each $t \geq 0$, N_t is a random variable. The collection of random variables $(N_t)_{t\geq 0}$ is a continuous-time, integer-valued stochastic process, called a counting process. Since N_t counts events in $[0, t]$, as t increases, the number of events N_t increases.

Introduction to Stochastic Processes with R, First Edition. Robert P. Dobrow.

Counting Process

A *counting process* $(N_t)_{t\geq 0}$ is a collection of non-negative, integer-valued random variables such that if $0 \leq s \leq t$, then $N_s \leq N_t$.

Unlike a Markov chain, which is a *sequence* of random variables, a counting process forms an uncountable collection, since it is indexed over a continuous time interval.

Figure 6.1 shows the path of a counting process in which events occur at times t_1, t_2, t_3, t_4, t_5. As shown, the path of a counting process is a right-continuous step function. If $0 \leq s < t$, then $N_t - N_s$ is the number of events in the interval $(s, t]$.

Note that in the discussion of the Poisson process, we use the word *event* in a loose, generic sense, and not in the rigorous sense of probability theory where an event is a subset of the sample space.

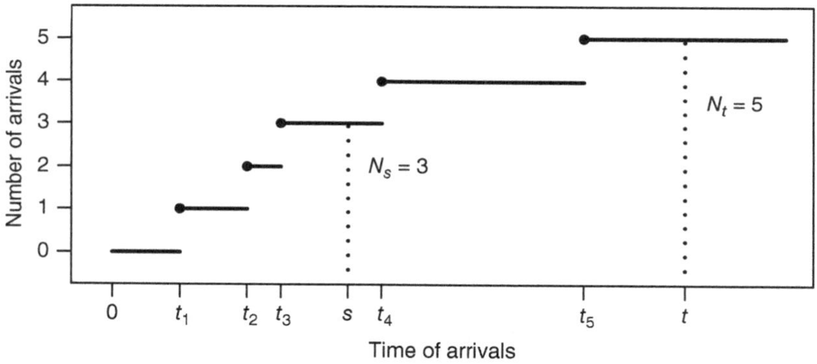

Figure 6.1 Counting process.

There are several ways to characterize the Poisson process. One can focus on (i) the *number* of events that occur in fixed intervals, (ii) *when* events occur, and the times between those events, or (iii) the probabilistic behavior of individual events on infinitesimal intervals. This leads to three equivalent definitions of a Poisson process, each of which gives special insights into the stochastic model.

Poisson Process—Definition 1

A *Poisson process with parameter* λ is a counting process $(N_t)_{t\geq 0}$ with the following properties:

1. $N_0 = 0$.
2. For all $t > 0$, N_t has a Poisson distribution with parameter λt.

3. (Stationary increments) For all $s, t > 0$, $N_{t+s} - N_s$ has the same distribution as N_t. That is,

$$P(N_{t+s} - N_s = k) = P(N_t = k) = \frac{e^{-\lambda t}(\lambda t)^k}{k!}, \text{ for } k = 0, 1, \ldots$$

4. (Independent increments) For $0 \le q < r \le s < t$, $N_t - N_s$ and $N_r - N_q$ are independent random variables.

The stationary increments property says that the distribution of the number of arrivals in an interval only depends on the *length* of the interval.

The independent increments property says that the number of arrivals on disjoint intervals are independent random variables.

Since N_t has a Poisson distribution, $E(N_t) = \lambda t$. That is, we expect about λt arrivals in t time units. Thus, the *rate* of arrivals is $E(N_t)/t = \lambda$.

Example 6.1 Starting at 6 a.m., customers arrive at Martha's bakery according to a Poisson process at the rate of 30 customers per hour. Find the probability that more than 65 customers arrive between 9 and 11 a.m.

Solution Let $t = 0$ represent 6 a.m. Then, the desired probability is $P(N_5 - N_3 > 65)$. By stationary increments,

$$\begin{aligned} P(N_5 - N_3 > 65) &= P(N_2 > 65) = 1 - P(N_2 \le 65) \\ &= 1 - \sum_{k=0}^{65} P(N_2 = k) \\ &= 1 - \sum_{k=0}^{65} \frac{e^{-30(2)}(30(2))^k}{k!} = 0.2355. \end{aligned}$$

In R , the result is obtained by typing

```
> 1-ppois(65,2*30)
[1] 0.2355065
```

■

Example 6.2 Joe receives text messages starting at 10 a.m. at the rate of 10 texts per hour according to a Poisson process. Find the probability that he will receive exactly 18 texts by noon and 70 texts by 5 p.m.

Solution The desired probability is $P(N_2 = 18, N_7 = 70)$, with time as hours. If 18 texts arrive in $[0, 2]$ and 70 texts arrive in $[0, 7]$, then there are $70 - 18 = 52$ texts in

(2, 7]. That is,

$$\{N_2 = 18, N_7 = 70\} = \{N_2 = 18, N_7 - N_2 = 52\}.$$

The intervals [0, 2] and (2, 7] are disjoint, which gives

$$\begin{aligned}
P(N_2 = 18, N_7 = 70) &= P(N_2 = 18, N_7 - N_2 = 52) \\
&= P(N_2 = 18)\, P(N_7 - N_2 = 52) \\
&= P(N_2 = 18)\, P(N_5 = 52) \\
&= \left(\frac{e^{-2(10)}(2(10))^{18}}{18!}\right)\left(\frac{e^{-5(10)}(5(10))^{52}}{52!}\right) \\
&= 0.0045,
\end{aligned}$$

where the second equality is because of independent increments, and the third equality is because of stationary increments. The final calculation in R is

```
> dpois(18,2*10)*dpois(52,5*10)
[1] 0.004481021
```

Warning: It would be incorrect to write

$$P(N_2 = 18, N_7 - N_2 = 52) = P(N_2 = 18, N_5 = 52).$$

It is not true that $N_7 - N_2 = N_5$. The number of arrivals in (2, 7] is not necessarily equal to the number of arrivals in (0, 5]. What is true is that the *distribution* of $N_7 - N_2$ is equal to the distribution of N_5. Note that while $N_7 - N_2$ is independent of N_2, the random variable N_5 is not independent of N_2. Indeed, $N_5 \geq N_2$. ∎

Translated Poisson Process

Let $(N_t)_{t\geq 0}$ be a Poisson process with parameter λ. For fixed time $s > 0$, consider the *translated process* $(N_{t+s} - N_s)_{t\geq 0}$. The translated process is probabilistically equivalent to the original process.

Translated Poisson Process is a Poisson Process

Proposition 6.1. *Let $(N_t)_{t\geq 0}$ be a Poisson process with parameter λ. For $s > 0$, let*

$$\widetilde{N}_t = N_{t+s} - N_s, \quad \text{for } t \geq 0.$$

Then, $(\widetilde{N}_t)_{t\geq 0}$ is a Poisson process with parameter λ.

We have that $\left(\widetilde{N}_t\right)_{t\geq 0}$ is a counting process with $\widetilde{N}_0 = N_s - N_s = 0$. By stationary increments, $\widetilde{N}_t$ has the same distribution as N_t. And the new process inherits stationary and independent increments from the original. It follows that if $N_s = k$, the distribution of $N_{t+s} - k$ is equal to the distribution of N_t.

Example 6.3 On election day, people arrive at a voting center according to a Poisson process. On average, 100 voters arrive every hour. If 150 people arrive during the first hour, what is the probability that at most 350 people arrive before the third hour?

Solution Let N_t denote the number of arrivals in the first t hours. Then, $(N_t)_{t\geq 0}$ is a Poisson process with parameter $\lambda = 100$. Given $N_1 = 150$, the distribution of $N_3 - N_1 = N_3 - 150$ is equal to the distribution of N_2. This gives

$$\begin{aligned} P(N_3 \leq 350 | N_1 = 150) &= P(N_3 - 150 \leq 200 | N_1 = 150) \\ &= P(N_2 \leq 200) \\ &= \sum_{k=0}^{200} \frac{e^{-100\,(2)}(100(2))^k}{k!} \\ &= 0.519. \end{aligned}$$

■

6.2 ARRIVAL, INTERARRIVAL TIMES

For a Poisson process with parameter λ, let X denote the time of the first arrival. Then, $X > t$ if and only if there are no arrivals in $[0, t]$. Thus,

$$P(X > t) = P(N_t = 0) = e^{-\lambda t}, \quad \text{for } t > 0.$$

Hence, X has an exponential distribution with parameter λ.

The exponential distribution plays a central role in the Poisson process. What is true for the time of the first arrival is also true for the time between the first and second arrival, and for all *interarrival times*. A Poisson process is a counting process for which interarrival times are independent and identically distributed exponential random variables.

Poisson Process—Definition 2

Let $X_1, X_2, \ldots$ be a sequence of i.i.d. exponential random variables with parameter λ. For $t > 0$, let

$$N_t = \max\{n : X_1 + \cdots + X_n \leq t\},$$

with $N_0 = 0$. Then, $(N_t)_{t\geq 0}$ defines a Poisson process with parameter λ.

Let

$$S_n = X_1 + \cdots + X_n, \quad \text{for } n = 1, 2, \ldots$$

We call $S_1, S_2, \ldots$ the *arrival times* of the process, where S_k is the time of the kth arrival. Furthermore,

$$X_k = S_k - S_{k-1}, \quad \text{for } k = 1, 2, \ldots$$

is the *interarrival time* between the $(k-1)$th and kth arrival, with $S_0 = 0$.

The relationship between the interarrival and arrival times for a Poisson process is illustrated in Figure 6.2.

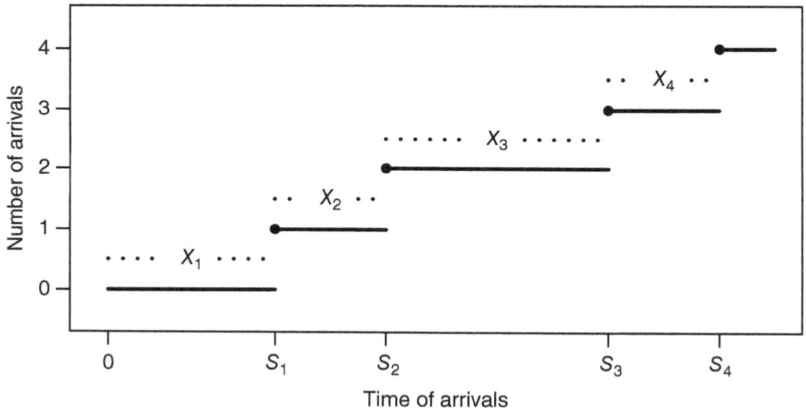

Figure 6.2 Arrival times $S_1, S_2, \ldots$, and interarrival times $X_1, X_2, \ldots$

In the next section, we show that Definitions 1 and 2 are equivalent. A benefit of Definition 2 is that it leads to a direct method for constructing, and simulating, a Poisson process:

1. Let $S_0 = 0$.
2. Generate i.i.d. exponential random variables $X_1, X_2, \ldots$
3. Let $S_n = X_1 + \cdots + X_n$, for $n = 1, 2, \ldots$
4. For each $k = 0, 1, \ldots$, let $N_t = k$, for $S_k \leq t < S_{k+1}$.

Two realizations of a Poisson process with $\lambda = 0.1$ obtained by this method on the interval $[0, 100]$ are shown in Figure 6.3.

The importance of the exponential distribution to the Poisson process lies in its unique memoryless property, a topic from probability that merits review.

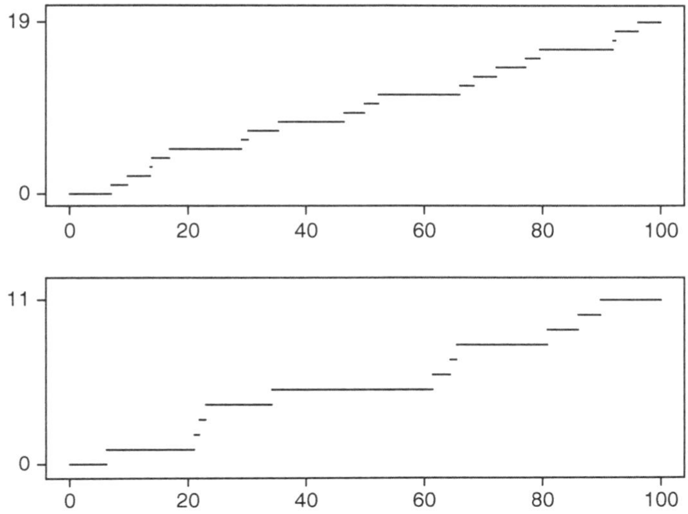

Figure 6.3 Realizations of a Poisson process with $\lambda = 0.1$.

Memorylessness

To illustrate, assume that Amy and Zach each want to take a bus. Buses arrive at a bus stop according to a Poisson process with parameter $\lambda = 1/30$. That is, the times between buses have an exponential distribution, and buses arrive, on average, once every 30 minutes. Unlucky Amy gets to the bus stop just as a bus pulls out of the station. Her waiting time for the next bus is about 30 minutes. Zach arrives at the bus stop 10 minutes after Amy. Remarkably, the time that Zach waits for a bus also has an exponential distribution with parameter $\lambda = 1/30$. Memorylessness means that their waiting time distributions are the same, and they will both wait, on average, the same amount of time!

To prove it true, observe that Zach waits more than t minutes if and only if Amy waits more than $t + 10$ minutes, given that a bus does not come in the first 10 minutes. Let A and Z denote Amy and Zach's waiting times, respectively. Amy's waiting time is exponentially distributed. Hence,

$$P(Z > t) = P(A > t + 10 | A > 10) = \frac{P(A > t + 10)}{P(A > 10)}$$
$$= \frac{e^{-(t+10)/30}}{e^{-10/30}} = e^{-t/30} = P(A > t),$$

from which it follows that A and Z have the same distribution. See the `R` code and Figure 6.4 for the results of a simulation.

Of course, there is nothing special about $t = 10$. Memorylessness means that regardless of how long you have waited, the distribution of the time you still have to wait is the same as the original waiting time.

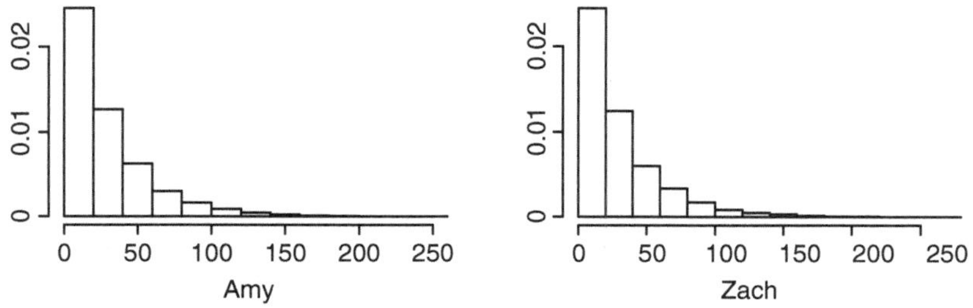

Figure 6.4 Waiting time distributions for Amy and Zach. Zach arrives 10 minutes after Amy. By memorylessness, the distributions are the same.

R : Bus Waiting Times

```
# buswaiting.R
> trials <- 5000
> amy <- numeric(trials)
> zach <- numeric(trials)
> for (i in 1:trials) {
 + bus <- rexp(1,1/30)
 + amy[i] <- bus
 + while (bus < 10) { bus <- bus + rexp(1,1/30) }
 + zach[i] <- bus-10 }
> mean(amy)
[1] 29.8043
> mean(zach)
[1] 30.39833
> hist(amy,xlab="Amy",prob=T,ylab="",main="")
> hist(zach,xlab="Zach",prob=T,ylab="",main="")
```

The exponential distribution is the only continuous distribution that is memoryless. (The geometric distribution has the honors for the discrete case.) Here is the general statement of the property.

Memorylessness

A random variable X is *memoryless* if, for all $s, t > 0$,

$$P(X > s + t | X > s) = P(X > t).$$

Results for the *minimum* of independent exponential random variables are particularly useful when working with the Poisson process. We highlight two properties that arise in many settings.

Minimum of Independent Exponential Random Variables

Let $X_1, \ldots, X_n$ be independent exponential random variables with respective parameters $\lambda_1, \ldots, \lambda_n$. Let $M = \min(X_1, \ldots, X_n)$.

1. For $t > 0$,
$$P(M > t) = e^{-t(\lambda_1+\cdots+\lambda_n)}. \tag{6.1}$$
That is, M has an exponential distribution with parameter $\lambda_1 + \cdots + \lambda_n$.
2. For $k = 1, \ldots, n$,
$$P(M = X_k) = \frac{\lambda_k}{\lambda_1 + \cdots + \lambda_n}. \tag{6.2}$$

Proof.

1. For $t > 0$,
$$\begin{aligned} P(M > t) &= P(X_1 > t, \ldots, X_n > t) = P(X_1 > t) \cdots P(X_n > t) \\ &= e^{-\lambda_1 t} \cdots e^{-\lambda_n t} = e^{-t(\lambda_1+\cdots+\lambda_n)}. \end{aligned}$$
2. For $1 \le k \le n$, conditioning on X_k gives
$$\begin{aligned} P(M = X_k) &= P(\min(X_1, \ldots, X_n) = X_k) \\ &= P(X_1 \ge X_k, \ldots, X_n \ge X_k) \\ &= \int_0^\infty P(X_1 \ge t, \ldots, X_n \ge t | X_k = t)\lambda_k e^{-\lambda_k t}\, dt \\ &= \int_0^\infty P(X_1 \ge t, \ldots, X_{k-1} \ge t, X_{k+1} \ge t, \ldots, X_n \ge t)\lambda_k e^{-\lambda_k t}\, dt \\ &= \int_0^\infty P(X_1 \ge t) \cdots P(X_{k-1} \ge t) P(X_{k+1} \ge t) \cdots P(X_n \ge t)\lambda_k e^{-\lambda_k t}\, dt \\ &= \lambda_k \int_0^\infty e^{-t(\lambda_1+\cdots+\lambda_n)}\, dt = \frac{\lambda_k}{\lambda_1 + \cdots + \lambda_n}. \end{aligned}$$

∎

Example 6.4 A Boston subway station services the red, green, and orange lines. Subways on each line arrive at the station according to three independent Poisson processes. On average, there is one red train every 10 minutes, one green train every 15 minutes, and one orange train every 20 minutes.

(i) When you arrive at the station what is the probability that the first subway that arrives is for the green line?

(ii) How long will you wait, on average, before some train arrives?

(iii) You have been waiting 20 minutes for a red train and have watched three orange trains go by. What is the expected additional time you will wait for your subway?

Solution

(i) Let X_G, X_R, and X_O denote, respectively, the times of the first green, red, and orange subways that arrive at the station. The event that the first subway is green is the event that X_G is the minimum of the three independent random variables. The desired probability is

$$P(\min(X_G, X_R, X_O) = X_G) = \frac{1/15}{1/10 + 1/15 + 1/20} = \frac{4}{13} = 0.31.$$

(ii) The time of the first train arrival is the minimum of X_G, X_R, and X_O, which has an exponential distribution with parameter

$$\frac{1}{10} + \frac{1}{15} + \frac{1}{20} = \frac{13}{60}.$$

Thus, you will wait, on average $60/13 = 4.615$ minutes. A *quick* simulation gives

```
> sim <- replicate(10000,
+    min(rexp(1,1/10),rexp(1,1/15),rexp(1,1/20)))
> mean(sim)
[1] 4.588456
```

(iii) Your waiting time is independent of the orange arrivals. By memorylessness of interarrival times, the additional waiting time for the red line has the same distribution as the original waiting time. You will wait, on average, 10 more minutes. ■

For a Poisson process, each arrival time S_n is a sum of n i.i.d. exponential interarrival times. A sum of i.i.d. exponential variables has a gamma distribution.

Arrival Times and Gamma Distribution

For $n = 1, 2, \ldots$, let S_n be the time of the nth arrival in a Poisson process with parameter λ. Then, S_n has a gamma distribution with parameters n and λ. The density function of S_n is

$$f_{S_n}(t) = \frac{\lambda^n t^{n-1} e^{-\lambda t}}{(n-1)!}, \quad \text{for } t > 0.$$

Mean and variance are

$$E(S_n) = \frac{n}{\lambda} \quad \text{and} \quad Var(S_n) = \frac{n}{\lambda^2}.$$

For a general gamma distribution, the parameter n does not have to be an integer. When it is, the distribution is sometimes called an *Erlang distribution*. Observe that if $n = 1$, the gamma distribution reduces to the exponential distribution with parameter λ.

Two ways to derive the gamma distribution for sums of i.i.d. exponentials are (i) by induction on n, using the fact that $S_n = S_{n-1} + X_n$, and (ii) by moment-generating functions. We leave both derivations to the exercises.

Example 6.5 The times when goals are scored in hockey are modeled as a Poisson process in Morrison (1976). For such a process, assume that the average time between goals is 15 minutes.

(i) In a 60-minute game, find the probability that a fourth goal occurs in the last 5 minutes of the game.

(ii) Assume that at least three goals are scored in a game. What is the mean time of the third goal?

Solution The parameter of the hockey Poisson process is $\lambda = 1/15$.

(i) The desired probability is

$$P(55 < S_4 \leq 60) = \frac{1}{6} \int_{55}^{60} (1/15)^4 t^3 e^{-t/15}\, dt = 0.068.$$

In R, the probability is found by typing

```
> pgamma(60,4,1/15)-pgamma(55,4,1/15)
[1] 0.06766216
```

(ii) The desired expectation is

$$\begin{aligned} E(S_3|S_3 < 60) &= \frac{1}{P(S_3 < 60)} \int_0^{60} t f_{S_3}(t)\, dt \\ &= \frac{1}{P(S_3 < 60)} \int_0^{60} t \frac{(1/15)^3 t^2 e^{-t/15}}{2}\, dt \\ &= \frac{25.4938}{0.7619} = 33.461 \text{ minutes.} \end{aligned}$$

■

6.3 INFINITESIMAL PROBABILITIES

A third way to define the Poisson process is based on an infinitesimal description of the distribution of points (e.g., events) in small intervals.

To state the new definition, we use *little-oh* notation. Write $f(h) = o(h)$ to mean that

$$\lim_{h \to 0} \frac{f(h)}{h} = 0.$$

More generally, say that a function f is *little-oh of* g, and write $f(h) = o(g(h))$, to mean that

$$\lim_{h \to 0} \frac{f(h)}{g(h)} = 0.$$

Little-oh notation is often used when making order of magnitude statements about a function, or in referencing the remainder term of an approximation.

For example, the Taylor series expansion of e^h with remainder term is

$$e^h = 1 + h + \frac{h^2}{2} + \frac{h^3}{6} + \cdots = 1 + h + R(h),$$

where $R(h) = e^z h^2/2$, for some $z \in (-h, h)$. Since $R(h)/h = e^z h/2 \to 0$, as $h \to 0$, we can write

$$e^h = 1 + h + o(h).$$

Note that if two functions f and g are little-oh of h, then $f(h) + g(h) = o(h)$, since $(f(h) + g(h))/h \to 0$, as $h \to 0$. Similarly, if $f(h) = o(h)$, then $cf(h) = o(h)$, for any constant c. If $f(h) = o(1)$, then $f(h) \to 0$, as $h \to 0$.

Poisson Process—Definition 3

A *Poisson process with parameter* λ is a counting process $(N_t)_{t \geq 0}$ with the following properties:

1. $N_0 = 0$.
2. The process has stationary and independent increments.
3. $P(N_h = 0) = 1 - \lambda h + o(h)$.
4. $P(N_h = 1) = \lambda h + o(h)$.
5. $P(N_h > 1) = o(h)$.

Properties 3–5 essentially ensure that there cannot be infinitely many arrivals in a finite interval, and that in an infinitesimal interval there may occur at most one event.

It is straightforward to show that Definition 3 is a consequence of Definition 1. If N_h has a Poisson distribution with parameter λh, then

$$P(N_h = 0) = e^{-\lambda h} = 1 - \lambda h + o(h),$$
$$P(N_h = 1) = e^{-\lambda h}\lambda h = (1 - \lambda h + o(h))\lambda h = \lambda h + o(h),$$

and

$$\begin{aligned} P(N_h > 1) &= 1 - P(N_h = 0) - P(N_h = 1) \\ &= 1 - (1 - \lambda h + o(h)) - (\lambda h + o(h)) \\ &= o(h). \end{aligned}$$

The converse, that Definition 3 implies Definition 1, is often shown by deriving and solving a differential equation. We will forego the rigorous proof, but give a heuristic explanation, which offers insight into the nature of Poisson arrivals.

Assume that Definition 3 holds. We need to show that N_t has a Poisson distribution with parameter λt.

Consider N_t, the number of points in the interval $[0, t]$. Partition $[0, t]$ into n subintervals each of length t/n. Properties 3–5 imply that for sufficiently large n, each subinterval will contain either 0 or 1 point with high probability. The chance that a small subinterval contains 2 or more points is negligible. See Figure 6.5.

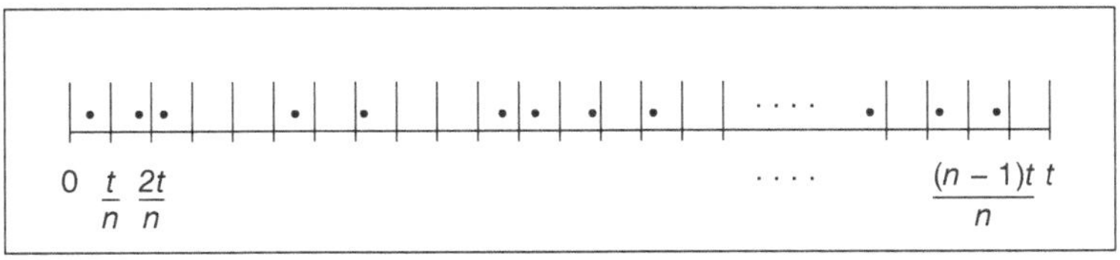

Figure 6.5 A partition of $[0, t]$ where each subinterval contains 0 or 1 point.

A subinterval has the form $((k-1)t/n, kt/n]$ for some $k = 1, \dots, n$. By stationary increments,

$$P(N_{kt/n} - N_{(k-1)t/n} = 1) = P(N_{t/n} = 1) = \frac{\lambda t}{n} + o\left(\frac{t}{n}\right), \quad \text{for all } k.$$

Furthermore, by independent increments, whether or not a point is contained in a particular subinterval is independent of the points in any other subinterval.

Hence, for large n, the outcomes in each subinterval can be considered a sequence of n i.i.d. Bernoulli trials, where the probability p_n that a point is contained in a subinterval is

$$p_n = \frac{\lambda t}{n} + o\left(\frac{t}{n}\right).$$

Thus, the number of points in $[0, t]$, being the sum of n i.i.d. Bernoulli trials, has a binomial distribution with parameters n and p_n.

The result that N_t has a Poisson distribution is obtained by letting n tend to infinity and appealing to the Poisson approximation of the binomial distribution. Since

$$np_n = n\left[\frac{\lambda t}{n} + \mathrm{o}\left(\frac{t}{n}\right)\right] = \lambda t + n\left[\mathrm{o}\left(\frac{t}{n}\right)\right] \to \lambda t > 0, \quad \text{as } n \to \infty,$$

the approximation holds, which gives that N_t has a Poisson distribution with parameter λt. Note that for the $\mathrm{o}(t/n)$ term, since

$$\lim_{1/n \to 0} \frac{\mathrm{o}(t/n)}{t/n} = 0, \quad \text{then equivalently} \quad \lim_{n\to\infty} n[\mathrm{o}(t/n)] = 0.$$

A statement and proof of the Poisson approximation of the binomial distribution is given in Appendix B, Section B.4.

Equivalence of Poisson Definitions

We have shown that Definitions 1 and 3 are equivalent. Here, we show that Definitions 1 and 2 are equivalent.

Assume Definition 2. That is, let $X_1, X_2, \ldots$ be an i.i.d. sequence of exponential random variables with parameter λ. For each n, let $S_n = X_1 + \cdots + X_n$, with $S_0 = 0$, and let $N_t = \max\{n : S_n \le t\}$. We show N_t has a Poisson distribution with parameter λt.

Observe that for $k \ge 0$, $N_t = k$ if and only if the kth arrival occurs by time t and the $(k+1)$th arrival occurs after t. That is, $S_k \le t < S_k + X_{k+1}$. Since S_k is a function of $X_1, \ldots, X_k$, S_k and X_{k+1} are independent random variables, and their joint density is the product of their marginal densities. This gives

$$f_{S_k, X_{k+1}}(s, x) = f_{S_k}(s) f_{X_{k+1}}(x) = \left(\frac{\lambda^k s^{k-1} e^{-\lambda s}}{(k-1)!}\right) \lambda e^{-\lambda x}, \quad \text{for } s, x > 0.$$

For $k \ge 0$,

$$\begin{aligned}
P(N_t = k) &= P(S_k \le t \le S_k + X_{k+1}) \\
&= P(S_k \le t, X_{k+1} \ge t - S_k) \\
&= \int_0^t \int_{t-s}^{\infty} f_{S_k, X_{k+1}}(s, x)\, dx\, ds \\
&= \int_0^t \int_{t-s}^{\infty} \left(\frac{\lambda^k s^{k-1} e^{-\lambda s}}{(k-1)!}\right) \lambda e^{-\lambda x}\, dx\, ds \\
&= \frac{\lambda^k}{(k-1)!} \int_0^t \left(s^{k-1} e^{-\lambda s}\right) e^{-\lambda(t-s)}\, ds \\
&= \frac{e^{-\lambda t} \lambda^k}{(k-1)!} \int_0^t s^{k-1}\, ds = \frac{e^{-\lambda t} (\lambda t)^k}{k!},
\end{aligned}$$

which gives the desired Poisson distribution.

The fact that the interarrival times of a Poisson process are memoryless means that the pattern of arrivals from an arbitrary time s onward behaves the same as the pattern of arrivals from time 0 onward. It follows that the number of arrivals in the interval $(s, s+t]$ has the same distribution as the number of arrivals in $(0, t]$. Stationary, as well as independent, increments are direct consequences.

Conversely, assume Definition 1. Consider the distribution of the first interarrival time X_1. As shown at the beginning of Section 6.2 the distribution is exponential with parameter λ.

For X_2, consider $P(X_2 > t|X_1 = s)$. If the first arrival occurs at time s, and the second arrival occurs more than t time units later, then there are no arrivals in the interval $(s, s+t]$. Conversely, if there are no arrivals in $(s, s+t]$ and $X_1 = s$, then $X_2 > t$. Thus,

$$\begin{aligned} P(X_2 > t|X_1 = s) &= P(N_{s+t} - N_s = 0|X_1 = s) \\ &= P(N_{s+t} - N_s = 0) \\ &= P(N_t = 0) = e^{-\lambda t}, \quad \text{for } t > 0. \end{aligned}$$

The second equality is because of independent increments. It follows that X_1 and X_2 are independent, and the distribution of X_2 is exponential with parameter λ.

For the general case, consider $P(X_{k+1} > t|X_1 = x_1, X_2 = x_2, \ldots, X_k = x_k)$. If the kth arrival occurs at time $x_1 + \cdots + x_k$ and $X_{k+1} > t$, then there are no arrivals in the interval $(x_1 + \cdots + x_k, x_1 + \cdots + x_k + t]$. See Figure 6.6.

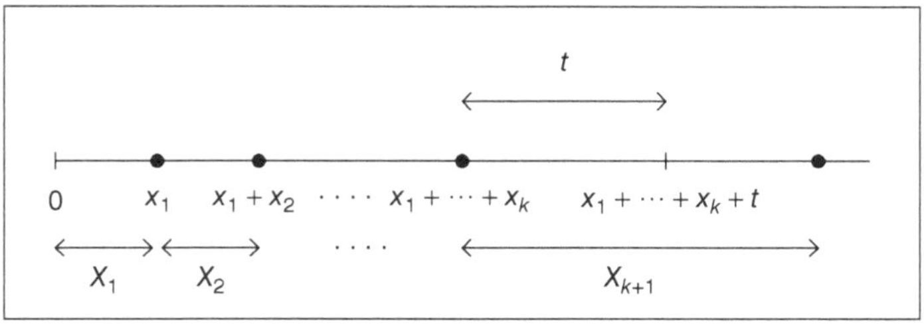

Figure 6.6 With $X_{k+1} > t$, there are no arrivals between $x_1 + \cdots + x_k$ and $x_1 + \cdots + x_{k+t}$.

We have that

$$\begin{aligned} &P(X_{k+1} > t|X_1 = x_1, \ldots, X_k = x_k) \\ &\quad = P(N_{x_1+\cdots+x_k+t} - N_{x_1+\cdots+x_k} = 0|X_1 = x_1, \ldots, X_k = x_k) \\ &\quad = P(N_{x_1+\cdots+x_k+t} - N_{x_1+\cdots+x_k} = 0) \\ &\quad = P(N_t = 0) = e^{-\lambda t}, \quad \text{for } t > 0. \end{aligned}$$

Hence, X_{k+1} is independent of $X_1, \ldots, X_k$, and the distribution of X_{k+1} is exponential with parameter λ.

It follows that $X_1, X_2, \ldots$ is an i.i.d. sequence of exponential random variables with parameter λ, which gives Definition 2.

The equivalence of all three definitions has been established.

6.4 THINNING, SUPERPOSITION

According to the United Nations Population Division, the worldwide sex ratio at birth is 108 boys to 100 girls. Thus, the probability that any birth is a boy is

$$p = \frac{108}{108 + 100} = 0.519.$$

That this probability is greater than one-half is said to be nature's way of balancing the fact that boys have a slightly higher risk than girls of not surviving birth.

Assume that babies are born on a maternity ward according to a Poisson process $(N_t)_{t\geq 0}$ with parameter λ. How can the number of male births and the number of female births be described?

Babies' sex is independent of each other. We can think of a male birth as the result of a coin flip whose heads probability is p. Assume that there are n births by time t. Then, the number of male births by time t is the number of heads in n i.i.d. coin flips, which has a binomial distribution with parameters n and p. Similarly the number of female births in $[0, t]$ has a binomial distribution with parameters n and $1 - p$.

Let M_t denote the number of male births by time t. Similarly define the number of female births F_t. Thus, $M_t + F_t = N_t$. The joint probability mass function of (M_t, F_t) is

$$\begin{aligned}
P(M_t = m, F_t = f) &= P(M_t = m, F_t = f, N_t = m + f)\\
&= P(M_t = m, F_t = f | N_t = m + f)P(N_t = m + f)\\
&= P(M_t = m | N_t = m + f)P(N_t = m + f)\\
&= \frac{(m+f)!}{m!f!}p^m(1-p)^f\frac{e^{-\lambda t}(\lambda t)^{m+f}}{(m+f)!}\\
&= \frac{p^m(1-p)^f e^{-\lambda t(p+(1-p))}(\lambda t)^{m+f}}{m!f!}\\
&= \left(\frac{e^{-\lambda pt}(\lambda pt)^m}{m!}\right)\left(\frac{e^{-\lambda(1-p)t}(\lambda(1-p)t)^f}{f!}\right),
\end{aligned}$$

for $m,f = 0, 1, \ldots$ This shows that M_t and F_t are independent Poisson random variables with parameters λpt and $\lambda(1-p)t$, respectively. In fact, each process $(M_t)_{t\geq 0}$ and $(F_t)_{t\geq 0}$ is a Poisson process, called a *thinned process*. It is not hard to show that both processes inherit stationary and independent increments from the original Poisson birth process.

The birth example with two thinned processes illustrates a general result.

Thinned Poisson Process

Let $(N_t)_{t\geq 0}$ be a Poisson process with parameter λ. Assume that each arrival, independent of other arrivals, is marked as a *type k* event with probability p_k, for $k = 1, \dots, n$, where $p_1 + \cdots + p_n = 1$. Let $N_t^{(k)}$ be the number of type k events in $[0, t]$. Then, $\left(N_t^{(k)}\right)_{t\geq 0}$ is a Poisson process with parameter λp_k. Furthermore, the processes

$$\left(N_t^{(1)}\right)_{t\geq 0}, \dots, \left(N_t^{(n)}\right)_{t\geq 0}$$

are independent. Each process is called a *thinned Poisson process*.

Example 6.6 Consider the male and female birth processes. Assume that births occur on a maternity ward at the average rate of 2 births per hour.

(i) On an 8-hour shift, what is the expectation and standard deviation of the number of female births?

(ii) Find the probability that only girls were born between 2 and 5 p.m.

(iii) Assume that five babies were born on the ward yesterday. Find the probability that two are boys.

Solution Let $(N_t)_{t\geq 0}$, $(M_t)_{t\geq 0}$, and $(F_t)_{t\geq 0}$ denote the overall birth, male, and female processes, respectively.

(i) Female births form a Poisson process with parameter

$$\lambda(1-p) = 2(0.481) = 0.962.$$

The number of female births on an 8-hour shift F_8 has a Poisson distribution with expectation

$$E(F_8) = \lambda(1-p)8 = 2(0.481)8 = 7.696,$$

and standard deviation

$$SD(F_8) = \sqrt{7.696} = 2.774.$$

(ii) The desired probability is $P(M_3 = 0, F_3 > 0)$. By independence,

$$\begin{aligned} P(M_3 = 0, F_3 > 0) &= P(M_3 = 0)P(F_3 > 0) \\ &= e^{-2(0.519)3}\left(1 - e^{-2(0.481)3}\right) \\ &= e^{-3.114}\left(1 - e^{-2.886}\right) = 0.042. \end{aligned}$$

(iii) Conditional on there being five births in a given interval, the number of boys in that interval has a binomial distribution with parameters $n = 5$ and $p = 0.519$. The desired probability is

$$\frac{5!}{2!3!}(0.519)^2(0.481)^3 = 0.30.$$

■

Related to the thinned process is the *superposition* process obtained by merging, or adding, independent Poisson processes. We state the following intuitive result without proof:

Superposition Process

Assume that $\left(N_t^{(1)}\right)_{t\geq 0}, \ldots, \left(N_t^{(n)}\right)_{t\geq 0}$ are n independent Poisson processes with respective parameters $\lambda_1, \ldots, \lambda_n$. Let $N_t = N_t^{(1)} + \cdots + N_t^{(n)}$, for $t \geq 0$. Then, $(N_t)_{t\geq 0}$ is a Poisson process with parameter $\lambda = \lambda_1 + \cdots + \lambda_n$.

See Figure 6.7 to visualize the superposition of three independent Poisson processes.

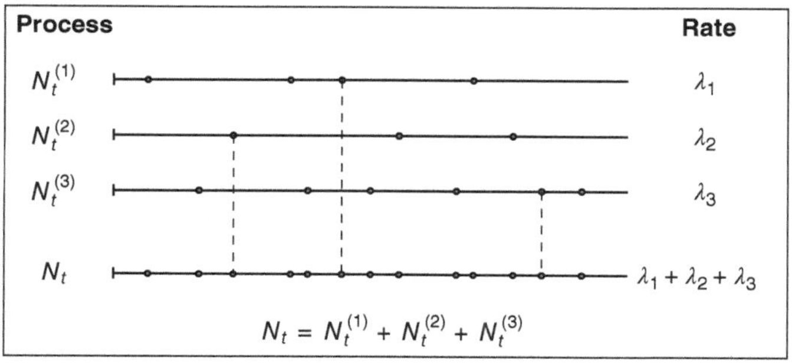

Figure 6.7 The N_t process is the superposition of $N_t^{(1)}$, $N_t^{(2)}$, and $N_t^{(3)}$.

Example 6.7 (Oh my!) In the land of Oz, sightings of lions, tigers, and bears each follow a Poisson process with respective parameters, $\lambda_L, \lambda_T, \lambda_B$, where the time unit is hours. Sightings of the three species are independent of each other.

(i) Find the probability that Dorothy will not see any animal in the first 24 hours from when she arrives in Oz.

(ii) Dorothy saw three animals one day. Find the probability that each species was seen.

Solution

(i) The process of animal sightings $(N_t)_{t\geq 0}$ is the superposition of three independent Poisson processes. Thus, it is a Poisson process with parameter $\lambda_L + \lambda_T + \lambda_B$. The desired probability is

$$P(N_{24} = 0) = e^{-24(\lambda_L+\lambda_T+\lambda_B)}.$$

(ii) Let L_t, T_t, and B_t be the numbers of lions, tigers, and bears, respectively, seen by time t. The desired probability is

$$\begin{aligned}
P(L_{24} = 1, B_{24} = 1, T_{24} = 1|N_{24} = 3) \\
&= \frac{P(L_{24} = 1, B_{24} = 1, T_{24} = 1, N_{24} = 3)}{P(N_{24} = 3)} \\
&= \frac{P(L_{24} = 1, B_{24} = 1, T_{24} = 1)}{P(N_{24} = 3)} \\
&= \frac{P(L_{24} = 1)P(B_{24} = 1)P(T_{24} = 1)}{P(N_{24} = 3)} \\
&= \frac{(e^{-24\lambda_L}24\lambda_L)(e^{-24\lambda_T}24\lambda_T)(e^{-24\lambda_B}24\lambda_B)}{e^{-24(\lambda_L+\lambda_T+\lambda_B}(24(\lambda_L + \lambda_T + \lambda_B))^3/3!} \\
&= \frac{6\lambda_L\lambda_B\lambda_T}{(\lambda_L + \lambda_B + \lambda_T)^3}.
\end{aligned}$$

■

Embedding and the Birthday Problem

Sometimes discrete problems can be solved by *embedding* them in continuous ones. The methods in this section were popularized in Blom and Holst (1991) for solving discrete balls-and-urn models, which involve sampling with and without replacement. We illustrate the method on the famous birthday problem. (If your probability class did not cover the birthday problem, you should ask for your money back.)

The classic birthday problem asks, "How many people must be in a room before the probability that some share a birthday, ignoring year and leap days, is at least 50%?"

The probability that two people have the same birthday is 1 minus the probability that no one shares a birthday, which is

$$p_k = 1 - \prod_{i=1}^{k} \frac{366 - i}{365} = 1 - \frac{365!}{(365 - k)!365^k}.$$

One finds that $p_{22} = 0.476$ and $p_{23} = 0.507$. Thus, 23 people are needed.

Consider this variant of the birthday problem, assuming a random person's birthday is uniformly distributed on the 365 days of the year. People enter a room one by one. How many people are in the room the first time that two people share the same birthday? Let K be the desired number. We show how to find the mean and standard deviation of K by embedding.

Consider a continuous-time version of the previous question. People enter a room according to a Poisson process $(N_t)_{t\geq 0}$ with rate $\lambda = 1$. Each person is independently *marked* with one of 365 birthdays, where all birthdays are equally likely. The procedure creates 365 thinned Poisson processes, one for each birthday. Each of the 365 processes are independent, and their superposition gives the process of people entering the room. See Figure 6.8.

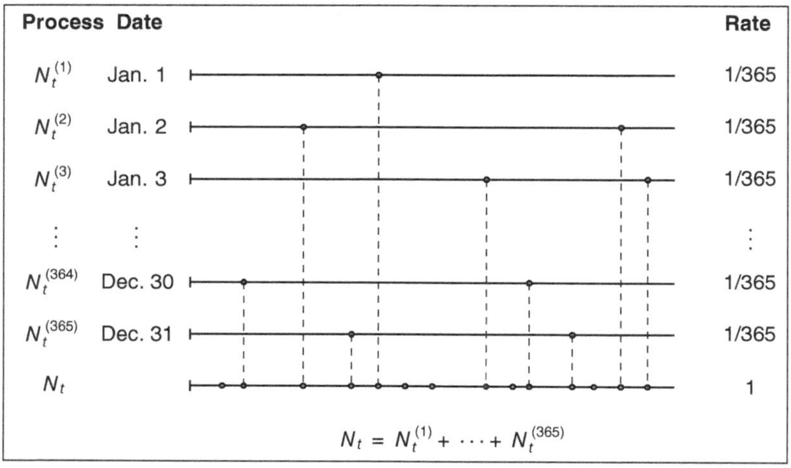

Figure 6.8 Embedding the birthday problem in a superposition of Poisson processes.

Let $X_1, X_2, \ldots,$ be the interarrival sequence for the process of people entering the room. The X_i are i.i.d. exponential random variables with mean 1. Let T be the first time when two people in the room share the same birthday. Then,

$$T = \sum_{i=1}^{K} X_i. \tag{6.3}$$

Equation (6.3) relates the interarrival times $X_1, X_2, \ldots,$ the continuous time T, and the discrete count K.

The X_i are independent of K. The random variable T is represented as a random sum of random variables. By results for such sums (see Example 1.27),

$$E[T] = E[K]\ E[X_1] = E[K].$$

For each $k = 1, \ldots, 365$, let Z_k be the time when the second person marked with birthday k enters the room. Then, the first time two people in the room have the same birthday is $T = \min_{1\leq k\leq 365} Z_k$. Each Z_k, being the arrival time of the second event of a Poisson process, has a gamma distribution with parameters $n = 2$ and $\lambda = 1/365$, with density

$$f(t) = \frac{te^{-t/365}}{365^2}, \quad \text{for } t > 0,$$

and cumulative distribution function

$$P(Z_1 \leq t) = \int_0^t \frac{se^{-s/365}}{365^2}\, ds = 1 - \frac{e^{-t/365}(365 + t)}{365}.$$

This gives

$$\begin{aligned} P(T > t) &= P(\min(Z_1, \ldots, Z_{365}) > t) \\ &= P(Z_1 > t, \ldots, Z_{365} > t) \\ &= P(Z_1 > t)^{365} = \left(1 + \frac{t}{365}\right)^{365} e^{-t}, \quad \text{for } t > 0. \end{aligned}$$

The desired birthday expectation is

$$E(K) = E(T) = \int_0^\infty P(T > t)\, dt = \int_0^\infty \left(1 + \frac{t}{365}\right)^{365} e^{-t}\, dt. \tag{6.4}$$

The second equality makes use of a general result for the expectation of a positive, continuous random variable. (For reference, see Exercise 1.21.) The last integral of Equation (6.4) is difficult to solve exactly. It can be estimated using a Taylor series approximation. A numerical software package finds $E(K) = 24.617$ and $Var(K) = 779.23$, with standard deviation 27.91.

We invite the reader to use embedding to find the expected number of people needed for *three* people to share the same birthday. See Exercise 6.19.

6.5 UNIFORM DISTRIBUTION

It is common to think of a Poisson process as modeling a *completely random* distribution of events, or points, on the positive number line. Although it is not possible to have a uniform distribution on $[0, \infty)$, or any unbounded interval, there is nevertheless a strong connection between a Poisson process and the uniform distribution.

If a Poisson process contains exactly n events in an interval $[0, t]$, then the unordered *locations*, or times, of those events are uniformly distributed on the interval.

For the case of one event, the assertion is easily shown. Consider the distribution of the time of the first arrival, conditional on there being one arrival by time t. For $0 \le s \le t$,

$$\begin{aligned} P(S_1 \le s | N_t = 1) &= \frac{P(S_1 \le s, N_t = 1)}{P(N_t = 1)} = \frac{P(N_s = 1, N_t = 1)}{P(N_t = 1)} \\ &= \frac{P(N_s = 1, N_t - N_s = 0)}{P(N_t = 1)} \\ &= \frac{P(N_s = 1)P(N_{t-s} = 0)}{P(N_t = 1)} \\ &= \frac{e^{-\lambda s}\lambda s e^{-\lambda(t-s)}}{e^{-\lambda t}\lambda t} = \frac{s}{t}, \end{aligned}$$

which is the cumulative distribution function of the uniform distribution on $[0, t]$.

To discuss the case of more than one arrival in $[0, t]$, we introduce the topic of *order statistics*. Let $U_1, \ldots, U_n$ be an i.i.d. sequence of random variables uniformly distributed on $[0, t]$. Their joint density function is

$$f_{U_1,\ldots,U_n}(u_1, \ldots, u_n) = \frac{1}{t^n}, \quad \text{for } 0 \le u_1, \ldots, u_n \le t.$$

Arrange the U_i in increasing order $U_{(1)} \le U_{(2)} \le \cdots \le U_{(n)}$, where $U_{(k)}$ is the kth smallest of the U_i. The ordered sequence $(U_{(1)}, \ldots, U_{(n)})$ is called the *order statistics* of the original sequence.

The joint density function of the order statistics is

$$f_{U_{(1)},\ldots,U_{(n)}}(u_1, \ldots, u_n) = \frac{n!}{t^n}, \text{ for } 0 \le u_1 < \cdots < u_n \le t.$$

We will not prove this rigorously, but give an intuitive argument. Assume that

$$U_{(1)} = u_1, \ldots, U_{(n)} = u_n, \quad \text{for } 0 < u_1 < \cdots < u_n < t.$$

Consider a sample of n independent uniform random variables on $[0, t]$. There are $n!$ such samples that would give rise to these order statistic values, as there are $n!$ orderings of the n distinct numbers $u_1, \ldots, u_n$. The value of the joint density for each of these uniform samples is equal to $1/t^n$. Hence, the infinitesimal probability

$$\begin{aligned} f_{U_{(1)},\ldots,U_{(n)}}(u_1, \ldots, u_n)du_1 \cdots du_n &= n! f_{U_1,\ldots,U_n}(u_1, \ldots, u_n)du_1 \cdots du_n \\ &= \frac{n!}{t^n} du_1 \cdots du_n, \end{aligned}$$

which establishes the claim.

We can now describe the joint distribution of the arrival times in a Poisson process, conditional on the number of arrivals.

Arrival Times and Uniform Distribution

Let $S_1, S_2, \ldots,$ be the arrival times of a Poisson process with parameter λ. Conditional on $N_t = n$, the joint distribution of $(S_1, \ldots, S_n)$ is the distribution of the order statistics of n i.i.d. uniform random variables on $[0, t]$. That is, the joint density function of $S_1, \ldots, S_n$ is

$$f(s_1, \ldots, s_n) = \frac{n!}{t^n}, \text{ for } 0 < s_1 < \cdots < s_n < t. \tag{6.5}$$

Equivalently, let $U_1, \ldots, U_n$ be an i.i.d. sequence of random variables uniformly distributed on $[0, t]$. Then, conditional on $N_t = n$,

$$(S_1, \ldots, S_n) \text{ and } (U_{(1)}, \ldots, U_{(n)})$$

have the same distribution.

Proof. For jointly distributed random variables $S_1, \ldots, S_n$ with joint density f,

$$f(s_1, \ldots, s_n) = \lim_{\epsilon_1 \to 0} \cdots \lim_{\epsilon_n \to 0} \frac{P(s_1 \leq S_1 \leq s_1 + \epsilon_1, \ldots, s_n \leq S_n \leq s_n + \epsilon_n)}{\epsilon_1 \cdots \epsilon_n}.$$

To establish Equation (6.5), assume that $0 < s_1 < s_2 < \cdots < s_n < t$ and consider the event

$$\{s_1 \leq S_1 \leq s_1 + \epsilon_1, \ldots, s_n \leq S_n \leq s_n + \epsilon_n\},$$

given that there are exactly n arrivals in $[0, t]$. For $\epsilon_1, \ldots, \epsilon_n$ sufficiently small, this is the event that each of the intervals $(s_i, s_i + \epsilon_i]$ contains exactly one arrival, and no arrivals occur elsewhere in $[0, t]$. By stationary and independent increments,

$$\begin{aligned}
P(s_1 &\leq S_1 \leq s_1 + \epsilon_1, \ldots, s_n \leq S_n \leq s_n + \epsilon_n | N_t = n) \\
&= \frac{P(N_{s_1+\epsilon_1} - N_{s_1} = 1, \ldots, N_{s_n+\epsilon_n} - N_{s_n} = 1, N_t = n)}{P(N_t = n)} \\
&= \frac{\lambda\epsilon_1 e^{-\lambda\epsilon_1} \cdots \lambda\epsilon_n e^{-\lambda\epsilon_n} e^{-\lambda(t-\epsilon_1-\cdots-\epsilon_n)}}{e^{-\lambda t}(\lambda t)^n / n!} \\
&= \frac{n!\epsilon_1 \cdots \epsilon_n}{t^n}.
\end{aligned}$$

Dividing by $\epsilon_1 \cdots \epsilon_n$, and letting each $\epsilon_i \to 0$, gives the result. ∎

There is a lot of information contained in Equation (6.5). From this, one can obtain joint distributions of any subset of the arrival times, including the marginal distributions. For instance, the *last (first)* arrival time, conditional on there being n arrivals in $[0, t]$, has the same distribution as the maximum (minimum) of n independent uniform random variables on $(0, t)$.

Example 6.8 Starting at time $t = 0$, patrons arrive at a restaurant according to a Poisson process with rate 20 customers per hour.

(i) Find the probability that the 60th customer arrives in the interval $[2.9, 3]$.

(ii) If 60 people arrive at the restaurant by time $t = 3$, find the probability that the 60th customer arrives in the interval $[2.9, 3]$.

Solution

(i) The time of the 60th arrival S_{60} has a gamma distribution with parameters $n = 60$ and $\lambda = 20$. The desired probability is $P(2.9 < S_{60} < 3)$. In R, type

```
> pgamma(3,60,20)-pgamma(2.9,60,20)
[1] 0.1034368
```

(ii) Given $N_3 = 60$, the arrival time of the 60th customer has the same distribution as the maximum M of 60 i.i.d. random variables uniformly distributed on $(0, 3)$. The desired probability is

$$\begin{aligned}
P(2.9 < S_{60} < 3|N_3 = 60) &= P(2.9 < M < 3) = 1 - P(M \leq 2.9) \\
&= 1 - P(U_1 \leq 2.9, \dots, U_{60} \leq 2.9) \\
&= 1 - P(U_1 \leq 2.9)^{60} \\
&= 1 - \left(\frac{2.9}{3}\right)^{60} \\
&= 1 - 0.131 = 0.869.
\end{aligned}$$

■

Example 6.9 Concert-goers arrive at a show according to a Poisson process with parameter λ. The band starts playing at time t. The kth person to arrive in $[0, t]$ waits $t - S_k$ time units for the start of the concert, where S_k is the kth arrival time. Find the expected total waiting time of concert-goers who arrive before the band starts.

Solution The desired expectation is $E\left(\sum_{k=1}^{N_t}(t - S_k)\right)$. Conditioning on N_t,

$$\begin{aligned}
E\left(\sum_{k=1}^{N_t}(t - S_k)\right) &= \sum_{n=1}^{\infty} E\left(\sum_{k=1}^{n}(t - S_k)|N_t = n\right) P(N_t = n) \\
&= \sum_{n=1}^{\infty} E\left(tn - \sum_{k=1}^{n} S_k|N_t = n\right) P(N_t = n) \\
&= \sum_{n=1}^{\infty}\left(tn - E\left(\sum_{k=1}^{n} S_k|N_t = n\right)\right) P(N_t = n) \\
&= \lambda t^2 - \sum_{n=1}^{\infty} E\left(\sum_{k=1}^{n} S_k|N_t = n\right) P(N_t = n).
\end{aligned}$$

Conditional on n arrivals in $[0, t]$, $S_1 + \cdots + S_n$ has the same distribution as the sum of the uniform order statistics. Furthermore, $\sum_{k=1}^{n} U_{(k)} = \sum_{k=1}^{n} U_k$. This gives

$$\begin{aligned}\sum_{n=1}^{\infty} E\left(\sum_{k=1}^{n} S_k | N_t = n\right) P(N_t = n) &= \sum_{n=1}^{\infty} E\left(\sum_{k=1}^{n} U_{(k)}\right) P(N_t = n)\\ &= \sum_{n=1}^{\infty} E\left(\sum_{k=1}^{n} U_k\right) P(N_t = n)\\ &= \sum_{n=1}^{\infty} \frac{nt}{2} P(N_t = n)\\ &= \frac{\lambda t^2}{2}.\end{aligned}$$

The desired expectation is

$$E\left(\sum_{k=1}^{N_t} (t - S_k)\right) = \lambda t^2 - \frac{\lambda t^2}{2} = \frac{\lambda t^2}{2}.$$

■

Example 6.10 Students enter a campus building according to a Poisson process $(N_t)_{t \geq 0}$ with parameter λ. The times spent by each student in the building are i.i.d. random variables with continuous cumulative distribution function $F(t)$. Find the probability mass function of the number of students in the building at time t, assuming there are no students in the building at time 0.

Solution Let B_t denote the number of students in the building at time t. Conditioning on N_t,

$$\begin{aligned}P(B_t = k) &= \sum_{n=k}^{\infty} P(B_t = k | N_t = n) P(N_t = n)\\ &= \sum_{n=k}^{\infty} P(B_t = k | N_t = n) \frac{e^{-\lambda t}(\lambda t)^n}{n!}.\end{aligned}$$

Assume that n students enter the building by time t, with arrival times $S_1, \ldots, S_n$. Let Z_k be the length of time spent in the building by the kth student, for $1 \leq k \leq n$. Then, $Z_1, \ldots, Z_n$ are i.i.d. random variables with cdf F, and students leave the building at times $S_1 + Z_1, \ldots, S_n + Z_n$. There are k students in the building at time t if and only if k of the departure times $S_1 + Z_1, \dots, S_n + Z_n$ exceed t. This gives

$$\begin{aligned}
P(B_t = k | N_t = n) &= P(k \text{ of the } S_1 + Z_1, \dots, S_n + Z_n \text{ exceed } t | N_t = n) \\
&= P(k \text{ of the } U_{(1)} + Z_1, \dots, U_{(n)} + Z_n \text{ exceed } t) \\
&= P(k \text{ of the } U_1 + Z_1, \dots, U_n + Z_n \text{ exceed } t) \\
&= \binom{n}{k} p^k (1-p)^{n-k},
\end{aligned}$$

where

$$p = P(U_1 + Z_1 > t) = \frac{1}{t}\int_0^t P(Z_1 > t - x)\, dx = \frac{1}{t}\int_0^t [1 - F(x)]\, dx.$$

This gives

$$\begin{aligned}
P(B_t = k) &= \sum_{n=k}^{\infty} \binom{n}{k} p^k (1-p)^{n-k} e^{-\lambda t} \frac{(\lambda t)^n}{n!} \\
&= \frac{p^k (\lambda t)^k}{k!} \sum_{n=k}^{\infty} \frac{(1-p)^{n-k} (\lambda t)^{n-k}}{(n-k)!} \\
&= \frac{p^k (\lambda t)^k}{k!} e^{\lambda(1-p)t} \\
&= \frac{e^{-\lambda pt} (\lambda pt)^k}{k!}, \quad \text{for } k = 0, 1, \dots
\end{aligned}$$

That is, B_t has a Poisson distribution with parameter λp, where

$$p = \int_0^t [1 - F(x)]\, dx.$$

■

Simulation

Results for arrival times and the uniform distribution offer a new method for simulating a Poisson process with parameter λ on an interval $[0, t]$:

1. Simulate the number of arrivals N in $[0, t]$ from a Poisson distribution with parameter λt.
2. Generate N i.i.d. random variables uniformly distributed on $(0, t)$.
3. Sort the variables in increasing order to give the Poisson arrival times.

R: Simulating a Poisson Process on $[0, t]$

Following is a simulation of the arrival times of a Poisson process with parameter $\lambda = 1/2$ on $[0, 30]$.

```
# poissonsim.R
> t <- 30
> lambda <- 1/2
> N <- rpois(1,lambda*t)
> unifs <- runif(N,0,t))
> arrivals <- sort(unifs)
> arrivals
 [1]  8.943  9.835 11.478 12.039 16.009 17.064 17.568
 [8] 17.696 18.663 22.961 24.082 24.440 28.250
```

6.6 SPATIAL POISSON PROCESS

The spatial Poisson process is a model for the distribution of events, or *points*, in two- or higher-dimensional space. Such processes have been used to model the location of trees in a forest, galaxies in the night sky, and cancer clusters across the United States. For $d \geq 1$ and $A \subseteq \mathbb{R}^d$, let N_A denote the number of points in the set A. Write $|A|$ for the *size* of A (e.g., area in $\mathbb{R}^2$, volume in $\mathbb{R}^3$).

Spatial Poisson Process

A collection of random variables $(N_A)_{A \subseteq \mathbb{R}^d}$ is a spatial Poisson process with parameter λ if

1. for each bounded set $A \subseteq \mathbb{R}^d$, N_A has a Poisson distribution with parameter $\lambda|A|$.
2. whenever A and B are disjoint sets, N_A and N_B are independent random variables.

Observe how the spatial Poisson process generalizes the regular one-dimensional Poisson process. Property 1 gives the analogue of stationary increments, where the *size* of an interval is the length of the interval. Property 2 is the counterpart of independent increments.

Example 6.11 A spatial Poisson process in the plane has parameter $\lambda = 1/2$. Find the probability that a disk of radius 2 centered at $(3, 4)$ contains exactly 5 points.

Solution Let C denote the disk. Then, $|C| = \pi r^2 = 4\pi$. The desired probability is

$$P(N_C = 5) = \frac{e^{-\lambda|C|}(\lambda|C|)^5}{5!} = \frac{e^{-2\pi}(2\pi)^5}{5!} = 0.152.$$

■

The uniform distribution arises for the spatial process in a similar way to how it does for the one-dimensional Poisson process. Given a bounded set $A \subseteq \mathbb{R}^d$, then conditional on there being n points in A, the locations of the points are uniformly distributed in A. For this reason, a spatial Poisson process is sometimes called a model of *complete spatial randomness.*

To simulate a spatial Poisson process with parameter λ on a bounded set A, first simulate the number of points N in A according to a Poisson distribution with parameter $\lambda|A|$. Then, generate N points uniformly distributed in A.

Four realizations of a spatial Poisson process with parameter $\lambda = 100$ on the square $[0, 1] \times [0, 1]$ are shown in Figure 6.9. The circle C inside the square is centered at $(0.7, 0.7)$ with radius $r = 0.2$. The simulation was repeated 100,000 times, counting the number of points in the circle at each iteration. Table 6.1 shows the numerical results. The distribution of counts is seen to be close to the expected counts for a Poisson distribution with mean $\lambda|C| = 100\pi(0.2)^2 = 12.567$.

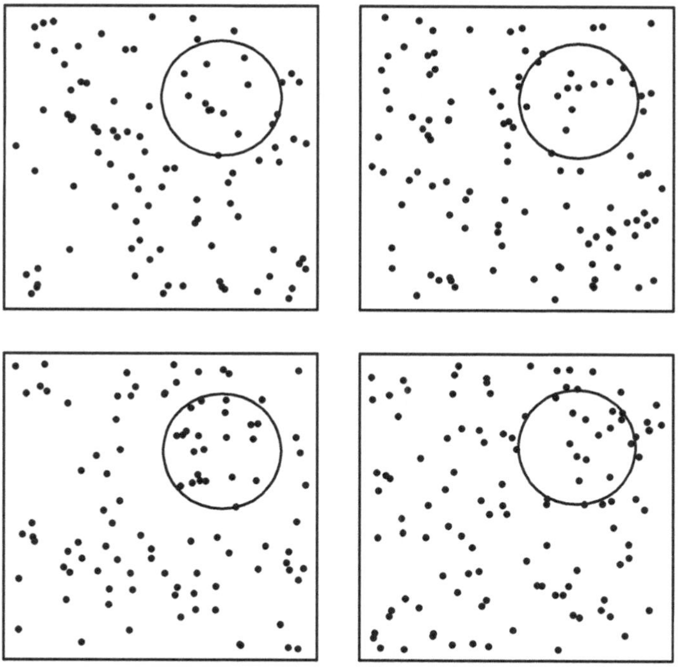

Figure 6.9 Samples of a spatial Poisson process.

TABLE 6.1 Number of Points in a Circle of Radius $r = 0.2$ for a Spatial Poisson Process with $\lambda = 100$.

Counts	0–4	5–9	10–14	15–19	20–24	25–29
Observed	522	19200	52058	24975	3135	106
Expected	510.0	19130.0	522215.6	24941.5	3075.00	125.9

Simulation is based on 100,000 trials.

R: Simulating a Spatial Poisson Process

```
# spatialPoisson.R
> lambda <- 100
> squarearea <- 1
> trials <- 100000
> simlist <- numeric(trials)
> for (i in 1:trials) {
+  N <- rpois(1,lambda*squarearea)
+  xpoints <- runif(N,0,1)
+  ypoints <- runif(N,0,1)
+  ct <- sum(((xpoints-0.7)^2+(ypoints-0.7)^2)<=0.2^2)
+  simlist[i] <- ct } # number of points in circle
> mean(simlist)
[1] 12.57771
> var(simlist)
[1] 12.57435
> # Compare with theoretical mean and variance
> lambda*pi*(0.2)^2
[1] 12.56637
```

A spatial Poisson process is a special case of a *point process*, which is a general model for the distribution of points in space. There is an abundance of applications of point processes, which include models that incorporate clustering, attraction, repulsion, time and space dependence, and so on. Often one wants to measure how close or far a given point pattern is from complete spatial randomness, that is, from a spatial Poisson process. A common measure is the *nearest-neighbor distance* defined to be the distance between an arbitrary point and the point of the process closest to it.

Consider a spatial Poisson process in $\mathbb{R}^2$ with parameter λ. Let x denote a fixed point in the plane. Let D be the distance from x to its nearest neighbor. The event $\{D > t\}$ occurs if and only if there are no points in the circle centered at x of radius t. Let C_x denote such a circle. Then,

$$P(D > t) = P\left(N_{C_x} = 0\right) = e^{-\lambda|C_x|} = e^{-\lambda\pi t^2}, \quad \text{for } t > 0.$$

Differentiating gives the density function for the nearest-neighbor distance

$$f_D(t) = e^{-\lambda\pi t^2} 2\lambda\pi t, \quad \text{for } t > 0,$$

with mean and variance

$$E(D) = \frac{1}{2\sqrt{\lambda}} \quad \text{and} \quad Var(D) = \frac{4-\pi}{4\pi\lambda}.$$

Example 6.12 (South Carolina swamp forest) Jones et al. (1994) introduce data for the locations of 630 trees, including 91 cypress trees, in a swamp hardwood forest in South Carolina for the purpose of studying tree population dynamics. See Figure 6.10. The data are contained in a 200×50 m^2 area. The average nearest-neighbor distance for all tree locations is 1.990 m. For the cypress trees it is 5.08 m.

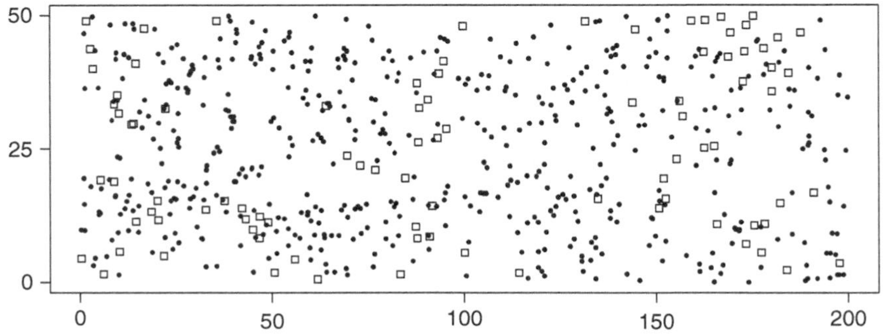

Figure 6.10 Plot of tree locations in a hardwood swamp in South Carolina. Squares are locations of cypress trees and dots locations of any other species. *Source:* Dixon (2012).

Researchers are interested in whether there is evidence of clustering. In a spatial Poisson process, with points distributed at the rate of 630 per 10,000 m^2, the expected nearest-neighbor distance is

$$E(D) = \frac{1}{2\sqrt{630/10000}} = 1.992 \text{ m},$$

with standard error

$$SD(D)/\sqrt{n} = \sqrt{\frac{4-\pi}{4\pi 630/10000}} \Big/ \sqrt{630} = 0.041.$$

For the 91 cypress trees, a model of complete spatial randomness would yield

$$E(D) = \frac{1}{2\sqrt{91/100000}} = 5.241 \text{ m},$$

with standard error

$$SD(D)/\sqrt{n} = \sqrt{\frac{4-\pi}{4\pi 630/10000}}\Big/\sqrt{91} = 1.041 \text{ m}.$$

As measured by nearest-neighbor distance, the data do not show evidence of clustering. We note that researchers were able to detect some small evidence of clustering for these data by using more sophisticated spatial statistic tools. ■

6.7 NONHOMOGENEOUS POISSON PROCESS

In a Poisson process, arrivals occur at a constant rate, independent of time. However, for many applications this is an unrealistic assumption. Consider lunch time at a college cafeteria. The doors open at 11 a.m. Students arrive at an increasing rate until the noon peak hour. Then, the rate stays constant for 2 hours, after which it declines until 3 p.m., when the cafeteria closes.

Such activity can be modeled by a nonhomogeneous Poisson process with rate $\lambda = \lambda(t)$, which depends on t. Such a rate *function* is called the *intensity function*.

Nonhomogeneous Poisson Process

A counting process $(N_t)_{t\geq 0}$ is a *nonhomogeneous Poisson process with intensity function* $\lambda(t)$, if

1. $N_0 = 0$.
2. For all $t > 0$, N_t has a Poisson distribution with mean

$$E(N_t) = \int_0^t \lambda(x)\, dx.$$

3. For $0 \leq q < r \leq s < t$, $N_r - N_q$ and $N_t - N_s$ are independent random variables.

A nonhomogeneous Poisson process has independent increments, but not necessarily stationary increments. It can be shown that for $0 < s < t$, $N_t - N_s$ has a Poisson distribution with parameter $\int_s^t \lambda(x)\, dx$. If $\lambda(t) = \lambda$ is constant, we obtain the regular Poisson process with parameter λ.

Example 6.13 Students arrive at the cafeteria for lunch according to a nonhomogeneous Poisson process. The arrival rate increases linearly from 100 to 200 students per hour between 11 a.m. and noon. The rate stays constant for the next 2 hours, and then decreases linearly down to 100 from 2 to 3 p.m. Find the probability that there are at least 400 people in the cafeteria between 11:30 a.m. and 1:30 p.m.

Solution The intensity function is

$$\lambda(t) = \begin{cases} 100 + 100t, & 0 \le t \le 1, \\ 200, & 1 < t \le 3, \\ 500 - 100t, & 3 \le t < 4, \end{cases}$$

where t represents hours past 11 a.m. See Figure 6.11.

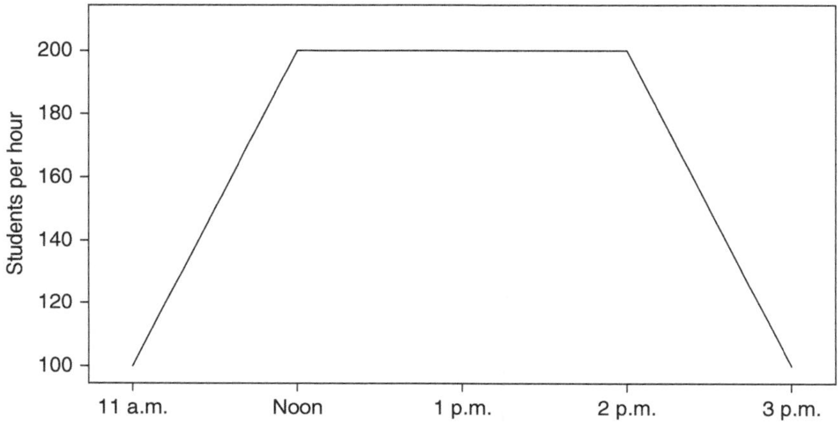

Figure 6.11 Intensity function.

The desired probability is $P(N_{2.5} - N_{0.5} \ge 400)$, where $N_{2.5} - N_{0.5}$ has a Poisson distribution with mean

$$E(N_{2.5} - N_{0.5}) = \int_{0.5}^{2.5} \lambda(t)\, dt = \int_{0.5}^{1} (100 + 100t)\, dt + \int_{1}^{2.5} 200\, dt = 387.5.$$

Then,

$$P(N_{2.5} - N_{0.5} \ge 400) = 1 - \sum_{k=0}^{399} \frac{e^{-387.5}(387.5)^k}{k!} = 0.269.$$

■

Example 6.14 In reliability engineering one is concerned with the probability that a system is working during an interval of time. A common model for failure times is a nonhomogeneous Poisson process with intensity function of the form

$$\lambda(t) = \alpha\beta t^{\beta-1},$$

where $\alpha, \beta > 0$ are parameters, and t represents the age of the system. At $\beta = 1$, the model reduces to a homogeneous Poisson process with parameter α. If the system starts at $t = 0$, the expected number of failures after t time units is

$$E(N_t) = \int_0^t \lambda(x)\, dx = \int_0^t \alpha\beta x^{\beta-1}\, dx = \alpha t^{\beta}.$$

Because of the power law form of the mean failure time, the process is sometimes called a *power law Poisson process*.

Of interest, is the *reliability* $R(t)$, defined as the probability that a system, which starts at time t, is operational up through time $t + c$, for some constant c, that is, the probability of no failures in the interval $(t, t + c]$. This gives

$$R(t) = P(N_{t+c} - N_t = 0) = e^{-\int_t^{t+c} \lambda(x)\, dx} = e^{-\int_t^{t+c} \alpha\beta x^{\beta-1}\, dx} = e^{-\alpha((t+c)^{\beta} - t^{\beta})}.$$

■

6.8 PARTING PARADOX

> How wonderful that we have met with a paradox. Now we have some hope of making progress.
>
> –Niels Bohr

The following classic is based on Feller (1968). Buses arrive at a bus stop according to a Poisson process. The time between buses, on average, is 10 minutes. Lisa gets to the bus stop at time t. How long can she expect to wait for a bus?

Here are two possible answers:

(i) By memorylessness, the time until the next bus is exponentially distributed with mean 10 minutes. Lisa will wait, on average, 10 minutes.

(ii) Lisa arrives at some time between two consecutive buses. The expected time between consecutive buses is 10 minutes. By symmetry, her expected waiting time should be half that, or 5 minutes.

Paradoxically, *both* answers have some truth to them! On the one hand, the time until the next bus will be shown to have an exponential distribution with mean 10 minutes. But the *backwards* time to the previous bus is almost exponential as well, with mean close to 10 minutes. Thus, the time when Lisa arrives at the bus stop is a point in an interval whose length is about 20 minutes. And the argument in (ii) essentially holds. By symmetry, her expected waiting time should be half that, or 10 minutes.

The surprising result is that the interarrival time of the buses before and after Lisa's arrival is about 20 minutes. And yet the expected interarrival time for buses is 10 minutes!

R : Waiting Time Paradox

A Poisson process with parameter $\lambda = 1/10$ is generated on $[0, 200]$. Lisa arrives at the bus stop at time $t = 50$. Simulation shows her average wait time is about 10 minutes.

```
# waitingparadox.R
> mytime <- 50
> lambda <- 1/10
> trials <- 10000
> simlist <- numeric(trials)
> for (i in 1:trials) {
+  N <- rpois(1,300*lambda)
+  arrivals <- sort(runif(N,0,300))
+  wait <- arrivals[arrivals > mytime][1] - mytime
+  simlist[i] <- wait }
> mean(simlist)
[1] 10.04728
```

To explain the paradox, consider the process of bus arrivals. The rate of one arrival per 10 minutes is an average. The time between buses is random, and buses may arrive one right after the other, or there may be a long time between consecutive buses. When Lisa gets to the bus stop, she is more likely to get there during a longer interval between buses than a shorter interval.

To illustrate the idea, pick a random number between 1 and 200. Do it now before reading on. Now look at Figure 6.12, which gives arrival times for a Poisson process with parameter $\lambda = 1/10$ on $[0, 200]$. Find your number. Is your number in a short interval (length less than 10) or a long interval (length greater than 10)? Most readers will find their number in a long interval.

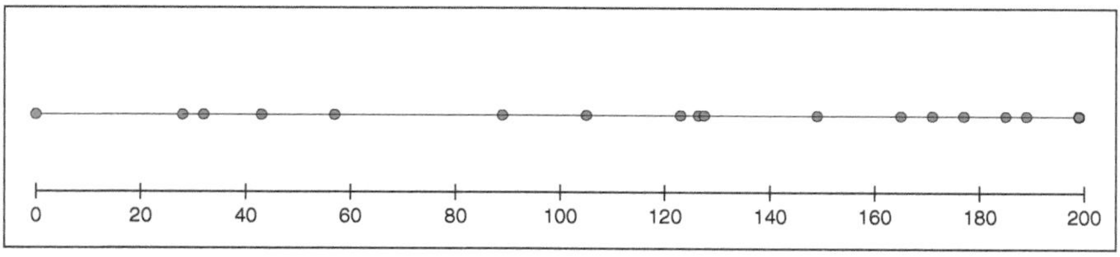

Figure 6.12 Pick a number from 0 to 200. Is your number in a short or long interval?

This example illustrates the phenomenon of *length-biased* or *size-biased* sampling. If you reach into a bag containing pieces of string of different lengths and pick a string *at random*, you tend to pick a longer rather than a shorter piece. Bus interarrival times are analogous to lengths of string.

Here is another example of size-biased sampling. Suppose you want to estimate how much time people spend exercising at the gym. If you go to the gym to survey people at random, you are likely to get a biased estimate, as you are more likely to sample people who work out a lot. Those who rarely go to the gym are not likely to be there when you go!

For the bus waiting problem, the expected length of an interarrival time, *which contains a fixed time* t, is larger—about twice as large—than the average interval length between buses.

Here is an exact analysis. Consider a fixed $t > 0$. The time of the last bus before t is S_{N_t}. The time of the next bus after t is S_{N_t+1}. The expected length of the interval containing t is

$$E\left(S_{N_t+1} - S_{N_t}\right) = E\left(S_{N_t+1}\right) - E\left(S_{N_t}\right).$$

For $E\left(S_{N_t+1}\right)$, condition on N_t. Consider

$$E\left(S_{N_t+1}|N_t = k\right) = E(S_{k+1}|N_t = k) = E(S_{k+1}) = \frac{k+1}{\lambda}.$$

The second equality is because the $(k+1)$th arrival occurs after time t, and is thus independent of N_t. It follows that $E\left(S_{N_t+1}|N_t\right) = (N_t + 1)/\lambda$. By the law of total expectation,

$$E\left(S_{N_t+1}\right) = E\left(E\left(S_{N_t+1}|N_t\right)\right) = E\left(\frac{N_t+1}{\lambda}\right) = \frac{\lambda t + 1}{\lambda} = t + \frac{1}{\lambda}. \tag{6.6}$$

For $E\left(S_{N_t}\right)$, we have that $E\left(S_{N_t}|N_t = k\right) = E(S_k|N_t = k)$. Conditional on $N_t = k$, the kth arrival time has the same distribution as the maximum of k i.i.d. uniform random variables distributed on $(0, t)$. We leave it to the reader to show that this expectation is equal to $tk/(k+1)$. That is, $E\left(S_{N_t}|N_t = k\right) = tk/(k+1)$ and thus

$$E\left(S_{N_t}|N_t\right) = tN_t/(N_t + 1) = t - t/(N_t + 1).$$

This gives

$$E\left(S_{N_t}\right) = E\left(E\left(S_{N_t}|N_t\right)\right) = E\left(t - \frac{t}{N_t+1}\right) = t - tE\left(\frac{1}{N_t+1}\right). \tag{6.7}$$

We find

$$\begin{aligned} E\left(\frac{1}{N_t+1}\right) &= \sum_{k=0}^{\infty}\left(\frac{1}{k+1}\right)\frac{e^{-\lambda t}(\lambda t)^k}{k!} \\ &= \frac{e^{-\lambda t}}{\lambda t}\sum_{k=0}^{\infty}\frac{(\lambda t)^{k+1}}{(k+1)!} = \frac{e^{-\lambda t}}{\lambda t}\sum_{k=1}^{\infty}\frac{(\lambda t)^k}{k!} \\ &= \frac{e^{-\lambda t}}{\lambda t}(e^{\lambda t} - 1) = \frac{1 - e^{-\lambda t}}{\lambda t}. \end{aligned}$$

Together with Equation (6.7),

$$E\left(S_{N_t}\right) = t - \frac{1}{\lambda} + \frac{e^{-\lambda t}}{\lambda}.$$

Finally with Equation (6.6), the expected length of the interval that contains t is

$$E\left(S_{N_t+1} - S_{N_t}\right) = \left(t + \frac{1}{\lambda}\right) - \left(t - \frac{1}{\lambda} + \frac{e^{-\lambda t}}{\lambda}\right) = \frac{2 - e^{-\lambda t}}{\lambda} \approx \frac{2}{\lambda},$$

for large (or even moderate) t.

EXERCISES

6.1 Let $(N_t)_{t\geq 0}$ be a Poisson process with parameter $\lambda = 1.5$. Find the following:

(a) $P(N_1 = 2, N_4 = 6)$

(b) $P(N_4 = 6|N_1 = 2)$

(c) $P(N_1 = 2|N_4 = 6)$

6.2 Let $(N_t)_{t\geq 0}$ be a Poisson process with parameter $\lambda = 2$. Find the following:

(a) $E(N_3 N_4)$

(b) $E(X_3 X_4)$

(c) $E(S_3 S_4)$

6.3 Calls are received at a company call center according to a Poisson process at the rate of five calls per minute.

(a) Find the probability that no call occurs over a 30-second period.

(b) Find the probability that exactly four calls occur in the first minute, and six calls occur in the second minute.

(c) Find the probability that 25 calls are received in the first 5 minutes and six of those calls occur in the first minute.

6.4 Starting at 9 a.m., patients arrive at a doctor's office according to a Poisson process. On average, three patients arrive every hour.

(a) Find the probability that at least two patients arrive by 9:30 a.m.

(b) Find the probability that 10 patients arrive by noon and eight of them come to the office before 11 a.m.

(c) If six patients arrive by 10 a.m., find the probability that only one patient arrives by 9:15 a.m.

6.5 Let $(N_t)_{t\geq 0}$ be a Poisson process. Explain what is wrong with the following *proof* that N_3 is a constant.

$$\begin{aligned} E\left((N_3)^2\right) &= E(N_3 N_3) = E(N_3(N_6 - N_3)) \\ &= E(N_3)E(N_6 - N_3) = E(N_3)E(N_3) \\ &= E(N_3)^2. \end{aligned}$$

Thus, $\text{Var}(N_3) = E\left((N_3)^2\right) - E(N_3)^2 = 0$, which gives that N_3 is a constant with probability 1.

6.6 Occurrences of landfalling hurricanes during an El Niño event are modeled as a Poisson process in Bove et al. (1998). The authors assert that "During an El Niño year, the probability of two or more hurricanes making landfall in the United States is 28%." Find the rate of the Poisson process.

6.7 Ben, Max, and Yolanda are at the front of three separate lines in the cafeteria waiting to be served. The serving times for the three lines follow independent Poisson processes with respective parameters 1, 2, and 3.

(a) Find the probability that Yolanda is served first.

(b) Find the probability that Ben is served before Yolanda.

(c) Find the expected waiting time for the first person served.

6.8 Starting at 6 a.m., cars, buses, and motorcycles arrive at a highway toll booth according to independent Poisson processes. Cars arrive about once every 5 minutes. Buses arrive about once every 10 minutes. Motorcycles arrive about once every 30 minutes.

(a) Find the probability that in the first 20 minutes, exactly three vehicles—two cars and one motorcycle—arrive at the booth.

(b) At the toll booth, the chance that a driver has exact change is 1/4, independent of vehicle. Find the probability that no vehicle has exact change in the first 10 minutes.

(c) Find the probability that the seventh motorcycle arrives within 45 minutes of the third motorcycle.

(d) Find the probability that at least one other vehicle arrives at the toll booth between the third and fourth car arrival.

6.9 Show that the geometric distribution is memoryless.

6.10 Assume that $X_1, X_2, \ldots$ is an i.i.d. sequence of exponential random variables with parameter λ. Let $S_n = X_1 + \cdots + X_n$. Show that S_n has a gamma distribution with parameters n and λ

(a) by moment-generating functions,

(b) by induction on n.

6.11 Show that a continuous probability distribution that is memoryless must be exponential. Hint: For $g(t) = P(X > t)$, show that $g(t) = (g(1))^t$ for all positive, rational t.

6.12 Starting at noon, diners arrive at a restaurant according to a Poisson process at the rate of five customers per minute. The time each customer spends eating at the restaurant has an exponential distribution with mean 40 minutes, independent of other customers and independent of arrival times. Find the distribution, as well as the mean and variance, of the number of diners in the restaurant at 2 p.m.

6.13 Assume that $(N_t)_{t\geq 0}$ is a Poisson process with parameter λ. Find the conditional distribution of N_s given $N_t = n$, for

(a) $s < t$,

(b) $s > t$.

6.14 Red cars arrive at an intersection according to a Poisson process with parameter r. Blue cars arrive, independently of red cars, according to a Poisson process with parameter b. Let X be the number of blue cars which arrive between two successive red cars. Show that X has a geometric distribution.

6.15 Failures occur for a mechanical process according to a Poisson process. Failures are classified as either major or minor. Major failures occur at the rate of 1.5 failures per hour. Minor failures occur at the rate of 3.0 failures per hour.

(a) Find the probability that two failures occur in 1 hour.

(b) Find the probability that in half an hour, no major failures occur.

(c) Find the probability that in 2 hours, at least two major failures occur or at least two minor failures occur.

6.16 Accidents occur at a busy intersection according to a Poisson process at the rate of two accidents per week. Three out of four accidents involve the use of alcohol.

(a) What is the probability that three accidents involving alcohol will occur next week?

(b) What is the probability that at least one accident occurs tomorrow?

(c) If six accidents occur in February (four weeks), what is the probability that less than half of them involve alcohol?

6.17 Let $(N_t)_{t\geq 0}$ be a Poisson process with parameter λ and arrival times $S_1, S_2, \ldots$ Evaluate the expected sum of squares of the arrival times before time t,

$$E\left(\sum_{n=1}^{N_t} S_n^2\right).$$

6.18 The planets of the Galactic Empire are distributed in space according to a spatial Poisson process at an approximate density of one planet per cubic parsec. From the Death Star, let X be the distance to the nearest planet.

(a) Find the probability density function of X.

(b) Find the mean distance from the Death Star to the nearest planet.

6.19 Members of a large audience are asked to state their birthdays, one at a time. How many people will be asked before three persons are found to have the same birthday? Use embedding to estimate the expected number. You will need to use numerical methods to evaluate the resulting integral.

6.20 Consider a spatial point process in $\mathbb{R}^2$ with parameter λ. Assume that A is a bounded set in $\mathbb{R}^2$ which contains exactly one point of the process. Given $B \subseteq A$, find the probability that B contains one point.

6.21 For a Poisson process with parameter λ show that for $s < t$, the correlation between N_s and N_t is

$$\text{Corr}(N_s, N_t) = \sqrt{\frac{s}{t}}.$$

6.22 At the Poisson Casino, two dice are rolled at random times according to a Poisson process with parameter λ. Find the probability that in $[0, t]$ every pair of dice rolled comes up 7.

6.23 Oak and maple trees are each located in the arboretum according to independent spatial Poisson processes with parameters λ_O and λ_M, respectively.

(a) In a region of x square meters, find the probability that both species of trees are present.

(b) In the arboretum, there is a circular pond of radius 100 m. Find the probability that within 20 m of the pond there are only oaks.

6.24 Tom is bird watching in the arboretum. The times when he sights a meadowlark occur in accordance with a Poisson process with parameter λ. The times when he sights a sparrow occur as a Poisson process with parameter μ. Assume that meadowlark and sparrow sightings are independent.

(a) Find the probability that a meadowlark is seen first.

(b) Find the probability the one bird is seen by time $t = 1$.

(c) Find the probability that one sparrow and two meadowlarks are seen by time $t = 2$.

6.25 Computers in the lab fail, on average, twice a day, according to a Poisson process. Last week, 10 computers failed. Find the expected time of the last failure, and give an approximate time of day when the last failure occurred.

6.26 If $(N_t)_{t \geq 0}$ is a Poisson process with parameter λ, find the probability generating function of N_t.

6.27 In a small parliamentary election, votes are counted according to a Poisson process at the rate of 60 votes per minute. There are six political parties, whose popularity among the electorate is shown by this distribution.

A	B	C	D	E	F
0.05	0.30	0.10	0.10	0.25	0.20

(a) In the first 2 minutes of the vote tally, 40 people had voted for parties E and F. Find the probability that more than 100 votes were counted in the first 2 minutes.

(b) If vote counting starts at 3 p.m., find the probability that the first vote is counted by the first second after 3 p.m.

(c) Find the probability that the first vote for party C is counted before a vote for B or D.

6.28 Tornadoes hit a region according to a Poisson process with $\lambda = 2$. The number of insurance claims filed after any tornado has a Poisson distribution with mean 30. The number of tornadoes is independent of the number of insurance claims. Find the expectation and standard deviation of the total number of claims filed by time t.

6.29 Job offers for a recent college graduate arrive according to a Poisson process with mean two per month. A job offer is acceptable if the salary offered is at least \$35,000. Salary offers follow an exponential distribution with mean \$25,000. Find the probability that an acceptable job offer will be received within 3 months.

6.30 See Example 6.9. Find the variance of total waiting time of concert-goers who arrive before the band starts.

6.31 See Example 6.9. Find the expected *average* waiting time of concert-goers who arrive before the band starts.

6.32 Investors purchase \$1,000 bonds at the random times of a Poisson process with parameter λ. If the interest rate is r, then the *present value* of an investment purchased at time t is $1000e^{-rt}$. Show that the expected total present value of the bonds purchased by time t is $1000\lambda(1 - e^{-rt})/r$.

6.33 Let $S_1, S_2, \ldots$ be the arrival times of a Poisson process with parameter λ. Given the time of the nth arrival, find the expected time $E(S_1 | S_n)$ of the first arrival.

6.34 Describe in words the random variable S_{N_t}. Find the distribution of S_{N_t} by giving the cumulative distribution function.

6.35 See the definitions for the spatial and nonhomogeneous Poisson processes. Define a nonhomogeneous, spatial Poisson process in $\mathbb{R}^2$. Consider such a process $(N_A)_{A \subseteq \mathbb{R}^2}$ with intensity function

$$\lambda(x, y) = e^{-(x^2+y^2)}, \text{ for } -\infty < x, y < \infty.$$

Let C denote the unit circle, that is, the circle of radius 1 centered at the origin. Find $P(N_C = 0)$.

6.36 Starting at 9 a.m., customers arrive at a store according to a nonhomogeneous Poisson process with intensity function $\lambda(t) = t^2$, for $t > 0$, where the time unit is hours. Find the probability mass function of the number of customers who enter the store by noon.

6.37 A *compound Poisson process* $(C_t)_{t\geq 0}$ is defined as

$$C_t = \sum_{k=1}^{N_t} X_k,$$

where $(N_t)_{t\geq 0}$ is a Poisson process, and $X_1, X_2, \ldots$ is an i.i.d. sequence of random variables that are independent of $(N_t)_{t\geq 0}$.

Assume that automobile accidents at a dangerous intersection occur according to a Poisson process at the rate of 3 accidents per week. Furthermore, the number of people seriously injured in an accident has a Poisson distribution with mean 2. Show that the process of serious injuries is a compound Poisson process, and find the mean and standard deviation of the number of serious injuries over 1 year's time.

6.38 A *mixed Poisson process*, also called a *Cox process* or *doubly stochastic process*, arises from a Poisson process where the parameter Λ is itself a random variable. If $(N_t)_{t\geq 0}$ is a mixed Poisson process, then the conditional distribution of N_t given $\Lambda = \lambda$ is Poisson with parameter λt. Assume that for such a process Λ has an exponential distribution with parameter μ. Find the probability mass function of N_t.

6.39 See Exercise 6.38. Assume that $(N_t)_{t\geq 0}$ is a mixed Poisson process with rate parameter uniformly distributed on (0, 1). Find $P(N_1 = 1)$.

6.40 Assume that $(N_t)_{t\geq 0}$ is a mixed Poisson process whose rate Λ has a gamma distribution with parameters n and λ. Show that

$$P(N_t = k) = \binom{n+k-1}{k} \left(\frac{\lambda}{\lambda + t}\right)^n \left(\frac{t}{\lambda + t}\right)^k, \quad \text{for } k = 0, 1, \ldots$$

6.41 R : Goals occur in a soccer game according to a Poisson process. The average total number of goals scored for a 90-minute match is 2.68. Assume that two teams are evenly matched. Use simulation to estimate the probability both teams will score the same number of goals. Compare with the theoretical result.

6.42 R : Simulate the restaurant results of Exercise 6.12.

6.43 R : Simulate a spatial Poisson process with $\lambda = 10$ on the box of volume 8 with vertices at the 8 points $(\pm 1, \pm 1, \pm 1)$. Estimate the mean and variance of the number of points in the ball centered at the origin of radius 1. Compare with the exact values.

6.44 `R` : See Exercise 6.32. Simulate the expected total present value of the bonds if the interest rate is 4%, the Poisson parameter is $\lambda = 50$, and $t = 10$. Compare with the exact value.

6.45 `R` : Simulate the birthday problem of Exercise 6.19.

7

CONTINUOUS-TIME MARKOV CHAINS

> Life defies our phrases, it is infinitely continuous and subtle and shaded, whilst our verbal terms are discrete, rude, and few.
>
> –William James

7.1 INTRODUCTION

In this chapter, we extend the Markov chain model to continuous time. A continuous-time process allows one to model not only the transitions between states, but also the duration of time in each state. The central Markov property continues to hold—given the present, past and future are independent.

The Markov property is a form of memorylessness. This leads to the exponential distribution. In a continuous-time Markov chain, when a state is visited, the process stays in that state for an exponentially distributed length of time before moving to a new state. If one just watches the sequence of states that are visited, ignoring the length of time spent in each state, the process looks like a discrete-time Markov chain.

One of the Markov chains introduced in this book was the three-state weather chain of Example 2.3, with state space {rain, snow, clear}. Consider a continuous-time extension. Assume that rainfall lasts, on average, 3 hours at a time. When it snows, the duration, on average, is 6 hours. And the weather stays clear, on average, for

Introduction to Stochastic Processes with R, First Edition. Robert P. Dobrow.

12 hours. Furthermore, changes in weather states are described by the stochastic transition matrix

$$\widetilde{P} = \begin{array}{c} \\ \text{Rain} \\ \text{Snow} \\ \text{Clear} \end{array} \begin{array}{c} \begin{array}{ccc} \text{Rain} & \text{Snow} & \text{Clear} \end{array} \\ \begin{pmatrix} 0 & 1/2 & 1/2 \\ 3/4 & 0 & 1/4 \\ 1/4 & 3/4 & 0 \end{pmatrix} \end{array}. \tag{7.1}$$

To elaborate, assume that it is currently snowing. Under this model, it snows for an exponential length of time with parameter $\lambda_s = 1/6$. (Remember that the parameter of an exponential distribution is the reciprocal of the mean.) Then, the weather changes to either rain or clear, with respective probabilities 3/4 and 1/4. If it switches to rain, it rains for an exponential length of time with parameter $\lambda_r = 1/3$. Then, the weather changes to either snow or clear with equal probability, and so on. Figure 7.1 gives an example of how the process unfolds over 50 hours.

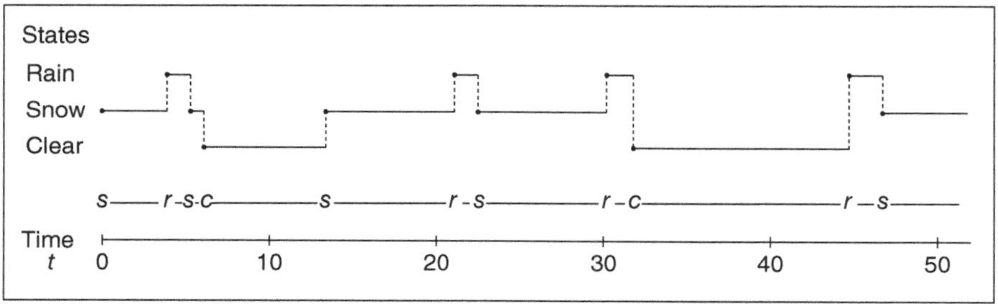

Figure 7.1 Realization of a continuous-time weather chain.

Let X_t denote the weather at time t. Then, $(X_t)_{t\geq 0}$ is a continuous-time Markov chain. The $\widetilde{P}$ matrix, exponential time parameters $(\lambda_r, \lambda_s, \lambda_c) = (1/3, 1/6, 1/12)$, and initial distribution completely specify the process. That is, they are sufficient for computing all probabilities of the form $P\left(X_{t_1} = i_1, \ldots, X_{t_n} = i_n\right)$, for $n \geq 1$, states $i_1, \ldots, i_n$, and times $t_1, \ldots, t_n \geq 0$.

Markov Transition Function

The formal treatment of continuous-time Markov chains begins with the defining Markov property.

Markov Property

A continuous-time stochastic process $(X_t)_{t\geq 0}$ with discrete state space S is a *continuous-time Markov chain* if

$$P(X_{t+s} = j | X_s = i, X_u = x_u, 0 \leq u < s) = P(X_{t+s} = j | X_s = i),$$

for all $s, t, \geq 0$, $i, j, x_u \in S$, and $0 \leq u < s$.

The process is said to be *time-homogeneous* if this probability does not depend on s. That is,

$$P(X_{t+s} = j | X_s = i) = P(X_t = j | X_0 = i), \text{ for } s \geq 0. \tag{7.2}$$

The Markov chains we treat in this book are all time-homogeneous. For each $t \geq 0$, the transition probabilities in Equation (7.2) can be arranged in a matrix function $\boldsymbol{P}(t)$ called the *transition function*, with

$$P_{ij}(t) = P(X_t = j | X_0 = i).$$

Note that for the weather chain, the $\widetilde{\boldsymbol{P}}$ matrix in Equation (7.1) is *not* the transition function. The $\widetilde{\boldsymbol{P}}$ matrix gives the probabilities of moving from state to state in a discretized process in which time has been ignored.

The transition function $\boldsymbol{P}(t)$ has similar properties as that of the transition matrix for a discrete-time Markov chain. For instance, the Chapman–Kolmogorov equations hold.

Chapman–Kolmogorov Equations

For a continuous-time Markov chain $(X_t)_{t\geq 0}$ with transition function $\boldsymbol{P}(t)$,

$$\boldsymbol{P}(s+t) = \boldsymbol{P}(s)\boldsymbol{P}(t),$$

for $s, t \geq 0$. That is,

$$P_{ij}(s+t) = [P(s)P(t)]_{ij} = \sum_k P_{ik}(s)P_{kj}(t), \text{ for states } i, j, \text{ and } s, t \geq 0.$$

Proof. By conditioning on X_s,

$$\begin{aligned} P_{ij}(s+t) &= P(X_{s+t} = j | X_0 = i) \\ &= \sum_k P(X_{s+t} = j | X_s = k, X_0 = i) P(X_s = k | X_0 = i) \\ &= \sum_k P(X_{s+t} = j | X_s = k) P(X_s = k | X_0 = i) \\ &= \sum_k P(X_t = j | X_0 = k) P(X_s = k | X_0 = i) \\ &= \sum_k P_{ik}(s) P_{kj}(t) = [P(s)P(t)]_{ij}, \end{aligned}$$

where the third equality is because of the Markov property, and the fourth equality is by time-homogeneity. ■

Example 7.1 (Poisson process) A Poisson process $(N_t)_{t\geq 0}$ with parameter λ is a continuous-time Markov chain. The Markov property holds as a consequence of stationary and independent increments. For $0 \leq i \leq j$,

$$\begin{aligned}
P_{ij}(t) = P(N_{t+s} = j|N_s = i) &= \frac{P(N_{t+s} = j, N_s = i)}{P(N_s = i)} \\
&= \frac{P(N_{t+s} - N_s = j - i, N_s = i)}{P(N_s = i)} \\
&= P(N_{t+s} - N_s = j - i) \\
&= P(N_t = j - i) = \frac{e^{-\lambda t}(\lambda t)^{j-i}}{(j-i)!}.
\end{aligned}$$

The transition function is

$$P(t) = \begin{array}{c} \\ 0 \\ 1 \\ 2 \\ 3 \\ \vdots \end{array} \begin{array}{c} \begin{array}{ccccc} 0 & 1 & 2 & 3 & \cdots \end{array} \\ \begin{pmatrix} e^{-\lambda t} & (\lambda t)e^{-\lambda t} & (\lambda t)^2 e^{-\lambda t}/2 & (\lambda t)^3 e^{-\lambda t}/6 & \cdots \\ 0 & e^{-\lambda t} & (\lambda t)e^{-\lambda t} & (\lambda t)^2 e^{-\lambda t}/2 & \cdots \\ 0 & 0 & e^{-\lambda t} & (\lambda t)e^{-\lambda t} & \cdots \\ 0 & 0 & 0 & e^{-\lambda t} & \cdots \\ \vdots & \vdots & \vdots & \vdots & \ddots \end{pmatrix} \end{array}.$$

■

Holding Times and Embedded Chains

By homogeneity, when a Markov chain visits state i its forward evolution from that time onward behaves the same as the process started in i at time $t = 0$. Time-homogeneity and the Markov property characterizes the distribution of the length of time that a continuous-time chain stays in state i before transitioning to a new state.

Holding Times are Exponentially Distributed

Let T_i be the *holding time* at state i, that is, the length of time that a continuous-time Markov chain started in i stays in i before transitioning to a new state. Then, T_i has an exponential distribution.

Proof. We show that T_i is memoryless. Let $s, t \geq 0$. For the chain started in i, the event $\{T_i > s\}$ is equal to the event that $\{X_u = i$, for $0 \leq u \leq s\}$. Since $\{T_i > s + t\}$ implies $\{T_i > s\}$,

$$P(T_i > s + t|X_0 = i) = P(T_i > s + t, T_i > s|X_0 = i)$$

$$
\begin{aligned}
&= P(T_i > s + t | X_0 = i, T_i > s) P(T_i > s | X_0 = i) \\
&= P(T_i > s + t | X_u = i, \quad \text{for } 0 \le u \le s) P(T_i > s | X_0 = i) \\
&= P(T_i > s + t | X_s = i) P(T_i > s | X_0 = i) \\
&= P(T_i > t | X_0 = i) P(T_i > s | X_0 = i),
\end{aligned}
$$

where the next to last equality is because of the Markov property, and the last equality is because of homogeneity. This gives that T_i is memoryless. The result follows since the exponential distribution is the only continuous distribution that is memoryless. ∎

For each i, let q_i be the parameter of the exponential distribution for the holding time T_i. We assume that $0 < q_i < \infty$. Technically, a continuous-time process can be defined where $q_i = 0$ or $+\infty$. In the former case, when i is visited the process never leaves, and i is called an *absorbing state*. In the latter case, the process leaves i immediately upon entering i. This would allow for infinitely many transitions in a finite interval. Such a process is called *explosive*.

The evolution of a continuous-time Markov chain which is neither absorbing nor explosive can be described as follows. Starting from i, the process stays in i for an exponentially distributed length of time, on average $1/q_i$ time units. Then, it hits a new state $j \neq i$, with some probability p_{ij}. The process stays in j for an exponentially distributed length of time, on average $1/q_j$ time units. It then hits a new state $k \neq j$, with probability p_{jk}, and so on.

The transition probabilities p_{ij} describe the discrete transitions from state to state. If we ignore time, and just watch state to state transitions, we see a sequence $Y_0, Y_1, \ldots$, where Y_n is the nth state visited by the continuous process. The sequence $Y_0, Y_1, \ldots$ is a discrete-time Markov chain called the *embedded chain*.

Let $\widetilde{\boldsymbol{P}}$ be the transition matrix for the embedded chain. That is, $\widetilde{P}_{ij} = p_{ij}$. Then, $\widetilde{\boldsymbol{P}}$ is a stochastic matrix whose diagonal entries are 0.

Example 7.2 (Poisson process) For a Poisson process with parameter λ, the holding time parameters are constant. That is, $q_i = \lambda$, for $i = 0, 1, 2, \ldots$ The process moves from 0 to 1 to 2, and so on. The transition matrix of the embedded chain is

$$
\widetilde{\boldsymbol{P}} = \begin{array}{c} \\ 0 \\ 1 \\ 2 \\ 3 \\ 4 \\ \vdots \end{array}
\begin{array}{c} \begin{array}{cccccc} 0 & 1 & 2 & 3 & 4 & \cdots \end{array} \\
\begin{pmatrix}
0 & 1 & 0 & 0 & 0 & \cdots \\
0 & 0 & 1 & 0 & 0 & \cdots \\
0 & 0 & 0 & 1 & 0 & \cdots \\
0 & 0 & 0 & 0 & 1 & \cdots \\
0 & 0 & 0 & 0 & 0 & \cdots \\
\vdots & \vdots & \vdots & \vdots & \vdots & \ddots
\end{pmatrix} \end{array}.
$$

∎

Example 7.3 (Two-state chain) A two-state continuous-time Markov chain is specified by two holding time parameters as depicted in the transition graph in Figure 7.2. Edges of the graph are labeled with holding time rates, not probabilities.

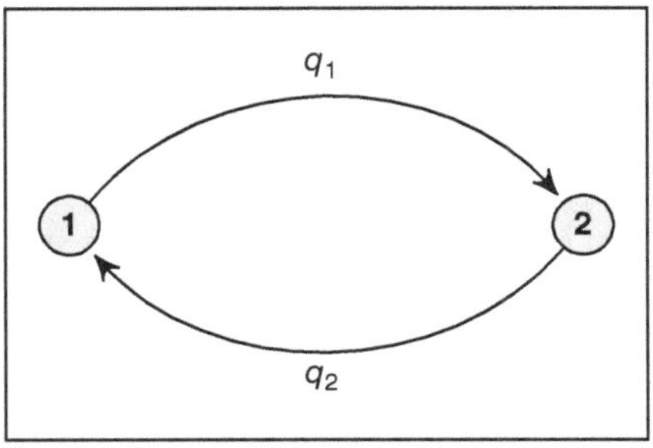

Figure 7.2 Two-state process.

The embedded chain transition matrix is

$$\widetilde{P} = \begin{matrix} & \begin{matrix} 1 & 2 \end{matrix} \\ \begin{matrix} 1 \\ 2 \end{matrix} & \begin{pmatrix} 0 & 1 \\ 1 & 0 \end{pmatrix} \end{matrix}.$$

■

The process stays in state 1 an exponential length of time with parameter q_1 before moving to 2. It stays in 2 an exponential length of time with parameter q_2 before moving to 1, and so on.

7.2 ALARM CLOCKS AND TRANSITION RATES

A continuous-time Markov chain can also be described by specifying *transition rates* between pairs of states. Central to this approach is the notion of the *exponential alarm clock*.

Imagine that for each state i, there are independent alarm clocks associated with each of the states that the process can visit after i. If j can be hit from i, then the alarm clock associated with (i, j) will ring after an exponentially distributed length of time with parameter q_{ij}. When the process first hits i, the clocks are started simultaneously. The first alarm that rings determines the next state to visit. If the (i, j) clock rings first and the process moves to j, a new set of exponential alarm clocks are started, with transition rates $q_{j1}, q_{j2}, \ldots$ Again, the first alarm that rings determines the next state hit, and so on.

The q_{ij} are called the *transition rates* of the continuous-time process. From the transition rates, we can obtain the holding time parameters and the embedded chain transition probabilities.

Consider the process started in i. The clocks are started, and the first one that rings determines the next transition. The time of the first alarm is the minimum of independent exponential random variables with parameters $q_{i1}, q_{i2}, \ldots$ Recall results for the exponential distribution and the minimum of independent exponentials, as given in Equations (6.1) and (6.2). The minimum has an exponential distribution with parameter $\sum_k q_{ik}$. That is, the chain stays at i for an exponentially distributed amount of time with parameter $\sum_k q_{ik}$. From the discussion of holding times, the exponential length of time that the process stays in i has parameter q_i. That is,

$$q_i = \sum_k q_{ik}.$$

The interpretation is that the rate that the process leaves state i is equal to the sum of the rates from i to each of the next states.

From i, the chain moves to j if the (i,j) clock rings first, which occurs with probability $q_{ij}/\sum_k q_{ik} = q_{ij}/q_i$. Thus, for the embedded chain transition probabilities

$$p_{ij} = \frac{q_{ij}}{\sum_k q_{ik}} = \frac{q_{ij}}{q_i}.$$

Example 7.4 The general three-state continuous-time Markov chain is described by the transition graph in Figure 7.3. In terms of the transition rates, holding time parameters are

$$(q_1, q_2, q_3) = (q_{12} + q_{13}, q_{21} + q_{23}, q_{31} + q_{32}),$$

with embedded chain transition matrix

$$\widetilde{P} = \begin{matrix} & \begin{matrix} 1 & \quad 2 & \quad 3 \end{matrix} \\ \begin{matrix} 1 \\ 2 \\ 3 \end{matrix} & \begin{pmatrix} 0 & q_{12}/q_1 & q_{13}/q_1 \\ q_{21}/q_2 & 0 & q_{23}/q_2 \\ q_{31}/q_3 & q_{32}/q_3 & 0 \end{pmatrix} \end{matrix}.$$

■

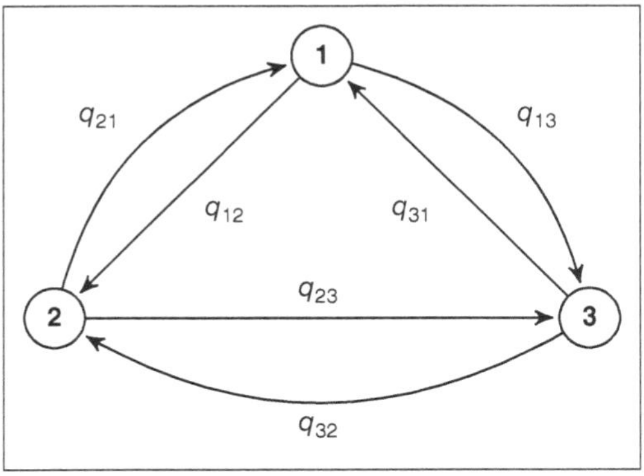

Figure 7.3 Transition rates for three-state chain.

Example 7.5 (Registration line) It is time for students to register for classes, and a line is forming at the registrar's office for those who need assistance. It takes the registrar an exponentially distributed amount of time to service each student, at the rate of one student every 5 minutes. Students arrive at the office and get in line according to a Poisson process at the rate of one student every 4 minutes. Line size is capped at 4 people. If an arriving student finds that there are already 4 people in line, then they try again later. As soon as there is at least one person in line, the registrar starts assisting the first available student. The arrival times of the students are independent of the registrar's service time.

Let X_t be the number of students in line at time t. Then, $(X_t)_{t\geq 0}$ is a continuous-time Markov chain on $\{0, 1, 2, 3, 4\}$.

If there is no one in line, then the size of the line increases to 1 when a student arrives. If there are 4 people in line, then the number decreases to 3 when the registrar finishes assisting the student they are meeting with. If there are 1, 2, or 3 students in line, then the line size can either decrease or increase by 1. If a student arrives at the registrar's office before the registrar has finished serving the student being helped, then the line increases by 1. If the registrar finishes serving the student being helped before another student arrives, the line decreases by 1.

Imagine that when there is 1 person in line two exponential alarm clocks are started—one for student arrivals, with rate 1/4, the other for the registrar's service time, with rate 1/5. If the arrival time clock rings first, the line increases by one. If the service clock rings first, the line decreases by one. The same dynamics hold if there are 2 or 3 people in line. The process is described by the transition graph in Figure 7.4.

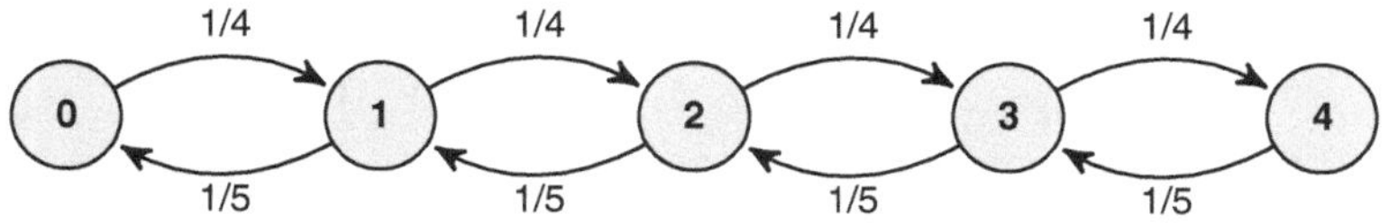

Figure 7.4 Transition rate graph for registration line Markov chain.

The holding time parameters are

$$(q_0, q_1, q_2, q_3, q_4) = \left(\frac{1}{4}, \frac{9}{20}, \frac{9}{20}, \frac{9}{20}, \frac{1}{5}\right),$$

with embedded chain transition matrix

$$\widetilde{P} = \begin{array}{c} \\ 0 \\ 1 \\ 2 \\ 3 \\ 4 \end{array} \begin{array}{c} \begin{array}{ccccc} 0 & 1 & 2 & 3 & 4 \end{array} \\ \begin{pmatrix} 0 & 1 & 0 & 0 & 0 \\ 4/9 & 0 & 5/9 & 0 & 0 \\ 0 & 4/9 & 0 & 5/9 & 0 \\ 0 & 0 & 4/9 & 0 & 5/9 \\ 0 & 0 & 0 & 1 & 0 \end{pmatrix} \end{array}.$$

Since students arrive in line at a faster rate than the registrar's service time, the line tends to grow over time. See a realization of the process on $[0, 60]$ in Figure 7.5.

■

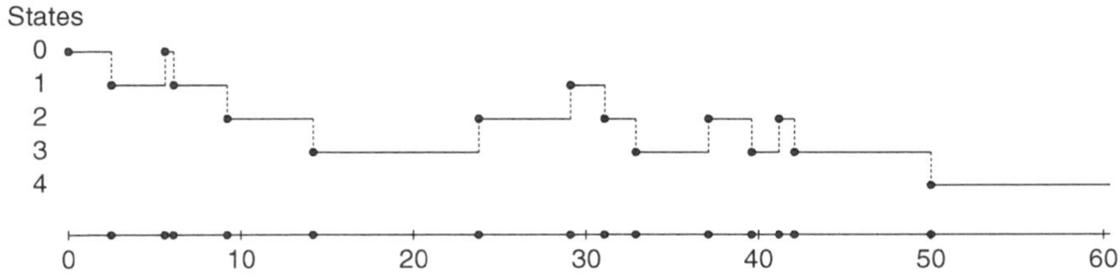

Figure 7.5 Realization of the size of the line at the registrar's office.

7.3 INFINITESIMAL GENERATOR

In continuous time, transition rates play a fundamental role when working with Markov chains. Since the derivative of a function describes its rate of change, it is not surprising that the derivative of the transition function $\boldsymbol{P}'(t)$ is most important.

Assume that $(X_t)_{t\geq 0}$ is a continuous-time Markov chain with transition function $\boldsymbol{P}(t)$. Assume the transition function is differentiable. Note that

$$P_{ij}(0) = \begin{cases} 1, & \text{if } i = j, \\ 0, & \text{if } i \neq j. \end{cases}$$

If $X_t = i$, then the instantaneous transition rate of hitting $j \neq i$ is

$$\begin{aligned}
\lim_{h\to 0^+} \frac{E(\text{Number of transitions to } j \text{ in } (t, t+h])}{h} &= \lim_{h\to 0^+} \frac{P(X_{t+h} = j | X_t = i)}{h} \\
&= \lim_{h\to 0^+} \frac{P(X_h = j | X_0 = i)}{h} \\
&= \lim_{h\to 0^+} \frac{P_{ij}(h)}{h} \\
&= \lim_{h\to 0^+} \frac{P_{ij}(h) - P_{ij}(0)}{h} \\
&= P'_{ij}(0).
\end{aligned}$$

The first equality is because for h sufficiently small, the number of transitions to j in $(t, t+h]$ is either 0 or 1. Let $\boldsymbol{Q} = \boldsymbol{P}'(0)$. The off-diagonal entries of $\boldsymbol{Q}$ are the instantaneous transition rates, which are the transition rates q_{ij} introduced in the last section. That is, $Q_{ij} = q_{ij}$, for $i \neq j$. In the language of infinitesimals, if $X_t = i$, then the chance that $X_{t+dt} = j$ is $q_{ij}\ dt$.

The diagonal entries of $\boldsymbol{Q}$ are

$$Q_{ii} = P'_{ii}(0) = \lim_{h\to 0^+} \frac{P_{ii}(h) - P_{ii}(0)}{h} \lim_{h\to 0^+} \frac{P_{ii}(h) - 1}{h}$$

$$= \lim_{h\to 0^+} -\frac{\sum_{j\neq i} P_{ij}(h)}{h} = -\sum_{j\neq i} \lim_{h\to 0^+} \frac{P_{ij}(h)}{h}$$

$$= -\sum_{j\neq i} Q_{ij} = -\sum_{j\neq i} q_{ij} = -q_i.$$

The $\boldsymbol{Q}$ matrix is called the *generator* or *infinitesimal generator*. It is the most important matrix for continuous-time Markov chains. Here, we derived $\boldsymbol{Q}$ from the transition function $\boldsymbol{P}(t)$. However, in a modeling context one typically starts with $\boldsymbol{Q}$, identifying the transition rates q_{ij} based on the qualitative and quantitative dynamics of the process. The transition function and related quantities are derived from $\boldsymbol{Q}$.

Clearly, the generator is not a stochastic matrix. Diagonal entries are negative, entries can be greater than 1, and rows sum to 0.

Example 7.6 The infinitesimal generator matrix for the registration line chain of Example 7.5 is

$$\boldsymbol{Q} = \begin{array}{c} \\ 0 \\ 1 \\ 2 \\ 3 \\ 4 \end{array} \begin{array}{c} \begin{matrix} 0 & 1 & 2 & 3 & 4 \end{matrix} \\ \begin{pmatrix} -1/4 & 1/4 & 0 & 0 & 0 \\ 1/5 & -9/20 & 1/4 & 0 & 0 \\ 0 & 1/5 & -9/20 & 1/4 & 0 \\ 0 & 0 & 1/5 & -9/20 & 1/4 \\ 0 & 0 & 0 & 1/5 & -1/5 \end{pmatrix} \end{array}.$$

■

Example 7.7 The generator for a Poisson process with parameter λ is

$$\boldsymbol{Q} = \begin{array}{c} \\ 0 \\ 1 \\ 2 \\ 3 \\ 4 \\ \vdots \end{array} \begin{array}{c} \begin{matrix} 0 & 1 & 2 & 3 & 4 & \cdots \end{matrix} \\ \begin{pmatrix} -\lambda & \lambda & 0 & 0 & 0 & \cdots \\ 0 & -\lambda & \lambda & 0 & 0 & \cdots \\ 0 & 0 & -\lambda & \lambda & 0 & \cdots \\ 0 & 0 & 0 & -\lambda & \lambda & \cdots \\ 0 & 0 & 0 & 0 & -\lambda & \cdots \\ \vdots & \vdots & \vdots & \vdots & \vdots & \ddots \end{pmatrix} \end{array}.$$

■

For a continuous-time Markov chain, the transition probabilities of the embedded chain can be derived from first principles from the transition function and generator matrix. From i, the probability that if a transition occurs at time t the process moves to a different state $j \neq i$ is

$$\lim_{h\to 0^+} P(X_{t+h} = j | X_t = i, X_{t+h} \neq i) = \lim_{h\to 0^+} P(X_h = j | X_0 = i, X_h \neq i)$$

$$= \lim_{h\to 0^+} \frac{P(X_h = j, X_h \neq i, X_0 = i)}{P(X_h \neq i, X_0 = i)}$$

$$= \lim_{h\to 0^+} \frac{P(X_h = j|X_0 = i)}{P(X_h \neq i|X_0 = i)}$$
$$= \lim_{h\to 0^+} \frac{P_{ij}(h)/h}{1 - P_{ii}(h)/h}$$
$$= \frac{q_{ij}}{q_i},$$

which is independent of t. This gives the transition probability p_{ij} of the embedded chain. It also establishes the relationship between instantaneous transition rates, holding time parameters, and the embedded chain transition probabilities.

Instantaneous Rates, Holding Times, Transition Probabilities

For a continuous-time Markov chain, let q_{ij}, q_i, and p_{ij} be defined as above. For $i \neq j$,

$$q_{ij} = q_i p_{ij}.$$

For discrete-time Markov chains, there is no generator matrix and the probabilistic properties of the stochastic process are captured by the transition matrix $\boldsymbol{P}$. For continuous-time Markov chains the generator matrix $\boldsymbol{Q}$ gives a complete description of the dynamics of the process. The distribution of any finite subset of the X_t, and all probabilistic quantities of the stochastic process, can, in principle, be obtained from the infinitesimal generator and the initial distribution.

Forward, Backward Equations

The transition function $\boldsymbol{P}(t)$ can be computed from the generator $\boldsymbol{Q}$ by solving a system of differential equations.

Kolmogorov Forward, Backward Equations

A continuous-time Markov chain with transition function $\boldsymbol{P}(t)$ and infinitesimal generator $\boldsymbol{Q}$ satisfies the *forward equation*

$$\boldsymbol{P}'(t) = \boldsymbol{P}(t)\boldsymbol{Q} \tag{7.3}$$

and the *backward equation*

$$\boldsymbol{P}'(t) = \boldsymbol{Q}\boldsymbol{P}(t). \tag{7.4}$$

Equivalently, for all states i and j,

$$P'_{ij}(t) = \sum_k P_{ik}(t)q_{kj} = -P_{ij}(t)q_j + \sum_{k\neq j} P_{ik}(t)q_{kj}$$

and

$$P'_{ij}(t) = \sum_k q_{ik} P_{kj}(t) = -q_i P_{ij}(t) + \sum_{k \neq i} q_{ik} P_{kj}(t).$$

Proof. The equations are a consequence of the Chapman–Kolmogorov property. For the forward equation, for $h, t \geq 0$,

$$\begin{aligned}\frac{\boldsymbol{P}(t+h) - \boldsymbol{P}(t)}{h} &= \frac{\boldsymbol{P}(t)\boldsymbol{P}(h) - \boldsymbol{P}(t)}{h} \\ &= \boldsymbol{P}(t)\left(\frac{\boldsymbol{P}(h) - \boldsymbol{I}}{h}\right) \\ &= \boldsymbol{P}(t)\left(\frac{\boldsymbol{P}(h) - \boldsymbol{P}(0)}{h}\right).\end{aligned}$$

Taking limits as $h \to 0^+$ gives $\boldsymbol{P}'(t) = \boldsymbol{P}(t)\boldsymbol{Q}$. The backward equation is derived similarly, starting with $\boldsymbol{P}(t+h) = \boldsymbol{P}(h)\boldsymbol{P}(t)$. ■

Example 7.8 (Poisson process) The transition probabilities for the Poisson process with parameter λ,

$$P_{ij}(t) = \frac{e^{-\lambda t}(\lambda t)^{j-i}}{(j-i)!}, \text{ for } j \geq i,$$

were derived in Example 7.1. They satisfy the Kolmogorov forward equations

$$\begin{aligned}P'_{ii}(t) &= -\lambda P_{ii}(t), \\ P'_{ij}(t) &= -\lambda P_{ij}(t) + \lambda P_{i,j-1}, \text{ for } j = i+1, i+2, \ldots,\end{aligned}$$

and the backward equations

$$\begin{aligned}P'_{ii}(t) &= -\lambda P_{ii}(t), \\ P'_{ij}(t) &= -\lambda P_{ij}(t) + \lambda P_{j+1,i}, \text{ for } j = i+1, i+2, \ldots\end{aligned}$$

■

Example 7.9 (Two-state process) For a continuous-time process with generator

$$\boldsymbol{Q} = \begin{array}{c} \\ 1 \\ 2 \end{array} \!\!\begin{array}{c} \begin{array}{cc} 1 & 2 \end{array} \\ \begin{pmatrix} -\lambda & \lambda \\ \mu & -\mu \end{pmatrix} \end{array}$$

the forward equations give

$$\begin{aligned} P'_{11}(t) &= -P_{11}(t)q_1 + P_{12}(t)q_{21} \\ &= -\lambda P_{11}(t) + (1 - P_{11}(t))\mu \\ &= \mu - (\lambda + \mu)P_{11}(t), \end{aligned}$$

using the fact that the first row of the matrix $\boldsymbol{P}(t)$ sums to 1. The solution to the linear differential equation is

$$P_{11}(t) = \frac{\mu}{\lambda + \mu} + \frac{\lambda}{\lambda + \mu}e^{-(\lambda+\mu)t}.$$

Also,

$$P'_{22}(t) = -P_{22}(t)q_2 + P_{21}(t)q_{12} = \lambda - (\lambda + \mu)P_{22}(t),$$

with solution

$$P_{22}(t) = \frac{\lambda}{\lambda + \mu} + \frac{\mu}{\lambda + \mu}e^{-(\lambda+\mu)t}.$$

The transition function is

$$\boldsymbol{P}(t) = \frac{1}{\lambda + \mu}\begin{array}{c} \begin{matrix} \qquad 1 \qquad & \qquad\qquad 2 \qquad \end{matrix} \\ \begin{pmatrix} \mu + \lambda e^{-(\lambda+\mu)t} & \lambda - \lambda e^{-(\lambda+\mu)t} \\ \mu - \mu e^{-(\lambda+\mu)t} & \lambda + \mu e^{-(\lambda+\mu)t} \end{pmatrix} \end{array}. \tag{7.5}$$

■

Matrix Exponential

The Kolmogorov backward equation $\boldsymbol{P}'(t) = \boldsymbol{QP}(t)$ is a matrix equation, which bears a striking resemblance to the nonmatrix differential equation $p'(t) = qp(t)$, where p is a differentiable function and q is a constant. If $p(0) = 1$, the latter has the unique solution

$$p(t) = e^{tq}, \text{ for } t \geq 0.$$

If you are not familiar with solutions to matrix differential equations it might be tempting to try to solve the backward equation by analogy, and write

$$\boldsymbol{P}(t) = e^{t\boldsymbol{Q}}, \text{ for } t \geq 0,$$

since $\boldsymbol{P}(0) = \boldsymbol{I}$. Remarkably, this is exactly correct, as long as $e^{t\boldsymbol{Q}}$ is defined properly.

Matrix Exponential

Let $\boldsymbol{A}$ be a $k \times k$ matrix. The *matrix exponential* $e^{\boldsymbol{A}}$ is the $k \times k$ matrix

$$e^{\boldsymbol{A}} = \sum_{n=0}^{\infty} \frac{1}{n!}\boldsymbol{A}^n = \boldsymbol{I} + \boldsymbol{A} + \frac{1}{2}\boldsymbol{A}^2 + \frac{1}{6}\boldsymbol{A}^3 + \cdots .$$

The matrix exponential is the matrix version of the exponential function and reduces to the ordinary exponential function e^x when $\boldsymbol{A}$ is a 1×1 matrix. The matrix $e^{\boldsymbol{A}}$ is well-defined as its defining series converges for all square matrices $\boldsymbol{A}$.

The matrix exponential satisfies many familiar properties of the exponential function. These include

1. $e^{\boldsymbol{0}} = \boldsymbol{I}$.
2. $e^{\boldsymbol{A}}e^{-\boldsymbol{A}} = \boldsymbol{I}$.
3. $e^{(s+t)\boldsymbol{A}} = e^{s\boldsymbol{A}}e^{t\boldsymbol{A}}$.
4. If $\boldsymbol{AB} = \boldsymbol{BA}$, then $e^{\boldsymbol{A}+\boldsymbol{B}} = e^{\boldsymbol{A}}e^{\boldsymbol{B}} = e^{\boldsymbol{B}}e^{\boldsymbol{A}}$.
5. $\frac{d}{dt}e^{t\boldsymbol{A}} = \boldsymbol{A}e^{t\boldsymbol{A}} = e^{t\boldsymbol{A}}\boldsymbol{A}$.

For a continuous-time Markov chain with generator $\boldsymbol{Q}$, the matrix exponential $e^{t\boldsymbol{Q}}$ is the unique solution to the forward and backward equations. Letting $\boldsymbol{P}(t) = e^{t\boldsymbol{Q}}$ gives

$$\boldsymbol{P}'(t) = \frac{d}{dt}e^{t\boldsymbol{Q}} = \boldsymbol{Q}e^{t\boldsymbol{Q}} = e^{t\boldsymbol{Q}}\boldsymbol{Q} = \boldsymbol{P}(t)\boldsymbol{Q} = \boldsymbol{Q}\boldsymbol{P}(t).$$

Transition Function and Generator

For a continuous-time Markov chain with transition function $\boldsymbol{P}(t)$ and infinitesimal generator $\boldsymbol{Q}$,

$$\boldsymbol{P}(t) = e^{t\boldsymbol{Q}} = \sum_{n=0}^{\infty} \frac{1}{n!}(t\boldsymbol{Q})^n = \boldsymbol{I} + t\boldsymbol{Q} + \frac{t^2}{2}\boldsymbol{Q}^2 + \frac{t^3}{6}\boldsymbol{Q}^3 + \cdots \tag{7.6}$$

Computing the matrix exponential is often numerically challenging. Finding accurate and efficient algorithms is still a topic of current research. Furthermore, the transition function is difficult to obtain in closed form for all but the most specialized models. For applied problems, numerical approximation methods are often needed. The `R` package `expm` contains the function `expm(mat)` for computing the matrix exponential of a numerical matrix `mat`.

R: Computing the Transition Function

For the registration line Markov chain of Example 7.5, we find the transition function $\boldsymbol{P}(t)$ for $t = 2.5$.

```
# matrixexp.R
> install.packages("expm")
> library(expm)
> Q <- matrix(c(-1/4,1/4,0,0,0,1/5,-9/20,1/4,0,0,0,
```

```
+ 1/5,-9/20,1/4,0,0,0,1/5,-9/20,1/4,0,0,0,1/5,-1/5),
+ nrow=5,byrow=T)
> Q
      0     1     2     3     4
0 -0.25  0.25  0.00  0.00  0.00
1  0.20 -0.45  0.25  0.00  0.00
2  0.00  0.20 -0.45  0.25  0.00
3  0.00  0.00  0.20 -0.45  0.25
4  0.00  0.00  0.00  0.20 -0.20
> P  <- function(t) expm(t*Q)
> P(2.5)
      0     1     2     3     4
0 0.610 0.290 0.081 0.016 0.003
1 0.232 0.443 0.238 0.071 0.017
2 0.052 0.190 0.435 0.238 0.085
3 0.008 0.045 0.191 0.446 0.310
4 0.001 0.008 0.054 0.248 0.688
```

Diagonalization*

If the $\boldsymbol{Q}$ matrix is diagonalizable, then so is $e^{t\boldsymbol{Q}}$, and the transition function can be expressed in terms of the eigenvalues and eigenvectors of $\boldsymbol{Q}$. Write $\boldsymbol{Q} = \boldsymbol{SDS}^{-1}$, where

$$\boldsymbol{D} = \begin{pmatrix} \lambda_1 & 0 & \cdots & 0 \\ 0 & \lambda_2 & \cdots & 0 \\ \vdots & \vdots & \ddots & \vdots \\ 0 & 0 & \cdots & \lambda_k \end{pmatrix}$$

is a diagonal matrix whose diagonal entries are the eigenvalues of $\boldsymbol{Q}$, and $\boldsymbol{S}$ is an invertible matrix whose columns are the corresponding eigenvectors. This gives

$$\begin{aligned} e^{t\boldsymbol{Q}} &= \sum_{n=0}^{\infty} \frac{1}{n!}(t\boldsymbol{Q})^n = \sum_{n=0}^{\infty} \frac{t^n}{n!}(\boldsymbol{SDS}^{-1})^n \\ &= \sum_{n=0}^{\infty} \frac{t^n}{n!}\boldsymbol{SD}^n\boldsymbol{S}^{-1} = \boldsymbol{S}\left(\sum_{n=0}^{\infty} \frac{t^n}{n!}\boldsymbol{D}^n\right)\boldsymbol{S}^{-1} \\ &= \boldsymbol{S}e^{t\boldsymbol{D}}\boldsymbol{S}^{-1}, \end{aligned}$$

where

$$e^{t\boldsymbol{D}} = \sum_{n=0}^{\infty} \frac{t^n}{n!}\boldsymbol{D}^n = \sum_{n=0}^{\infty} \frac{t^n}{n!} \begin{pmatrix} \lambda_1^n & 0 & \cdots & 0 \\ 0 & \lambda_2^n & \cdots & 0 \\ \vdots & \vdots & \ddots & \vdots \\ 0 & 0 & \cdots & \lambda_k^n \end{pmatrix}$$

$$= \begin{pmatrix} \sum_{n=0}^{\infty} (t\lambda_1)^n/n! & 0 & \cdots & 0 \\ 0 & \sum_{n=0}^{\infty} (t\lambda_2)^n/n! & \cdots & 0 \\ \vdots & \vdots & \ddots & \vdots \\ 0 & 0 & \cdots & \sum_{n=0}^{\infty} (t\lambda_k)^n/n! \end{pmatrix}$$

$$= \begin{pmatrix} e^{t\lambda_1} & 0 & \cdots & 0 \\ 0 & e^{t\lambda_2} & \cdots & 0 \\ \vdots & \vdots & \ddots & \vdots \\ 0 & 0 & \cdots & e^{t\lambda_k} \end{pmatrix}.$$

Example 7.10 For the two-state chain, the generator

$$Q = \begin{pmatrix} -\lambda & \lambda \\ \mu & -\mu \end{pmatrix}$$

is diagonalizable with eigenvalues 0 and $-(\lambda + \mu)$, and corresponding eigenvectors $\begin{pmatrix} 1 \\ 1 \end{pmatrix}$ and $\begin{pmatrix} -\lambda \\ \mu \end{pmatrix}$. This gives

$$Q = SDS^{-1} = \begin{pmatrix} 1 & -\lambda \\ 1 & \mu \end{pmatrix} \begin{pmatrix} 0 & 0 \\ 0 & -(\lambda+\mu) \end{pmatrix} \begin{pmatrix} \mu/(\lambda+\mu) & \lambda/(\lambda+\mu) \\ -1/(\lambda+\mu) & 1/(\lambda+\mu) \end{pmatrix}.$$

The transition function is

$$\begin{aligned} P(t) = e^{tQ} &= Se^{tD}S^{-1} \\ &= \begin{pmatrix} 1 & -\lambda \\ 1 & \mu \end{pmatrix} \begin{pmatrix} 1 & 0 \\ 0 & e^{-t(\lambda+\mu)} \end{pmatrix} \begin{pmatrix} \mu/(\lambda+\mu) & \lambda/(\lambda+\mu) \\ -1/(\lambda+\mu) & 1/(\lambda+\mu) \end{pmatrix} \\ &= \frac{1}{\lambda+\mu} \begin{pmatrix} \mu + \lambda e^{-t(\lambda+\mu)} & \lambda - \lambda e^{-t(\lambda+\mu)} \\ \mu - \mu e^{-t(\lambda+\mu)} & \lambda + \mu e^{-t(\lambda+\mu)} \end{pmatrix}. \end{aligned}$$

This result was also shown in Example 7.9 as the solution to the Kolmogorov forward equations. ■

Example 7.11 (DNA evolution) Continuous-time Markov chains are used to study the evolution of DNA sequences. Numerous models have been proposed for the evolutionary changes on the genome as a result of mutation. Such models are often specified in terms of transition rates between base nucleotides adenine, guanine, cytosine, and thymine at a fixed chromosome location.

The *Jukes–Cantor* model assumes that all transition rates are the same. The infinitesimal generator, with parameter $r > 0$, is

$$Q = \begin{array}{c} \\ a \\ g \\ c \\ t \end{array} \begin{array}{c} \begin{array}{cccc} a & g & c & t \end{array} \\ \begin{pmatrix} -3r & r & r & r \\ r & -3r & r & r \\ r & r & -3r & r \\ r & r & r & -3r \end{pmatrix} \end{array}.$$

The generator is diagonalizable with linearly independent eigenvectors

$$\begin{pmatrix}-1\\0\\0\\1\end{pmatrix}, \begin{pmatrix}-1\\0\\1\\0\end{pmatrix}, \begin{pmatrix}-1\\1\\0\\0\end{pmatrix}, \begin{pmatrix}1\\1\\1\\1\end{pmatrix},$$

corresponding to eigenvalues $-4r, -4r, -4r$, and 0. This gives

$$\begin{aligned}
\boldsymbol{P}(t) &= e^{t\boldsymbol{Q}} = \boldsymbol{S}e^{t\boldsymbol{D}}\boldsymbol{S}^{-1}\\
&= \begin{pmatrix}-1 & -1 & -1 & 1\\ 0 & 0 & 1 & 1\\ 0 & 1 & 0 & 1\\ 1 & 0 & 0 & 1\end{pmatrix}\begin{pmatrix}e^{-4rt} & 0 & 0 & 0\\ 0 & e^{-4rt} & 0 & 0\\ 0 & 0 & e^{-4rt} & 0\\ 0 & 0 & 0 & 1\end{pmatrix}\begin{pmatrix}-\frac{1}{4} & -\frac{1}{4} & -\frac{1}{4} & \frac{3}{4}\\ -\frac{1}{4} & -\frac{1}{4} & \frac{3}{4} & -\frac{1}{4}\\ -\frac{1}{4} & \frac{3}{4} & -\frac{1}{4} & -\frac{1}{4}\\ \frac{1}{4} & \frac{1}{4} & \frac{1}{4} & \frac{1}{4}\end{pmatrix}\\
&= \frac{1}{4}\begin{pmatrix}1+3e^{-4rt} & 1-e^{-4rt} & 1-e^{-4rt} & 1-e^{-4rt}\\ 1-e^{-4rt} & 1+3e^{-4rt} & 1-e^{-4rt} & 1-e^{-4rt}\\ 1-e^{-4rt} & 1-e^{-4rt} & 1+3e^{-4rt} & 1-e^{-4rt}\\ 1-e^{-4rt} & 1-e^{-4rt} & 1-e^{-4rt} & 1+3e^{-4rt}\end{pmatrix}.
\end{aligned}$$

The Jukes–Cantor model does not distinguish between base types. Nucleotides a and g are purines, c and t are pyrimidines. Changes from purine to purine or pyrimidine to pyrimidine are called *transitions*. Changes from purine to pyrimidine or vice versa are called *transversions*.

The *Kimura* model, which includes two parameters r and s, distinguishes between transitions and transversions. The generator is

$$\boldsymbol{Q} = \begin{array}{c} \\ a\\ g\\ c\\ t\end{array}\begin{array}{c}\begin{array}{cccc} a & g & c & t\end{array}\\ \begin{pmatrix}-(r+2s) & r & s & s\\ r & -(r+2s) & s & s\\ s & s & -(r+2s) & r\\ s & s & r & -(r+2s)\end{pmatrix}\end{array}.$$

The matrix is diagonalizable. The respective matrices of eigenvalues and eigenvectors are

$$\boldsymbol{D} = \begin{pmatrix}-2(r+s) & 0 & 0 & 0\\ 0 & -2(r+s) & 0 & 0\\ 0 & 0 & -4s & 0\\ 0 & 0 & 0 & 0\end{pmatrix} \quad \text{and} \quad \boldsymbol{S} = \begin{pmatrix}-1 & 0 & -1 & 1\\ 0 & -1 & 1 & 1\\ 1 & 0 & -1 & 1\\ 0 & 1 & 1 & 1\end{pmatrix}.$$

The transition function $\boldsymbol{P}(t) = e^{t\boldsymbol{Q}} = \boldsymbol{S}e^{t\boldsymbol{D}}\boldsymbol{S}^{-1}$ is

$$\boldsymbol{P}_{xy}(t) = \begin{cases} \left(1 + e^{-4st} - 2e^{-2(r+s)t}\right)/4, & \text{if } xy \in \{ag, ga, ct, tc\}, \\ \left(1 - e^{-4st}\right)/4, & \text{if } xy \in \{ac, at, gc, gt, ca, cg, ta, tg\}, \\ \left(1 + e^{-4st} + 2e^{-2(r+s)t}\right)/4, & \text{if } xy \in \{aa, gg, cc, tt\}. \end{cases}$$

Estimating Mutation Rate and Evolutionary Distance

Continuous-time Markov models are used by biologists and geneticists to estimate the evolutionary distance between species, as well as the related mutation rate. The following application is based on Durrett (2002).

A statistic for estimating evolutionary distance from DNA sequences is the number of locations in the sequences that differ. Given two DNA strands and a common nucleotide site, consider the probability q that the nucleotides in the two strands are identical given that the most recent common ancestor occurred t time units ago. Assume at that time that the nucleotide at the given site was a. Then, the probability that the two sequences are identical, by the Jukes–Cantor model, is

$$\begin{aligned} q &= (P_{aa}(t))^2 + (P_{ag}(t))^2 + (P_{ac}(t))^2 + (P_{at}(t))^2 \\ &= \left(\frac{1}{4} + \frac{3}{4}e^{-4rt}\right)^2 + 3\left(\frac{1}{4} - \frac{1}{4}e^{-4rt}\right)^2 \\ &= \frac{1}{4} + \frac{3}{4}e^{-8rt}, \end{aligned}$$

which is independent of the starting nucleotide. Thus the probability p that the two sites are different is

$$p = 1 - q = \frac{3}{4}\left(1 - e^{-8rt}\right).$$

Solving for rt gives

$$rt = -\frac{1}{8}\ln\left(1 - \frac{4p}{3}\right).$$

In the Jukes–Cantor model, nucleotides change at the rate of $3r$ transitions per time unit. In two DNA strands we expect $2(3rt) = 6rt$ substitutions over t years. Let K denote the number of substitutions that occur over t years. Then,

$$E(K) = 6rt = -\frac{3}{4}\ln\left(1 - \frac{4p}{3}\right).$$

To estimate the actual number of substitutions that occurred since the most recent common ancestor t time units ago, take

$$\widehat{K} = -\frac{3}{4}\ln\left(1 - \frac{4\widehat{p}}{3}\right),$$

where $\widehat{p}$ is the observed fraction of differences between two sequences.

The number of nucleotide substitutions per site between two sequences since their separation from the last common ancestor is called the *evolutionary distance* between the sequences. It is an important quantity for estimating the rate of evolution and the divergence time between species.

Durrett compares the sequence of mitochondrial RNA for the somatotropin gene (a growth hormone) for rats and humans. The observed proportion of differences between the two sequences at a fixed site in the gene is $\widehat{p} = 0.366$. Hence, the estimate of the number of substitutions at that site that have occurred since divergence of the two species is

$$\widehat{K} = -\frac{3}{4} \ln\left(1 - \frac{4(0.366)}{3}\right) = 0.502.$$

Using the fact that rats and humans diverged about 80 million years ago, and choosing 0.502 as the number of substitutions per nucleotide, the mutation rate at that position is estimated as $0.502/(8 \times 10^7) = 6.275 \times 10^{-9}$ per year. ■

Example 7.12 (Using symbolic software) Symbolic software systems, such as *Mathematica* and *Maple*, work in exact integer arithmetic, and can be used to find the matrix exponential when the generator matrix contains symbolic parameters. *Wolfram Alpha*, which is freely available on the web, has the command **MatrixExp** for computing the matrix exponential.

To find the transition function for the Jukes–Cantor model from the previous example, we type the following command from our web browser

> **MatrixExp**$[t\{\{-3r, r, r, r\}, \{r, -3r, r, r\}, \{r, r, -3r, r\}, \{r, r, r, -3r\}\}]$,

which gives the output shown in Figure 7.6. ■

7.4 LONG-TERM BEHAVIOR

For continuous-time Markov chains, limiting and stationary distributions are defined similarly as for discrete time.

Limiting Distribution

A probability distribution $\boldsymbol{\pi}$ is the *limiting distribution* of a continuous-time Markov chain if for all states i and j,

$$\lim_{t \to \infty} \boldsymbol{P}_{ij}(t) = \pi_j.$$

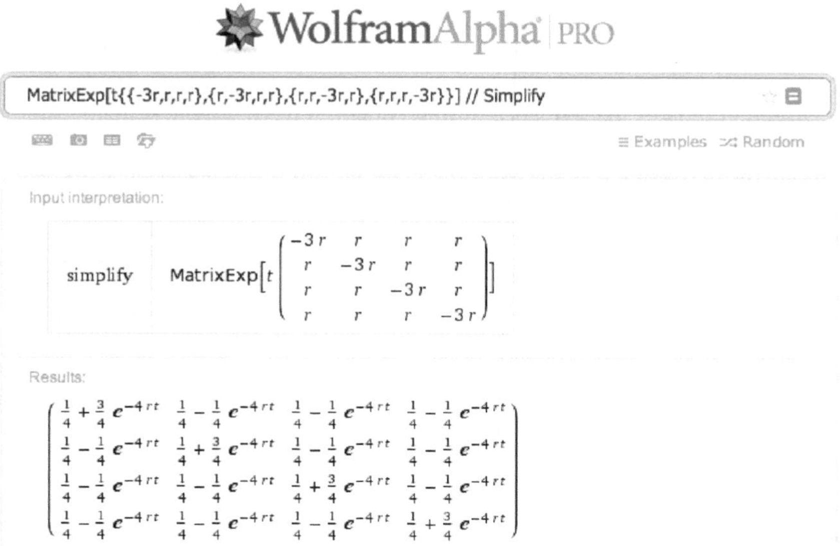

Figure 7.6 Computing the matrix exponential with *WolframAlpha. Source:* See Wolfram Alpha LLC (2015).

Stationary Distribution

A probability distribution $\boldsymbol{\pi}$ is a *stationary distribution* if

$$\boldsymbol{\pi} = \boldsymbol{\pi} \boldsymbol{P}(t), \text{ for } t \geq 0.$$

That is, for all states j,

$$\pi_j = \sum_i \pi_i P_{ij}(t), \text{ for } t \geq 0.$$

As in the discrete case, the limiting distribution, if it exists, is a stationary distribution. However, the converse is not necessarily true and depends on the class structure of the chain. For characterizing the states of a continuous-time Markov chain, notions of accessibility, communication, and irreducibility are defined as in the discrete case. A continuous-time Markov chain is *irreducible* if for all i and j, $P_{ij}(t) > 0$ for some $t > 0$.

In one regard, the classification of states is easier in continuous time since periodicity is not an issue. All states are essentially aperiodic, a consequence of the following lemma.

Lemma 7.1. *If $P_{ij}(t) > 0$, for some $t > 0$, then $P_{ij}(t) > 0$, for all $t > 0$.*

The result is intuitive since if $P_{ij}(t) > 0$ for some t, then there exists a path from i to j in the embedded chain, and for any time s there is positive probability of reaching j from i in s time units. We forego the complete proof, but show the result for forward time. Assume that $P_{ij}(t) > 0$ for some t. Then, for $s \geq 0$,

$$P_{ij}(t+s) = \sum_k P_{ik}(t)P_{kj}(s) \geq P_{ij}(t)P_{jj}(s) \geq P_{ij}(t)e^{-q_j s} > 0.$$

For the penultimate inequality, $e^{-q_j s}$ is the probability that there is no transition from j by time s. The latter event implies that the process started at j is at j at time s, whose probability is $P_{jj}(s)$.

A finite-state continuous-time Markov chain is irreducible if all the holding time parameters are positive. On the contrary, if $q_i = 0$ for some i, then i is an *absorbing state*. If we assume that all the holding time parameters are finite, then there are two possibilities: (i) the process is irreducible, all states communicate, and $P_{ij}(t) > 0$, for $t > 0$ and all i, j or (ii) the process contains one or more absorbing states.

The following fundamental limit theorem is given without proof. Note the analogies with the discrete-time results, for example, Theorems 3.6 and 3.8.

Fundamental Limit Theorem

Theorem 7.2. *Let $(X_t)_{t\geq 0}$ be a finite, irreducible, continuous-time Markov chain with transition function $\boldsymbol{P}(t)$. Then, there exists a unique stationary distribution $\boldsymbol{\pi}$, which is the limiting distribution. That is, for all j,*

$$\lim_{t\to\infty} P_{ij}(t) = \pi_j, \text{for all initial } i.$$

Equivalently,

$$\lim_{t\to\infty} \boldsymbol{P}(t) = \boldsymbol{\Pi},$$

where $\boldsymbol{\Pi}$ is a matrix all of whose rows are equal to $\boldsymbol{\pi}$.

Example 7.13 (Two-state process) For the continuous-time Markov chain on two states,

$$\lim_{t\to\infty} \boldsymbol{P}(t) = \lim_{t\to\infty} \frac{1}{\lambda+\mu}\begin{pmatrix} \mu + \lambda e^{-(\lambda+\mu)t} & \lambda - \lambda e^{-(\lambda+\mu)t} \\ \mu - \mu e^{-(\lambda+\mu)t} & \lambda + \mu e^{-(\lambda+\mu)t} \end{pmatrix} = \frac{1}{\lambda+\mu}\begin{pmatrix} \mu & \lambda \\ \mu & \lambda \end{pmatrix}.$$

The stationary distribution is

$$\boldsymbol{\pi} = \left(\frac{\mu}{\lambda+\mu}, \frac{\lambda}{\lambda+\mu}\right).$$

■

The following result links the stationary distribution with the generator.

Stationary Distribution and Generator Matrix

A probability distribution $\boldsymbol{\pi}$ is a stationary distribution of a continuous-time Markov chain with generator $\boldsymbol{Q}$ if and only if

$$\boldsymbol{\pi Q} = \mathbf{0}.$$

That is,

$$\sum_i \pi_i Q_{ij} = 0, \text{ for all } j.$$

Proof. Assume that $\boldsymbol{\pi} = \boldsymbol{\pi P}(t)$, for all $t \geq 0$. Take the derivative of both sides of the equation at $t = 0$ to get $\mathbf{0} = \boldsymbol{\pi P}'(0) = \boldsymbol{\pi Q}$. Conversely, assume that $\boldsymbol{\pi Q} = \mathbf{0}$. Then,

$$\mathbf{0} = \boldsymbol{\pi Q P}(t) = \boldsymbol{\pi P}'(t), \text{ for } t \geq 0,$$

by the Kolmogorov backward equation. Since the derivative is equal to $\mathbf{0}$, $\boldsymbol{\pi P}(t)$ is a constant. In particular, $\boldsymbol{P}(0) = \boldsymbol{I}$, and thus $\boldsymbol{\pi P}(t) = \boldsymbol{\pi P}(0) = \boldsymbol{\pi}$, for $t \geq 0$. ■

The stationary probability π_j can be interpreted as the long-term proportion of time that the chain spends in state j. This is analogous to the discrete-time case in which the stationary probability represents the long-term fraction of transitions that the chain visits a given state.

Example 7.14 (Eat, play, sleep) Jesse is a newborn baby who is always in one of three states: eat, play, and sleep. He eats on average for 30 minutes at a time; plays on average for 1 hour; and sleeps for about 3 hours. After eating, there is a 50–50 chance he will sleep or play. After playing, there is a 50–50 chance he will eat or sleep. And after sleeping, he always plays. Jesse's life is governed by a continuous-time Markov chain. What proportion of the day does Jesse sleep?

Solution The holding time parameters for the three-state chain (in hour units) are $(q_e, q_p, q_s) = (2, 1, 1/3)$. The embedded chain transition probabilities are

$$\widetilde{\boldsymbol{P}} = \begin{array}{c} \\ \text{Eat} \\ \text{Play} \\ \text{Sleep} \end{array} \begin{array}{c} \begin{array}{ccc} \text{Eat} & \text{Play} & \text{Sleep} \end{array} \\ \begin{pmatrix} 0 & 1/2 & 1/2 \\ 1/2 & 0 & 1/2 \\ 0 & 1 & 0 \end{pmatrix} \end{array}.$$

With $q_{ij} = q_i p_{ij}$, the generator matrix is

$$Q = \begin{array}{c} \\ \text{Eat} \\ \text{Play} \\ \text{Sleep} \end{array} \begin{array}{c} \begin{array}{ccc} \text{Eat} & \text{Play} & \text{Sleep} \end{array} \\ \begin{pmatrix} -2 & 1 & 1 \\ 1/2 & -1 & 1/2 \\ 0 & 1/3 & -1/3 \end{pmatrix} \end{array}.$$

The linear system $\boldsymbol{\pi Q} = \mathbf{0}$ gives

$$\begin{aligned} -2\pi_e + (1/2)\pi_p \qquad\qquad &= 0, \\ \pi_e - \qquad \pi_p + (1/3)\pi_s &= 0, \\ \pi_e + (1/2)\pi_p - (1/3)\pi_s &= 0, \end{aligned}$$

with solution

$$\boldsymbol{\pi} = (\pi_e, \pi_p, \pi_s) = \left(\frac{1}{14}, \frac{4}{14}, \frac{9}{14}\right) = (0.071, 0.286, 0.643).$$

Jesses spends almost two-thirds of his day sleeping.

Using R, we compute the transition function, and then take $\boldsymbol{P}(t)$ for large t (in this case $t = 100$) to find the approximate limiting distribution. ■

R: Eat, Play, Sleep

```
> install.packages("expm")
> library(expm)
> Q = matrix(c(-2,1,1,1/2,-1,1/2,0,1/3,-1/3),
+  nrow=3,byrow=T)
> P <- function(t) {expm(t*Q)}
> P(100)
          Eat     Play    Sleep
Eat   0.07143 0.28571 0.64286
Play  0.07143 0.28571 0.64286
Sleep 0.07143 0.28571 0.64286
```

Example 7.15 (DNA evolution) The Jukes–Cantor and Kimura models, introduced in Example 7.11, have uniform limiting distribution $\boldsymbol{\pi} = (1/4, 1/4, 1/4, 1/4)$. An objection to these models is that nucleotides are not distributed uniformly on the genome. For instance, frequencies for human DNA are approximately

a	g	c	t
0.292	0.207	0.207	0.292

The four-parameter *Felsenstein* model, introduced in 1981, has generator matrix

$$Q = \begin{array}{c} \\ a \\ g \\ c \\ t \end{array} \begin{array}{c} \begin{array}{cccc} a & g & c & t \end{array} \\ \begin{pmatrix} -\alpha(1-p_a) & \alpha p_g & \alpha p_c & \alpha p_t \\ \alpha p_a & \alpha(1-p_g) & \alpha p_c & \alpha p_t \\ \alpha p_a & \alpha p_g & -\alpha(1-p_c) & \alpha p_t \\ \alpha p_a & \alpha p_g & \alpha p_c & -\alpha(1-p_t) \end{pmatrix} \end{array},$$

where $p_a + p_g + p_c + p_t = 1$ and $\alpha > 0$. One checks that $\boldsymbol{\pi} = (p_a, p_g, p_c, p_t)$ satisfies $\boldsymbol{\pi Q} = \mathbf{0}$ and is the stationary distribution. The transition function, which we do not derive, is

$$\boldsymbol{P}_{ij}(t) = \begin{cases} (1 - e^{-\alpha t})p_j, & \text{if } i \neq j, \\ e^{-\alpha t} + (1 - e^{-\alpha t})p_j, & \text{if } i = j. \end{cases}$$

■

Absorbing States

An absorbing Markov chain is one in which there is at least one absorbing state. Let $(X_t)_{t\geq 0}$ be an absorbing continuous-time Markov chain on $\{1, \ldots, k\}$. For simplicity, assume the chain has one absorbing state a. As in the discrete case, the nonabsorbing states are *transient*. There is positive probability that the chain, started in a transient state, gets absorbed and never returns to that state. For transient state i, we derive an expression for a_i, the expected time until absorption.

Let T denote the set of transient states. Write the generator in canonical block matrix form

$$\boldsymbol{Q} = \begin{array}{c} \\ a \\ T \end{array} \begin{array}{c} \begin{array}{cc} a & T \end{array} \\ \begin{pmatrix} 0 & \mathbf{0} \\ * & \boldsymbol{V} \end{pmatrix} \end{array},$$

where $\boldsymbol{V}$ is a $(k-1) \times (k-1)$ matrix.

Mean Time Until Absorption

Theorem 7.3. *For an absorbing continuous-time Markov chain, define a square matrix* $\boldsymbol{F}$ *on the set of transient states, where* F_{ij} *is the expected time, for the chain started in i, that the process is in j until absorption. Then,*

$$\boldsymbol{F} = -\boldsymbol{V}^{-1}.$$

For the chain started in i, the mean time until absorption is,

$$a_i = \sum_j F_{ij}.$$

The matrix $\boldsymbol{F}$ *is called the* fundamental matrix.

Proof. The proof is based on conditioning on the first transition. For the chain started in i, consider the mean time F_{ij} spent in j before absorption.

Assume that $j \neq i$ and that the process first moves to $k \neq i$. The probability of moving from i to k is q_{ik}/q_i. If $k = a$ is the absorbing state, the time spent in j is 0. Otherwise, the mean time in j until absorption is F_{kj}. This gives

$$\begin{aligned}
F_{ij} &= \sum_{\substack{k\in T\\ k\neq i}} \frac{q_{ik}}{q_i} F_{kj} = \frac{1}{q_i}\sum_{\substack{k\in T\\ k\neq i}} V_{ik}F_{kj}\\
&= \frac{1}{q_i}[(VF)_{ij} - V_{ii}F_{ij}] = \frac{1}{q_i}[(VF)_{ij} + q_iF_{ij}]\\
&= \frac{1}{q_i}(VF)_{ij} + F_{ij}.
\end{aligned}$$

Thus $(VF)_{ij} = 0$. We have used that $V_{ij} = Q_{ij}$, for $i, j \in T$.

Assume that $j = i$. The mean time spent in i before transitioning to a new state is $1/q_i$. Conditioning on the next state,

$$\begin{aligned}
F_{ii} &= \frac{q_{ia}}{q_i}\left(\frac{1}{q_i}\right) + \sum_{\substack{k\in T\\ k\neq i}} \frac{q_{ik}}{q_i}\left(\frac{1}{q_i} + F_{ki}\right)\\
&= \frac{q_{ia}}{q_i^2} + \frac{1}{q_i^2}\sum_{\substack{k\in T\\ k\neq i}} q_{ik} + \frac{1}{q_i}\sum_{\substack{k\in T\\ k\neq i}} q_{ik}F_{ki}\\
&= \frac{-1}{q_i^2}Q_{ii} + \frac{1}{q_i}[(VF)_{ii} - V_{ii}F_{ii}]\\
&= \frac{1}{q_i} + \frac{1}{q_i}(VF)_{ii} + F_{ii}.
\end{aligned}$$

Hence, $(VF)_{ii} = -1$.

In summary,

$$(VF)_{ij} = \begin{cases} -1, & \text{if } i = j,\\ 0, & \text{if } i \neq j. \end{cases}$$

That is, $\boldsymbol{VF} = -\boldsymbol{I}$. and $\boldsymbol{F} = -\boldsymbol{V}^{-1}$.

For the chain started in i, the mean time until absorption is the sum of the mean times in each transient state j until absorption, which gives the final result. ∎

Example 7.16 (Multistate models) Multistate Markov models are used in medicine to model the course of diseases. A patient may advance into, or recover from, successively more severe stages of a disease until some terminal state. Each stage represents a state of an absorbing continuous-time Markov chain.

Bartolomeo et al. (2011) develops such a model to study the progression of liver disease among patients diagnosed with cirrhosis of the liver. The general form of the infinitesimal generator matrix for their three-parameter model is

$$Q = \begin{matrix} 1 \\ 2 \\ 3 \end{matrix} \overset{\begin{matrix} 1 & \qquad\qquad 2 & \qquad 3 \end{matrix}}{\begin{pmatrix} -(q_{12}+q_{13}) & q_{12} & q_{13} \\ 0 & -q_{23} & q_{23} \\ 0 & 0 & 0 \end{pmatrix}},$$

where state 1 represents cirrhosis, state 2 denotes liver cancer (hepatocellular carcinoma), and state 3 is death. The fundamental matrix is

$$F = -\begin{pmatrix} -(q_{12}+q_{13}) & q_{12} \\ 0 & -q_{23} \end{pmatrix}^{-1} = \overset{\begin{matrix} 1 & \qquad\qquad\qquad 2 \end{matrix}}{\begin{pmatrix} 1/(q_{12}+q_{13}) & q_{12}/(q_{23}(q_{12}+q_{13})) \\ 0 & 1/q_{23} \end{pmatrix}},$$

with mean absorption times

$$a_1 = \frac{1}{q_{12}+q_{13}} + \frac{q_{12}}{q_{23}(q_{12}+q_{13})} \quad \text{and} \quad a_2 = \frac{1}{q_{23}}.$$

From a sample of 1,925 patients diagnosed with cirrhosis, and data on the number of months at each stage of the disease, the authors estimate the parameters of the model for subgroups depending on age, gender, and other variables. Their mean parameter estimates are $\widehat{q_{12}} = 0.0151$, $\widehat{q_{13}} = 0.0071$, and $\widehat{q_{23}} = 0.0284$.

Plots of the transition probabilities between states are shown in Figure 7.7. The fundamental matrix is estimated to be

$$\widehat{F} = \begin{matrix} 1 \\ 2 \end{matrix} \overset{\begin{matrix} 1 & \quad 2 \end{matrix}}{\begin{pmatrix} 45.05 & 23.95 \\ 0.00 & 35.21 \end{pmatrix}}.$$

From the fundamental matrix, the estimated mean time to death for patients with liver cirrhosis is $45.05 + 23.95 = 69$ months. See the foregoing R code for a simulation of the mean time to death.

R: Simulation of Time to Absorption

```
# absorption.R
> trials <- 100000
> simlist <- numeric(trials)
> init <- 1   # initial state of liver cirrhosis
> for (i in 1:trials) {
+ state <- init
+ t <- 0
```

```
+ while (TRUE) {
+ if (state == 1) { q12 <- rexp(1,0.0151)
+   q13 <- rexp(1,0.0071) }
+   if (q12 < q13) {t <-t + q12
+   state <- 2}
+   else {t <- t + q13
+   break}
+ if (state == 2) {q23 <- rexp(1,0.0284)
+   t <- t + q23
+   break}
+ }
+ simlist[i] <- t  }
> mean(simlist)
[1] 69.01561
```

■

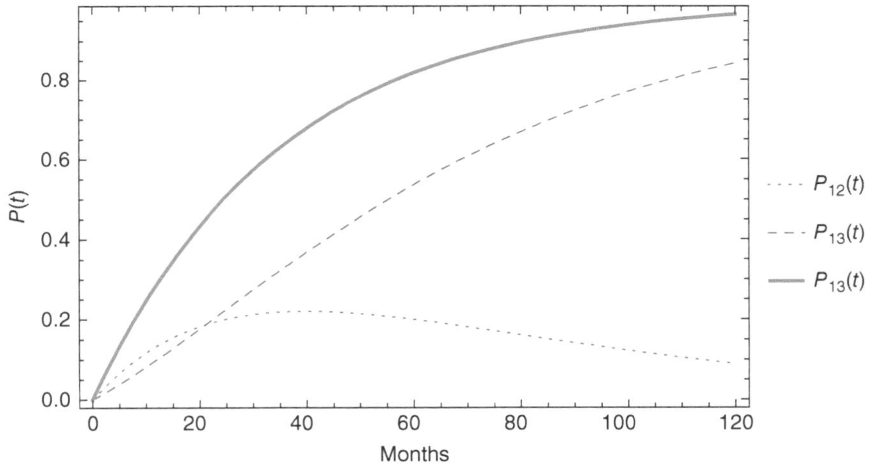

Figure 7.7 Estimated transition probabilities for stages of liver cirrhosis. *Source:* Bartolomeo et al. (2011).

Stationary Distribution of Embedded Chain

For a continuous-time Markov chain, the stationary distribution $\boldsymbol{\pi}$ is not the same as the stationary distribution of the embedded chain, which we denote as $\boldsymbol{\psi}$. Consider a three-state process with generator

$$\boldsymbol{Q} = \begin{matrix} \\ 1 \\ 2 \\ 3 \end{matrix} \begin{matrix} \begin{matrix} 1 & \;\;2 & \;\;3 \end{matrix} \\ \begin{pmatrix} -2 & 1 & 1 \\ 2 & -4 & 2 \\ 4 & 4 & -8 \end{pmatrix} \end{matrix}.$$

The unique solution of $\boldsymbol{\pi Q} = \mathbf{0}$ is $\boldsymbol{\pi} = (4/7, 2/7, 1/7)$, which is the stationary distribution of the continuous-time chain. However, the embedded chain has transition matrix

$$\widetilde{\boldsymbol{P}} = \begin{matrix} \\ 1 \\ 2 \\ 3 \end{matrix} \begin{matrix} \begin{matrix} 1 & 2 & 3 \end{matrix} \\ \begin{pmatrix} 0 & 1/2 & 1/2 \\ 1/2 & 0 & 1/2 \\ 1/2 & 1/2 & 0 \end{pmatrix} \end{matrix},$$

with uniform stationary distribution $\boldsymbol{\psi} = (1/3, 1/3, 1/3)$.

Note the difference in interpretation of the two distributions. The stationary probability π_j is the long-term proportion of *time* that the process spends in state j. On the other hand, the embedded chain stationary probability ψ_j is the long-term proportion of transitions that the process makes into state j.

Each stationary distribution can be derived from the other. From $\boldsymbol{\pi Q} = \mathbf{0}$, we have that

$$\pi_j q_j = \sum_{i \neq j} \pi_i q_{ij}, \text{ for all } j. \tag{7.7}$$

Define $\widetilde{\psi}_j = \pi_j q_j$. Then, Equation (7.7) gives

$$\widetilde{\psi}_j = \sum_{i \neq j} \widetilde{\psi}_i p_{ij}.$$

Thus $\widetilde{\psi}$ satisfies $\widetilde{\psi}\widetilde{\boldsymbol{P}} = \widetilde{\psi}$. To obtain a stationary distribution for the embedded chain, normalize $\widetilde{\psi}$. Let

$$\psi_j = \frac{\widetilde{\psi}_j}{\sum_k \widetilde{\psi}_k} = \frac{\pi_j q_j}{\sum_k \pi_k q_k}.$$

Then, $\boldsymbol{\psi} = \boldsymbol{\psi}\widetilde{\boldsymbol{P}}$, and $\boldsymbol{\psi}$ is the stationary distribution of the embedded Markov chain.

Having derived $\boldsymbol{\psi}$ from $\boldsymbol{\pi}$, we can also derive $\boldsymbol{\pi}$ from $\boldsymbol{\psi}$. Since $\psi_j = C\pi_j q_j$, where C is an appropriate normalizing constant, we have that $\pi_j = \psi_j/(Cq_j)$. This gives,

$$\pi_j = \frac{\psi_j / q_j}{\sum_k \psi_k / q_k}, \text{ for all } j.$$

Example 7.17 The embedded chain of a continuous-time process has transition matrix

$$\widetilde{\boldsymbol{P}} = \begin{matrix} \\ 1 \\ 2 \\ 3 \end{matrix} \begin{matrix} \begin{matrix} 1 & 2 & 3 \end{matrix} \\ \begin{pmatrix} 0 & 1 & 0 \\ 1/3 & 0 & 2/3 \\ 0 & 1 & 0 \end{pmatrix} \end{matrix}.$$

Assume that the process stays at state 1 on average 5 minutes before moving to 2. From state 2, it stays on average 2 minutes before moving to a new state. From 3, it stays on average 4 minutes before transitioning to 2. Find the stationary distribution of the continuous-time chain.

Solution Solve $\boldsymbol{\psi}\widetilde{\boldsymbol{P}} = \boldsymbol{\psi}$ to get $\boldsymbol{\psi} = (1/6, 1/2, 1/3)$. Holding time parameters are $(q_1, q_2, q_3) = (1/5, 1/2, 1/4)$. The stationary distribution $\boldsymbol{\pi}$ is proportional to

$$(\psi_1/q_1, \psi_2/q_2, \psi_3/q_3) = (5/6, 2/2, 4/3),$$

which gives $\boldsymbol{\pi} = (5/19, 6/19, 8/19)$. ■

Global Balance

Assume that $\boldsymbol{\pi}$ is the stationary distribution of a continuous-time Markov chain. From $\boldsymbol{\pi Q} = \mathbf{0}$, we have that

$$\sum_{i \neq j} \pi_i q_{ij} = \pi_j q_j, \text{ for all } j. \tag{7.8}$$

The holding time parameter q_j is the transition rate from j. Since π_j is the long-term proportion of time the process visits j, the right-hand side of Equation (7.8) is the long-term rate that the process leaves j. Also, $\pi_i q_{ij}$ is the long-term rate of transitioning from i to j. Thus, the left-hand side of Equation (7.8) is the long-term rate that the process enters j.

Equations (7.8) are known as the *global balance equations*. They say that in stationarity the rates in and out of any state are the same.

Example 7.18 In Example 7.17, we have that

$$(\pi_1 q_1, \pi_2 q_2, \pi_3 q_3) = (1/19, 3/19, 2/19).$$

By global balance, in the long-term, every 19 minutes the process sees one transition to and from state 1, three transitions to and from state 2, and two transitions to and from state 3.

We illustrate one of these facts by simulating the number of transitions to and from state 2 for the first 19 minutes of the stationary continuous-time chain.

R: Simulating Global Balance

```
# globalbalance.R
> trials <- 100000
> simin2 <- simout2 <- numeric(trials)
> for (i in 1:trials) {
+ state <- sample(1:3,1, prob=c(5/19,6/19,8/19))
+ t <- 0  # time
+ in2  <- 0  # counter for transitions to 2
+ out2 <- 0  # counter for transitions from 2
+ while (t < 19) {
```

```
+ if (state == 1) {t <- t+rexp(1,1/5)
+   if (t > 19) {break}
+   state <- 2
+   in2 <- in2 + 1 }
+ if (state == 3) {t <- t + rexp(1,1/4)
+   if (t > 19) {break}
+   state <- 2
+   in2 <- in2 + 1}
+ if (state ==2) { r1 <- rexp(1, (1/2)*(1/3))
+   r3 <- rexp(1, (1/2)*(2/3))
+   if (r1 < r3) { t <- t + r1
+   if (t > 19) {break}
+   out2 <- out2 + 1
+   state <- 1} else { t <-  t + r3
+   if (t > 19) {break}
+   out2 <- out2 + 1
+   state <- 3}}
+ }
+ simin2[i] <- in2
+ simout2[i] <- out2}
> mean(simin2)   # mean transitions to 2
[1] 2.99994
> mean(simout2)  # mean transitions from 2
[1] 3.00264
```

■

7.5 TIME REVERSIBILITY

Intuitively, a continuous-time Markov chain is *time reversible* if the process in forward time is indistinguishable from the process in reversed time. A consequence is that for all states i and j, the long-term *forward* transition rate from i to j is equal to the long-term *backward* rate from j to i. This is a stronger condition than global balance, which says that the long-term rate from a given state is equal to the long-term rate into that state.

Time Reversibility

A continuous-time Markov chain with generator $\boldsymbol{Q}$ and unique stationary distribution $\boldsymbol{\pi}$ is said to be *time reversible* if

$$\pi_i q_{ij} = \pi_j q_{ji}, \text{ for all } i,j. \tag{7.9}$$

Equations (7.9) give the *local balance*, or *detailed balance*, equations. They say that the long-term transition rate from i to j is equal to the long-term transition rate from j to i.

The local balance equations are a property of reversible chains, which can be used to find the stationary distribution. If a probability distribution λ satisfies

$$\lambda_i q_{ij} = \lambda_j q_{ji}, \text{ for all } i, j,$$

then the continuous-time chain is time reversible and $\boldsymbol{\pi} = \boldsymbol{\lambda}$ is the unique stationary distribution of the chain. Summing over i gives

$$\sum_i \lambda_i q_{ij} = \lambda_j \sum_i q_{ji} = 0.$$

That is, $\boldsymbol{\lambda Q} = \mathbf{0}$.

Example 7.19 The general, nine-parameter, time reversible model for DNA substitutions has generator matrix

$$Q = \begin{array}{c} \\ a \\ g \\ c \\ t \end{array} \begin{array}{c} \begin{array}{cccc} a & g & c & t \end{array} \\ \begin{pmatrix} - & \alpha p_g & \beta p_c & \gamma p_t \\ \alpha p_a & - & \delta p_c & \epsilon p_t \\ \beta p_a & \delta p_g & - & \eta p_t \\ \gamma p_a & \epsilon p_g & \eta p_c & - \end{pmatrix} \end{array},$$

where $p_a + p_g + p_c + p_t = 1$, and $\alpha, \beta, \gamma, \delta, \epsilon, \eta > 0$. Diagonal entries are chosen so that rows sum to 0. One checks that that $p_x q_{xy} = q_{yx} p_y$ for $x, y \in \{a, g, c, t\}$. It follows that the unique stationary distribution is $\boldsymbol{\pi} = (p_a, p_g, p_c, p_t)$. ■

A continuous-time Markov chain is time reversible if and only if its embedded discrete-time chain is time reversible. From Equation (7.9), we obtain that

$$\pi_i q_i p_{ij} = \pi_j q_j p_{ji}, \text{ for all } i, j.$$

Dividing both sides by the normalizing constant $\sum_k \pi_k q_k$ gives

$$\psi_i p_{ij} = \psi_j p_{ji}, \text{ for all } i, j.$$

That is, $\boldsymbol{\psi}$ satisfies the detailed balance equations for the embedded chain. In stationarity, the frequency of transitions from i to j is the same as that from j to i.

Tree Theorem

A *tree* is an undirected graph in which any pair of vertices is connected by exactly one path. Equivalently, a tree is a connected graph that has no cycles. Figure 7.8 shows three examples of trees.

Consider an undirected transition graph for a continuous-time Markov chain where vertices i and j are joined by an edge if $q_{ij} > 0$ or $q_{ji} > 0$. The next theorem, from

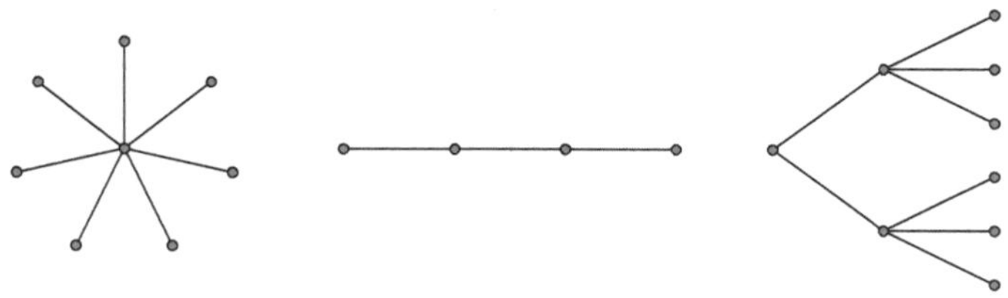

Figure 7.8 Trees.

Kelly (1994), gives a sufficient condition for time reversibility based on the structure of the graph.

Markov Processes on Trees are Time Reversible

Theorem 7.4. *Assume that the transition graph of an irreducible continuous-time Markov chain is a tree. Then, the process is time reversible.*

The result is shown by means of the following lemma, which is of independent interest. It states that the long-term transition rate out of any nonempty subset of states is equal to the long-term transition rate into that set.

Lemma 7.5. *Let S denote the state space of a continuous-time Markov chain with stationary distribution $\boldsymbol{\pi}$ and generator $\boldsymbol{Q}$. For any nonempty subset of the state space $A \subseteq S$,*

$$\sum_{i \in A} \sum_{j \notin A} \pi_i q_{ij} = \sum_{i \in A} \sum_{j \notin A} \pi_j q_{ji}. \tag{7.10}$$

Proof of Lemma 7.5. The left-hand side of Equation (7.10) is equal to

$$\sum_{i \in A} \sum_{j \in S} \pi_i q_{ij} - \sum_{i \in A} \sum_{j \in A} \pi_i q_{ij} = -\sum_{i \in A} \sum_{j \in A} \pi_i q_{ij},$$

since the rows of $\boldsymbol{Q}$ sum to 0. The right-hand side of Equation (7.10) is equal to

$$\sum_{i \in A} \sum_{j \in S} \pi_j q_{ji} - \sum_{i \in A} \sum_{j \in A} \pi_j q_{ji} = -\sum_{i \in A} \sum_{j \in A} \pi_i q_{ij},$$

since $\boldsymbol{\pi}$ is the stationary distribution and $\boldsymbol{\pi Q} = \boldsymbol{0}$. ■

Proof of Theorem 7.4. Assume that the transition graph is a tree. We show that

$$\pi_i q_{ij} = \pi_j q_{ji}, \text{ for all } i, j.$$

If there is no edge between i and j in the transition graph, then $q_{ij} = q_{ji} = 0$. Assume that there is an edge between i and j. A tree has the property, since it contains no cycles, that removal of any edge cuts the graph into two disconnected components. Let A denote the set of states connected to i, after removing the edge between i and j. Since the only vertex connected to i in the original graph, which is not in S, is j,

$$\sum_{k\in A}\sum_{l\notin A} \pi_k q_{kl} = \pi_i q_{ij}$$

and

$$\sum_{k\in A}\sum_{l\notin A} \pi_l q_{lk} = \pi_j q_{ji}.$$

By Lemma 7.5, the result is proven. ∎

Theorem 7.4 gives a sufficient condition for a continuous-time Markov chain to be time reversible. However, the condition is not necessary. Following is an example of a reversible Markov chain whose undirected transition graph is not a tree.

Example 7.20 The continuous-time Markov chain with symmetric generator matrix

$$\boldsymbol{Q} = \begin{pmatrix} -2 & 1 & 1 \\ 1 & -2 & 1 \\ 1 & 1 & -2 \end{pmatrix}$$

has stationary distribution $\boldsymbol{\pi} = (1/3, 1/3, 1/3)$. One checks that the local balance equations are satisfied and the process is time reversible. However, the undirected transition graph is a triangle. ∎

Birth-and-Death Process

Birth-and-death processes form a large class of time reversible, continuous-time Markov chains, which arise in many applications. For these processes, transitions only occur to neighboring states. *Births* occur from i to $i+1$ at the rate λ_i. *Deaths* occur from i to $i-1$ at the rate μ_i.

The generator matrix for the general birth-and-death process on $\{0, 1, \dots\}$ is

$$\boldsymbol{Q} = \begin{matrix} \\ 0 \\ 1 \\ 2 \\ 3 \\ \vdots \end{matrix}\begin{matrix} \begin{matrix} 0 & 1 & 2 & 3 & \cdots \end{matrix} \\ \begin{pmatrix} -\lambda_0 & \lambda_0 & 0 & 0 & \cdots \\ \mu_1 & -(\lambda_1+\mu_1) & \lambda_1 & 0 & \cdots \\ 0 & \mu_2 & -(\lambda_2+\mu_2) & \lambda_2 & \cdots \\ 0 & 0 & \mu_3 & -(\lambda_3+\mu_3) & \cdots \\ \vdots & \vdots & \vdots & \vdots & \ddots \end{pmatrix}\end{matrix}.$$

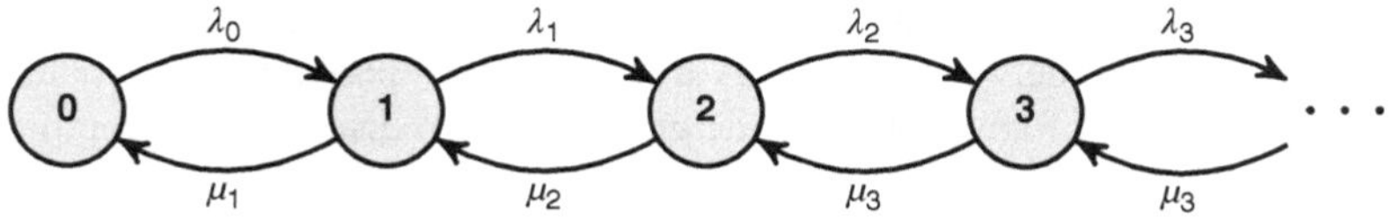

Figure 7.9 Birth-and-death process.

See Figure 7.9 for the directed transition graph. The undirected graph is a path, which is a tree. By Theorem 7.4, the chain is reversible.

The local balance equations for a birth-and-death process are

$$\pi_i \lambda_i = \pi_{i+1}\mu_{i+1}, \text{ for } i = 0, 1, \ldots$$

Solving for the stationary distribution,

$$\pi_1 = \pi_0 \frac{\lambda_0}{\mu_1},$$

$$\pi_2 = \pi_1 \frac{\lambda_1}{\mu_2} = \pi_0 \frac{\lambda_0 \lambda_1}{\mu_1 \mu_2},$$

and so on, giving

$$\pi_k = \pi_0 \frac{\lambda_0 \cdots \lambda_{k-1}}{\mu_1 \cdots \mu_k} = \pi_0 \prod_{i=1}^{k} \frac{\lambda_{i-1}}{\mu_i}, \text{ for } k = 0, 1, \ldots, \tag{7.11}$$

where we use the convention, for $k = 0$, that an empty product is equal to 1. For the components of $\boldsymbol{\pi}$ to sum to 1, we need

$$1 = \sum_{k=0}^{\infty} \pi_k = \pi_0 \sum_{k=0}^{\infty} \prod_{i=1}^{k} \frac{\lambda_{i-1}}{\mu_i}. \tag{7.12}$$

A necessary and sufficient condition for the stationary distribution to exist is that

$$\sum_{k=0}^{\infty} \prod_{i=1}^{k} \frac{\lambda_{i-1}}{\mu_i} < \infty,$$

in which case we can solve for π_0 in Equation (7.12).

Stationary Distribution for Birth-and-Death Process

For a birth-and-death process with birth rates λ_i and death rates μ_i, for $i = 1, 2, \ldots$, assume that

$$\sum_{k=0}^{\infty} \prod_{i=1}^{k} \frac{\lambda_{i-1}}{\mu_i} < \infty.$$

Then, the unique stationary distribution $\boldsymbol{\pi}$ is

$$\pi_k = \pi_0 \prod_{i=1}^{k} \frac{\lambda_{i-1}}{\mu_i}, \text{ for } k = 1, 2, \dots,$$

where

$$\pi_0 = \left(\sum_{k=0}^{\infty} \prod_{i=1}^{k} \frac{\lambda_{i-1}}{\mu_i} \right)^{-1}.$$

Example 7.21 (Continuous-time random walk) Consider a continuous-time version of simple random walk on $\{0, 1, 2, \dots\}$ with reflecting boundaries. From 0, the walk moves to 1 after an exponentially distributed length of time with rate λ. From $i > 0$, transitions to the left occur at the rate μ, and transitions to the right occur at the rate λ. Find the stationary distribution.

Solution The walk is a birth-and-death process with constant birth rate $\lambda_i = \lambda$ and death rate $\mu_i = \mu$. For the stationary distribution,

$$\pi_k = \pi_0 \prod_{i=1}^{k} \frac{\lambda}{\mu} = \pi_0 \left(\frac{\lambda}{\mu} \right)^k, \text{ for } k = 0, 1, \dots,$$

with

$$\pi_0 = \left(\sum_{k=0}^{\infty} \left(\frac{\lambda}{\mu} \right)^k \right)^{-1} = 1 - \frac{\lambda}{\mu},$$

provided that $\lambda < \mu$. In that case,

$$\pi_k = (1 - \lambda/\mu)(\lambda/\mu)^k, \text{ for } k = 0, 1, \dots$$

For $\lambda < \mu$, the stationary distribution is the geometric distribution on $\{0, 1, \dots\}$ with parameter $1 - \lambda/\mu$. ■

Example 7.22 (Yule process) The Yule process arises in biology to describe the growth of a population where each individual gives birth to an offspring at a constant rate λ independently of other individuals. Let X_t denote the size of the population at time t. If $X_t = i$, then a new individual is born when one of the i members of the population gives birth, which occurs at rate $i\lambda$. A Yule process is a birth-and-death process with birth rate $\lambda_i = i\lambda$ and death rate $\mu_i = 0$. In a Yule process, all states are transient and no limiting distribution exists.

Assume that the initial size of the population is 1. The Yule process satisfies the Kolmogorov forward equations

$$P'_{1,j}(t) = -\lambda j P_{1,j}(t) + \lambda(j-1)P_{1,j-1}(t), \text{ for } j = 1, 2, \dots,$$

with

$$P_{1,1}(0) = 1 \quad \text{and} \quad P_{1,j} = 0, \text{ for } j \geq 2.$$

The solution to the system of differential equations, which may be verified directly, is

$$P_{1,j}(t) = e^{-\lambda t}(1 - e^{-\lambda t})^{j-1}, \text{ for } j = 1, 2 \ \dots ,$$

which is a geometric distribution with parameter $e^{-\lambda t}$.

For the process started with i individuals, the transition function is

$$P_{ij}(t) = \binom{j-1}{i-1} e^{-\lambda it}(1 - e^{-\lambda t})^{j-i}, \text{ for } j \geq i, \tag{7.13}$$

which is a negative binomial distribution. See Exercise 7.16. ■

Example 7.23 An academic support center has N tutors who help students with their homework. Students arrive at the center according to a Poisson process at rate λ. Each tutor takes an exponential length of time to work with students. Tutors' service times and student arrival times are independent. If all the tutors are busy when a student arrives at the center, the student will leave. Let X_t denote the number of tutors who are busy at time t. Find the stationary distribution.

Solution Assume that $i < N$ tutors are busy at time t. The number of busy tutors will increase by one if a student arrives, which occurs at rate λ. That is, $q_{i,i+1} = \lambda$, for $i = 0, \dots, N-1$.

On the other hand, if $i > 0$ tutors are busy, the number of busy tutors decreases by one if one of the i busy tutors finishes with their student. Each tutor's service time is exponentially distributed with parameter μ. Thus, the first time one of the i tutors is free is the minimum of i exponential random variables with parameter μ, which has an exponential distribution with parameter $i\mu$. This gives $q_{i,i-1} = i\mu$, for $i = 1, \dots, N$.

The process is a finite birth-and-death process with constant birth rate and linear death rate. The generator for $N = 4$ tutors is

$$Q = \begin{matrix} & \begin{matrix} 0 & 1 & 2 & 3 & 4 \end{matrix} \\ \begin{matrix} 0 \\ 1 \\ 2 \\ 3 \\ 4 \end{matrix} & \begin{pmatrix} -\lambda & \lambda & 0 & 0 & 0 \\ \mu & -(\lambda+\mu) & \lambda & 0 & 0 \\ 0 & 2\mu & -(\lambda+2\mu) & \lambda & 0 \\ 0 & 0 & 3\mu & -(\lambda+3\mu) & \lambda \\ 0 & 0 & 0 & 4\mu & -4\mu \end{pmatrix} \end{matrix}.$$

For general N, the stationary distribution is

$$\pi_k = \pi_0 \prod_{i=1}^{k} \frac{\lambda}{i\mu} = \pi_0 \frac{(\lambda/\mu)^k}{k!}, \text{ for } k = 0, 1, \ldots, N,$$

with

$$\pi_0 = \left(\sum_{k=0}^{N} \prod_{i=1}^{k} \frac{\lambda}{i\mu} \right)^{-1} = \left(\sum_{k=0}^{N} \frac{(\lambda/\mu)^k}{k!} \right)^{-1}.$$

The distribution is a truncated Poisson distribution on $\{0, 1, \ldots, N\}$ with parameter λ/μ. For large N, the distribution is an approximate Poisson distribution, and the long-term expected number of busy tutors is λ/μ. ■

Common birth-and-death processes are listed in Table 7.1.

TABLE 7.1 Types of Birth-and-Death Processes

Type	Birth Rate	Death Rate
Pure birth	λ_i	$\mu_i = 0$
Poisson process	$\lambda_i = \lambda$	$\mu_i = 0$
Pure death	$\lambda_i = 0$	μ_i
Linear process	$\lambda_i = i\lambda, i > 0$	$\mu_i = i\mu$
Yule process	$\lambda_i = \lambda i, i, \lambda > 0$	$\mu_i = 0$
Linear with immigration	$\lambda_i = i\lambda + \alpha, i, \alpha > 0$	$\mu_i = i\mu$

7.6 QUEUEING THEORY

Queueing theory is the study of waiting lines, or queues. In the terminology of queueing theory, *customers* arrive at a facility for *service*. If the service is not immediate, they wait for service, and leave the system when the service is complete. The framework is very general and could describe a diner waiting to be seated at a restaurant, a computer program waiting to be run, and a machine waiting to be repaired.

The general queueing model can be quite broad with many parameters, which describe things such as the distribution of arrival times, the distribution of service times, the number of servers, the capacity of the system, the stages of service, and how customers line up to be served.

A standard notation of the form A/B/n is used to describe a queueing model, where A denotes the arrival time distribution, B the service time distribution, and n the number of servers.

The M/M/1 queue is a basic model. The M stands for Markov or memoryless. In this model, both arrival and service times have exponential distributions, and there is one server. The M/M/1 queue is a birth-and-death process with constant birth and death rates.

The queueing models we explore in this section are continuous-time Markov chains where X_t denotes the number of customers in the system at time t.

A central result in queueing theory is Little's formula, which is deceptively simple and remarkably diverse.

Little's Formula

In a queueing system, let L denote the long-term average number of customers in the system, λ the rate of arrivals, and W the long-term average time that a customer is in the system. Then,

$$L = \lambda W.$$

The power of Little's formula is that it applies to a very broadly defined queueing system. We will not prove the formula but justify it intuitively with a *proof by picture*. The argument is based on Gross et al. (2008).

Consider a realization of a queueing system between the time when a customer first enters the system and when the system is next empty. Assume that four customers enter the system in the time interval $[0, t]$ with arrival and departure times

Customer	Arrival Time	Departure Time
1	t_1	t_5
2	t_2	t_4
3	t_3	t_7
4	t_6	t

See Figure 7.10. Let A be the area of the shaded region. Little's formula is obtained by computing A in two ways.

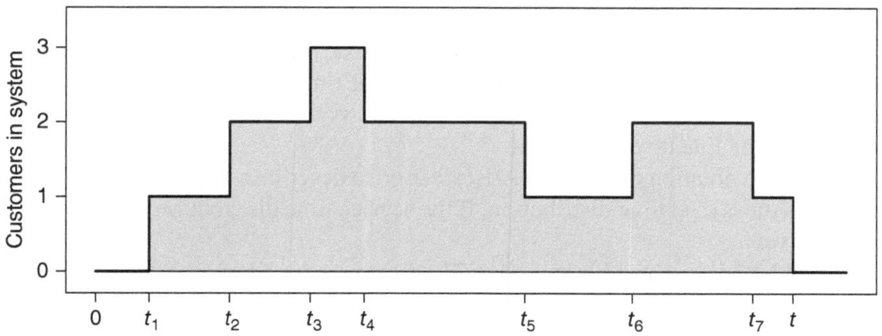

Figure 7.10 Little's formula is obtained by finding the area of the shaded region two ways.

The average length of time W that a customer spends in the system is

$$W = \frac{(t_5 - t_1) + (t_4 - t_2) + (t_7 - t_3) + (t - t_6)}{4}$$
$$= \frac{(t - t_1) + (t_5 - t_2) + (t_7 - t_6) + (t_4 - t_3)}{\text{Number of customers in } [0, t]}$$
$$= \frac{A}{\text{Number of customers in } [0, t]}. \tag{7.14}$$

In addition, the average number of customers L in the system is

$$L = \frac{1}{t}[1(t_2 - t_1) + 2(t_3 - t_2) + 3(t_4 - t_3) + 2(t_5 - t_4)$$
$$+1(t_6 - t_5) + 2(t_7 - t_6) + 1(t - t_7)]$$
$$= \frac{A}{t}. \tag{7.15}$$

Equations (7.14) and (7.15) give

$$L = \frac{A}{t} = \frac{W \times \text{Number of customers in } [0, t]}{t}.$$

For large t, this gives $L = \lambda W$.

Example 7.24 (At the carwash) Cars arrive at a drive-through carwash according to a Poisson process at the rate of nine customers per hour. The time to wash a car has an exponential distribution with mean 5 minutes. Many questions can be asked.

1. How many cars, on average, are at the carwash?
2. How long, on average, is a customer at the carwash?
3. How long, on average, does a customer wait to be served?
4. What is the expected number of cars waiting to be served?

Solution Let X_t denote the number of cars in the system at hour t. The process is an M/M/1 queueing system, which is a birth-and-death process with constant birth and death rates. The arrival rate is $\lambda = 9$. Since the average time to wash a car is one-twelfth of an hour, the service rate is $\mu = 12$. The limiting distribution probabilities (see Example 7.21) are

$$\pi_k = \left(1 - \frac{\lambda}{\mu}\right)\left(\frac{\lambda}{\mu}\right)^k = \left(\frac{1}{4}\right)\left(\frac{3}{4}\right)^k, \text{ for } k = 0, 1, \ldots,$$

which is a geometric distribution with parameter $p = 1 - \lambda/\mu = 1/4$.

1. The long-term expected number of cars at the carwash is the mean of the geometric distribution. A geometric distribution with parameter p, which takes values on $\{0, 1, \dots\}$, has expectation $(1-p)/p$. The desired expectation is

$$\frac{\lambda/\mu}{1-\lambda/\mu} = \frac{\lambda}{\mu-\lambda} = 3.$$

On average, there will be three cars at the carwash. In the notation of Little's formula, $L = 3$.

2. By Little's formula, the long-term average time W that a customer is at the carwash is

$$W = \frac{L}{\lambda} = \frac{\lambda/(\mu-\lambda)}{\lambda} = \frac{1}{\mu-\lambda} = \frac{1}{3}.$$

A customer will be at the carwash, on average, for 20 minutes.

3. Let W_q denote the long-term average time a customer spends in the queue waiting to be served. Let W_s be the average time it takes for a car to be washed. Then, $W = W_q + W_s$. The service time for a car to be washed is the mean of an exponential distribution with parameter μ. Thus, $W_s = 1/\mu$. This gives

$$W_q = W - W_s = \frac{1}{\mu-\lambda} - \frac{1}{\mu} = \frac{\lambda}{\mu(\lambda-\mu)} = \frac{1}{4}.$$

A customer waits, on average, 15 minutes to be served.

4. We can consider the process restricted to just the queue as its own queueing system. Little's formula applies with $L_q = \lambda W_q$, where L_q denotes the long-term average number of cars waiting to be served. This gives

$$L_q = \lambda W_q = \frac{\lambda^2}{\mu(\lambda-\mu)} = \frac{9}{4}.$$

On average, there are 2.25 cars in the queue. ■

M/M/c Queue

An M/M/c queue has c servers. Consider the dynamics of the process. If there are $0 < k \le c$ customers then k servers are busy. The number of customers will decrease by one, the first time one of the servers completes their service. The time until that happens is the minimum of k independent exponential random variables, which has an exponential distribution with parameter $k\mu$. If there are more than c customers in the system, then the time until the first service time is complete is the minimum of c independent exponential random variables, and thus is exponentially distributed with parameter $c\mu$.

The M/M/c queue is a birth-and-death process with parameters $\lambda_i = \lambda$, for all i, and

$$\mu_i = \begin{cases} i\mu, & \text{for } i = 1, \dots, c, \\ c\mu, & \text{for } i = c+1, c+2, \dots \end{cases}$$

We have that

$$\sum_{k=0}^{\infty}\prod_{i=1}^{k}\frac{\lambda_i}{\mu_i} = \sum_{k=0}^{c-1}\prod_{i=1}^{k}\frac{\lambda}{i\mu} + \sum_{k=c}^{\infty}\left(\prod_{i=1}^{c}\frac{\lambda}{i\mu}\right)\left(\prod_{i=c+1}^{k}\frac{\lambda}{c\mu}\right)$$

$$= \sum_{k=0}^{c-1}\left(\frac{\lambda}{\mu}\right)^k\frac{1}{k!} + \frac{1}{c!}\sum_{k=c}^{\infty}\left(\frac{\lambda}{\mu}\right)^k\left(\frac{1}{c}\right)^{k-c}$$

$$= \sum_{k=0}^{c-1}\left(\frac{\lambda}{\mu}\right)^k\frac{1}{k!} + \frac{(\lambda/\mu)^c}{c!}\sum_{k=c}^{\infty}\left(\frac{\lambda}{c\mu}\right)^{k-c}.$$

The infinite sum converges for $0 < \lambda < c\mu$, in which case the stationary distribution $\boldsymbol{\pi}$ exists, with

$$\pi_0 = \left(\sum_{k=0}^{c-1}\left(\frac{\lambda}{\mu}\right)^k\frac{1}{k!} + \frac{(\lambda/\mu)^c}{c!}\left(\frac{1}{1-\lambda/c\mu}\right)\right)^{-1}.$$

The stationary probabilities are

$$\pi_k = \begin{cases} \dfrac{\pi_0}{k!}\left(\dfrac{\lambda}{\mu}\right)^k, & \text{for } 0 \le k < c,\\[2ex] \dfrac{\pi_0}{c^{k-c}c!}\left(\dfrac{\lambda}{\mu}\right)^k, & \text{for } k \ge c. \end{cases}$$

Example 7.25 (At the hair salon) A hair salon has five chairs. Customers arrive at the salon at the rate of 6 per hour. The hair stylists each take, on average, half an hour to service a customer, independent of arrival times.

1. Jill, the owner, wants to know the long-term probability that no customers are in the salon.
2. Danny, a potential customer, wants to know the average waiting time for a haircut.
3. Leslie, a hair stylist, wants to know the long-term expected number of customers in the salon.

Solution The system is an M/M/5 queue with $\lambda = 6$, $c = 5$, and $\mu = 2$.

1. The long-term probability that no customers are in the salon is

$$\pi_0 = \left(\sum_{k=0}^{4}\frac{3^k}{k!} + \frac{3^5}{5!}\left(\frac{1}{1-6/10}\right)\right)^{-1} = \frac{16}{343} = 0.0466.$$

2. The long-time average waiting time in the queue, in the notation of Little's formula, is $W_q = L_q/\lambda$. To find L_q, the expected number of customers in the queue, observe that there are k people in the queue if and only if there are $k + c$ customers in the system. This gives

$$\begin{aligned}
L_q &= \sum_{k=c}^{\infty}(k-c)\pi_k = \sum_{k=c}^{\infty}(k-c)\frac{\pi_0}{c^{k-c}c!}\left(\frac{\lambda}{\mu}\right)^k \\
&= \frac{\pi_0}{c!}\left(\frac{\lambda}{\mu}\right)^c \sum_{k=c}^{\infty}(k-c)\frac{1}{c^{k-c}}\left(\frac{\lambda}{\mu}\right)^{k-c} \\
&= \frac{\pi_0}{c!}\left(\frac{\lambda}{\mu}\right)^c \sum_{k=0}^{\infty}k\left(\frac{\lambda}{c\mu}\right)^k \\
&= \frac{\pi_0}{c!}\left(\frac{\lambda}{\mu}\right)^c \frac{\lambda}{c\mu}\left(\frac{1}{1-\lambda/c\mu}\right)^2 \\
&= \frac{\pi_0 3^5}{5!}\left(\frac{6}{10}\right)\left(\frac{1}{1-6/10}\right)^2 \\
&= 0.35423.
\end{aligned}$$

By Little's formula, the expected waiting time in the queue is

$$W_q = \frac{L_q}{\lambda} = \frac{0.35423}{6} = 0.059,$$

or about 3.6 minutes.

3. The long-term expected waiting time in the system is

$$W = W_q + W_s = W_q + \frac{1}{\mu} = 0.059 + 0.5 = 0.559,$$

or about 33.54 minutes. The expected number of customers in the system, by Little's formula, is

$$L = \lambda W = 6(0.559) = 3.354.$$

Note that the last result can also be obtained by finding the expectation with respect to the stationary distribution

$$L = \sum_{k=0}^{\infty} k\pi_k.$$

■

7.7 POISSON SUBORDINATION

The times when transitions occur for a continuous-time Markov chain are exponentially distributed with holding time rates $q_1, q_2, \ldots$ In this section, we present an interesting representation of a continuous-time Markov chain in which all holding time rates are the same, but where transitions from a state to itself are allowed. The representation is remarkably useful for simulation and numerical computation.

Consider a finite-state, irreducible, discrete-time Markov chain $Y_0, Y_1, \ldots$ with transition matrix $\boldsymbol{R}$. Let $(N_t)_{t\geq 0}$ be a Poisson process with parameter λ, which is independent of the Markov chain. Define a continuous-time process $(X_t)_{t\geq 0}$ by $X_t = Y_{N_t}$. That is, transitions for the X_t process occur at the arrival times of the Poisson process. From state i, the process holds an exponentially distributed amount of time with parameter λ and then transitions to j with probability R_{ij}.

The process $(X_t)_{t\geq 0}$ is a continuous-time Markov chain whose transition function $\boldsymbol{P}(t)$ has a surprisingly simple form. By conditioning on N_t,

$$\begin{aligned} P_{ij}(t) &= P(X_t = j|X_0 = i) \\ &= \sum_{k=0}^{\infty} P(X_t = j|N_t = k, X_0 = i)P(N_t = k|X_0 = i) \\ &= \sum_{k=0}^{\infty} P(Y_k = j|N_t = k, X_0 = i)P(N_t = k) \\ &= \sum_{k=0}^{\infty} P(Y_k = j|Y_0 = i)P(N_t = k) \\ &= \sum_{k=0}^{\infty} R_{ij}^k \frac{e^{-\lambda t}(\lambda t)^k}{k!}. \end{aligned}$$

We say that the $(X_t)_{t\geq 0}$ Markov chain is *subordinated to a Poisson process.*

Not only can we construct a continuous-time Markov chain from a discrete-time chain and a Poisson process, but conversely many continuous-time Markov chains can be represented as a chain subordinated to a Poisson process.

Consider a continuous-time Markov chain with generator $\boldsymbol{Q}$ and holding time parameters $q_1, q_2, \ldots$ Assume the parameters are uniformly bounded. That is, there exists a constant λ such that $q_i \leq \lambda$, for all i. This will always be the case if the chain is finite, and we can take $\lambda = \max_i q_i$. Let

$$\boldsymbol{R} = \frac{1}{\lambda}\boldsymbol{Q} + \boldsymbol{I}.$$

The matrix $\boldsymbol{R}$ is a stochastic matrix. Entries are non-negative and rows sum to 1. The transition function can be given in terms of $\boldsymbol{R}$, as

$$\begin{aligned} \boldsymbol{P}(t) = e^{t\boldsymbol{Q}} &= e^{-\lambda t}e^{t\boldsymbol{Q}}e^{\lambda t} = e^{-\lambda t}e^{t(\boldsymbol{Q}+\lambda \boldsymbol{I})} \\ &= e^{-\lambda t}\sum_{k=0}^{\infty} \frac{1}{k!}t^k(\boldsymbol{Q} + \lambda \boldsymbol{I})^k \\ &= \sum_{k=0}^{\infty} \left(\frac{1}{\lambda}\boldsymbol{Q} + \boldsymbol{I}\right)^k \frac{e^{-\lambda t}(\lambda t)^k}{k!} \end{aligned}$$

$$= \sum_{k=0}^{\infty} \boldsymbol{R}^k \frac{e^{-\lambda t}(\lambda t)^k}{k!}.$$

The continuous-time chain is represented as a Markov chain subordinated to a Poisson process. Other names for Poisson subordination are *randomization* and *uniformization*.

Note that the $\boldsymbol{R}$ matrix is *not* the matrix of the embedded Markov chain. The entries of the embedded Markov matrix are

$$\widetilde{P}_{ij} = \begin{cases} q_{ij}/q_i, & \text{for } i \neq j, \\ 0, & \text{for } i = j, \end{cases}$$

while the entries of the $\boldsymbol{R}$ matrix are

$$R_{ij} = \begin{cases} q_{ij}/\lambda, & \text{for } i \neq j, \\ 1 - q_i/\lambda, & \text{for } i = j. \end{cases}$$

Poisson subordination can be described as follows. From a given state i, wait an exponential length of time with rate λ. Then, flip a coin whose heads probability is q_i/λ. If heads, transition to a new state according to the $\boldsymbol{R}$ matrix. If tails, stay at i and repeat. Thus, holding time parameters are constant, and transitions, or pseudo-transitions, from a state to itself are allowed.

To illustrate, consider a continuous-time Markov chain on $\{a, b, c\}$ with generator

$$\boldsymbol{Q} = \begin{matrix} & \begin{matrix} a & b & c \end{matrix} \\ \begin{matrix} a \\ b \\ c \end{matrix} & \begin{pmatrix} -2 & 1 & 1 \\ 1 & -3 & 2 \\ 5 & 1 & -6 \end{pmatrix} \end{matrix}.$$

Choose $\lambda = \max\{2, 3, 6\} = 6$. Then,

$$\boldsymbol{R} = \frac{1}{\lambda}\boldsymbol{Q} + \boldsymbol{I} = \frac{1}{6}\begin{pmatrix} -2 & 1 & 1 \\ 1 & -3 & 2 \\ 5 & 1 & -6 \end{pmatrix} + \begin{pmatrix} 1 & 0 & 0 \\ 0 & 1 & 0 \\ 0 & 0 & 1 \end{pmatrix} = \begin{pmatrix} 2/3 & 1/6 & 1/6 \\ 1/6 & 1/2 & 1/3 \\ 5/6 & 1/6 & 0 \end{pmatrix}.$$

See Figure 7.11 for a comparison of the dynamics of the original chain with the subordinated process.

Long-Term Behavior, Simulation, Computation

For a Markov chain subordinated to a Poisson process, the discrete $\boldsymbol{R}$-chain has the same stationary distribution as the original chain. We have that

$$\boldsymbol{\pi Q} = \boldsymbol{\pi}\lambda(\boldsymbol{R} - \boldsymbol{I}) = \lambda\boldsymbol{\pi R} - \lambda\boldsymbol{\pi}.$$

Thus, $\boldsymbol{\pi Q} = \boldsymbol{0}$ if and only if $\boldsymbol{\pi R} = \boldsymbol{\pi}$.

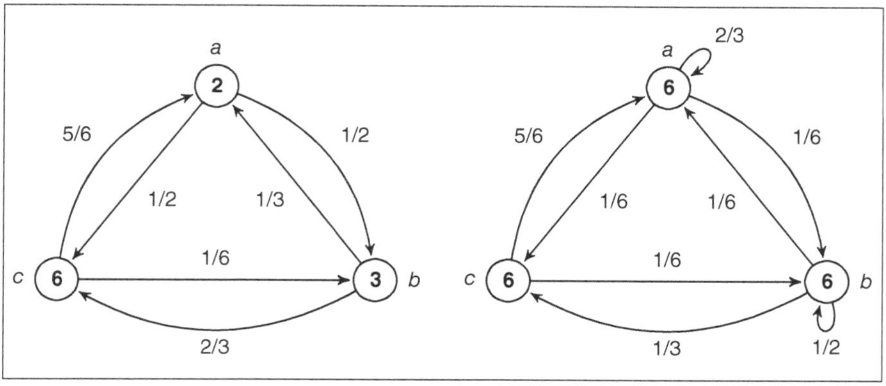

Figure 7.11 Both graphs describe the same Markov chain. Nodes are labeled with holding time parameters, edges are labeled with transition probabilities. The graph on the right shows the chain subordinated to a Poisson process. Holding time parameters are constant, and transitions to the same state are allowed.

Example 7.26 Consider a Markov chain $(X_t)_{t\geq 0}$ with generator

$$\boldsymbol{Q} = \begin{pmatrix} -2 & 1 & 1 \\ 1 & -3 & 2 \\ 0 & 1 & -1 \end{pmatrix}.$$

Letting $\lambda = \max\{q_1, q_2, q_3\} = 3$ gives

$$\boldsymbol{R} = \frac{1}{3}\boldsymbol{Q} + \boldsymbol{I} = \begin{pmatrix} 1/3 & 1/3 & 1/3 \\ 1/3 & 0 & 2/3 \\ 0 & 1/3 & 2/3 \end{pmatrix}.$$

Observe that $\boldsymbol{\pi} = (1/8, 2/8, 5/8)$ is the stationary distribution for both the original chain and the discrete-time $\boldsymbol{R}$-chain. ■

Poisson subordination leads to an efficient method to simulate a continuous-time Markov chain, since transitions rates are constant and thus not dependent on the current state. Here, we simulate the distribution of $X_{1.5}$ for the Markov chain in Example 7.26.

R: Simulation of the Distribution of $X_{1.5}$

```
# Psubordination.R
> Q
   1  2  3
1 -2  1  1
2  1 -3  2
3  0  1 -1
```

```
> lambda <- 3
> R <- (1/lambda)*Q+diag(3)
> R
          1         2         3
1 0.3333333 0.3333333 0.3333333
2 0.3333333 0.0000000 0.6666667
3 0.0000000 0.3333333 0.6666667
> trials <- 100000
> simlist <- numeric(trials)
> for (i in 1:trials) {
 +   s <- 0  # time
 +   state <- 1
 +   newstate <- 1
 +   while(s < 1.5) {
 +     state <- newstate
 +     s <- s+rexp(1,lambda)
 +     newstate <- sample(1:3,1,prob=r[state,]) }
 +     simlist[i] <- state
 +          }
> table(simlist)/trials
simlist
      1       2       3
0.16274 0.24958 0.58768
> expm(1.5*Q)[1,]  #   Compare with exact values
      1     2     3
1 0.163 0.249 0.588
```

Another benefit of Poisson subordination is that it gives a numerically stable method for computing a Markov transition function, as compared with the matrix exponential. For large matrices, the matrix exponential is notoriously difficult to compute. The infinite sum $\sum_{k=0}^{\infty}(t\boldsymbol{Q})^k/k!$ converges slowly. The matrix $\boldsymbol{Q}$ contains positive and negative numbers, some of which may be greater than 1, and thus $\boldsymbol{Q}^k$ may contain large positive and negative numbers, a source of numerical instability.

On the other hand, with Poisson subordination the transition function

$$P_{ij}(t) = \sum_{k=0}^{\infty} R_{ij}^k \frac{e^{-\lambda t}(\lambda t)^k}{k!} \tag{7.16}$$

can be computed numerically by truncating the infinite sum to a desired level of accuracy. Consider the approximation

$$\widehat{P_{ij}(t)} = \sum_{k=0}^{N} R_{ij}^k \frac{e^{-\lambda t}(\lambda t)^k}{k!}, \tag{7.17}$$

for some N. The absolute error of the approximation is

$$\begin{aligned}|P_{ij}(t) - \widehat{P_{ij}(t)}| &= \left| \sum_{k=0}^{\infty} R_{ij}^k \frac{e^{-\lambda t}(\lambda t)^k}{k!} - \sum_{k=0}^{N} R_{ij}^k \frac{e^{-\lambda t}(\lambda t)^k}{k!} \right| \\ &= \sum_{k=N+1}^{\infty} R_{ij}^k \frac{e^{-\lambda t}(\lambda t)^k}{k!} \\ &\leq \sum_{k=N+1}^{\infty} \frac{e^{-\lambda t}(\lambda t)^k}{k!} \\ &= P(Y > N),\end{aligned}$$

where Y is a Poisson random variable with parameter λt. The inequality holds since $\boldsymbol{R}^k$ is a stochastic matrix, all of whose entries are between 0 and 1.

To obtain an absolute error in the approximation of at most ϵ, choose N such that $P(Y > N) \leq \epsilon$.

The following example is small enough so that exact calculations are possible and numerical approximations are not necessary. However, it illustrates the general method.

Example 7.27 Consider a birth-and-death process on $\{0, \dots, 5\}$ with constant death rate $\mu_i = 2$ and linear birth rates $\lambda_i = i$, for $i = 1, \dots, 4$. The generator is

$$\boldsymbol{Q} = \begin{matrix} & \begin{matrix} 0 & 1 & 2 & 3 & 4 & 5 \end{matrix} \\ \begin{matrix} 0 \\ 1 \\ 2 \\ 3 \\ 4 \\ 5 \end{matrix} & \begin{pmatrix} -1 & 1 & 0 & 0 & 0 & 0 \\ 2 & -4 & 2 & 0 & 0 & 0 \\ 0 & 2 & -5 & 3 & 0 & 0 \\ 0 & 0 & 2 & -6 & 4 & 0 \\ 0 & 0 & 0 & 2 & -7 & 5 \\ 0 & 0 & 0 & 0 & 2 & -2 \end{pmatrix} \end{matrix}.$$

Find $\boldsymbol{P}(1.5)$ to within four-digit accuracy.

Solution Let $\lambda = 7$, and set

$$\boldsymbol{R} = \frac{1}{\lambda}\boldsymbol{Q} + \boldsymbol{I} = \begin{matrix} & \begin{matrix} 0 & 1 & 2 & 3 & 4 & 5 \end{matrix} \\ \begin{matrix} 0 \\ 1 \\ 2 \\ 3 \\ 4 \\ 5 \end{matrix} & \begin{pmatrix} 6/7 & 1/7 & 0 & 0 & 0 & 0 \\ 2/7 & 3/7 & 2/7 & 0 & 0 & 0 \\ 0 & 2/7 & 2/7 & 3/7 & 0 & 0 \\ 0 & 0 & 2/7 & 1/7 & 4/7 & 0 \\ 0 & 0 & 0 & 2/7 & 0 & 5/7 \\ 0 & 0 & 0 & 0 & 2/7 & 5/7 \end{pmatrix} \end{matrix}.$$

Choose N such that $P(Y > N) < 0.5 \times 10^{-4}$, where Y is a Poisson random variable with parameter $7 \times 1.5 = 10.5$. In R, we find that

```
> 1-ppois(24,10.5)
[1] 9.933169e-05
> 1-ppois(25,10.5)
[1] 3.921504e-05
```

Thus, truncate the infinite series at $N = 25$. This gives

$$\widehat{P(1.5)} = \sum_{k=0}^{25} \boldsymbol{R}^k \frac{e^{-\lambda(1.5)}((1.5)t)^k}{k!}$$

$$= \begin{pmatrix} 0.50706 & 0.18466 & 0.09985 & 0.06535 & 0.05600 & 0.08703 \\ 0.36932 & 0.15278 & 0.10063 & 0.08480 & 0.09944 & 0.19299 \\ 0.19970 & 0.10063 & 0.08741 & 0.10025 & 0.15741 & 0.35457 \\ 0.08713 & 0.05653 & 0.06683 & 0.10241 & 0.19880 & 0.48826 \\ 0.03734 & 0.03315 & 0.05247 & 0.09940 & 0.21813 & 0.55948 \\ 0.02321 & 0.02573 & 0.04728 & 0.09765 & 0.22379 & 0.58230 \end{pmatrix}.$$

For this example, we can compare the numerical approximation with the matrix exponential computation $\boldsymbol{P}(1.5) = e^{1.5\boldsymbol{Q}}$, which can be found with software. The transition function, to five decimal places, is

$$\boldsymbol{P}(1.5) = \begin{pmatrix} 0.50708 & 0.18467 & 0.09985 & 0.06535 & 0.05601 & 0.08704 \\ 0.36933 & 0.15278 & 0.10064 & 0.08480 & 0.09945 & 0.19300 \\ 0.19970 & 0.10064 & 0.08741 & 0.10025 & 0.15742 & 0.35458 \\ 0.08714 & 0.05653 & 0.06683 & 0.10241 & 0.19881 & 0.48828 \\ 0.03734 & 0.03315 & 0.05247 & 0.09940 & 0.21814 & 0.55950 \\ 0.02321 & 0.02573 & 0.04728 & 0.09766 & 0.2238 & 0.58232 \end{pmatrix}.$$

By inspection, we see that our approximation is accurate to the desired level of accuracy.

By contrast, if we truncate the infinite sum in the matrix exponential to 25 terms we get

$$e^{1.5\boldsymbol{Q}} \approx \sum_{k=0}^{25} (1.5\boldsymbol{Q})^k / k!$$

$$= \begin{pmatrix} -4.06 & 22.01 & -69.10 & 170.51 & -284.73 & 166.39 \\ 44.01 & -208.29 & 661.43 & -1629.30 & 2723.19 & -1590.05 \\ -138.20 & 661.43 & -2099.20 & 5174.13 & -8648.70 & 5051.54 \\ 227.34 & -1086.20 & 3449.42 & -8503.51 & 14216.90 & -8302.93 \\ -189.82 & 907.73 & -2882.90 & 7108.44 & -11884.80 & 6942.40 \\ 44.37 & -212.01 & 673.54 & -1660.59 & 2776.96 & -1621.28 \end{pmatrix},$$

which is not close to converging. In fact, it takes almost twice as many terms in this case before reaching the desired level of accuracy. ∎

EXERCISES

7.1 A continuous-time Markov chain has generator matrix

$$\boldsymbol{Q} = \begin{array}{c} \\ a \\ b \\ c \end{array}\begin{array}{c} \begin{array}{ccc} a & b & c \end{array} \\ \begin{pmatrix} -1 & 1 & 0 \\ 1 & -2 & 1 \\ 2 & 2 & -4 \end{pmatrix} \end{array}.$$

Exhibit (i) the transition matrix of the embedded Markov chain and (ii) the holding time parameter for each state.

7.2 A Markov chain on $\{1, 2, 3, 4\}$ has nonzero transition rates

$$q_{12} = q_{23} = q_{31} = q_{41} = 1 \quad \text{and} \quad q_{14} = q_{32} = q_{34} = q_{43} = 2.$$

(a) Exhibit the (i) generator, (ii) holding time parameters, and (iii) transition matrix for the embedded Markov chain.

(b) If the chain is at state 1, how long on average will it take before moving to a new state?

(c) If the chain is at state 3, how long on average will it take before moving to state 4?

(d) Over the long term, what proportion of visits will be to state 2?

7.3 A three-state Markov chain has distinct holding time parameters a, b, and c. From each state, the process is equally likely to transition to the other two states. Exhibit the generator matrix and find the stationary distribution.

7.4 During lunch hour, customers arrive at a fast-food restaurant at the rate of 120 customers per hour. The restaurant has one line, with three workers taking food orders at independent service stations. Each worker takes an exponentially distributed amount of time—on average 1 minute—to service a customer. Let X_t denote the number of customers in the restaurant (in line and being serviced) at time t. The process $(X_t)_{t\geq 0}$ is a continuous-time Markov chain. Exhibit the generator matrix.

7.5 For the fast-food restaurant chain of the previous exercise, assume that customers turn away from the store if all three service stations are busy. Let Y_t denote the number of service stations busy at time t. Then, $(Y_t)_{t\geq 0}$ is a continuous-time Markov chain. Exhibit the generator matrix.

7.6 A Markov chain $(X_t)_{t\geq 0}$ on $\{1, 2, 3, 4\}$ has generator matrix

$$\boldsymbol{Q} = \begin{pmatrix} -2 & 1 & 1 & 0 \\ 1 & -3 & 1 & 1 \\ 2 & 2 & -4 & 0 \\ 1 & 2 & 3 & 6 \end{pmatrix}.$$

Use technology as needed for the following:

(a) Find the long-term proportion of time that the chain visits state 1.

(b) For the chain started in state 2, find the long-term probability that the chain visits state 3.

(c) Find $P(X_1 = 3|X_0 = 1)$.

(d) Find $P(X_5 = 1, X_2 = 4|X_1 = 3)$.

7.7 A Markov chain has generator matrix

$$\boldsymbol{Q} = \begin{pmatrix} -1 & 1 & 0 \\ 0 & -2 & 2 \\ 3 & 0 & -3 \end{pmatrix}.$$

(a) Exhibit the Kolmogorov backward equations.

(b) Find the transition function by diagonalizing the generator and finding the matrix exponential.

(c) Find the transition function using a symbolic software system such as *Wolfram Alpha*.

7.8 Consider the Jukes–Cantor model for DNA nucleotide substitution in Example 7.11. Find the transition function $\boldsymbol{P}(t)$ by solving the backward, or forward, equations.

7.9 Consider the Felsenstein DNA model of Example 7.15.

(a) Show that $\boldsymbol{\pi} = (p_a, p_c, p_c, p_t)$ is the stationary distribution.

(b) For $\alpha = 2$ and the parameter values given in the example, find the probability that an a nucleotide mutates to a c nucleotide in 1.5 time units.

7.10 Unlike the Felsenstein DNA model, introduced in Example 7.15, the Hasegawa, Kishino, Yao model distinguishes between transitions and transversions. The generator matrix is

$$\boldsymbol{Q} = \begin{array}{c} \\ a \\ g \\ c \\ t \end{array} \begin{array}{c} \begin{array}{cccc} a & g & c & t \end{array} \\ \begin{pmatrix} -\alpha p_g - \beta p_r & \alpha p_g & \beta p_c & \beta p_t \\ \alpha p_a & -\alpha p_a - \beta p_r & \beta p_c & \beta p_t \\ \beta p_a & \beta p_g & -\alpha p_t - \beta p_s & \alpha p_t \\ \beta p_a & \beta p_g & \alpha p_c & -\alpha p_c - \beta p_s \end{pmatrix} \end{array},$$

where $p_r = p_c + p_t$, $p_s = p_a + p_g$, $p_a + p_g + p_c + p_t = 1$, and $\alpha, \beta > 0$. Show that $\boldsymbol{\pi} = (p_a, p_g, p_c, p_t)$ is the stationary distribution of the chain.

7.11 Let A be a square matrix. Show that

$$\frac{d}{dt}e^{tA} = \boldsymbol{A}e^{tA} = e^{tA}\boldsymbol{A}.$$

7.12 The following result from linear algebra relates the determinant and trace of a matrix $\boldsymbol{A}$:

$$\det e^{\boldsymbol{A}} = e^{\operatorname{tr} \boldsymbol{A}}.$$

Prove this for the case that $\boldsymbol{A}$ is diagonalizable.

7.13 Assume that $\boldsymbol{\pi}$ is the limiting distribution of a continuous-time chain. Show that $\boldsymbol{\pi}$ is a stationary distribution. (*Hint:* start with the forward equation.)

7.14 For the Markov chain with transition rate graph shown in Figure 7.12, find

(a) the generator matrix,

(b) the stationary distribution of the continuous-time Markov chain,

(c) the transition matrix of the embedded chain,

(d) the stationary distribution of the embedded chain.

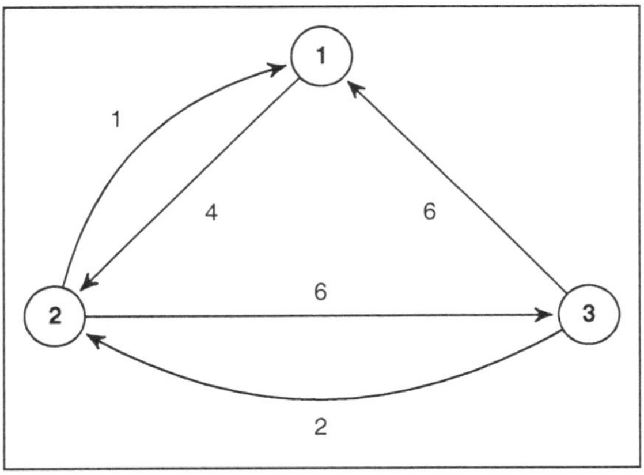

Figure 7.12

7.15 Let $(N_t)_{t\geq 0}$ be a Poisson process with parameter $\lambda = 1$. Define the process $X_t = N_t \bmod 4$, for $t \geq 0$. Then, $(X_t)_{t\geq 0}$ is a continuous-time Markov chain on $\{0, 1, 2, 3\}$.

(a) Exhibit the generator matrix.

(b) Use a symbolic software system such as *Wolfram Alpha* to find the transition function $\boldsymbol{P}(t)$.

7.16 For a Yule process started with i individuals, derive the transition probabilities

$$P_{ij}(t) = \binom{j-1}{i-1} e^{-\lambda it}(1 - e^{-\lambda t})^{j-i}, \text{ for } j \geq i.$$

Hint: Relate the process to a Yule process started with one individual.

7.17 Each individual in a population gives birth to an offspring at the rate of 1.5 per unit time independently of other individuals. If the population starts with 4 individuals, find the mean and variance of the size of the population at time $t = 8$.

7.18 For a general birth-and-death process with birth rates λ_i and death rates μ_i, let T_i denote the time, from state i, for the process to hit state $i + 1$.

(a) Show that

$$E[T_i] = \frac{1}{\lambda_i} + \frac{\mu_i}{\lambda_i} E[T_{i-1}], \text{ for } i = 1, 2, \ldots$$

Hint: Condition on whether the first transition is a birth or a death.

(b) Solve $E[T_i]$ for the case $\lambda_i = \lambda$ and $\mu_i = \mu$, for all i.

7.19 Taxis arrive at a taxi stand according to a Poisson process with parameter λ. Customers arrive, independently of taxis, at the rate μ. If there are no taxis when a customer arrives at the stand they will leave. Assume that $\lambda < \mu$. What is the long-term probability that an arriving customer gets a taxi?

7.20 The M/M/∞ queue has infinitely many servers. Show that the limiting distribution is Poisson and find the mean number of customers in the system.

7.21 For an M/M/∞ queue with $\lambda = \mu = 1$, find the mean time until state 4 is hit for the process started in state 1.

7.22 Tom is taking an online exam in which there are four questions of increasing difficulty. Tom needs to complete each question before moving on to the next question. It takes on average 10, 20, 30, and 40 minutes, respectively, to answer each question. The times spent for each question are independent and exponentially distributed.

(a) Find the probability that after 45 minutes Tom has completed the exam.

(b) Find the probability that after 45 minutes Tom is still on the third question.

7.23 A facility has four machines, with two repair workers to maintain them. Individual machines fail on average every 10 hours. It takes an individual repair worker on average 4 hours to fix a machine. Repair and failure times are independent and exponentially distributed.

(a) Find the generator matrix.

(b) In the long term, how many machines are typically operational?

(c) If all four machines are initially working, find the probability that only two machines are working after 5 hours.

7.24 Customers arrive at a busy food truck according to a Poisson process with parameter λ. If there are i people already in line, the customer will join the line with probability $1/(i + 1)$. Assume that the chef at the truck takes, on average, α minutes to process an order.

(a) Find the long-term average number of people in line.

(b) Find the long-term probability that there are at least two people in line.

7.25 Over the long term, a continuous-time Markov chain on $\{1, 2, 3, 4\}$ makes 10% of its transitions to 1 and 30% each to 2, 3, and 4, respectively. From state i, it stays on average i minutes before moving to a new state, for $i = 1, 2, 3, 4$.

(a) Find the stationary distribution of the embedded discrete-time chain.

(b) Find the stationary distribution of the continuous-time chain.

7.26 A facility has three machines and three mechanics. Machines break down at the rate of one per 24 hours. Breakdown times are exponentially distributed. The time it takes a mechanic to fix a machine is exponentially distributed with mean 6 hours. Only one mechanic can work on a failed machine at any given time. Let X_t be the number of machines working at time t. Find the long-term probability that all machines are working.

7.27 Recall the discrete-time Ehrenfest dog–flea model of Example 3.7. In the continuous-time version, there are N fleas distributed between two dogs. Fleas jump from one dog to another independently at rate λ. Let X_t denote the number of fleas on the first dog.

(a) Show that the process is a birth-and-death process. Give the birth and death rates.

(b) Find the stationary distribution.

(c) Assume that fleas jump at the rate of 2 per minute. If there are 10 fleas on Cooper and no fleas on Lisa, how long, on average, will it take for Lisa to get 4 fleas?

7.28 A linear birth-and-death process with immigration (see Table 7.1) has parameters $\lambda = 3$, $\mu = 4$, and $\alpha = 2$. Find the stationary distribution.

7.29 Consider an absorbing, continuous-time Markov chain with possibly more than one absorbing states.

(a) Argue that the continuous-time chain is absorbed in state a if and only if the embedded discrete-time chain is absorbed in state a.

(b) Let

$$\boldsymbol{Q} = \begin{matrix} \\ 1 \\ 2 \\ 3 \\ 4 \\ 5 \end{matrix}\begin{matrix} \begin{matrix} 1 & \;\;2 & \;\;3 & \;\;4 & \;\;5 \end{matrix} \\ \begin{pmatrix} 0 & 0 & 0 & 0 & 0 \\ 1 & -3 & 2 & 0 & 0 \\ 0 & 2 & -4 & 2 & 0 \\ 0 & 0 & 2 & -5 & 3 \\ 0 & 0 & 0 & 0 & 0 \end{pmatrix} \end{matrix}$$

be the generator matrix for a continuous-time Markov chain. For the chain started in state 2, find the probability that the chain is absorbed in state 5.

7.30 Cars arrive at a toll booth according to a Poisson process at the rate of two cars per minute. The time taken by the attendant to collect the toll is exponentially distributed with mean 20 seconds.

(a) Find the long-term mean number of cars in line at the toll booth.

(b) Find the long-term probability that there are more than three cars at the toll booth.

7.31 See the last exercise. Assume that the toll booth has two attendants on duty. Find the long-term probability there are no cars at the toll booth.

7.32 Calls come in to a computer help center at the rate of 15 calls per hour. There are three tech support workers on duty, and the times they take to provide assistance are exponentially distributed with a mean of 10 minutes.

(a) Find the average number of callers waiting to be helped.

(b) Find the average amount of time that a caller spends waiting.

7.33 Consider an M/M/1 queue. Assume that the arrival and service time rates are both increased by a factor of k. What effect does this have on

(a) the long-term expected number of customers in the system?

(b) the long-term expected time that a customer is in the system?

7.34 You are a frequent customer at a coffee shop, where you typically wait 3 minutes to be served. Furthermore, on average you spend \$4 per visit. Over many months you estimate that on a typical day there are 20 customers in the shop, which is open from 6 a.m. to 10 p.m. Estimate the shop's total revenue per day.

7.35 Consider a continuous-time Markov chain with generator

$$\boldsymbol{Q} = \begin{pmatrix} -1 & 0 & 1 \\ 1 & -2 & 1 \\ 1 & 3 & -4 \end{pmatrix}.$$

Represent the process as a Markov chain subordinated to a Poisson process. Exhibit the transition function $\boldsymbol{P}(t)$ in terms of $\boldsymbol{R}$.

7.36 A discrete-time Markov chain has transition matrix

$$\boldsymbol{P} = \begin{pmatrix} p & q \\ p & q \end{pmatrix},$$

for $p + q = 1$. Extend the process to continuous time by allowing transitions to occur at the points of a Poisson process with parameter λ. Find the transition function $\boldsymbol{P}(t)$.

7.37 R: Consider a continuous-time Markov chain with generator

$$\boldsymbol{Q} = \begin{pmatrix} -4 & 1 & 2 & 1 \\ 2 & -3 & 0 & 1 \\ 3 & 3 & -9 & 3 \\ 4 & 2 & 0 & -6 \end{pmatrix}.$$

(a) To use Poisson subordination to estimate $\boldsymbol{P}(0.8)$ to three significant digits, how many terms of the series are needed?

(b) Use (a) and compute $\boldsymbol{P}(0.8)$.

(c) Use `R`'s matrix exponential command to find $\boldsymbol{P}(0.8)$ and check that your result in (a) is accurate to three digits.

7.38 `R`: Simulate Tom's online exam in Exercise 7.22 and estimate the probabilities in that problem.

7.39 `R`: Simulate an M/M/∞ queue and verify the result of Exercise 7.20, choosing your own values for λ and μ.

7.40 `R`: A multistate Markov model for the progression of HIV infection is developed in Hendricks et al. (1996). A sample of 467 HIV-positive men were monitored between 1984 and 1993. A 7-state absorbing chain is developed where stages 1–6 represent a range of values of an immunological marker for HIV, and state 7 corresponds to AIDS. The researchers give the following estimates for the monthly transition rates of the model.

λ_{12}	λ_{21}	λ_{23}	λ_{32}	λ_{34}	λ_{43}	λ_{45}	λ_{47}	λ_{54}	λ_{56}	λ_{57}	λ_{65}	λ_{67}
0.055	0.008	0.060	0.008	0.039	0.008	0.033	0.006	0.009	0.029	0.007	0.002	0.042

(a) Find the mean time to develop AIDS from each HIV state.

(b) Starting from the first stage of HIV infection, estimate the probability of developing AIDS within k years, for $k = 5, 10, 15, 20$.

8

BROWNIAN MOTION

> Observe what happens when sunbeams are admitted into a building and shed light on its shadowy places. You will see a multitude of tiny particles mingling in a multitude of ways ... their dancing is an actual indication of underlying movements of matter that are hidden from our sight.... It originates with the atoms which move of themselves [i.e., spontaneously]. Then those small compound bodies that are least removed from the impetus of the atoms are set in motion by the impact of their invisible blows and in turn cannon against slightly larger bodies. So the movement mounts up from the atoms and gradually emerges to the level of our senses, so that those bodies are in motion that we see in sunbeams, moved by blows that remain invisible.
>
> —Lucretius, *On the Nature of Things*, 60 B.C.

8.1 INTRODUCTION

Brownian motion is a stochastic process, which is rooted in a physical phenomenon discovered almost 200 years ago. In 1827, the botanist Robert Brown, observing pollen grains suspended in water, noted the erratic and continuous movement of tiny particles ejected from the grains. He studied the phenomenon for many years, ruled out the belief that it emanated from some "life force" within the pollen, but could not explain the motion. Neither could any other scientist of the 19th century.

In 1905, Albert Einstein solved the riddle in his paper, *On the movement of small particles suspended in a stationary liquid demanded by the molecular-kinetic theory of heat*. Einstein explained the movement by the continual bombardment of the

Introduction to Stochastic Processes with R, First Edition. Robert P. Dobrow.

immersed particles by the molecules in the liquid, resulting in "motions of such magnitude that these motions can easily be detected by a microscope." Einstein's theoretical explanation was confirmed 3 years later by empirical experiment, which led to the acceptance of the atomic nature of matter.

The description of the motion of dust particles in the classic poem *On the Nature of Things*, written by the Roman philosopher Lucretius over 2,000 years ago as an ancient proof of the existence of atoms, could have been a summary of Einstein's work!

Einstein showed that the position x of a particle at time t was described by the partial differential heat equation

$$\frac{\partial}{\partial t} f(x,t) = \frac{1}{2} \frac{\partial^2}{\partial x^2} f(x,t),$$

where $f(x, t)$ represents the density (number of particles per unit volume) at position x and time t. The solution to that equation is

$$f(x,t) = \frac{1}{\sqrt{2\pi t}} e^{-x^2/2t},$$

which is the probability density function of the normal distribution with mean 0 and variance t.

The mathematical object we call Brownian motion is a continuous-time, continuous-state stochastic process, also called the *Wiener process*, named after the American mathematician Norbert Wiener. The British mathematician Bertrand Russell influenced Wiener to take up the theory of Brownian motion as had been studied by Einstein. In his 1956 autobiography, Wiener writes,

> The Brownian motion was nothing new as an object of study by physicists. There were fundamental papers by Einstein and Smoluchowski that covered it. But whereas these papers concerned what was happening to any given particle at a specific time, or the long-time statistics of many particles, they did not concern themselves with the mathematical properties of the curve followed by a single particle.
>
> Here the literature was very scant, but it did include a telling comment by the French physicist Perrin in his book *Les Atomes*, where he said in effect that the very irregular curves followed by particles in the Brownian motion led one to think of the supposed continuous non-differentiable curves of the mathematicians.

Standard Brownian Motion

A continuous-time stochastic process $(B_t)_{t\geq 0}$ is a *standard Brownian motion* if it satisfies the following properties:

1. $B_0 = 0$.

2. *(Normal distribution)* For $t > 0$, B_t has a normal distribution with mean 0 and variance t.
3. *(Stationary increments)* For $s, t > 0$, $B_{t+s} - B_s$ has the same distribution as B_t. That is,

$$P(B_{t+s} - B_s \le z) = P(B_t \le z) = \int_{-\infty}^{z} \frac{1}{\sqrt{2\pi t}} e^{-x^2/2t}\, dx,$$

for $-\infty < z < \infty$.
4. *(Independent increments)* If $0 \le q < r \le s < t$, then $B_t - B_s$ and $B_r - B_q$ are independent random variables.
5. *(Continuous paths)* The function $t \mapsto B_t$ is continuous, with probability 1.

The normal distribution plays a central role in Brownian motion. The reader may find it helpful to review properties of the univariate and bivariate normal distributions in Appendix B, Section B.4. We write $X \sim \text{Normal}(\mu, \sigma^2)$ to mean that the random variable X is normally distributed with mean μ and variance σ^2.

Brownian motion can be thought of as the motion of a particle that diffuses randomly along a line. At each point t, the particle's position is normally distributed about the line with variance t. As t increases, the particle's position is more *diffuse*; see Figure 8.1.

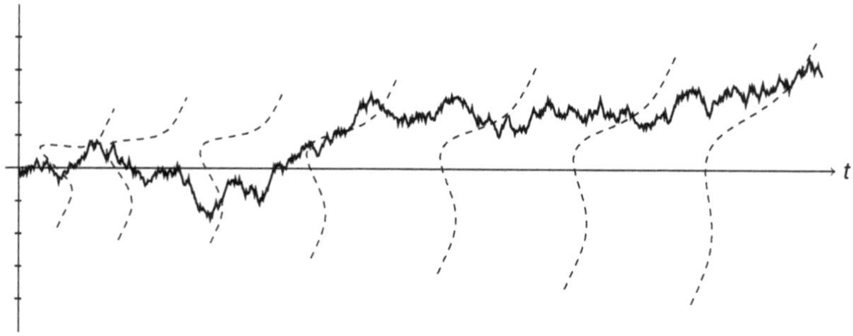

Figure 8.1 Brownian motion path. Superimposed on the graph are normal density curves with mean 0 and variance t.

It is not at all obvious that a stochastic process with the properties of Brownian motion actually exists. And Wiener's fundamental contribution was proving this existence. A rigorous derivation of Brownian motion is beyond the scope of this book, and requires measure theory and advanced analysis. The difficult part is showing the existence of a process that has stationary and independent increments together with continuous paths. The issue is touched upon at the end of the next section. First, however, we get our hands dirty with some calculations.

Computations involving Brownian motion are often tackled by exploiting stationary and independent increments. In the following examples, the reader may recognize similarities with the Poisson process, another stochastic process with stationary and independent increments. Unless stated otherwise, B_t denotes standard Brownian motion.

Example 8.1 For $0 < s < t$, find the distribution of $B_s + B_t$.

Solution Write $B_s + B_t = 2B_s + (B_t - B_s)$. By independent increments, B_s and $B_t - B_s$ are independent random variables, and thus $2B_s$ and $B_t - B_s$ are independent. The sum of independent normal variables is normal. Thus, $B_s + B_t$ is normally distributed with mean $E(B_s + B_t) = E(B_s) + E(B_t) = 0$, and variance

$$\begin{aligned} Var(B_s + B_t) &= Var(2B_s + (B_t - B_s)) = Var(2B_s) + Var(B_t - B_s) \\ &= 4Var(B_s) + Var(B_{t-s}) = 4s + (t - s) \\ &= 3s + t. \end{aligned}$$

The second equality is because the variance of a sum of independent random variables is the sum of their variances. The third equality uses stationary increments. We have that $B_s + B_t \sim \text{Normal}(0, 3s + t)$. ■

Example 8.2 A particle's position is modeled with a standard Brownian motion. If the particle is at position 1 at time $t = 2$, find the probability that its position is at most 3 at time $t = 5$.

Solution The desired probability is

$$\begin{aligned} P(B_5 \le 3 | B_2 = 1) &= P(B_5 - B_2 \le 3 - B_2 | B_2 = 1) \\ &= P(B_5 - B_2 \le 2 | B_2 = 1) \\ &= P(B_5 - B_2 \le 2) \\ &= P(B_3 \le 2) = 0.876. \end{aligned}$$

The third equality is because $B_5 - B_2$ and B_2 are independent. The penultimate equality is by stationary increments. The desired probability in R is

```
> pnorm(2,0,sqrt(3))
[1] 0.8758935
```

Note that in R commands involving the normal distribution are parameterized by standard deviation, not variance. ■

Example 8.3 Find the covariance of B_s and B_t.

Solution For the covariance,

$$\text{Cov}(B_s, B_t) = E(B_s B_t) - E(B_s)E(B_t) = E(B_s B_t).$$

For $s < t$, write $B_t = (B_t - B_s) + B_s$, which gives

$$\begin{aligned} E(B_s B_t) &= E(B_s(B_t - B_s + B_s)) \\ &= E(B_s(B_t - B_s)) + E\left(B_s^2\right) \\ &= E(B_s)E(B_t - B_s) + E\left(B_s^2\right) \\ &= 0 + Var(B_s) = s. \end{aligned}$$

Thus, $\text{Cov}(B_s, B_t) = s$. For $t < s$, by symmetry $\text{Cov}(B_s, B_t) = t$. In either case,

$$\text{Cov}(B_s, B_t) = \min\{s, t\}.$$

■

Simulating Brownian Motion

Consider simulating Brownian motion on $[0, t]$. Assume that we want to generate n variables at equally spaced time points, that is $B_{t_1}, B_{t_2}, \dots, B_{t_n}$, where $t_i = it/n$, for $i = 1, 2, \dots, n$. By stationary and independent increments, with $B_{t_0} = B_0 = 0$,

$$B_{t_i} = B_{t_{i-1}} + \left(B_{t_i} - B_{t_{i-1}}\right) \stackrel{d}{=} B_{t_{i-1}} + X_i,$$

where X_i is normally distributed with mean 0 and variance $t_i - t_{i-1} = t/n$, and is independent of $B_{t_{i-1}}$. The notation $X \stackrel{d}{=} Y$ means that random variables X and Y have the same distribution.

This leads to a recursive simulation method. Let $Z_1, Z_2, \dots, Z_n$ be independent and identically distributed standard normal random variables. Set

$$B_{t_i} = B_{t_{i-1}} + \sqrt{t/n}Z_i, \text{ for } i = 1, 2, \dots, n.$$

This gives

$$B_{t_i} = \sqrt{\frac{t}{n}}(Z_1 + \cdots + Z_n).$$

In R, the cumulative sum command

```
> cumsum(rnorm(n,0,sqrt(t/n)))
```

generates the Brownian motion variables $B_{t/n}, B_{2t/n}, \dots, B_t$.

Simulations of Brownian motion on [0, 1] are shown in Figure 8.2. The paths were drawn by simulating $n = 1,000$ points in [0, 1] and then connecting the dots.

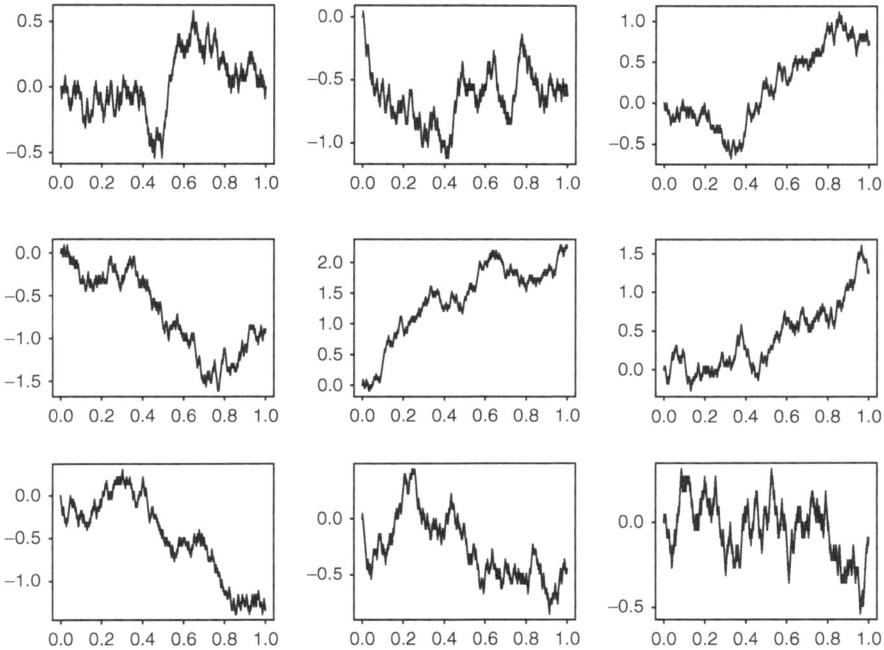

Figure 8.2 Sample paths of Brownian motion on [0, 1].

R: Simulating Brownian Motion

```
# bm.R
> n <- 1000
> t <- 1
> bm <- c(0, cumsum(rnorm(n,0,sqrt(t/n))))
> steps <- seq(0,t,length=n+1)
> plot(steps,bm,type="l")
```

More generally, to simulate $B_{t_1}, B_{t_2}, \ldots, B_{t_n}$, for time points $t_1 < t_2 < \cdots < t_n$, set

$$B_{t_i} = B_{t_{i-1}} + \sqrt{t_i - t_{i-1}} Z_i, \text{ for } i = 1, 2, \ldots, n,$$

with $t_0 = 0$.

Sample Space for Brownian Motion and Continuous Paths*

Consider the fifth defining property of Brownian motion: the function $t \mapsto B_t$ is continuous.

A continuous-time stochastic process $(X_t)_{-\infty<t<\infty}$ is a collection of random variables defined on a common sample space, or probability space, Ω. A random variable is really a function on a probability space. If $X : \Omega \to \mathbb{R}$ is a random variable, then X takes values depending on the outcome $\omega \in \Omega$. Thus, we can write $X = X(\omega)$ to emphasize the dependence on the outcome ω, although we usually suppress ω for simplicity.

In the context of stochastic processes, $X_t = X_t(\omega)$ is a function of two variables: t and ω. For fixed $t \in \mathbb{R}$, X_t is a random variable. Letting $X_t(\omega)$ vary as $\omega \in \Omega$ generates the different values of the process at the fixed time t. On the other hand, for fixed $\omega \in \Omega$, $X_t(\omega)$ is a function of t. Letting t vary generates a *sample path* or *realization*. One can think of these realizations as *random functions*.

For instance, each of the graphs in Figure 8.2, is one realization of such a random function. Using function notation, we could write $f(t) = B_t(\omega)$, for $-\infty < t < \infty$. For fixed ω, it makes sense to ask whether f is continuous, differentiable, etc.

A more precise statement of Property 5 is that

$$P(\omega \in \Omega : B_t(\omega) \text{ is a continuous function of } t) = 1.$$

Implicit in this statement is the existence of (i) the probability space Ω and (ii) a suitable probability function, or *probability measure*, P, which is consistent with the other defining properties of Brownian motion.

A probability space Ω is easy enough to identify. Let Ω be the set of all continuous functions on $[0, \infty)$. Each $\omega \in \Omega$ is a continuous function. Then, $B_t(\omega) = \omega_t$, the value of ω evaluated at t. This is the easiest way to insure that B_t has continuous sample paths. The hard part is to construct a probability function P on the set of continuous functions, which is consistent with the properties of Brownian motion. This was precisely Norbert Wiener's contribution. That probability function, introduced by Wiener in 1923, is called *Wiener measure*.

8.2 BROWNIAN MOTION AND RANDOM WALK

Continuous-time, continuous-state Brownian motion is intimately related to discrete-time, discrete-state random walk. Brownian motion can be constructed from simple symmetric random walk by suitably scaling the values of the walk while simultaneously speeding up the steps of the walk.

Let $X_1, X_2, \ldots$ be an i.i.d. sequence with each X_i taking values ± 1 with probability 1/2 each. Set $S_0 = 0$ and for integer $t > 0$, let $S_t = X_1 + \cdots + X_t$. Then, $S_0, S_1, S_2, \ldots$ is a simple symmetric random walk with $E(S_t) = 0$ and $Var(S_t) = t$ for $t = 0, 1, \ldots$. As a sum of i.i.d. random variables, for large t, S_t is approximately normally distributed by the central limit theorem.

The random walk has independent increments. For integers $0 < q < r < s < t$, $S_t - S_s = X_{s+1} + \cdots + X_t$, and $S_r - S_q = X_{q+1} + \cdots + X_r$. Since $X_{s+1}, \ldots, X_t$ is independent of $X_{q+1}, \ldots, X_r$, it follows that $S_t - S_s$ and $S_r - S_q$ are independent random variables. Furthermore, for all integers $0 < s < t$, the distribution of $S_t - S_s$ and S_{t-s}

is the same since they are both a function of $t - s$ distinct X_i. Thus, the walk has stationary increments.

The simple random walk is a discrete-state process. To obtain a continuous-time process with continuous sample paths, values are connected by linear interpolation. Recall that $\lfloor x \rfloor$ is the *floor* of x, or integer part of x, which is the largest integer not greater than x. Extending the definition of S_t to real $t \geq 0$, let

$$S_t = \begin{cases} X_1 + \cdots + X_t, & \text{if } t \text{ is an integer,} \\ S_{\lfloor t \rfloor} + X_{\lfloor t \rfloor + 1} (t - \lfloor t \rfloor), & \text{otherwise.} \end{cases}$$

Observe that if k is a positive integer, then for $k \leq t \leq k + 1$, S_t is the linear interpolation of the points (k, S_k) and $(k + 1, S_{k+1})$. See Figure 8.3 to visualize the construction. We have $E(S_t) = 0$ and

$$\begin{aligned} Var(S_t) &= Var(S_{\lfloor t \rfloor} + X_{\lfloor t \rfloor + 1} (t - \lfloor t \rfloor)) \\ &= Var(S_{\lfloor t \rfloor}) + (t - \lfloor t \rfloor)^2 Var(X_{\lfloor t \rfloor + 1}) \\ &= \lfloor t \rfloor + (t - \lfloor t \rfloor)^2 \approx t, \end{aligned}$$

as $0 \leq t - \lfloor t \rfloor < 1$.

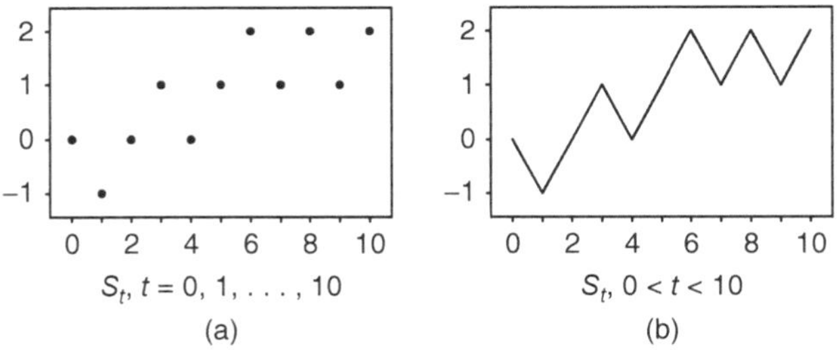

Figure 8.3 (a) Realization of a simple random walk. (b) Walk is extended to a continuous path by linear interpolation.

The process is now scaled both vertically and horizontally. Let $S_t^{(n)} = S_{nt}/\sqrt{n}$, for $n = 1, 2, \ldots$ On any interval, the new process has n times as many steps as the original walk. And the height at each step is shrunk by a factor of $1/\sqrt{n}$. The construction is illustrated in Figure 8.4.

The scaling preserves mean and variance, as $E\left(S_t^{(n)}\right) = 0$ and $Var\left(S_t^{(n)}\right) = Var(S_{nt})/n \approx t$. Sample paths are continuous and for each n, the process retains independent and stationary increments. Considering the central limit theorem, it is reasonable to think that as $n \to \infty$, the process converges to Brownian motion. This, in fact, is the case, a result proven by Monroe Donsker in 1951.

We have not precisely said what *convergence* actually means since we are not talking about convergence of a sequence of numbers, nor even convergence of random variables, but rather convergence of stochastic processes. We do not

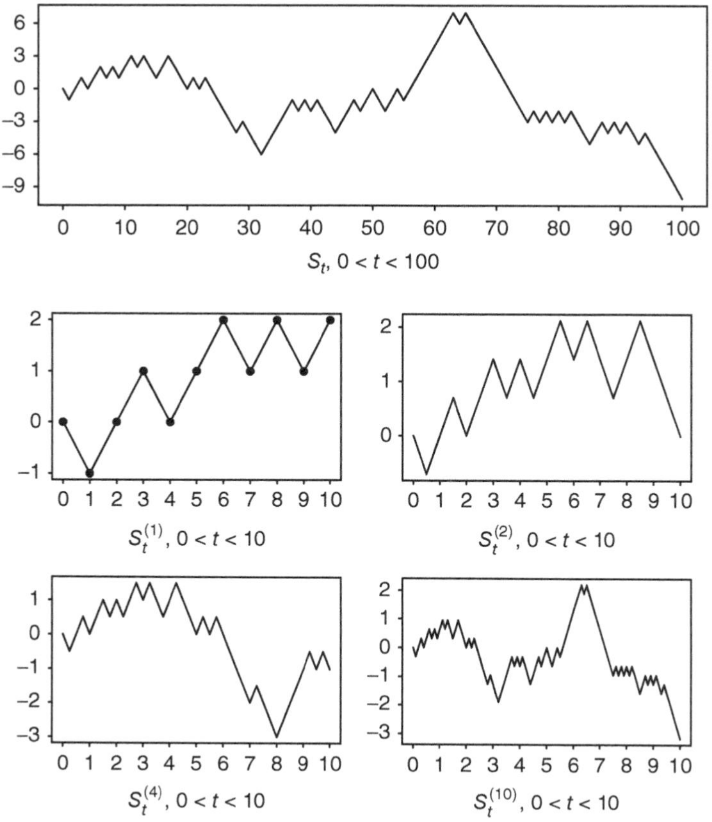

Figure 8.4 Top graph is a simple symmetric random walk for 100 steps. Bottom graphs show the scaled process $S_t^{(n)} = S_{nt}/\sqrt{n}$, for $n = 1, 2, 4, 10$.

give a rigorous statement. Nevertheless, the reader can trust their intuition that for large n, $\left(S_{nt}/\sqrt{n}\right)_{t\geq 0}$ behaves like a standard Brownian motion process $(B_t)_{t\geq 0}$.

Invariance Principle*

The construction of Brownian motion from simple symmetric random walk can be generalized so that we start with any i.i.d. sequence $X_1, \ldots, X_n$ with mean 0 and variance 1. Let $S_n = X_1 + \cdots + X_n$. Then, $S_{nt}/\sqrt{n}$ converges to B_t, as $n \to \infty$. This is known as *Donsker's invariance principle*. The word *invariance* is used because all random walks with increments that have mean 0 and variance 1, regardless of distribution, give the same Brownian motion limit.

A consequence of the invariance principle and continuity is that *functions* of the discrete process $S_{nt}/\sqrt{n}$ converge to the corresponding function of

Brownian motion, as $n \to \infty$. If g is a bounded, continuous function, whose domain is the set of continuous functions on $[0, 1]$, then $g\left(S_{nt}/\sqrt{n}\right) \approx g(B_t)$, for large n.

Functions whose domain is a set of functions are called *functionals*. The invariance principle means that properties of random walk and of functionals of random walk can be derived by considering analogous properties of Brownian motion, and vice versa.

For example, assume that f is a continuous function on $[0, 1]$. Let $g(f) = f(1)$, the evaluation of $f(t)$ at $t = 1$. Then, $g(S_{nt}/\sqrt{n}) = S_n/\sqrt{n}$ and $g(B_t) = B_1$. By the invariance principle, $S_n/\sqrt{n} \to B_1$, as $n \to \infty$. The random variable B_1 is normally distributed with mean 0 and variance 1, and we have thus recaptured the central limit theorem from Donsker's invariance principle.

Example 8.4 For a simple symmetric random walk, consider the maximum value of the walk in the first n steps. Let $g(f) = \max_{0\le t\le 1} f(t)$. By the invariance principle,

$$\lim_{n\to\infty} g\left(\frac{S_{tn}}{\sqrt{n}}\right) = \lim_{n\to\infty} \max_{0\le t\le 1} \frac{S_{tn}}{\sqrt{n}} = \lim_{n\to\infty} \max_{0\le k\le n} \frac{S_k}{\sqrt{n}} = g\left(B_t\right) = \max_{0\le t\le 1} B_t.$$

This gives $\max_{0\le k\le n} S_k \approx \sqrt{n}\max_{0\le t\le 1} B_t$, for large n.

In Section 8.4, it is shown that the random variable $M = \max_{0\le t\le 1} B_t$ has density function

$$f_M(x) = \sqrt{\frac{2}{\pi}} e^{-x^2/2}, \text{ for } x > 0.$$

Mean and standard deviation are

$$E(M) = \sqrt{\frac{2}{\pi}} \approx 0.80 \quad \text{and} \quad SD(M) = \frac{\pi - 2}{\pi} \approx 0.60.$$

With these results, we see that in the first n steps of simple symmetric random walk, the maximum value is about $(0.80)\sqrt{n}$ give or take $(0.60)\sqrt{n}$. In $n = 10,000$ steps, the probability that a value greater than 200 is reached is

$$\begin{aligned} P\left(\max_{0\le k\le n} S_k > 200\right) &= P\left(\max_{0\le k\le n} \frac{S_k}{100} > 2\right) \\ &= P\left(\max_{0\le k\le n} \frac{S_k}{\sqrt{n}} > 2\right) \\ &\approx P(M > 2) \\ &= \int_2^\infty \sqrt{\frac{2}{\pi}} e^{-x^2/2}\, dx = 0.0455. \end{aligned}$$

■

R: Maximum for Simple Symmetric Random Walk

```
# maxssrw.R
> n <- 10000
> sim <- replicate(10000,
  + max(cumsum(sample(c(-1,1),n,replace=T))))
> mean(sim)
[1] 79.7128
> sd(sim)
[1] 60.02429
> sim <- replicate(10000,
  + if(max(cumsum(sample(c(-1,1),n,rep=T)))>200)
  1 else 0)
> mean(sim) # P(max > 200)
[1] 0.0461
```

8.3 GAUSSIAN PROCESS

The normal distribution is also called the *Gaussian distribution*, in honor of Carl Friedrich Gauss, who introduced the distribution more than 200 years ago. The bivariate and multivariate normal distributions extend the univariate distribution to higher finite-dimensional spaces. A *Gaussian process* is a continuous-time stochastic process, which extends the Gaussian distribution to infinite dimensions. In this section, we show that Brownian motion is a Gaussian process and identify the conditions for when a Gaussian process is Brownian motion.

(Gauss, considered by historians to have been one of the greatest mathematicians of all time, first used the normal distribution as a model for measurement error in celestial observations, which led to computing the orbit of the planetoid Ceres. The story of the discovery of Ceres, and the contest to compute its orbit, is a mathematical page-turner. See *The Discovery of Ceres: How Gauss Became Famous* by Teets and Whitehead (1999).)

We first define the multivariate normal distribution and establish some of its key properties.

Multivariate Normal Distribution

Random variables $X_1, \dots, X_k$ have a *multivariate normal distribution* if for all real numbers $a_1, \dots, a_k$, the linear combination

$$a_1X_1 + \cdots + a_kX_k$$

has a univariate normal distribution. A multivariate normal distribution is completely determined by its *mean vector*

$$\boldsymbol{\mu} = (\mu_1, \ldots, \mu_k) = (E(X_1), \ldots, E(X_k))$$

and *covariance matrix* $\boldsymbol{V}$, where

$$V_{ij} = \text{Cov}(X_i, X_j), \text{for } 1 \leq i, j \leq k.$$

The joint density function of the multivariate normal distribution is

$$f(\boldsymbol{x}) = \frac{1}{(2\pi)^{k/2}|\boldsymbol{V}|^{1/2}} \exp\left(-\frac{1}{2}(\boldsymbol{x} - \boldsymbol{\mu})^T \boldsymbol{V}^{-1}(\boldsymbol{x} - \boldsymbol{\mu})\right),$$

where $\boldsymbol{x} = (x_1, \ldots, x_k)$ and $|\boldsymbol{V}|$ is the determinant of $\boldsymbol{V}$.

The multivariate normal distribution has the remarkable property that all marginal and conditional distributions are normal. If $X_1, \ldots, X_k$ have a multivariate normal distribution, then the X_i are normally distributed, joint distributions of subsets of the X_i have multivariate normal distributions, and conditional distributions given subsets of the X_i are normal.

If $X_1, \ldots, X_k$ are independent normal random variables, then their joint distribution is multivariate normal. For jointly distributed normal random variables, independence is equivalent to being uncorrelated. That is, if X and Y are jointly distributed normal random variables, then X and Y are independent if and only if $E(XY) = E(X)E(Y)$.

A *Gaussian process* extends the multivariate normal distribution to stochastic processes.

Gaussian Process

A *Gaussian process* $(X_t)_{t\geq 0}$ is a continuous-time stochastic process with the property that for all $n = 1, 2, \ldots$ and $0 \leq t_1 < \cdots < t_n$, the random variables $X_{t_1}, \ldots, X_{t_n}$ have a multivariate normal distribution.

A Gaussian process is completely determined by its *mean function* $E(X_t)$, for $t \geq 0$, and *covariance function* $\text{Cov}(X_s, X_t)$, for $s, t \geq 0$.

Standard Brownian motion is a Gaussian process. The following characterization gives conditions for when a Gaussian process is a standard Brownian motion.

Gaussian Process and Brownian Motion

A stochastic process $(B_t)_{t\geq 0}$ is a standard Brownian motion if and only if it is a Gaussian process with the following properties:

1. $B_0 = 0$.
2. *(Mean function)* $E(B_t) = 0$, for all t.
3. *(Covariance function)* $\text{Cov}(B_s, B_t) = \min\{s, t\}$, for all s, t.
4. *(Continuous paths)* The function $t \mapsto B_t$ is continuous, with probability 1.

Proof. Let $(B_t)_{t\geq 0}$ be a standard Brownian motion. Consider $(B_{t_1}, B_{t_2}, \dots, B_{t_k})$, for $0 < t_1 < t_2 < \cdots < t_k$. For constants $a_1, a_2, \dots, a_k$, we need to show that $a_1 B_{t_1} + a_2 B_{t_2} + \cdots + a_k B_{t_k}$ has a univariate normal distribution. By independent increments, $B_{t_1}, B_{t_2} - B_{t_1}, \dots, B_{t_k} - B_{t_{k-1}}$ are independent normal random variables, whose joint distribution is multivariate normal. Write

$$\begin{aligned} &a_1 B_{t_1} + a_2 B_{t_2} + \cdots + a_k B_{t_k} \\ &\quad = (a_1 + \cdots + a_k) B_{t_1} + (a_2 + \cdots + a_k)(B_{t_2} - B_{t_1}) \\ &\qquad + \cdots + (a_{k-1} + a_k)(B_{t_{k-1}} - B_{t_{k-2}}) + a_k (B_{t_k} - B_{t_{k-1}}), \end{aligned}$$

which is a linear combination of $B_{t_1}, B_{t_2} - B_{t_1}, \dots, B_{t_k} - B_{t_{k-1}}$, and thus has a univariate normal distribution.

The mean and covariance functions for Brownian motion are

$$E(B_t) = 0 \quad \text{and} \quad \text{Cov}(B_s, B_t) = \min\{s, t\}.$$

It follows that standard Brownian motion is the unique Gaussian process with these mean and covariance functions.

Conversely, assume that $(B_t)_{t\geq 0}$ is a Gaussian process that satisfies the stated properties. We need to show the process has stationary and independent increments.

Since the process is Gaussian, for $s, t \geq 0$, $B_{t+s} - B_s$ is normally distributed with mean $E(B_{t+s} - B_s) = E(B_{t+s}) - E(B_s) = 0$, and variance

$$Var(B_{t+s} - B_s) = Var(B_{t+s}) + Var(B_s) - 2\text{Cov}(B_{t+s}, B_s) = (t + s) + s - 2s = t.$$

Thus, $B_{t+s} - B_s$ has the same distribution as B_t, which gives stationary increments.

For $0 \leq q < r \leq s < t$,

$$\begin{aligned} E((B_r - B_q)(B_t - B_s)) &= E(B_r B_t) - E(B_r B_s) - E(B_q B_t) + E(B_q B_s) \\ &= \text{Cov}(B_r, B_t) - \text{Cov}(B_r, B_s) - \text{Cov}(B_q, B_t) + \text{Cov}(B_q, B_s) \\ &= r - r - q + q = 0. \end{aligned}$$

Thus, $B_r - B_q$ and $B_t - B_s$ are uncorrelated. Since $B_r - B_q$ and $B_t - B_s$ are normally distributed, it follows that they are independent.

We have shown that $(B_t)_{t\geq 0}$ is a standard Brownian motion. ■

Example 8.5 For $a > 0$, let $X_t = B_{at}/\sqrt{a}$, for $t \geq 0$. Show that $(X_t)_{t\geq 0}$ is a standard Brownian motion.

Solution For $0 \leq t_1 < \cdots < t_k$ and real numbers $a_1, \ldots, a_k$,

$$\sum_{i=1}^{k} a_i X_{t_i} = \sum_{i=1}^{k} \frac{a_i}{\sqrt{a}} B_{at_i},$$

which has a univariate normal distribution, since $(B_t)_{t\geq 0}$ is a Gaussian process. Thus, $(X_t)_{t\geq 0}$ is a Gaussian process. Clearly, $X_0 = 0$. The mean function is

$$E(X_t) = E\left(\frac{1}{\sqrt{a}} B_{at}\right) = \frac{1}{\sqrt{a}} E(B_{at}) = 0.$$

The covariance function is

$$\begin{aligned}\text{Cov}(X_s, X_t) &= \text{Cov}\left(\frac{1}{\sqrt{a}} B_{as}, \frac{1}{\sqrt{a}} B_{at}\right) = \frac{1}{a}\text{Cov}(B_{as}, B_{at}) \\ &= \frac{1}{a}\min\{as, at\} = \min\{s, t\}.\end{aligned}$$

Finally, path continuity of $(X_t)_{t\geq 0}$ follows from the path continuity of standard Brownian motion, as the function $t \mapsto B_{at}/\sqrt{a}$ is continuous for all $a > 0$, with probability 1. ■

Nowhere Differentiable Paths

The property illustrated in Example 8.5 shows that Brownian motion preserves its character after rescaling. For instance, given a standard Brownian motion on [0, 1], if we look at the process on, say, an interval of length one-trillionth ($= 10^{-12}$) then after resizing by a factor of $1/\sqrt{10^{-12}} = 1{,}000{,}000$, what we see is indistinguishable from the original Brownian motion!

This highlights the invariance, or *fractal*, structure of Brownian motion sample paths. It means that the jagged character of these paths remains jagged at all time scales. This leads to the remarkable fact that Brownian motion sample paths are *nowhere differentiable*. It is hard to even contemplate a function that is continuous at every point on its domain, but not differentiable at any point. Indeed, for many years, mathematicians believed that such a function was impossible, until Karl Weierstrass, considered the founder of modern analysis, demonstrated their existence in 1872.

The proof that Brownian motion is nowhere differentiable requires advanced analysis. Here is a heuristic argument. Consider the formal derivative

$$\frac{d}{dt}B_t = \lim_{h\to 0} \frac{B_{t+h} - B_t}{h}.$$

By stationary increments, $B_{t+h} - B_t$ has the same distribution as B_h, which is normal with mean 0 and variance h. Thus, the difference quotient $(B_{t+h} - B_t)/h$ is normally distributed with mean 0 and variance $1/h$. As h tends to 0, the variance tends to infinity. Since the difference quotient takes arbitrarily large values, the limit, and hence the derivative, does not exist.

8.4 TRANSFORMATIONS AND PROPERTIES

In addition to invariance under rescaling, Brownian motion satisfies numerous reflection, translation, and symmetry properties.

Transformations of Brownian Motion

Let $(B_t)_{t\geq 0}$ be a standard Brownian motion. Then, each of the following transformations is a standard Brownian motion.

1. *(Reflection)* $(-B_t)_{t\geq 0}$.
2. *(Translation)* $(B_{t+s} - B_s)_{t\geq 0}$, for all $s \geq 0$.
3. *(Rescaling)* $(a^{-1/2}B_{at})_{t\geq 0,}$ for all $a > 0$.
4. *(Inversion)* The process $(X_t)_{t\geq 0}$ defined by $X_0 = 0$ and $X_t = tB_{1/t}$, for $t > 0$.

We leave the proofs that reflection and translation are standard Brownian motions as exercises. Rescaling was shown in Example 8.5. For inversion, let $t_1 < \cdots < t_k$. For constants $a_1, \ldots, a_k$,

$$a_1X_{t_1} + \cdots + a_kX_{t_k} = a_1t_1B_{1/t_1} + \cdots + a_kt_kB_{1/t_k}$$

is normally distributed and thus $(X_t)_{t\geq 0}$ is a Gaussian process. The mean function is $E(X_t) = E(tB_{1/t}) = tE(B_{1/t}) = 0$, for $t > 0$. Covariance is

$$\begin{aligned}\text{Cov}(X_s, X_t) = & \text{ Cov}(sB_{1/s}, tB_{1/t}) = E(sB_{1/s}tB_{1/t}) \\ & = stE(B_{1/s}B_{1/t}) = st\,\text{Cov}(B_{1/s}, B_{1/t}) \\ & = st\min\{1/s, 1/t\} = \min\{t, s\}.\end{aligned}$$

Continuity, for all $t > 0$, is inherited from $(B_t)_{t\geq 0}$. What remains is to show the process is continuous at $t = 0$. We do not prove this rigorously. Suffice it to show that

$$0 = \lim_{t\to 0} X_t = \lim_{t\to 0} tB_{1/t}, \text{ with probability 1}$$

is equivalent to

$$0 = \lim_{s\to\infty} \frac{B_s}{s}, \text{ with probability 1,}$$

and the latter is intuitive by the strong law of large numbers.

For real x, the process defined by $X_t = x + B_t$, for $t \geq 0$, is called *Brownian motion started at* x. For such a process, $X_0 = x$ and X_t is normally distributed with mean function $E(X_t) = x$, for all t. The process retains all other defining properties of standard Brownian motion: stationary and independent increments, and continuous sample paths.

See Figure 8.5 for examples of transformations of Brownian motion.

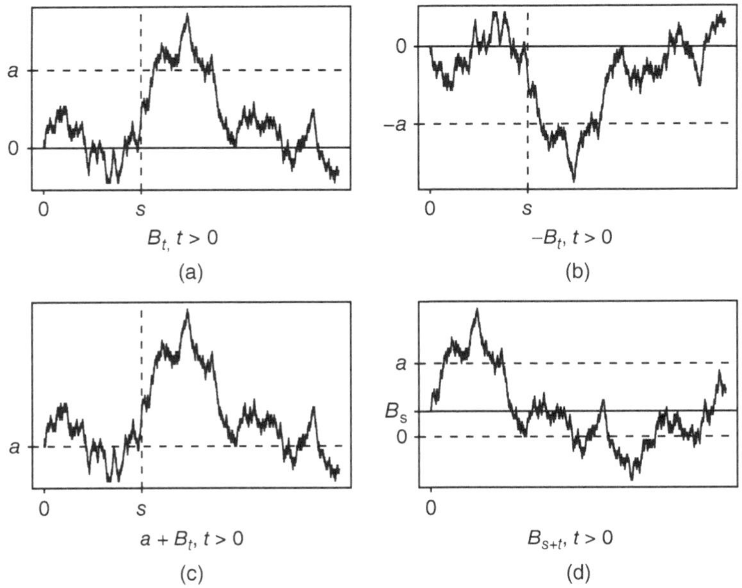

Figure 8.5 Transformations of Brownian motion. (a) Standard Brownian motion. (b) Reflection across t-axis. (c) Brownian motion started at a. (d) Translation.

Example 8.6 Let $(X_t)_{t\geq 0}$ be a Brownian motion process started at $x = 3$. Find $P(X_2 > 0)$.

Solution Write $X_t = B_t + 3$. Then,

$$P(X_2 > 0) = P(B_2 + 3 > 0) = P(B_2 > -3) = \int_{-3}^{\infty} \frac{1}{\sqrt{4\pi}} e^{-x^2/4}\, dx = 0.983.$$

In R, type

```
> 1-pnorm(-3,0,sqrt(2))
[1] 0.9830526
```

■

Markov Process

Brownian motion satisfies the Markov property that conditional on the present, past and future are independent. This is a consequence of independent increments. Brownian motion is an example of a *Markov process*. A continuous-state stochastic process $(X_t)_{t\geq 0}$ is a Markov process if

$$P(X_{t+s} \leq y | X_u, 0 \leq u \leq s) = P(X_{t+s} \leq y | X_s), \tag{8.1}$$

for all $s, t \geq 0$ and real y. The process is *time-homogeneous* if the probability in Equation (8.1) does not depend on s. That is,

$$P(X_{t+s} \leq y | X_s) = P(X_t \leq y | X_0).$$

For a continuous-state Markov process, the *transition function*, or *transition kernel*, $K_t(x, y)$ plays the role that the transition matrix plays for a discrete-state Markov chain. The function $K_t(x, \cdot)$ is the conditional density of X_t given $X_0 = x$.

If $(X_t)_{t\geq 0}$ is Brownian motion started at x, then X_t is normally distributed with mean x and variance t. The transition kernel is

$$K_t(x, y) = \frac{1}{\sqrt{2\pi t}} e^{-(y-x)^2/2t}.$$

The transition kernel of a Markov process satisfies the Chapman–Kolmogorov equations. For continuous-state processes, integrals replace sums. The equations are

$$K_{s+t}(x, y) = \int_{-\infty}^{\infty} K_s(x, z) K_t(z, y)\, dz, \quad \text{for all } s, t, \tag{8.2}$$

as

$$\begin{aligned}
\int_{-\infty}^{y} K_{s+t}(x, w)\, dw &= P(X_{s+t} \leq y | X_0 = x) \\
&= \int_{-\infty}^{\infty} P(X_{s+t} \leq y | X_s = z, X_0 = x) K_s(x, z)\, dz \\
&= \int_{-\infty}^{\infty} P(X_t \leq y | X_0 = z) K_s(x, z)\, dz \\
&= \int_{-\infty}^{\infty} \left(\int_{-\infty}^{y} K_t(z, w)\, dw \right) K_s(x, z)\, dz \\
&= \int_{-\infty}^{y} \left(\int_{-\infty}^{\infty} K_s(x, z) K_t(z, w)\, dz \right) dw.
\end{aligned}$$

Taking derivatives with respect to y gives Equation (8.2).

First Hitting Time and Strong Markov Property

Brownian motion satisfies the strong Markov property. Recall that Brownian motion translated left or right by a constant is also a Brownian motion. For $s > 0$, with $B_s = x$, the process $(B_{t+s})_{t\geq 0}$ is a Brownian motion started at x.

By the strong Markov property this also holds for some random times as well. If S is a *stopping time*, $(B_{t+S})_{t\geq 0}$ is a Brownian motion process. See Section 3.9 to reference the strong Markov property for discrete-time Markov chains.

A common setting is when Brownian motion first hits a particular state or set of states. Let $T_a = \min\{t : B_t = a\}$ be the *first hitting time* that Brownian motion hits level a. See Figure 8.6. The random variable T_a is a stopping time. Moving forward from T_a, the translated process is a Brownian motion started at a.

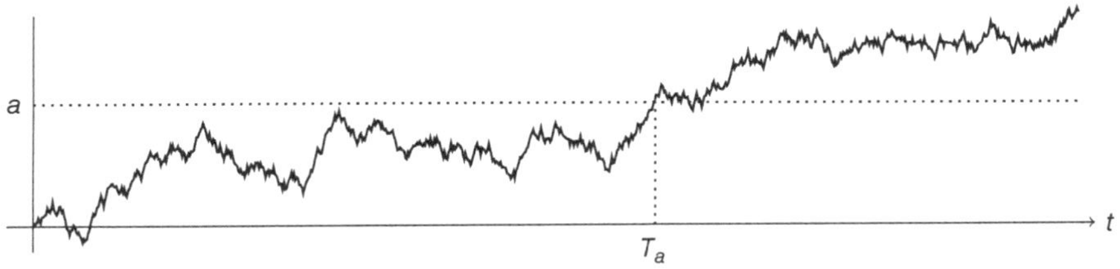

Figure 8.6 T_a is the first time Brownian motion hits level a.

The strong Markov property is used to find the distribution of T_a. Consider standard Brownian motion. At any time t, B_t is equally likely to be above or below the line $y = 0$. Assume that $a > 0$. For Brownian motion started at a, the process is equally likely to be above or below the line $y = a$. This gives,

$$P(B_t > a | T_a < t) = P(B_t > 0) = \frac{1}{2},$$

and thus,

$$\frac{1}{2} = P(B_t > a | T_a < t) = \frac{P(B_t > a, T_a < t)}{P(T_a < t)} = \frac{P(B_t > a)}{P(T_a < t)}.$$

The last equality is because the event $\{B_t > a\}$ implies $\{T_a < t\}$ by continuity of sample paths. We have that

$$\begin{aligned} P(T_a < t) &= 2P(B_t > a) & (8.3)\\ &= 2\int_a^{\infty} \frac{1}{\sqrt{2\pi t}} e^{-x^2/2t}\, dx \\ &= 2\int_{a/\sqrt{t}}^{\infty} \frac{1}{\sqrt{2\pi}} e^{-x^2/2}\, dx. & (8.4)\end{aligned}$$

If $a < 0$, the argument is similar with $1/2 = P(B_t < a | T_a < t)$. In either case,

$$P(T_a < t) = 2\int_{|a|/\sqrt{t}}^{\infty} \frac{1}{\sqrt{2\pi}} e^{-x^2/2}\, dx.$$

Differentiating with respect to t gives the probability density function of the first hitting time.

First Hitting Time Distribution

For a standard Brownian motion, let T_a be the first time the process hits level a. The density function of T_a is

$$f_{T_a}(t) = \frac{|a|}{\sqrt{2\pi t^3}} e^{-a^2/2t}, \text{ for } t > 0. \tag{8.5}$$

Example 8.7 A particle moves according to Brownian motion started at $x = 1$. After $t = 3$ hours, the particle is at level 1.5. Find the probability that the particle reaches level 2 sometime in the next hour.

Solution For $t \geq 3$, the translated process is a Brownian motion started at $x = 1.5$. The event that the translated process reaches level 2 in the next hour, is equal to the event that a standard Brownian motion first hits level $a = 2 - 1.5 = 0.5$ in $[0, 1]$. The desired probability is

$$P(T_{0.5} \leq 1) = \int_0^1 \frac{0.5}{\sqrt{2\pi t^3}} e^{-(0.5)^2/2t}\, dt = 0.617.$$

■

The first hitting time distribution has some surprising properties. Consider

$$\begin{aligned} P(T_a < \infty) &= \lim_{t\to\infty} P(T_a < t) = \lim_{t\to\infty} 2\int_{|a|/\sqrt{t}}^{\infty} \frac{1}{\sqrt{2\pi}} e^{-x^2/2}\, dx \\ &= 2\int_0^{\infty} \frac{1}{\sqrt{2\pi}} e^{-x^2/2}\, dx = 2\left(\frac{1}{2}\right) = 1. \end{aligned} \tag{8.6}$$

Brownian motion hits level a, with probability 1, for all a, no matter how large.

On the contrary,

$$E(T_a) = \int_0^{\infty} \frac{t|a|}{\sqrt{2\pi t^3}} e^{-a^2/2t}\, dt = +\infty.$$

The expected time to first reach level a is infinite. This is true for all a, no matter how small!

Reflection Principle and the Maximum of Brownian Motion

Brownian motion reflected at a first hitting time is a Brownian motion. This property is known as the *reflection principle* and is a consequence of the strong Markov property.

For a standard Brownian motion, and first hitting time T_a, consider the process $\left(\tilde{B}_t\right)_{t\geq 0}$ defined by

$$\tilde{B}_t = \begin{cases} B_t, & \text{if } 0 \leq t \leq T_a, \\ 2a - B_t, & \text{if } t \geq T_a. \end{cases}$$

This is called *Brownian motion reflected at* T_a. The construction is shown in Figure 8.7. The reflected portion $a - (B_t - a) = 2a - B_t$ is a Brownian motion process started at a by the strong Markov property and the fact that $-B_t$ is a standard Brownian motion. Concatenating the front of the original process $(B_t)_{0\leq t\leq T_a}$ to the reflected piece preserves continuity.

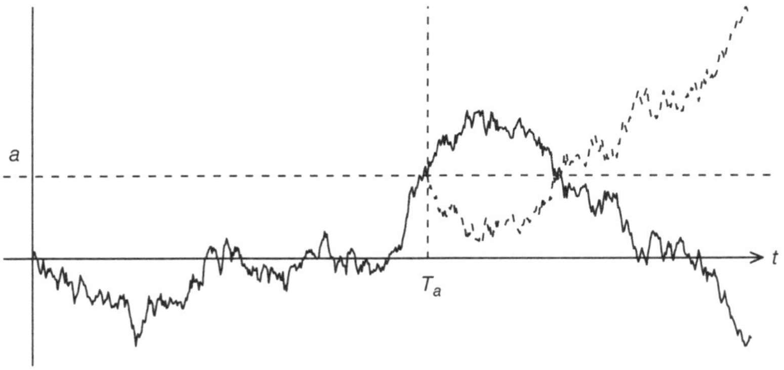

Figure 8.7 Reflection principle.

Another way of thinking of the reflection principle is that it establishes a one-to-one correspondence between paths that exceed level a at time t and paths that are below a at time t and have hit a by time t.

The reflection principle is applied in the following derivation of the distribution of $M_t = \max_{0\leq s\leq t} B_s$, the maximum value of Brownian motion on $[0, t]$.

Let $a > 0$. If at time t, B_t exceeds a, then the maximum on $[0, t]$ is greater than a. That is, $\{B_t > a\}$ implies $\{M_t > a\}$. This gives

$$\begin{aligned} \{M_t > a\} &= \{M_t > a, B_t > a\} \cup \{M_t > a, B_t \leq a\} \\ &= \{B_t > a\} \cup \{M_t > a, B_t \leq a\}. \end{aligned}$$

As the union is disjoint, $P(M_t > a) = P(B_t > a) + P(M_t > a, B_t \leq a)$.

Consider a sample path that has crossed a by time t and is at most a at time t. By the reflection principle, the path corresponds to a reflected path that is at least a at time t. This gives that $P(M_t > a, B_t \leq a) = P\left(\tilde{B}_t \geq a\right) = P(B_t > a)$. Thus,

$$P(M_t > a) = 2P(B_t > a) = \int_a^\infty \sqrt{\frac{2}{\pi t}} e^{-x^2/2t}\, dx, \text{ for } a > 0.$$

The distribution of M_t is equal to the distribution of $|B_t|$, the absolute value of a normally distributed random variable with mean 0 and variance t.

Since we have already found the distribution of T_a, a different derivation of the distribution of M_t is possible, using the fact that $M_t > a$ if and only if the process hits a by time t, that is $T_a < t$. This gives

$$P(M_t > a) = P(T_a < t) = \int_0^t \frac{a}{\sqrt{2\pi s^3}} e^{-a^2/2s}\, ds \tag{8.7}$$

$$= \int_a^\infty \sqrt{\frac{2}{\pi t}} e^{-x^2/2t}\, dx, \text{ for } a > 0. \tag{8.8}$$

The last equality is achieved by the change of variables $a^2/s = x^2/t$.

Example 8.8 A laboratory instrument takes annual temperature measurements. Measurement errors are assumed to be independent and normally distributed. As precision decreases over time, errors are modeled as standard Brownian motion. For how many years can the lab be guaranteed that there is at least 90% probability that all errors are less than 4 degrees?

Solution The problem asks for the largest t such that $P(M_t \leq 4) \geq 0.90$. We have

$$0.90 \leq P(M_t \leq 4) = 1 - P(M_t > 4) = 1 - 2P(B_t > 4) = 2P(B_t \leq 4) - 1.$$

This gives

$$0.95 \leq P(B_t \leq 4) = P\left(Z \leq \frac{4}{\sqrt{t}}\right),$$

where Z is a standard normal random variable. The 95th percentile of the standard normal distribution is 1.645. Solving $4/\sqrt{t} = 1.645$ gives

$$t = \left(\frac{4}{1.645}\right)^2 = 5.91 \text{ years.}$$

■

Zeros of Brownian Motion and Arcsine Distribution

Brownian motion reaches level x, no matter how large x, with probability 1. Brownian motion also returns to the origin infinitely often. In fact, on any interval $(0, \epsilon)$, no matter how small ϵ, the process crosses the t-axis infinitely many times.

The times when the process crosses the t-axis are the *zeros* of Brownian motion. That Brownian motion has infinitely many zeros in any interval of the form $(0, \epsilon)$ is a consequence of the following.

Zeros of Brownian Motion

Theorem 8.1. *For $0 \le r < t$, let $z_{r,t}$ be the probability that standard Brownian motion has at least one zero in (r, t). Then,*

$$z_{r,t} = \frac{2}{\pi} \arccos\left(\sqrt{\frac{r}{t}}\right).$$

With $r = 0$, the result gives that standard Brownian motion has at least one zero in $(0, \epsilon)$ with probability

$$z_{0,\epsilon} = (2/\pi) \arccos(0) = (2/\pi)(\pi/2) = 1.$$

That is, $B_t = 0$, for some $0 < t < \epsilon$. By the strong Markov property, for Brownian motion restarted at t, there is at least one zero in (t, ϵ), with probability 1. Continuing in this way, there are infinitely many zeros in $(0, \epsilon)$.

The *arcsine distribution* arises in the proof of Theorem 8.1 and other results related to the zeros of Brownian motion. The distribution is a special case of the beta distribution, and has cumulative distribution function

$$F(t) = \frac{2}{\pi} \arcsin\left(\sqrt{t}\right), \text{ for } 0 \le t \le 1. \tag{8.9}$$

The arcsine density function is

$$f(t) = F'(t) = \frac{1}{\pi\sqrt{t(1-t)}}, \text{ for } 0 \le t \le 1.$$

The density is bimodal and symmetric, as shown in Figure 8.8.

Proof of Theorem 8.1. Conditioning on B_r,

$$\begin{aligned} z_{r,t} &= P(B_s = 0 \text{ for some } s \in (r,t)) \\ &= \int_{-\infty}^{\infty} P(B_s = 0 \text{ for some } s \in (r,t) | B_r = x) \frac{1}{\sqrt{2\pi r}} e^{-x^2/2r}\, dx. \end{aligned} \tag{8.10}$$

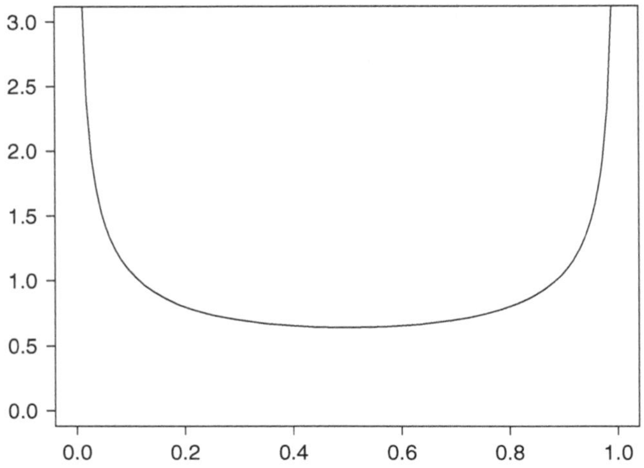

Figure 8.8 Arcsine distribution.

Assume that $B_r = x < 0$. The probability that $B_s = 0$ for some $s \in (r, t)$ is equal to the probability that for the process started in x, the maximum on $(0, t - r)$ is greater than 0. By translation, the latter is equal to the probability that for the process started in 0, the maximum on $(0, t - r)$ is greater than x. That is,

$$P(B_s = 0 \text{ for some } s \in (r, t) | B_r = x) = P(M_{t-r} > x).$$

For $x > 0$, consider the reflected process $-B_s$ started in $-x$. In either case, with Equations (8.7) and (8.10),

$$\begin{aligned} z_{r,t} &= \int_{-\infty}^{\infty} P(M_{t-r} > |x|) \frac{1}{\sqrt{2\pi r}} e^{-x^2/2r} \, dx \\ &= \int_{-\infty}^{\infty} \int_0^{t-r} \frac{1}{\sqrt{2\pi s^3}} |x| e^{-x^2/2s} \, ds \frac{1}{\sqrt{2\pi r}} e^{-x^2/2r} \, dx \\ &= \frac{1}{\pi} \int_0^{t-r} \frac{1}{\sqrt{rs^3}} \int_0^{\infty} x e^{-x^2(r+s)/2rs} \, dx \, ds \\ &= \frac{1}{\pi} \int_0^{t-r} \frac{1}{\sqrt{rs^3}} \int_0^{\infty} e^{-z(r+s)/rs} \, dz \, ds \\ &= \frac{1}{\pi} \int_0^{t-r} \frac{1}{\sqrt{rs^3}} \left(\frac{rs}{r+s} \right) ds \\ &= \frac{1}{\pi} \int_{r/t}^{1} \frac{1}{\sqrt{x(1-x)}} \, dx. \end{aligned}$$

The last equality is by the change of variables $r/(r+s) = x$. The last expression is an arcsine probability, which, by Equation (8.9), gives

$$z_{r,t} = \frac{2}{\pi}\left(\arcsin\left(\sqrt{1}\right) - \arcsin\left(\sqrt{\frac{r}{t}}\right)\right)$$
$$= 1 - \frac{2}{\pi}\arcsin\left(\sqrt{\frac{r}{t}}\right) = \frac{2}{\pi}\arccos\left(\sqrt{\frac{r}{t}}\right).$$

■

Last Zero Standing

Corollary 8.2. *Let L_t be the last zero in $(0, t)$. Then,*

$$P(L_t \le x) = \frac{2}{\pi}\arcsin\left(\sqrt{\frac{x}{t}}\right), \textit{ for } 0 < x < t.$$

Proof of corollary. The last zero occurs by time $x < t$ if and only if there are no zeros in (x, t). This occurs with probability

$$1 - z_{x,t} = 1 - \frac{2}{\pi}\arccos\left(\sqrt{\frac{x}{t}}\right) = \frac{2}{\pi}\arcsin\left(\sqrt{\frac{x}{t}}\right).$$

■

Example 8.9 (Fluctuations in Coin Tossing)

> We shall encounter theoretical conclusions which not only are unexpected but actually come as a shock to intuition and common sense. They will reveal that commonly accepted notions concerning chance fluctuations are without foundation and that the implications of the law of large numbers are widely misconstrued.
>
> —William Feller

Feller, the author of the quotation and of the classic and ground-breaking probability textbook *An Introduction to Probability Theory and Its Applications*, was discussing fluctuations in coin tossing and random walk. As a discrete process random walk is often studied with counting and combinatorial tools. Because of the connection between random walk and Brownian motion many discrete results can be obtained by taking suitable limits and invoking the invariance principle.

Consider a fair coin-tossing game between two players, A and B. If the coin lands heads, A pays B one dollar. If tails, B pays A. The coin is flipped $n = 10,000$ times. To test your "intuition and common sense," when would you guess is the *last* time the players are even? That is, at what point will the remaining duration of the game see one player always ahead?

Perhaps you think that in an evenly matched game there will be frequent changes in the lead and thus the last time the players are even is likely to be close to n, near the end of the game?

Let $\tilde{L}_n$ be the last time, in n plays, that the players are tied. This is the last zero for simple symmetric random walk on $\{0, 1, \dots, n\}$. Before continuing, we invite the reader to sketch their guesstimate of the distribution of $\tilde{L}_n$.

We simulated the coin-flipping game 5,000 times, generating the histogram of $\tilde{L}_n$ in Figure 8.9. The distribution is symmetric. It is equally likely that the last zero of the random walk is either k or $n - k$. Furthermore, the probabilities near the ends are the greatest. There is a surprisingly large probability of one player gaining the lead early in the game, and keeping the lead throughout. After just 20% of the game, there is almost 30% probability that one player will be in the lead for the remainder.

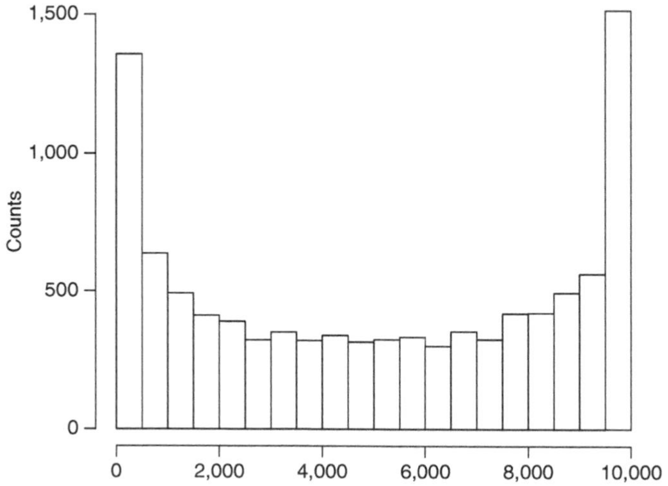

Figure 8.9 Last time players are tied in 10,000 coin flips.

It is no accident, of course, that the histogram bears a striking resemblance to the arcsine density curve in Figure 8.8. Let $0 < t < 1$. The probability that the last zero of the random walk occurs by step tn, that is, after $100t$ percent of the walk, is approximately the probability that the last zero of Brownian motion on $[0, 1]$ occurs by time t. For large n,

$$P\left(\tilde{L}_n \leq tn\right) \approx P(L_1 \leq t) = \frac{2}{\pi} \arcsin\left(\sqrt{t}\right).$$

Simulated probabilities for the random walk and theoretical values for Brownian motion are compared in Table 8.1.

TABLE 8.1 Random Walk and Brownian Motion Probabilities for the Last Zero ($n = 10,000$)

t	0.1	0.2	0.3	0.4	0.5	0.6	0.7	0.8	0.9
$P\left(\tilde{L}_n \leq tn\right)$	0.207	0.297	0.367	0.434	0.499	0.565	0.632	0.704	0.795
$P(L_1 \leq t)$	0.205	0.295	0.369	0.436	0.500	0.564	0.631	0.705	0.795

R: Random Walk and Coin Tossing

```
# coinflips.R
trials <- 10000
simlist <- numeric(trials)
for (i in 1:trials) {
 rw <- c(0,cumsum(sample(c(-1,1),(trials-1),
 replace=T)))
 simlist[i] <- tail(which(rw==0),1)
  }
mean(simlist)
hist(simlist,xlab="",ylab="Counts",main="")
```

■

8.5 VARIATIONS AND APPLICATIONS

Standard Brownian motion is often too simple a model for real-life applications. Many variations arise in practice. Brownian motion started in x has a constant mean function. A common variant of Brownian motion has linear mean function as well as an additional variance parameter.

Brownian Motion with Drift

For real μ and $\sigma > 0$, the process defined by

$$X_t = \mu t + \sigma B_t, \text{ for } t \geq 0,$$

is called *Brownian motion with drift parameter μ and variance parameter σ^2*.

Brownian motion with drift is a Gaussian process with continuous sample paths and independent and stationary increments. For $s, t > 0$, $X_{t+s} - X_t$ is normally distributed with mean μs and variance $\sigma^2 s$.

Example 8.10 Find the probability that Brownian motion with drift parameter $\mu = 0.6$ and variance $\sigma^2 = 0.25$ takes values between 1 and 3 at time $t = 4$.

Solution Write $X_t = (0.6)t + (0.5)B_t$. The desired probability is

$$P(1 \leq X_4 \leq 3) = P(1 \leq (0.6)4 + (0.5)B_4 \leq 3) = P(-2.8 \leq B_4 \leq 1.2)$$

$$= \int_{-2.8}^{1.2} \frac{1}{\sqrt{8\pi}} e^{-x^2/8} \, dx = 0.645.$$

■

Example 8.11 (Home team advantage) A novel application of Brownian motion to sports scores is given in Stern (1994). The goal is to quantify the home team advantage by finding the probability in a sports match that the home team wins the game given that they lead by l points after a fraction $0 \leq t \leq 1$ of the game is completed. The model is applied to basketball where scores can be reasonably approximated by a continuous distribution.

For $0 \leq t \leq 1$, let X_t denote the difference in scores between the home and visiting teams after $100t$ percent of the game has been completed. The process is modeled as a Brownian motion with drift, where the mean parameter μ is a measure of home team advantage. The probability that the home team wins the game, given that they have an l point lead at time $t < 1$, is

$$\begin{aligned} p(l,t) &= P(X_1 > 0 | X_t = l) = P(X_1 - X_t > -l) \\ &= P(X_{1-t} > -l) = P(\mu(1-t) + \sigma B_{1-t} > -l) \\ &= P\left(B_{1-t} < \frac{l + \mu(1-t)}{\sigma}\right) \\ &= P\left(B_t < \frac{\sqrt{t}[l + \mu(1-t)]}{\sigma\sqrt{1-t}}\right). \end{aligned}$$

The last equality is because B_t has the same distribution as $\sqrt{t/(1-t)}B_{1-t}$.

The model is applied to the results of 493 National Basketball Association games in 1992. Drift and variance parameters are fit from the available data with estimates $\hat{\mu} = 4.87$ and $\hat{\sigma} = 15.82$.

Table 8.2 gives the probability of a home team win for several values of l and t. Due to the home court advantage, the home team has a greater than 50% chance of winning even if it is behind by two points at halftime ($t = 0.50$). Even in the last

TABLE 8.2 Table for Basketball Data Probabilities $p(l,t)$ that the Home Team Wins the Game Given that they are in the Lead by l Points After a Fraction t of the Game is Completed

	Lead						
Time t	$l = -10$	$l = -5$	$l = -2$	$l = 0$	$l = 2$	$l = 5$	$l = 10$
0.00				0.62			
0.25	0.32	0.46	0.55	0.61	0.66	0.74	0.84
0.50	0.25	0.41	0.52	0.59	0.65	0.75	0.87
0.75	0.13	0.32	0.46	0.56	0.66	0.78	0.92
0.90	0.03	0.18	0.38	0.54	0.69	0.86	0.98
1.00	0.00	0.00	0.00		1.00	1.0	1.0

Source: Stern (1994).

five minutes of play ($t = 0.90$), home team comebacks from five points are not that unusual, according to the model, with probability 0.18.

We recommend this paper to the mathematically inclined sports fan. It is both accessible and readable. The author discusses model assumptions and limitations, the extent to which theoretical predictions follow empirical results, and an interesting extension from basketball to baseball. ■

Brownian Bridge

The two ends of a bridge are both secured to level ground. A Brownian bridge is a Brownian motion process conditioned so that the process ends at the same level as where it begins.

Brownian Bridge

From standard Brownian motion, the conditional process $(B_t)_{0\leq t\leq 1}$ given that $B_1 = 0$ is called a *Brownian bridge*. The Brownian bridge is *tied down* to 0 at the endpoints of [0, 1].

Examples of Brownian bridge are shown in Figure 8.10. Let $(X_t)_{t\geq 0}$ denote a Brownian bridge. For $0 \leq t \leq 1$, the distribution of X_t is equal to the conditional distribution of B_t given $B_1 = 0$. Since the conditional distributions of a Gaussian process are Gaussian, it follows that Brownian bridge is a Gaussian process. Continuity of sample paths, and independent and stationary increments are inherited from standard Brownian motion.

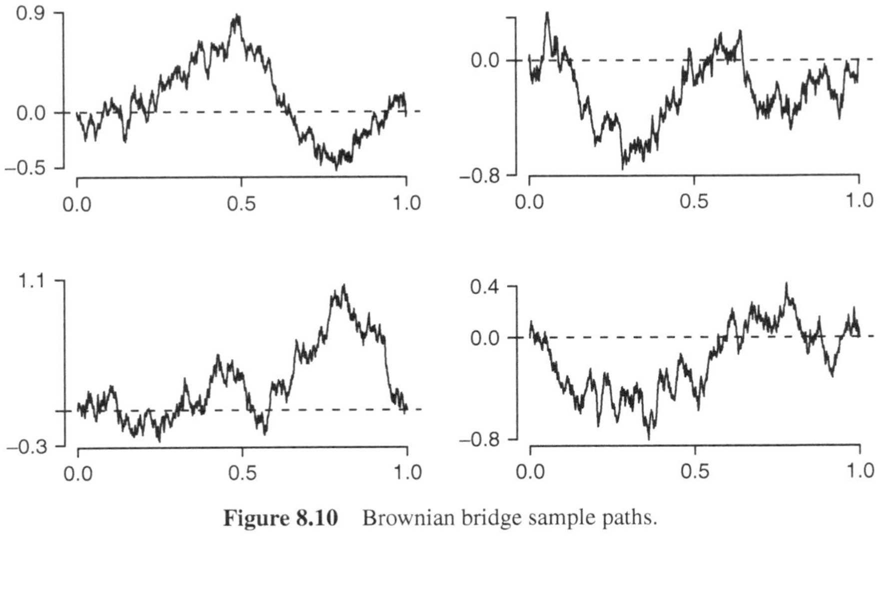

Figure 8.10 Brownian bridge sample paths.

To find the mean and covariance functions results are needed for bivariate normal distributions. We encourage the reader to work through Exercise 8.5(b) and show that for $0 < s < t$,

$$E(B_s|B_t = y) = \frac{sy}{t} \quad \text{and} \quad Var(B_s|B_t = y) = \frac{s(t-s)}{t}.$$

This gives the mean function of Brownian bridge

$$E(X_t) = E(B_t|B_1 = 0) = 0, \text{ for } 0 \le t \le 1.$$

For the covariance, $\text{Cov}(X_s, X_t) = E(X_sX_t)$. By the law of total expectation,

$$\begin{aligned} E(X_sX_t) &= E(E(X_sX_t)|X_t) = E(X_tE(X_s|X_t)) \\ &= E\left(X_t\frac{sX_t}{t}\right) = \frac{s}{t}E\left(X_t^2\right) = \frac{s}{t}E\left(B_t^2|B_1 = 0\right) \\ &= \frac{s}{t}Var(B_t|B_1 = 0) = \left(\frac{s}{t}\right)\frac{t(1-t)}{1} = s - st. \end{aligned}$$

By symmetry, for $t < s$, $E(X_sX_t) = t - st$. In either case, the covariance function is

$$\text{Cov}(X_s, X_t) = \min\{s, t\} - st.$$

Example 8.12 Let $X_t = B_t - tB_1$, for $0 \le t \le 1$. Show that $(X_t)_{0\le t\le 1}$ is a Brownian bridge.

Solution The process is a Gaussian process since $(B_t)_{t\ge 0}$ is a Gaussian process. Sample paths are continuous, with probability 1. It is suffice to show that the process has the same mean and covariance functions as a Brownian bridge.

The mean function is $E(X_t) = E(B_t - tB_1) = E(B_t) - tE(B_1) = 0$. The covariance function is

$$\begin{aligned} \text{Cov}(X_s, X_t) &= E(X_sX_t) = E((B_s - sB_1)(B_t - tB_1)) \\ &= E(B_sB_t) - tE(B_sB_1) - sE(B_tB_1) + stE(B_1^2) \\ &= \min\{s, t\} - ts - st + st = \min\{s, t\} - st, \end{aligned}$$

which is the covariance function of Brownian bridge.

The construction described in this example gives a direct method for simulating a Brownian bridge used to draw the graphs in Figure 8.10.

R: Simulating Brownian Bridge

```
# bbridge.R
> n <- 1000
> t <- seq(0,1,length=n)
```

```
> bm <- c(0,cumsum(rnorm(n-1,0,1)))/sqrt(n)
> bb <- bm - t*bm[n]
> plot(t,bb,type="l")
```

■

Example 8.13 (Animal tracking) In *Analyzing Animal Movements Using Brownian Bridges*, ecologists Horne et al. (2007) develop a Brownian bridge model for estimating the expected movement path of an animal. The model is based on a two-dimensional Brownian motion, where $Z_t^{a,b} = Z_t$ is defined to be the position in $\mathbb{R}^2$ of an animal at time $t \in [0, T]$, which starts at **a** and ends at **b**. Each Z_t is normally distributed with mean vector $E(Z_t) = \boldsymbol{a} + \frac{t}{T}(\boldsymbol{b} - \boldsymbol{a})$ and covariance matrix

$$\sigma_t^2 = \frac{t(T-t)}{T}\sigma_m^2 \boldsymbol{I},$$

where $\boldsymbol{I}$ is the identity matrix and σ_m^2 is a parameter related to the mobility of the animal. The probability that the animal is in region A at time t is $P(Z_t \in A)$. The model is applied to animal location data, often obtained through global positioning system telemetry, which allows for monitoring animal movements over great distances.

An objective of the researchers is to estimate the frequency of use of a region over the time of observation. Let $I_A(x)$ be the usual indicator function, which takes the value 1, if $x \in A$, and 0, otherwise. The *occupation time* for region A is defined as the random variable

$$\int_0^T I_A(Z_t)\, dt.$$

The expected fraction of time an animal occupies A is then

$$E\left(\frac{1}{T}\int_0^T I_A(Z_t)\, dt\right) = \frac{1}{T}\int_0^T P(Z_t \in A)\, dt,$$

a quantity used to estimate the home range of a male black bear in northern Idaho and the fall migration route of 11 caribou in Alaska.

The authors argue that the Brownian bridge movement model (BBMM) has the advantage over other methods in that BBMM assumes successive animal locations are not independent and explicitly incorporates the time between locations into the model. ■

Example 8.14 (Kolmogorov–Smirnov statistic) Following are 40 measurements, which take values between 0 and 1. We would like to test the claim that they are an i.i.d. sample from the uniform distribution on (0, 1). The Brownian bridge arises in the analysis of the *Kolmogorov–Smirnov test*, a common statistical method to test such claims.

0.100	0.296	0.212	0.385	0.993	0.870	0.070	0.815	0.123	0.588
0.332	0.035	0.758	0.362	0.453	0.047	0.134	0.389	0.147	0.424
0.060	0.003	0.800	0.011	0.085	0.674	0.196	0.715	0.342	0.519
0.675	0.799	0.768	0.721	0.315	0.009	0.109	0.835	0.044	0.152

Given a sample $X_1, \ldots, X_n$, define the *empirical distribution function*

$$F_n(t) = \frac{1}{n}\sum_{i=1}^{n} I_{\{X_i \le t\}},$$

where $I_{\{X_i \le t\}}$ is the indicator function equal to 1, if $X_i \le t$, and 0, otherwise. The empirical distribution function gives the proportion of values in the sample that are at most t. If the data are a sample from a population with cumulative distribution function F, then $F_n(t)$ is an estimate of $P(X_i \le t) = F(t)$.

If the data in our example is an i.i.d. sample from the uniform distribution on $(0, 1)$, then $F(t) = t$, for $0 \le t \le 1$, and we would expect $F_n(t) \approx t$. Figure 8.11 shows the empirical distribution function for these data plotted alongside the line $y = t$.

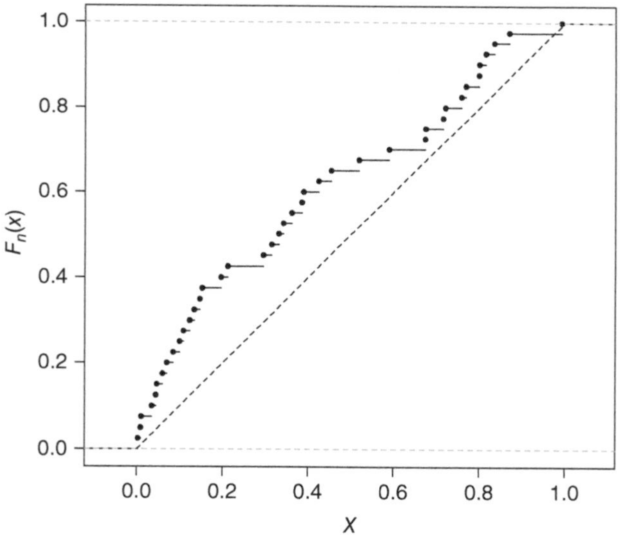

Figure 8.11 Empirical distribution function for sample data.

For a given cdf F, the *Kolmogorov–Smirnov statistic* is

$$D_n = \max_{0 \le t \le 1} |F_n(t) - F(t)|,$$

the maximum absolute distance between the empirical cdf and F. If the data are a sample from F, we expect $F_n(t) \approx F(t)$ and the value of D_n to be close to 0. Furthermore, large values of D_n are evidence against the hypothesis that the data are a sample from F. For the test of uniformity, $F(t) = t$, and

$$D_n = \max_{0 \le t \le 1} |F_n(t) - t|.$$

For our data, the Kolmogorov–Smirnov test statistic is $D_{40} = 0.223$. Is this large or small? Does it support or contradict the uniformity hypothesis? A statistician would ask: if the data do in fact come from a uniform distribution, what is the probability that D_{40} would be as large as 0.223? The probability $P(D_{40} > 0.223)$ is the *P-value* of the test. A small P-value means that it is unusual for D_{40} to be as large as 0.223, which would be evidence against uniformity. The distribution of D_n is difficult to obtain, which leads one to look for a good approximation.

If $X_1, \ldots, X_n$ is an i.i.d. sample from the uniform distribution on (0, 1), then for $0 < t < 1$,

$$F_n(t) - t = \sum_{i=1}^{n} \frac{I_{\{X_i \le t\}} - t}{n}$$

is a sum of i.i.d. random variables with common mean

$$E\left(\frac{I_{\{X_i \le t\}} - t}{n}\right) = \frac{P(X_i \le t) - t}{n} = \frac{t - t}{n} = 0$$

and variance

$$Var\left(\frac{I_{\{X_i \le t\}} - t}{n}\right) = \frac{1}{n^2} Var(I_{\{X_i \le t\}}) = \frac{P(X_i \le t)(1 - P(X_i \le t))}{n^2} = \frac{t(1-t)}{n^2}.$$

Thus, $F_n(t) - t$ has mean 0 and variance $t(1-t)/n$. For fixed $0 < t < 1$, the central limit theorem gives that $\sqrt{n}(F_n(t) - t)/\sqrt{t(1-t)}$ converges to a standard normal random variable, as $n \to \infty$. That is, for all real x,

$$\lim_{n \to \infty} P\left(\sqrt{n}(F_n(t) - t) \le x\right) = \int_{-\infty}^{x} \frac{1}{\sqrt{2\pi t(1-t)}} e^{-z^2/(2t(1-t))}\, dz = P(Y \le x),$$

where $Y \sim \text{Normal}(0, t(1-t))$.

By Donsker's invariance principle, it can be shown that the process $\sqrt{n}(F_n(t) - t)$, for $0 \le t \le 1$, converges to $B_t - tB_1$. The limiting process $(B_t - tB_1)_{0 \le t \le 1}$ is a Brownian bridge. The invariance principle further gives that

$$\sqrt{n} D_n = \sqrt{n} \max_{0 \le t \le 1} |F_n(t) - t|$$

converges to the maximum of a Brownian bridge. For large n, the distribution of D_n is approximately that of $M/\sqrt{n}$, where M is the maximum of a Brownian bridge.

We simulated the maximum of a Brownian bridge to find the P-value for our data, which is found to be

$$P\left(D_{40} > 0.223\right) \approx P(M/\sqrt{40} > 0.223) = P(M > 1.41) \approx 0.018.$$

The exact distribution of the maximum of a Brownian bridge is known. By using that distribution, the P-value is 0.0157. The P-value for the Kolmogorov–Smirnov test is obtained in R with the command ks.test.

The interpretation is that if the data were in fact uniformly distributed then the probability that D_{40} would be as large as 0.223 is less than 2%. Since the P-value is so small, we reject the claim and conclude that the data do not originate from a uniform distribution.

R: Test for Uniformity: Finding the *P*-value

```
# kstest.R
> trials <- 10000
> n <- 1000
> simlist <- numeric(trials)
> for (i in 1:trials) {
+ t <- seq(0,1,length=n)
# Brownian motion
+ bm <- c(0,cumsum(rnorm(n-1,0,1)))/sqrt(n)
+ bb <- bm-t*bm[n] # Brownian bridge
+ z <- max(bb) # maximum of Brownian bridge on [0,1]
+ simlist[i] <- if (z > 0.223*sqrt(40)) 1 else 0
}
> mean(simlist) # P-value = P(Z>1.41)
[1] 0.018
> ks.test(data,"punif",0,1)$p. value # exact P-value
[1] 0.015743
```

■

Geometric Brownian Motion

Geometric Brownian motion is a nonnegative process, which can be thought of as a stochastic model for exponential growth or decay. It is a favorite tool in mathematical finance, where it is used extensively to model stock prices.

Geometric Brownian Motion

Let $(X_t)_{t\geq 0}$ be a Brownian motion with drift parameter μ and variance parameter σ^2. The process $(G_t)_{t\geq 0}$ defined by

$$G_t = G_0 e^{X_t}, \text{ for } t \geq 0,$$

where $G_0 > 0$, is called *geometric Brownian motion.*

Taking logarithms, we see that $\ln G_t = \ln G_0 + X_t$ is normally distributed with mean

$$E(\ln G_t) = E(\ln G_0 + X_t) = \ln G_0 + \mu t$$

and variance

$$Var(\ln G_t) = Var(\ln G_0 + X_t) = Var(X_t) = \sigma^2 t.$$

A random variable whose logarithm is normally distributed is said to have a *lognormal distribution*. For each $t > 0$, G_t has a lognormal distribution.

We leave as an exercise the derivation of mean and variance for geometric Brownian motion

$$E(G_t) = G_0 e^{t(\mu+\sigma^2/2)} \quad \text{and} \quad Var(G_t) = G_0^2 e^{2t(\mu+\sigma^2/2)}(e^{t\sigma^2} - 1). \tag{8.11}$$

The exponential mean function shows that, on average, geometric Brownian motion exhibits exponential growth with growth rate $\mu + \sigma^2/2$.

Geometric Brownian motion arises as a model for quantities which can be expressed as the *product* of independent random multipliers. For $s, t \geq 0$, consider the ratio

$$\frac{G_{t+s}}{G_t} = \frac{G_0 e^{\mu(t+s)+\sigma X_{t+s}}}{G_0 e^{\mu t+\sigma X_t}} = e^{\mu s+\sigma(X_{t+s}-X_t)},$$

which has the same distribution as $e^{\mu s+\sigma X_s} = G_s/G_0$, because of stationary increments for the $(X_t)_{t\geq 0}$ process. For $0 \leq q < r \leq s < t$,

$$\frac{G_t}{G_s} = e^{\mu(t-s)+\sigma(X_t-X_s)} \quad \text{and} \quad \frac{G_r}{G_q} = e^{\mu(r-q)+\sigma(X_r-X_q)}$$

are independent random variables, because of independent increments for $(X_t)_{t\geq 0}$.

Let $Y_k = G_k/G_{k-1}$, for $k = 1, 2, \ldots$ Then, $Y_1, Y_2, \ldots$ is an i.i.d. sequence, and

$$G_n = \left(\frac{G_n}{G_{n-1}}\right)\left(\frac{G_{n-1}}{G_{n-2}}\right)\cdots\left(\frac{G_2}{G_1}\right)\left(\frac{G_1}{G_0}\right)G_0 = G_0 Y_1 Y_2 \cdots Y_{n-1} Y_n.$$

Example 8.15 Stock prices are commonly modeled with geometric Brownian motion. The process is attractive to economists because of several assumptions.

Historical data for many stocks indicate long-term exponential growth or decline. Prices cannot be negative and geometric Brownian motion takes only positive values. Let Y_t denote the price of a stock after t days. Since the price on a given day is probably close to the price on the next day (assuming normal market conditions), stock prices are not independent. However, the *percent changes in price* from day to day Y_t/Y_{t-1}, for $t = 1, 2, \ldots$ might be reasonably modeled as independent and identically distributed. This leads to geometric Brownian motion. In the context of stock prices, the standard deviation σ is called the *volatility*.

A criticism of the geometric Brownian motion model is that it does not account for extreme events like the stock market crash on October 19, 1987, when the world's stock markets lost more than 20% of their value within a few hours.

Assume that XYZ stock currently sells for \$80 a share and follows a geometric Brownian motion with drift parameter 0.10 and volatility 0.50. Find the probability that in 90 days the price of XYZ will rise to at least \$100.

Solution Let Y_t denote the price of XYZ after t years. Round 90 days as 1/4 of a year. Then,

$$\begin{aligned} P(Y_{0.25} \geq 100) &= P\left(80e^{\mu(0.25)+\sigma B_{0.25}} \geq 100\right) \\ &= P((0.1)(0.25) + (0.5)B_{0.25} \geq \ln 1.25) \\ &= P(B_{0.25} \geq 0.396) = 0.214. \end{aligned}$$

In R, type

```
> x <- (log(100/80)-(0.1)/4)/0.5
> 1-pnorm(x,0,sqrt(1/4))
[1] 0.214013
```

Simulations of the stock price are shown in Figure 8.12. ■

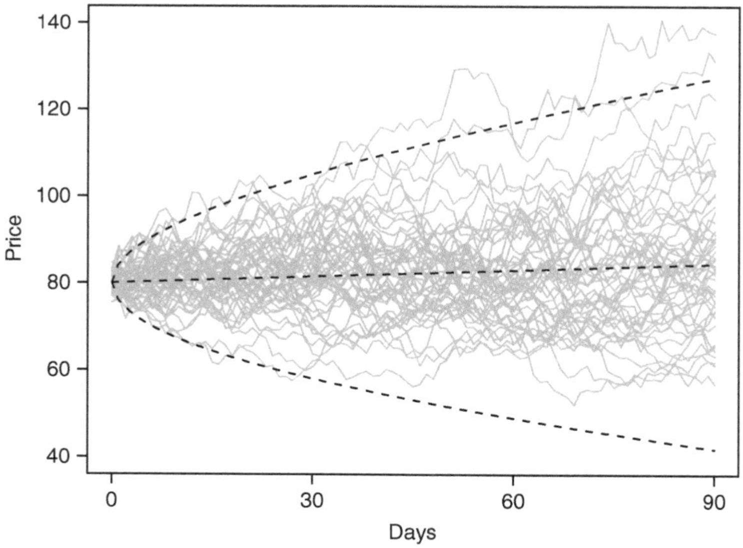

Figure 8.12 Fifty sample paths of a stock's price over 90 days modeled as geometric Brownian motion. Dotted lines are drawn at the mean function, and the mean plus or minus two standard deviations.

Example 8.16 (Financial options) An *option* is a contract that gives the buyer the right to buy shares of stock sometime in the future at a fixed price. In Example 8.13, we assumed that XYZ stock currently sells for \$80 a share. Assume that an XYZ option is selling for \$10. Under the terms of the option, in 90 days you may buy a share of XYZ stock for \$100.

If you decide to purchase the option, consider the payoff. Assume that in 90 days the price of XYZ is greater than \$100. Then, you can *exercise* the option, buy the stock for \$100, and turn around and sell XYZ for its current price. Your payoff would be $G_{90/365} - 100$, where $G_{90/365}$ is the price of XYZ in 90 days.

On the other hand, if XYZ sells for less than \$100 in 90 days, you would not exercise the option, and your payoff is nil. In either case, the payoff in 90 days is $\max\{G_{90/365} - 100, 0\}$. Your final profit would be the payoff minus the initial \$10 cost of the option.

Find the future expected payoff of the option, assuming the price of XYZ follows a geometric Brownian motion.

Solution Let G_0 denote the current stock price. Let t be the *expiration date*, which is the time until the option is exercised. Let K be the *strike price*, which is how much you can buy the stock for if you exercise the option. For XYZ, $G_0 = 80$, $t = 90/365$ (measuring time in years), and $K = 100$.

The goal is to find the expected payoff $E(\max\{G_t - K, 0\})$, assuming $(G_t)_{t\geq 0}$ is a geometric Brownian motion. Let $f(x)$ be the density function of a normal distribution with mean 0 and variance t. Then,

$$\begin{aligned}
E(\max\{G_t - K, 0\}) &= E\left(\max\{G_0 e^{\mu t+\sigma B_t} - K, 0\}\right) \\
&= \int_{-\infty}^{\infty} \max\{G_0 e^{\mu t+\sigma x} - K, 0\} f(x)\, dx \\
&= \int_{\beta}^{\infty} (G_0 e^{\mu t+\sigma x} - K) f(x)\, dx \\
&= G_0 e^{\mu t} \int_{\beta}^{\infty} e^{\sigma x} f(x)\, dx - KP\left(Z > \frac{\beta}{\sqrt{t}}\right),
\end{aligned}$$

where $\beta = \left(\ln\,(K/G_0) - \mu t\right)/\sigma$, and Z is a standard normal random variable.

By completing the square, the integral in the last expression is

$$\begin{aligned}
\int_{\beta}^{\infty} e^{\sigma x} f(x)\, dx &= \int_{\beta}^{\infty} e^{\sigma x} \frac{1}{\sqrt{2\pi t}} e^{-x^2/2t}\, dx \\
&= e^{\sigma^2 t/2} \int_{\beta}^{\infty} \frac{1}{\sqrt{2\pi t}} e^{-(x-\sigma t)^2/2t}\, dx
\end{aligned}$$

$$= e^{\sigma^2 t/2} \int_{(\beta-\sigma t)/\sqrt{t}}^{\infty} \frac{1}{\sqrt{2\pi}} e^{-x^2/2}\, dx$$

$$= e^{\sigma^2 t/2} P\left(Z > \frac{\beta - \sigma t}{\sqrt{t}} \right).$$

This gives

$$E(\max\{G_t - K, 0\})$$
$$= G_0 e^{t(\mu+\sigma^2/2)} P\left(Z > \frac{\beta - \sigma t}{\sqrt{t}} \right) - KP\left(Z > \frac{\beta}{\sqrt{t}} \right). \quad (8.12)$$

Assume that XYZ stock follows a geometric Brownian motion with drift parameter $\mu = 0.10$ and variance $\sigma^2 = 0.25$. Then,

$$\beta = \frac{\ln\ (K/G_0) - \mu t}{\sigma} = \frac{\ln\ (100/80) - (0.10)(90/365)}{0.5} = 0.3970,$$

which gives

$$E(\max\{G_{90/365} - 100, 0\})$$
$$= 80e^{(90/365)(0.10+0.25/2)} P\left(Z > \frac{0.397 - 0.5(90/365)}{\sqrt{90/365}} \right)$$
$$- 100P\left(Z > \frac{0.397}{\sqrt{90/365}} \right)$$
$$= 1.788.$$

Given that the initial cost of the option is \$10, your expected profit is $1.788 - 10 < 0$. So you can expect to lose money.

For this example, we set an arbitrary initial cost of the option. A fundamental question in finance is how such an option should be priced. This leads to the Black–Scholes model for option pricing, which is introduced in the next section. ■

8.6 MARTINGALES

A martingale is a stochastic process that generalizes the notion of a *fair game*. Assume that after n plays of a gambling game your winnings are x. Then, by *fair*, we mean that your expected future winnings should be x regardless of past history.

Martingale

A stochastic process $(Y_t)_{t\geq 0}$ is a *martingale*, if for all $t \geq 0$,

1. $E(Y_t|Y_r, 0 \leq r \leq s) = Y_s$, for all $0 \leq s \leq t$.
2. $E(|Y_t|) < \infty$.

A discrete-time martingale $Y_0, Y_1, \dots$ satisfies

1. $E(Y_{n+1}|Y_0, \dots, Y_n) = Y_n$, for all $n \geq 0$.
2. $E(|Y_n|) < \infty$.

A most important property of martingales is that they have constant expectation. By the law of total expectation,

$$E(Y_t) = E(E(Y_t|Y_r, 0 \leq r \leq s)) = E(Y_s),$$

for all $0 \leq s \leq t$. That is,

$$E(Y_t) = E(Y_0), \text{ for all } t.$$

Example 8.17 (Random walk) Show that simple symmetric random walk is a martingale.

Solution Let

$$X_i = \begin{cases} +1, & \text{with probability } 1/2, \\ -1, & \text{with probability } 1/2, \end{cases}$$

for $i = 1, 2, \dots$ For $n \geq 1$, let $S_n = X_1 + \cdots + X_n$, with $S_0 = 0$. Then,

$$\begin{aligned} E(S_{n+1}|S_0, \dots, S_n) &= E(X_{n+1} + S_n|S_0, \dots, S_n) \\ &= E(X_{n+1}|S_0, \dots, S_n) + E(S_n|S_0, \dots, S_n) \\ &= E(X_{n+1}) + S_n = S_n. \end{aligned}$$

The third equality is because X_{n+1} is independent of $X_1, \dots, X_n$, and thus independent of $S_0, S_1, \dots, S_n$. The fact that $E(S_n|S_0, \dots, S_n) = S_n$ is a consequence of a general property of conditional expectation, which states that if X is a random variable and g is a function, then $E(g(X)|X) = g(X)$.

The second part of the martingale definition is satisfied as

$$E(|S_n|) = E\left(\left|\sum_{i=1}^{n} X_i\right|\right) \leq E\left(\sum_{i=1}^{n} |X_i|\right) = \sum_{i=1}^{n} E(|X_i|) = n < \infty.$$

■

Since simple symmetric random walk is a martingale, the next example should not be surprising.

Example 8.18 (Brownian motion) Show that standard Brownian motion $(B_t)_{t\geq 0}$ is a martingale.

Solution We have that

$$\begin{aligned} E(B_t|B_r, 0 \leq r \leq s) &= E(B_t - B_s + B_s|B_r, 0 \leq r \leq s) \\ &= E(B_t - B_s|B_r, 0 \leq r \leq s) + E(B_s|B_r, 0 \leq r \leq s) \\ &= E(B_t - B_s) + B_s = B_s, \end{aligned}$$

where the second equality is because of independent increments. Also,

$$E(|B_t|) = \int_{-\infty}^{\infty} |x| \frac{1}{\sqrt{2\pi t}} e^{-x^2/2t}\, dx = \int_0^{\infty} x \sqrt{\frac{2}{\pi t}} e^{-x^2/2t}\, dx = \sqrt{\frac{2t}{\pi}} < \infty.$$

■

The following extension of the martingale definition is used frequently.

Martingale with Respect to Another Process

Let $(X_t)_{t\geq 0}$ and $(Y_t)_{t\geq 0}$ be stochastic processes. Then, $(Y_t)_{t\geq 0}$ is a *martingale with respect to* $(X_t)_{t\geq 0}$, if for all $t \geq 0$,

1. $E(Y_t|X_r, 0 \leq r \leq s) = Y_s,$ for all $0 \leq s \leq t$.
2. $E(|Y_t|) < \infty$.

The most common application of this is when Y_t is a function of X_t. That is, $Y_t = g(X_t)$ for some function g. It is useful to think of the conditioning random variables $(X_r)_{0\leq r\leq s}$ as representing past information, or history, of the process up to time s.

Following are several examples of martingales that are functions of Brownian motion.

Example 8.19 (Quadratic martingale) Let $Y_t = B_t^2 - t$, for $t \geq 0$. Show that $(Y_t)_{t\geq 0}$ is a martingale with respect to Brownian motion. This is called the *quadratic martingale*.

Solution For $0 \le s < t$,

$$\begin{aligned}
E(Y_t|B_r, 0 \le r \le s) &= E\left(B_t^2 - t|B_r, 0 \le r \le s\right) \\
&= E\left((B_t - B_s + B_s)^2|B_r, 0 \le r \le s\right) - t \\
&= E\left((B_t - B_s)^2 + 2(B_t - B_s)B_s + B_s^2|B_r, 0 \le r \le s\right) - t \\
&= E\left((B_t - B_s)^2\right) + 2B_sE(B_t - B_s) + B_s^2 - t \\
&= (t - s) + B_s^2 - t = B_s^2 - s = Y_s.
\end{aligned}$$

Furthermore,

$$E(|Y_t|) = E\left(|B_t^2 - t|\right) \le E\left(B_t^2 + t\right) = E\left(B_t^2\right) + t = 2t < \infty.$$

■

Example 8.20 Let $G_t = G_0e^{X_t}$ be a geometric Brownian motion, where $(X_t)_{t\ge 0}$ is Brownian motion with drift μ and variance σ^2. Let $r = \mu + \sigma^2/2$. Show that $e^{-rt}G_t$ is a martingale with respect to standard Brownian motion.

Solution For $0 \le s < t$,

$$\begin{aligned}
E\left(e^{-rt}G_t|B_r, 0 \le r \le s\right) &= e^{-rt}E\left(G_0e^{\mu t+\sigma B_t}|B_r, 0 \le r \le s\right) \\
&= e^{-rt}E\left(G_0e^{\mu(t-s)+\sigma(B_t-B_s)}e^{\mu s+\sigma B_s}|B_r, 0 \le r \le s\right) \\
&= e^{-rt}e^{\mu s+\sigma B_s}E\left(G_0e^{\mu(t-s)+\sigma(B_t-B_s)}\right) \\
&= e^{-rt}e^{\mu s+\sigma B_s}E(G_{t-s}) \\
&= e^{-t(\mu+\sigma^2/2)}e^{\mu s+\sigma B_s}G_0e^{(t-s)(\mu+\sigma^2/2)} \\
&= e^{-s(\mu+\sigma^2/2)}G_0e^{\mu s+\sigma B_s} \\
&= e^{-rs}G_s.
\end{aligned}$$

Also,

$$E(|e^{-rt}G_t|) = e^{-rt}E(G_t) = e^{-t(\mu+\sigma^2/2)}G_0e^{t(\mu+\sigma^2/2)} = G_0 < \infty, \text{ for all } t.$$

■

Example 8.21 (Black–Scholes) In Example 8.16, the expected payoff for a financial option was derived. This leads to the *Black–Scholes formula* for pricing options, a fundamental formula in mathematical finance.

The formula was first published by Fisher Black and Myron Scholes in 1973 and then further developed by Robert Merton. Merton and Scholes received the 1997 Nobel Prize in Economics for their work. The ability to price options and other financial *derivatives* opened up a massive global market for trading ever-more complicated financial instruments. Black–Scholes has been described both as a formula which "changed the world" and as "the mathematical equation that caused the banks to crash." See Stewart (2012).

There are several critical assumptions underlying the Black–Scholes formula. One is that stock prices follow a geometric Brownian motion. Another is that an investment should be *risk neutral*. What this means is that the expected return on an investment should be equal to the so-called risk-free rate of return, such as what is obtained by a short-term U.S. government bond.

Let r denote the risk-free interest rate. Starting with P dollars, because of compound interest, after t years of investing risk free your money will grow to $P(1+r)^t$ dollars. Under continuous compounding, the future value is $F = Pe^{rt}$. This gives the future value of your present dollars. On the other hand, assume that t years from now you will be given F dollars. To find its *present value* requires *discounting* the future amount by a factor of e^{-rt}. That is, $P = e^{-rt}F$.

Let G_t denote the price of a stock t years from today. Then, the present value of the stock price is $e^{-rt}G_t$. The Black–Scholes risk neutral assumption means that the discounted stock price process is a *fair game*, that is, a martingale. For any time $0 < s < t$, the expected present value of the stock t years from now given knowledge of the stock price up until time s should be equal to the present value of the stock price s years from now. In other words,

$$E(e^{-rt}G_t | G_x, 0 \le x \le s) = e^{-rs}G_s. \tag{8.13}$$

In Example 8.20, it was shown that Equation (8.13) holds for geometric Brownian motion if $r = \mu + \sigma^2/2$, or $\mu = r - \sigma^2/2$. The probability calculations for the Black–Scholes formula are obtained with this choice of μ. In the language of Black–Scholes, this gives the *risk-neutral probability* for computing the options price formula. The Black–Scholes formula for the price of an option is then the present value of the expected payoff of the option under the risk-neutral probability.

See Equation (8.12) for the future expected payoff of a financial option. The present value of the expected payoff is obtained by multiplying by the discount factor e^{-rt}. Replace μ with $r - \sigma^2$ to obtain the Black–Scholes option price formula

$$e^{-rt}E(\max\{G_t - K, 0\}) = G_0 P\left(Z > \frac{\alpha - \sigma t}{\sqrt{t}}\right) - e^{-rt}KP\left(Z > \frac{\alpha}{\sqrt{t}}\right),$$

where

$$\alpha = \frac{\ln\,(K/G_0) - (r - \sigma^2/2)t}{\sigma}.$$

For the XYZ stock example, $G_0 = 80$, $K = 100$, $\sigma^2 = 0.25$, and $t = 90/365$. Furthermore, assume $r = 0.02$ is the risk-free interest rate. Then,

$$\alpha = \frac{\ln\,(100/80) - (0.02 - 0.25/2)(90/365)}{0.5} = 0.498068,$$

and the Black–Scholes option price is

$$80P\left(Z > \frac{\alpha - 0.50(90/365)}{0.5}\right) - e^{-0.02(90/365)}(100)P\left(Z > \frac{\alpha}{0.5}\right) = \$2.426.$$

Remarks:

1. For a given expiration date and strike price, the Black–Scholes formula depends only on volatility σ and the risk-free interest rate r, and not on the drift parameter μ. In practice, volatility is often estimated from historical price data. The model assumes that σ and r are constant.
2. The original derivation of the Black–Scholes formula in the paper by Black and Scholes was based on deriving and solving a partial differential equation

$$\frac{\partial V}{\partial t} + \frac{\sigma^2 S^2}{2}\frac{\partial^2 V}{\partial S^2} + rS\frac{\partial V}{\partial S} = rV,$$

where $V = V(S, t)$ is the value of the option, a function of stock price S and time t. This is known as the *Black–Scholes equation*, and can be reduced to the simpler heat equation.
3. The Black–Scholes formula in the example is for a European call option, in which the option can only be exercised on the expiration date. In an American call option, one can exercise the option at any time before the expiration date. There are many types of options and financial derivatives whose pricing structure is based on Black–Scholes. ■

Optional Stopping Theorem

A martingale $(Y_t)_{t\geq 0}$ has constant expectation. For all $t \geq 0$, $E(Y_t) = E(Y_0)$. The property holds for all fixed, deterministic times, but not necessarily for *random times*. If T is a random variable, which takes values in the index set of a martingale, it is not necessarily true that $E(Y_T) = E(Y_0)$. For instance, let T be the first time that a standard Brownian motion hits level a. Then, $B_T = a = E(B_T)$. However, $E(B_t) = 0$, for all $t \geq 0$.

The optional stopping theorem gives conditions for when a random time T satisfies $E(Y_T) = E(Y_0)$. While that might not sound like a big deal, the theorem is remarkably powerful. As mathematicians David Aldous and Jim Fill write in *Reversible Markov chains and random walks on graphs*, "Modern probabilists regard the martingale optional stopping theorem as one of the most important results in their subject." (Aldous and Fill, 2002.)

In Section 3.9, we introduced the notion of a stopping time in the context of discrete-time Markov chains. For a stochastic process $(Y_t)_{t\geq 0}$, a nonnegative random variable T is a stopping time if for each t, the event $\{T \leq t\}$ can be determined from $\{Y_s, 0 \leq s \leq t\}$. That is, if the outcomes of Y_s are known for $0 \leq s \leq t$, then it can be determined whether or not $\{T \leq t\}$ occurs.

On the interval [0, 1], the first time Brownian motion hits level a is a stopping time. Whether or not a is first hit by time t can be determined from $\{B_s, 0 \le s \le t.\}$ On the other hand, the *last time* a is hit on [0, 1] is not a stopping time. To determine whether or not a was last hit by time $t < 1$ requires full knowledge of B_s, for all $s \in [0, 1]$.

Optional Stopping Theorem

Let $(Y_t)_{t\ge 0}$ be a martingale with respect to a stochastic process $(X_t)_{t\ge 0}$. Assume that T is a stopping time for $(X_t)_{t\ge 0}$. Then, $E(Y_T) = E(Y_0)$ if one of the following is satisfied.

1. T is bounded. That is, $T \le c$, for some constant c.
2. $P(T < \infty) = 1$ and $E(|Y_t|) \le c$, for some constant c, whenever $T > t$.

The proof of the optional stopping theorem is beyond the scope of this book. There are several versions of the theorem with alternate sets of conditions.

In the context of a fair gambling game, one can interpret the optional stopping theorem to mean that a gambler has no reasonable strategy for increasing their initial stake. Let Y_t denote the gambler's winnings at time t. If a gambler strategizes to stop play at time T, their expected winnings will be $E(Y_T) = E(Y_0) = Y_0$, the gambler's initial stake.

The word *martingale* has its origins in an 18th century French gambling strategy. Mansuy (2009) quotes the dictionary of the Académie Française, that "To play the martingale is to always bet all that was lost."

Consider such a betting strategy for a game where at each round a gambler can win or lose one franc with equal probability. After one turn, if you win you stop. Otherwise, bet 2 francs on the next round. If you win, stop. If you lose, bet 4 francs on the next round. And so on. Let T be the number of bets needed until you win. The random variable T is a stopping time. If the gambler wins their game after k plays, they gain

$$2^k - (1 + 2 + \cdots + 2^{k-1}) = 2^k - (2^k - 1) = 1 \text{ franc},$$

and, with probability 1, the gambler will eventually win some bet. Thus, T seems to be a winning strategy.

However, if Y_n denotes the gambler's winnings after n plays, then $E(Y_T) = 1 \ne 0 = E(Y_0)$. The random variable T does not satisfy the conditions of the optional sampling theorem. The reason this martingale gambling strategy is not *reasonable* is that it assumes the gambler has infinite capital.

The optional stopping theorem gives elegant and sometimes remarkably simple solutions to seemingly difficult problems.

Example 8.22 Let $a, b > 0$. For a standard Brownian motion, find the probability that the process hits level a before hitting level $-b$.

Solution Let p be the desired probability. Consider the time T that Brownian motion first hits either a or $-b$. That is, $T = \min\{t : B_t = a \text{ or } B_t = -b\}$. See Figure 8.13.

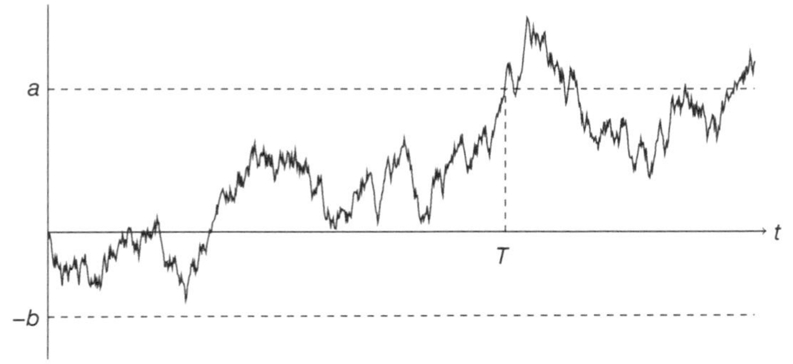

Figure 8.13 First hitting time T that Brownian motion hits level a or $-b$.

The random variable T is a stopping time. Furthermore, it satisfies the conditions of the optional stopping theorem. From Equation (8.6), the first hitting time T_a is finite with probability 1. By the strong Markov property, from a, the first time to hit $-b$ is also finite with probability 1. Thus, the first part of condition 2 is satisfied. Furthermore, for $t < T$, $B_t \in (-b, a)$. Thus, $|B_t| < \max\{a, b\}$, and the second part of condition 2 is satisfied.

Observe that $B_T = a$, with probability p, and $B_T = -b$, with probability $1 - p$. By the optional stopping theorem

$$0 = E(B_0) = E(B_T) = pa + (1 - p)(-b).$$

Solving for p gives $p = b/(a + b)$. ■

Example 8.23 (Expected hitting time) Apply the optional stopping theorem with the same stopping time as in Example 8.22, but with the quadratic martingale $B_t^2 - t$. This gives

$$E\left(B_T^2 - T\right) = E\left(B_0^2 - 0\right) = 0,$$

from which it follows that

$$E(T) = E\left(B_T^2\right) = a^2\left(\frac{b}{a+b}\right) + b^2\left(\frac{a}{a+b}\right) = ab.$$

We have thus discovered that the expected time that standard Brownian motion first hits the boundary of the region defined by the lines $y = a$ and $y = -b$ is ab. ■

Example 8.24 (Gambler's ruin) It was shown that discrete-time simple symmetric random walk $S_0, S_1, \ldots$ is a martingale. As with the quadratic martingale, the process $S_n^2 - n$ is a martingale. The results from the last two examples for Brownian motion can be restated for gambler's ruin on $\{-b, -b+1, \ldots, 0, \ldots, a-1, a\}$ starting at the origin. This gives the following:

1. The probability that the gambler gains a before losing b is $b/(a+b)$.
2. The expected duration of the game is ab. ■

Example 8.25 (Time to first hit the line $y = a - bt$) For $a, b > 0$, let $T = \min\{t : B_t = a - bt\}$ be the first time a standard Brownian motion hits the line $y = a - bt$. The random variable T is a stopping time, which satisfies the optional stopping theorem. This gives

$$0 = E(B_0) = E(B_T) = E(a - bT) = a - bE(T).$$

Hence, $E(T) = a/b$. For the line $y = 4 - (0.5)t$ in Figure 8.14, the mean time to first hit the line is $4/(0.5) = 8$. ■

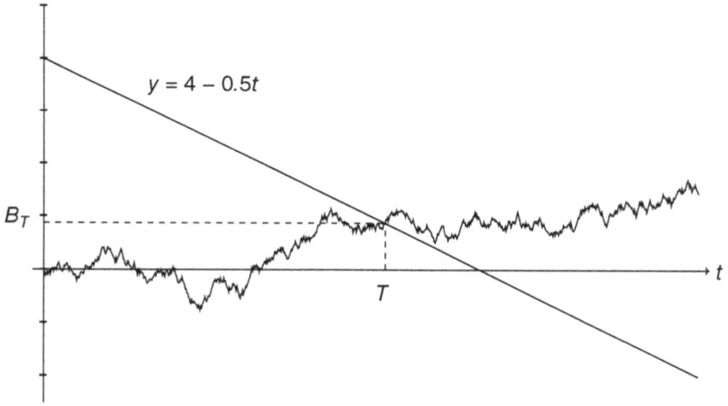

Figure 8.14 First time T Brownian motion hits the line $y = 4 - 0.5t$.

Example 8.26 (Time to reach a for Brownian motion with drift) Assume that $(X_t)_{t\geq 0}$ is a Brownian motion process with drift parameter μ and variance parameter σ^2, where $\mu > 0$. For $a > 0$, find the expected time that the process first hits level a.

Solution Let $T = \min\{t : X_t = a\}$ be the first time that the process hits level a. Write $X_t = \mu t + \sigma B_t$. Then, $X_t = a$ if and only if

$$B_t = \frac{a - \mu t}{\sigma} = \frac{a}{\sigma} - \left(\frac{\mu}{\sigma}\right) t.$$

Applying the result of Example 8.25, $E(T) = (a/\sigma)/(\mu/\sigma) = a/\mu$. ■

Example 8.27 (Variance of first hitting time) Assume that $(X_t)_{t\geq 0}$ is a Brownian motion process with drift μ and variance σ^2. Let $T = \min\{t : X_t = a\}$ be the first hitting time to reach level a. In the last example, the expectation of T was derived. Here, the variance of T is obtained using the quadratic martingale $Y_t = B_t^2 - t$.

Solution Since T is a stopping time with respect to B_t,

$$0 = E(Y_0) = E(Y_T) = E\left(B_T^2 - T\right) = E\left(B_T^2\right) - E(T).$$

Thus, $E\left(B_T^2\right) = E(T) = a/\mu$. Write $X_t = \mu t + \sigma B_t$. Then, $X_T = \mu T + \sigma B_T$, giving

$$B_T = \frac{X_T - \mu T}{\sigma} = \frac{a - \mu T}{\sigma}.$$

Thus,

$$\begin{aligned} Var(T) &= E\left((T - E(T))^2\right) = E\left(\left(T - \frac{a}{\mu}\right)^2\right) \\ &= \frac{1}{\mu^2}E\left((\mu T - a)^2\right) = \frac{\sigma^2}{\mu^2}E\left(\left(\frac{a - \mu T}{\sigma}\right)^2\right) \\ &= \frac{\sigma^2}{\mu^2}E\left(B_T^2\right) = \frac{\sigma^2}{\mu^2}\left(\frac{a}{\mu}\right) = \frac{a\sigma^2}{\mu^3}. \end{aligned}$$

■

R: Hitting Time Simulation for Brownian Motion with Drift

Let $(X_t)_{t\geq 0}$ be a Brownian motion with drift $\mu = 0.5$ and variance $\sigma^2 = 1$. Consider the first hitting time T of level $a = 10$. Exact results are

$$E(T) = a/\mu = 20 \quad \text{and } Var(T) = a\sigma^2/\mu^3 = 80.$$

Simulated results are based on 1,000 trials of a Brownian motion process on $[0, 80]$. See Figure 8.15 for several realizations.

```
# bmhitting.R
> mu <- 1/2
> sig <- 1
> a <- 10
> simlist <- numeric(1000)
> for (i in 1:1000) {
    t <- 80
    n <- 50000
    bm <- c(0,cumsum(rnorm(n,0,sqrt(t/n))))
```

```
     xproc <- mu*seq(0,t,t/n) + sig*bm
     simlist[i] <- which(xproc >= a)[1] * (t/n) }
> mean(simlist)
[1] 20.07139
> var(simlist)
[1] 81.31142
```

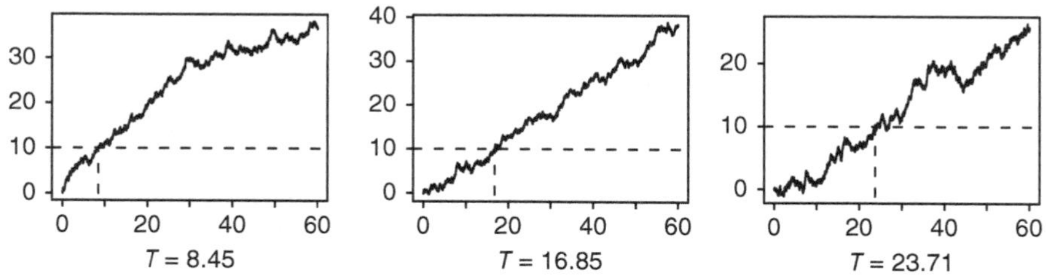

Figure 8.15 First time T to hit $a = 10$ for Brownian motion with drift $\mu = 0.5$ and variance $\sigma^2 = 1$.

EXERCISES

8.1 Show that

$$f(x,t) = \frac{1}{\sqrt{2\pi t}} e^{-x^2/(2t)},$$

satisfies the partial differential heat equation

$$\frac{\partial f}{\partial t} = \frac{1}{2} \frac{\partial^2 f}{\partial x^2}.$$

8.2 For standard Brownian motion, find

(a) $P(B_2 \leq 1)$

(b) $E(B_4 | B_1 = x)$

(c) $\text{Corr}(B_{t+s}, B_s)$

(d) $Var(B_4 | B_1)$

(e) $P(B_3 \leq 5 | B_1 = 2)$.

8.3 For standard Brownian motion started at $x = -3$, find

(a) $P(X_1 + X_2 > -1)$

(b) The conditional density of X_2 given $X_1 = 0$.

(c) $\text{Cov}(X_3, X_4)$

(d) $E(X_4 | X_1)$.

8.4 In a race between Lisa and Cooper, let X_t denote the amount of time (in seconds) by which Lisa is ahead when $100t$ percent of the race has been completed. Assume that $(X_t)_{0\le t\le 1}$ can be modeled by a Brownian motion with drift parameter 0 and variance parameter σ^2. If Lisa is leading by $\sigma/2$ seconds after three-fourths of the race is complete, what is the probability that she is the winner?

8.5 Consider standard Brownian motion. Let $0 < s < t$.

(a) Find the joint density of (B_s, B_t).

(b) Show that the conditional distribution of B_s given $B_t = y$ is normal, with mean and variance

$$E(B_s|B_t = y) = \frac{sy}{t} \quad \text{and } Var(B_s|B_t = y) = \frac{s(t-s)}{t}.$$

8.6 For $s > 0$, show that the translation $(B_{t+s} - B_s)_{t\ge 0}$ is a standard Brownian motion.

8.7 Show that the reflection $(-B_t)_{t\ge 0}$ is a standard Brownian motion.

8.8 Find the covariance function for Brownian motion with drift.

8.9 Let $W_t = B_{2t} - B_t$, for $t \ge 0$.

(a) Is $(W_t)_{t\ge 0}$ a Gaussian process?

(b) Is $(W_t)_{t\ge 0}$ a Brownian motion process?

8.10 Let $(B_t)_{t\ge 0}$ be a Brownian motion started in x. Let

$$X_t = B_t - t(B_1 - y), \text{ for } 0 \le t \le 1.$$

The process $(X_t)_{t\ge 0}$ is a Brownian bridge with start in x and end in y. Find the mean and covariance functions.

8.11 A standard Brownian motion crosses the t-axis at times $t = 2$ and $t = 5$. Find the probability that the process exceeds level $x = 1$ at time $t = 4$.

8.12 Show that Brownian motion with drift has independent and stationary increments.

8.13 Let $(X_t)_{t\ge 0}$ denote a Brownian motion with drift μ and variance σ^2. For $0 < s < t$, find $E(X_s X_t)$.

8.14 A Brownian motion with drift parameter $\mu = -1$ and variance $\sigma^2 = 4$ starts at $x = 1.5$. Find the probability that the process is positive at $t = 3$.

8.15 See Example 8.11 on using Brownian motion to model the home team advantage in basketball. In Stern (1994), Table 8.3 is given based on the outcomes of 493 basketball games played in 1992.

Here, X_t is the difference between the home team's score and the visiting team's score after $t(100)$ percent of the game is played. The data show the mean

TABLE 8.3 Results by Quarter of 493 NBA Games

Quarter	Variable	Mean	Standard Deviation
1	$X_{0.25}$	1.41	7.58
2	$X_{0.50} - X_{0.25}$	1.57	7.40
3	$X_{0.75} - X_{0.50}$	1.51	7.30
4	$X_1 - X_{0.75}$	0.22	6.99

and standard deviation of these differences at the end of each quarter. Why might the data support the use of a Brownian motion model? What aspects of the data give reason to doubt the Brownian motion model?

8.16 Let $(X_t)_{t\geq 0}$ and $(Y_t)_{t\geq 0}$ be independent, standard Brownian motions. Show that $Z_t = a(X_t - Y_t)$ defines a standard Brownian motion for some a. Find a.

8.17 For $a > 0$, show that for standard Brownian motion the first hitting time T_a has the same distribution as $1/X^2$, where X is a normal random variable with mean 0 and variance $1/a^2$.

8.18 Show that the first hitting time T_a has the same distribution as $a^2 T_1$.

8.19 Find the mean and variance of the maximum value of standard Brownian motion on $[0, t]$.

8.20 Use the reflection principle to show

$$P(M_t \geq a, B_t \leq a - b) = P(B_t \geq a + b), \quad \text{for } a, b > 0. \tag{8.14}$$

8.21 From standard Brownian motion, let X_t be the process defined by

$$X_t = \begin{cases} B_t, & \text{if } t < T_a, \\ a, & \text{if } t \geq T_a, \end{cases}$$

where T_a is the first hitting time of $a > 0$. The process $(X_t)_{t\geq 0}$ is called *Brownian motion absorbed at a*. The distribution of X_t has discrete and continuous parts.

(a) Show

$$P(X_t = a) = \frac{2}{\sqrt{2\pi t}} \int_a^\infty e^{-x^2/2t}\, dx.$$

(b) For $x < a$, show

$$P(X_t \leq x) = P(B_t \leq x) - P(B_t \leq x - 2a) = \frac{1}{\sqrt{2\pi t}} \int_{x-2a}^{x} e^{-z^2/2t}\, dz.$$

Hint: Use the result of Exercise 8.20.

8.22 Let Z be the smallest zero of standard Brownian motion past t. Show that

$$P(Z \leq z) = \frac{2}{\pi} \arccos \sqrt{\frac{t}{z}}, \quad \text{for } z > 0.$$

8.23 Let $0 < r < s < t$.

(a) Assume that standard Brownian motion is not zero in (r, s). Find the probability that standard Brownian motion is not zero in (r, t).

(b) Assume that standard Brownian motion is not zero in $(0, s)$. Find the probability that standard Brownian motion is not zero in $(0, t)$.

8.24 Derive the mean and variance of geometric Brownian motion.

8.25 The price of a stock is modeled with a geometric Brownian motion with drift $\mu = -0.25$ and volatility $\sigma = 0.4$. The stock currently sells for \$35. What is the probability that the price will be at least \$40 in 1 year?

8.26 For the stock price model in Exercise 8.25, assume that an option is available to purchase the stock in six months for \$40. Find the expected payoff of the option.

8.27 Assume that $Z_0, Z_1, \ldots$ is a branching process whose offspring distribution has mean μ. Show that Z_n/μ^n is a martingale.

8.28 An urn contains two balls—one red and one blue. At each discrete step, a ball is chosen at random from the urn. It is returned to the urn along with another ball of the same color. Let X_n denote the number of red balls in the urn after n draws. (Thus, $X_0 = 1$.) Let $R_n = X_n/(n+2)$ be the proportion of red balls in the urn after n draws. Show that $R_0, R_1, \ldots$ is a martingale with respect to $X_0, X_1, \ldots$ The process is called *Polya's Urn*.

8.29 Show that $\left(B_t^3 - 3tB_t\right)_{t\geq 0}$ is a martingale with respect to Brownian motion.

8.30 A Brownian motion with drift $\mu = 2$ and variance $\sigma^2 = 4$ is run until level $a = 3$ is first hit. The process is repeated 25 times. Find the approximate probability that the average first hitting time of 25 runs is between 1 and 2.

8.31 Consider standard Brownian motion started at $x = -3$.

(a) Find the probability of reaching level 2 before -7.

(b) When, on average, will the process leave the region between the lines $y = 2$ and $y = -7$?

8.32 Let $(N_t)_{t\geq 0}$ be a Poisson process with parameter λ. Let $X_t = N_t - \lambda t$, for $t \geq 0$. Show that X_t is a martingale with respect to N_t.

8.33 Let $X_1, X_2, \ldots$ be i.i.d. random variables with mean $\mu < \infty$. Let $Z_n = \sum_{i=1}^{n}(X_i - \mu)$, for $n = 0, 1, 2, \ldots$

(a) Show that $Z_0, Z_1, \ldots$ is a martingale with respect to $X_0, X_1, \ldots$

(b) Assume that N is a stopping time that satisfies the conditions of the optional stopping theorem. Show that

$$E\left(\sum_{i=1}^{N} X_i\right) = E(N)\mu.$$

This result is known as *Wald's equation.*

8.34 Let $(N_t)_{t\geq 0}$ be a Poisson process with parameter λ.

(a) Find the quantity $m(t)$ such that $M_t = (N_t - \lambda t)^2 - m(t)$ is a martingale.

(b) For fixed integer $k > 0$, let $T = \min\{t : N_t = k\}$ be the first time k arrivals occur for a Poisson process. Show that T is a stopping time that satisfies the conditions of the optional stopping theorem.

(c) Use the optional stopping theorem to find the standard deviation of T.

8.35 For $a > 0$, let T be the first time that standard Brownian motion exits the interval $(-a, a)$.

(a) Show that T is a stopping time that satisfies the optional stopping theorem.

(b) Find the expected time $E(T)$ to exit $(-a, a)$.

(c) Let $M_t = B_t^4 - 6tB_t^2 + 3t^2$. Then, $(M_t)_{t\geq 0}$ is a martingale, a fact that you do not need to prove. Use this to find the standard deviation of T.

8.36 Let $(X_t)_{t\geq 0}$ be a Brownian motion with drift $\mu \neq 0$ and variance σ^2. The goal of this exercise is to find the probability p that X_t hits a before $-b$, for $a, b > 0$.

(a) Let $Y_t = e^{-tc^2/2+cB_t}$, where B_t denotes standard Brownian motion. Show that $(Y_t)_{t\geq 0}$ is a martingale for constant $c \neq 0$.

(b) Let $T = \min\{t : X_t = a \text{ or } -b\}$. Then, T is a stopping time that satisfies the optional stopping theorem. Use (a) and appropriate choice of c to show

$$E\left(e^{-\frac{2\mu X_T}{\sigma^2}}\right) = 1.$$

(c) Show that

$$p = \frac{1 - e^{2\mu b/\sigma^2}}{e^{-2\mu a/\sigma^2} - e^{2\mu b/\sigma^2}}. \tag{8.15}$$

8.37 Consider a Brownian motion with drift μ and variance σ^2. Assume that $\mu < 0$. The process tends to $-\infty$, with probability 1. Let M be the maximum value reached.

(a) See Exercise 8.36. By letting $b \to \infty$ in Equation (8.15), show

$$P(M > a) = e^{2\mu a/\sigma^2}, \text{ for } a > 0.$$

Conclude that M has an exponential distribution.

(b) A particle moves according to a Brownian motion with drift $\mu = -1.6$ and variance $\sigma^2 = 0.4$. Find the mean and standard deviation of the largest level reached.

8.38 Let $T = \min\{t : B_t \notin (-a, a)\}$. Show that

$$E\left(e^{-\lambda T}\right) = \frac{2}{e^{a\sqrt{2\lambda}} + e^{-a\sqrt{2\lambda}}} = \frac{1}{\cosh(a\sqrt{2\lambda})}.$$

(a) Apply the optional stopping theorem to the exponential martingale in Exercise 8.36(a) to show that

$$E\left(e^{\sqrt{2\lambda}B_T}e^{-\lambda T}\right) = 1.$$

(b) Show that

$$P(B_T = a, T < x) = P(B_T = -a, T < x) = \frac{1}{2}P(T < x)$$

and conclude that B_T and T are independent, and thus establish the result.

8.39 R: Simulate a Brownian motion $(X_t)_{t\geq 0}$ with drift $\mu = 1.5$ and variance $\sigma^2 = 4$. Simulate the probability $P(X_3 > 4)$ and compare with the exact result.

8.40 R: Use the script file **bbridge.R** to simulate a Brownian bridge $(X_t)_{t\geq 0}$. Estimate the probability $P(X_{3/4} \leq 1/3)$. Compare with the exact result.

8.41 R: The price of a stock is modeled as a geometric Brownian motion with drift $\mu = -0.85$ and variance $\sigma^2 = 2.4$. If the current price is \$50, find the probability that the price is under \$40 in 2 years. Simulate the stock price, and compare with the exact value.

8.42 R: Simulate the mean and standard deviation of the maximum of standard Brownian motion on $[0, 1]$. Compare with the exact values.

8.43 R: Write a function for pricing an option using the Black–Scholes formula. Option price is a function of initial stock price, strike price, expiration date, interest rate, and volatility.

(a) Price an option for a stock that currently sells for \$400 and has volatility 40%. Assume that the option strike price is \$420 and the expiration date is 90 days. Assume a risk-free interest rate of 3%.

(b) For each of the five parameters in (a), how does varying the parameter, holding the other four fixed, effect the price of the option?

(c) For the option in (a), assume that volatility is not known. However, based on market experience, the option sells for \$30. Estimate the volatility.

9

A GENTLE INTRODUCTION TO STOCHASTIC CALCULUS*

The study of mathematics, like the Nile, begins in minuteness but ends in magnificence.
–Charles Caleb Colton

9.1 INTRODUCTION

Brownian motion paths are continuous functions. Continuous functions are integrable. Integration of Brownian motion opens the door to powerful calculus-based modeling tools, such as stochastic differential equations (SDEs). Stochastic calculus is an advanced topic, which requires measure theory, and often several graduate-level probability courses. Our goal in this section is to introduce the subject by emphasizing intuition, and whet your appetite for what is possible in this fascinating field.

We will make sense of integrals such as

$$\int_0^t B_s\, ds \quad \text{and} \quad \int_0^t B_s\, dB_s.$$

In the first integral, Brownian motion is integrated over the interval $[0, t]$. Think of the integral as representing the area under the Brownian motion curve on $[0, t]$. The fact that the integrand is random means that the integral is random, hence a random variable. As a function of t, it is a *random function*, that is, a stochastic process.

Introduction to Stochastic Processes with R, First Edition. Robert P. Dobrow.

If that is not strange enough, in the second integral Brownian motion appears in both the integrand and the *integrator*, where dB_s replaces the usual ds. Here, Brownian motion is integrated *with respect to* Brownian motion. To make sense of this will require new ideas, and even new rules, of calculus.

To start off, for $0 \leq a < b$, consider the integral

$$\int_a^b B_s \, ds.$$

For each outcome ω, $B_s(\omega)$ is a continuous function, and thus the integral

$$\int_a^b B_s(\omega) \, ds$$

is well defined in the usual sense as the limit of a Riemann sum. For a partition $a = t_0 < t_1 < \cdots < t_{n-1} < t_n = b$ of $[a, b]$, define the Riemann sum

$$I^{(n)}(\omega) = \sum_{k=1}^{n} B_{t_k^*}(\omega)(t_k - t_{k-1}),$$

where $t_k^* \in [t_{k-1}, t_k]$ is an arbitrary point in the subinterval $[t_{k-1}, t_k]$. The integral $\int_a^b B_s(\omega) \, ds$ is defined as the limit of the Riemann sum as n tends to infinity and the length of the longest subinterval of the partition converges to 0.

For each $n \geq 1$, the Riemann sum $I^{(n)}$ is a random variable, which is a linear combination of normal random variables. Since Brownian motion is a Gaussian process, $I^{(n)}$ is normally distributed. As this is true for all n, it is reasonable to expect that $\lim_{n\to\infty} I^{(n)}$ is normally distributed.

Let $I_t = \int_0^t B_s \, ds$, for $t \geq 0$. It can be shown that $(I_t)_{t\geq 0}$ is a Gaussian process with continuous sample paths. The mean function is

$$E(I_t) = E\left(\int_0^t B_s \, ds\right) = \int_0^t E(B_s) \, ds = 0,$$

where the interchange of expectation and integral can be justified.

For $s \leq t$, the covariance function is

$$\begin{aligned}
\text{Cov}(I_s, I_t) = E(I_s I_t) &= E\left(\int_0^s B_x \, dx \int_0^t B_y \, dy\right) \\
&= \int_0^s \int_0^t E(B_x B_y) \, dy \, dx = \int_0^s \int_0^t \min\{x, y\} \, dy \, dx \\
&= \int_0^s \int_0^x y \, dy \, dx + \int_0^s \int_x^t x \, dy \, dx \\
&= \frac{s^3}{6} + \left(\frac{ts^2}{2} - \frac{s^3}{3}\right) = \frac{3ts^2 - s^3}{6}.
\end{aligned}$$

Letting $s = t$ gives $Var(I_t) = t^3/3$.

Thus, the stochastic integral $\int_0^t B_s\,ds$ is a normally distributed random variable with mean 0 and variance $t^3/3$. The integral $\int_0^t B_s\,ds$ is called *integrated Brownian motion*. See Figure 9.1 for realizations when $t = 1$.

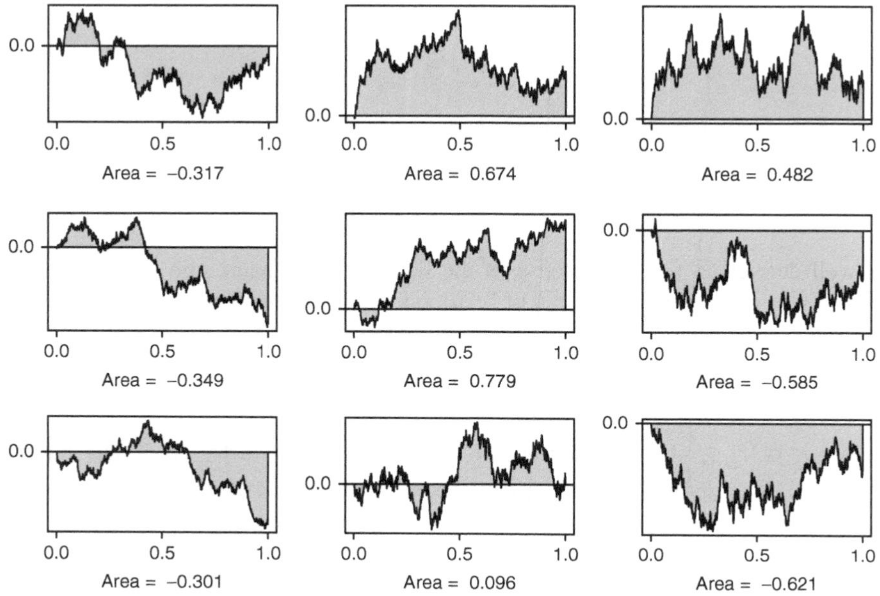

Figure 9.1 Realizations of the stochastic integral $\int_0^1 B_s\,ds$. The integral is normally distributed with mean 0 and variance 1/3.

We next introduce the *Riemann–Stieltjes integral* of g with respect to f

$$\int_0^t g(x)df(x),$$

where f and g are continuous, and nonrandom, functions. The integral is defined as the limit, as n tends to infinity, of the approximating sum

$$\sum_{k=1}^{n} g(t_k^*)\left(f(t_k) - f(t_{k-1})\right),$$

where $0 = t_0 < t_1 < \cdots < t_{n-1} < t_n = t$ is a partition of $[0, t]$, and $t_k^* \in [t_{k-1}, t_k]$. The definition generalizes the usual Riemann integral by letting $f(x) = x$. The integral can be interpreted as a *weighted* summation, or weighted average, of g, with weights determined by f.

If f is differentiable, with continuous derivative, then

$$\int_0^t g(x)df(x) = \int_0^t g(x)f'(x)\,dx,$$

which gives the usual Riemann integral. In probability, if X is a continuous random variable with density function f and cumulative distribution function F, and g is a function, the expectation $E(g(X))$ can be written as

$$E(g(X)) = \int_{-\infty}^{\infty} g(x)f(x)\,dx = \int_{-\infty}^{\infty} g(x)F'(x)\,dx = \int_{-\infty}^{\infty} g(x)dF(x),$$

that is, as a Riemann–Stieltjes integral of g with respect to the cumulative distribution function F.

Based on the Riemann–Stieltjes integral, we can define the integral of a function g with respect to Brownian motion

$$I_t = \int_0^t g(s)\,dB_s. \tag{9.1}$$

Technical conditions require that g be a bounded and continuous function, and satisfy $\int_0^\infty g^2(s)\,ds < \infty$. By analogy with the Riemann–Stieltjes integral, for the partition

$$0 = t_0 < t_1 < \cdots < t_{n-1} < t_n = t,$$

let

$$I_t^{(n)} = \sum_{k=1}^{n} g(t_k^*)\left(B_{t_k} - B_{t_{k-1}}\right),$$

where $t_k^* \in [t_{k-1}, t_k]$. Since $B_{t_k} - B_{t_{k-1}}$ is normally distributed with mean 0 and variance $t_k - t_{k-1}$, the approximating sum $I_t^{(n)}$ is normally distributed for all n. It can be shown that in the limit, as $n \to \infty$, the approximating sum converges to a normally distributed random variable, which we take to be the stochastic integral of Equation (9.1). Furthermore,

$$\begin{aligned} E(I_t) &= E\left(\lim_{n\to\infty} I_t^{(n)}\right) = \lim_{n\to\infty} E\left(\sum_{k=1}^{n} g(t_k^*)\left(B_{t_k} - B_{t_{k-1}}\right)\right) \\ &= \lim_{n\to\infty} \sum_{k=1}^{n} g(t_k^*)E\left(B_{t_k} - B_{t_{k-1}}\right) = 0. \end{aligned}$$

By independent increments,

$$Var\left(I_t^{(n)}\right) = \sum_{k=1}^{n} g^2(t_k^*)Var\left(B_{t_k} - B_{t_{k-1}}\right) = \sum_{k=1}^{n} g^2(t_k^*)(t_k - t_{k-1}).$$

The last expression is a Riemann sum whose limit, as n tends to infinity, is $\int_0^t g^2(s)\,ds$. In summary,

$$\int_0^t g(s)\,dB_s \sim \text{Normal}\left(0, \int_0^t g^2(s)\,ds\right). \tag{9.2}$$

In fact, it can be shown that $(I_t)_{t\geq 0}$ is a Gaussian process with continuous sample paths, independent increments, mean function 0, and covariance function

$$\mathrm{Cov}(I_s, I_t) = E\left(\int_0^s g(x)\, dB_x \int_0^t g(y)\, dB_y\right) = \int_0^{\min\{s,t\}} g^2(x)\, dx.$$

Example 9.1 Evaluate

$$\int_0^t dB_s.$$

Solution With $g(x) = 1$, the integral is normally distributed with mean 0 and variance $\int_0^t ds = t$. That is, the stochastic integral has the same distribution as B_t. Furthermore, the integral defines a continuous Gaussian process with mean 0 and covariance function

$$\int_0^{\min\{s,t\}} dx = \min\{s, t\}.$$

That is, $\left(\int_0^t dB_s\right)_{t\geq 0}$ is a standard Brownian motion, and

$$\int_0^t dB_s = B_t.$$

■

Example 9.2 Evaluate

$$\int_0^t e^s\, dB_s.$$

Solution The integral is normally distributed with mean 0 and variance

$$\int_0^t (e^s)^2\, ds = \int_0^t e^{2s}\, ds = \frac{1}{2}e^{2t}.$$

■

The stochastic integral

$$\int_a^b g(s)\, dB_s$$

has many familiar properties. Linearity holds. For functions g and h, for which the integral is defined, and constants α, β,

$$\int_a^b [\alpha g(s) + \beta h(s)]\, dB_s = \alpha \int_a^b g(s)\, dB_s + \beta \int_a^b h(s)\, dB_s.$$

For $a < c < b$,

$$\int_a^b g(s)\, dB_s = \int_a^c g(s)\, dB_s + \int_c^b g(s)\, dB_s.$$

The integral also satisfies an *integration-by-parts* formula. If g is differentiable,

$$\int_0^t g(s)\, dB_s = g(t)B_t - \int_0^t B_s g'(s)\, ds.$$

By letting $g(t) = 1$, we capture the identity

$$\int_0^t dB_s = B_t. \tag{9.3}$$

Example 9.3 Evaluate

$$\int_0^t s\, dB_s$$

in terms of integrated Brownian motion.

Solution Integration by parts gives

$$\int_0^t s\, dB_s = tB_t - \int_0^t B_s\, ds.$$

See Figure 9.2 for simulations of the process. ■

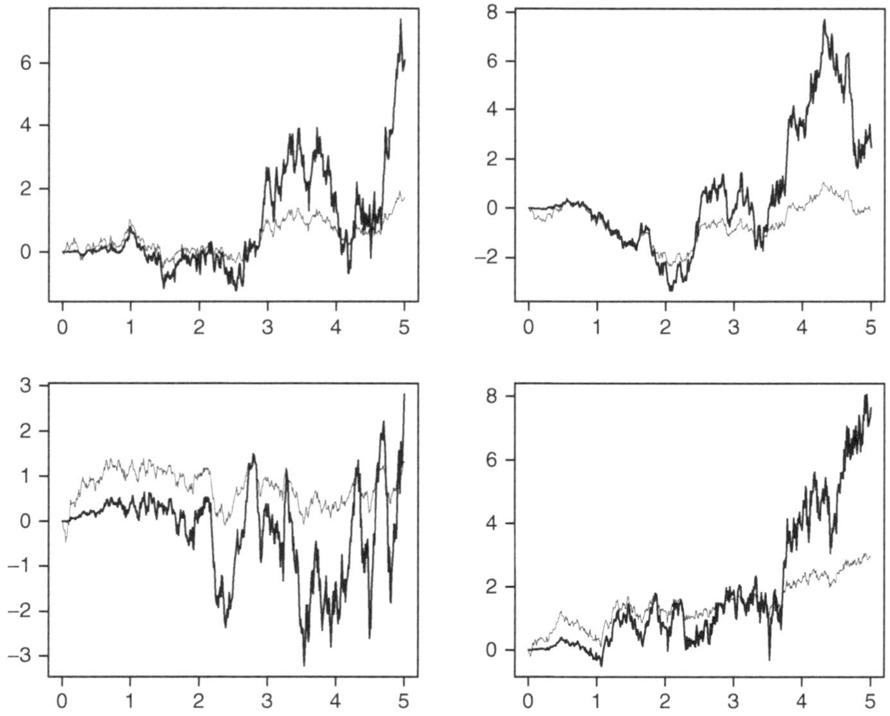

Figure 9.2 Simulations of $\left(\int_0^t s\, dB_s\right)_{0\le t\le 5}$. The light gray curve is the underlying standard Brownian motion.

White Noise

If Brownian motion paths were differentiable, Equation (9.3) could be written as

$$B_t = \int_0^t dB_s = \int_0^t \frac{dB_s}{ds}\, ds.$$

The "process" $W_t = dB_t/dt$ is called *white noise*. The reason for the quotation marks is because W_t is not a stochastic process in the usual sense, as Brownian motion derivatives do not exist. Nevertheless, Brownian motion is sometimes described as *integrated white noise*.

Consider the following formal treatment of the distribution of white noise. Letting Δ_t represent a small incremental change in t,

$$W_t \approx \frac{B_{t+\Delta_t} - B_t}{\Delta_t}.$$

The random variable W_t is approximately normally distributed with mean 0 and variance $1/\Delta_t$. We can think of W_t as the result of letting $\Delta_t \to 0$. White noise can be thought of as an *idealized* continuous-time Gaussian process, where W_t is normally distributed with mean 0 and variance $1/dt$. Furthermore, for $s \neq t$,

$$E(W_s W_t) = E\left(\frac{dB_s}{ds}\frac{dB_t}{dt}\right) = \frac{1}{ds\, dt} E\left((B_{s+ds} - B_s)(B_{t+dt} - B_t)\right) = 0,$$

by independent increments. Hence, W_s and W_t are independent, for all $s \neq t$.

It is hard to conceive of a real-world, time-indexed process in which all variables, no matter how close in time, are independent. Yet white noise is an extremely useful concept for real-world modeling, particularly in engineering, biology, physics, communication, and economics. In a physical context, white noise refers to sound that contains all frequencies in equal amounts, the analog of white light. See Figure 9.3. Applied to a time-varying signal g, the stochastic integral $\int_0^t g(s)\, dB_s$ can be interpreted as the output after the signal is transformed by white noise. For the case $g(s) = s$, see again Figure 9.2.

9.2 ITO INTEGRAL

We are now ready to consider a more general stochastic integral of the form

$$I_t = \int_0^t X_s\, dB_s,$$

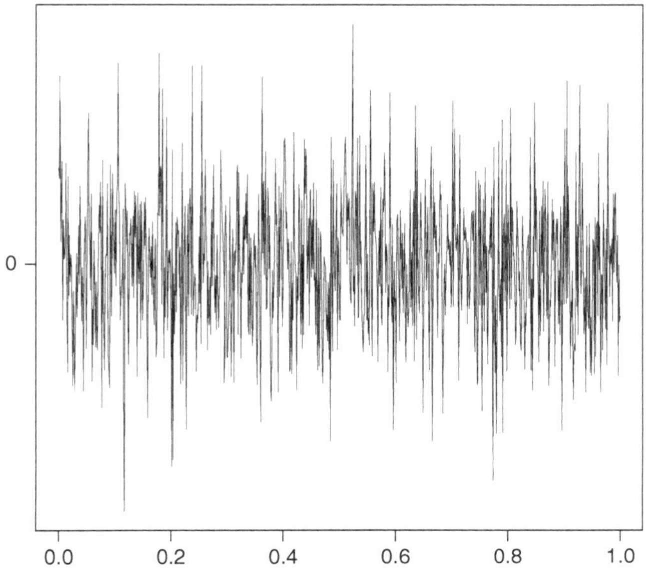

Figure 9.3 Simulation of white noise signal.

where $(X_t)_{t\geq 0}$ is a stochastic process, and $(B_t)_{t\geq 0}$ is standard Brownian motion. By analogy with what has come before, it would seem that a reasonable definition for the integral would be

$$\int_0^t X_s \, dB_s = \lim_{n\to\infty} \sum_{k=1}^{n} X_{t_k^*}\left(B_{t_k} - B_{t_{k-1}}\right), \tag{9.4}$$

for ever-finer partitions $0 = t_0 < t_1 < \cdots < t_{n-1} < t_n = t$, where $t_k^* \in [t_{k-1}, t_k]$. Unfortunately, the definition does not work. Unlike previous integrals, the choice of point t_k^* in the subinterval $[t_{k-1}, t_k]$ matters. The integral is not well-defined for arbitrary $t_k^* \in [t_{k-1}, t_k]$. Furthermore, the integral requires a precise definition of the meaning of the limit in Equation (9.4), as well as several regularity conditions for the process $(X_t)_{t\geq 0}$.

This brings us to the *Ito integral*, named after Kiyoshi Ito, a brilliant 20th century Japanese mathematician whose name is most closely associated with stochastic calculus. The Ito integral is based on taking each t_k^* to be the left endpoint[1] of the subinterval $[t_{k-1}, t_k]$. That is,

$$\int_0^t X_s \, dB_s = \lim_{n\to\infty} \sum_{k=1}^{n} X_{t_{k-1}}\left(B_{t_k} - B_{t_{k-1}}\right). \tag{9.5}$$

[1]A different type of stochastic integral, called the *Stratonovich integral*, is obtained by choosing $t_k^* = (t_{k-1} + t_k)/2$ to be the midpoint of the subinterval $[t_{k-1}, t_k]$.

The Ito integral requires

1. $\int_0^t E\left(X_s^2\right)\, ds < \infty$.
2. X_t does not depend on the values $\{B_s : s > t\}$ and only on $\{B_s : s \leq t\}$. We say that X_t is *adapted* to Brownian motion.
3. The limit in Equation (9.5) is defined in the *mean-square* sense. A sequence of random variables $X_0, X_1, \ldots$ is said to *converge to X in mean-square* if $\lim_{n\to\infty} E\Big((X_n - X)^2\Big) = 0$.

The Ito integral has many familiar properties, such as linearity. However, new rules of *stochastic calculus* will be needed for computations.

One of the most important properties of the Ito integral is that the process

$$\left(\int_0^t X_s\, dB_s\right)_{t\geq 0}$$

is a martingale with respect to Brownian motion.

The following properties of the Ito integral are summarized without proof.

Properties of the Ito Integral

The Ito integral

$$I_t = \int_0^t X_s\, dB_s$$

satisfies the following:

1. For processes $(X_t)_{t\geq 0}$ and $(Y_t)_{t\geq 0}$, and constants α, β,
$$\int_0^t \left(\alpha X_s + \beta Y_s\right)\, dB_s = \alpha \int_0^t X_s\, dB_s + \beta \int_0^t Y_s\, dB_s.$$
2. For $0 < r < t$,
$$\int_0^t X_s\, dB_s = \int_0^r X_s\, dB_s + \int_r^t X_s\, dB_s.$$
3. $$E(I_t) = 0.$$
4. $$Var(I_t) = E\left(\left(\int_0^t X_s\, dB_s\right)^2\right) = \int_0^t E\left(X_s^2\right)\, ds.$$
5. $(I_t)_{t\geq 0}$ is a martingale with respect to Brownian motion.

The Ito integral does not satisfy the usual integration-by-parts formula. Consider

$$\int_0^t B_s \, dB_s.$$

A formal application of integration by parts gives

$$\int_0^t B_s \, dB_s = B_t^2 - B_0^2 - \int_0^t B_s \, dB_s = B_t^2 - \int_0^t B_s \, dB_s,$$

which leads to $\int_0^t B_s \, dB_s = B_t^2/2$. However, this must be wrong as the Ito integral has mean 0, and thus $E\left(\int_0^t B_s \, dB_s\right) = 0$. However, $E\left(B_t^2/2\right) = t/2$.

To evaluate $\int_0^t B_s \, dB_s$, consider the approximating sum

$$\begin{aligned}
&\sum_{k=1}^{n} B_{t_{k-1}}\left(B_{t_k} - B_{t_{k-1}}\right)\\
&= \sum_{k=1}^{n}\left(\frac{1}{2}\left(B_{t_k} + B_{t_{k-1}}\right) - \frac{1}{2}\left(B_{t_k} - B_{t_{k-1}}\right)\right)\left(B_{t_k} - B_{t_{k-1}}\right)\\
&= \frac{1}{2}\sum_{k=1}^{n}\left(B_{t_k}^2 - B_{t_{k-1}}^2\right) - \frac{1}{2}\sum_{k=1}^{n}\left(B_{t_k} - B_{t_{k-1}}\right)^2\\
&= \frac{1}{2}B_t^2 - \frac{1}{2}\sum_{k=1}^{n}\left(B_{t_k} - B_{t_{k-1}}\right)^2.
\end{aligned}$$

It can be shown that $\sum_{k=1}^{n}\left(B_{t_k} - B_{t_{k-1}}\right)^2$ converges to the constant t in mean-square, that is,

$$\lim_{n\to\infty} E\left(\left(\sum_{k=1}^{n}(B_{t_k} - B_{t_{k-1}})^2 - t\right)^2\right) = 0.$$

This gives

$$\int_0^t B_s \, dB_s = \frac{1}{2}\left(B_t^2 - t\right),$$

which is a martingale. Recall that $B_t^2 - t$ is the quadratic martingale, shown in Example 8.19. Multiplying a martingale by a constant, does not change the martingale property.

ITO'S LEMMA

If one disqualifies the Pythagorean Theorem from contention, it is hard to think of a mathematical result which is better known and more widely applied in the world today than "Ito's Lemma." This result holds the same position in stochastic analysis that Newton's fundamental theorem holds in classical analysis. That is, it is the sine qua non of the subject.

–National Academy of Sciences

The most important result in stochastic calculus is Ito's Lemma, which is the stochastic version of the chain rule. It has been called the fundamental theorem of stochastic calculus.

Ito's Lemma

Let g be a real-valued function that is twice continuously differentiable. Then,

$$g(B_t) - g(B_0) = \int_0^t g'(B_s)\, dB_s + \frac{1}{2}\int_0^t g''(B_s)\, ds.$$

This is often written in shorthand differential form

$$dg(B_t) = g'(B_t)dB_t + \frac{1}{2}g''(B_t)dt.$$

Example 9.4 Let $g(x) = x^2$. By Ito's Lemma,

$$B_t^2 = \int_0^t 2B_s\, dB_s + \frac{1}{2}\int_0^t 2\, ds = 2\int_0^t B_s\, dB_s + t.$$

That is,

$$\int_0^t B_s\, dB_s = \frac{1}{2}B_t^2 - \frac{t}{2}.$$

■

Example 9.5 Evaluate $d(\sin B_t)$.

Solution Let $g(x) = \sin x$. By Ito's Lemma,

$$d(\sin B_t) = \cos B_t\, dB_t - \frac{1}{2}\sin B_t\, dt.$$

In integral form,

$$\sin B_t = \int_0^t \cos B_s\, dB_s - \frac{1}{2}\int_0^t \sin B_s\, ds.$$

■

Example 9.6 Evaluate

$$\int_0^t B_s^2\, dB_s \quad \text{and} \quad \int_0^t (B_s^2 - s)\, dB_s.$$

Solution

(i) Use Ito's Lemma with $g(x) = x^3$. This gives

$$B_t^3 = \int_0^t 3B_s^2 \, dB_s + \frac{1}{2}\int_0^t 6B_s \, ds.$$

Rearranging gives

$$\int_0^t B_s^2 \, dB_s = \frac{1}{3}B_t^3 - \int_0^t B_s \, ds.$$

(ii) By linearity of the Ito integral,

$$\begin{aligned}\int_0^t \left(B_s^2 - s\right) \, dB_s &= \int_0^t B_s^2 \, dB_s - \int_0^t s \, dB_s \\ &= \frac{1}{3}B_t^3 - \int_0^t B_s \, ds - \left(tB_t - \int_0^t B_s \, ds\right) \\ &= \frac{1}{3}B_t^3 - tB_t.\end{aligned}$$

The second equality is by integration by parts, which is valid for stochastic integrals with deterministic integrands.

Since Ito integrals are martingales, the process $\left(\frac{1}{3}B_t^3 - tB_t\right)_{t\geq 0}$ is a martingale. ■

Here is a heuristic derivation of Ito's Lemma. For a function g, consider its Taylor series expansion

$$g(t + dt) = g(t) + g'(t)dt + \frac{1}{2}g''(t)(dt)^2 + \cdots .$$

Higher-order terms, starting with $(dt)^2$, are negligible. Hence,

$$dg(t) = g(t + dt) - g(t) = g'(t)dt.$$

For a given function h,

$$g(h(t) + dh(t)) = g(h(t)) + g'(h(t))dh(t) + \frac{1}{2}g''(h(t))(dh(t))^2 + \cdots .$$

Under suitable regularity conditions, the higher-order terms drop out, giving the usual chain rule $dg(h) = g'(h)dh$.

Replacing $h(t)$ with B_t, the Taylor series expansion is

$$g(B_t + dB_t) = g(B_t) + g'(B_t)dB_t + \frac{1}{2}g''(B_t)(dB_t)^2 + \frac{1}{6}g'''(B_t)(dB_t)^3 + \cdots .$$

However, what is different for Brownian motion is that the term $(dB_t)^2$ is not negligible and cannot be eliminated. Intuitively, $dB_t = B_{t+dt} - B_t$ is a stochastic element with the same distribution as B_{dt}, which is normally distributed with mean 0 and variance dt. Thus, B_{dt} takes values on the order of the standard deviation $\sqrt{dt}$. This gives $(dB_t)^2 \approx \left(\sqrt{dt}\right)^2 = dt$. Thus, the $(dB_t)^2 = dt$ term is retained in the expansion.

Higher-order terms beyond the quadratic term are dropped from the expansion, as $(dB_t)^k \approx \left(\sqrt{dt}\right)^k = (dt)^{k/2}$, which is negligible for $k > 2$. This leaves

$$dg(B_t) = g(B_t + dB_t) - g(B_t) = g'(B_t)dB_t + \frac{1}{2}g''(B_t)dt,$$

the differential form of Ito's Lemma.

Here are the heuristic stochastic calculus rules for working with stochastic differentials:

$$(dt)^2 = 0, \quad (dt)(dB_t) = 0, \quad (dB_t)^2 = dt.$$

An extended version of Ito's Lemma allows g to be a function of both t and B_t. The extended result can be motivated by considering a second-order Taylor series expansion of g.

Extension of Ito's Lemma

Let $g(t, x)$ be a real-valued function whose second-order partial derivatives are continuous. Then,

$$g(t, B_t) - g(0, B_0) = \int_0^t \left(\frac{\partial}{\partial t} g(s, B_s) + \frac{1}{2} \frac{\partial^2}{\partial x^2} g(s, B_s) \right) ds + \int_0^t \frac{\partial}{\partial x} g(s, B_s)\, dB_s.$$

In shorthand differential form,

$$dg = \left(\frac{\partial g}{\partial t} + \frac{1}{2} \frac{\partial^2 g}{\partial x^2} \right) dt + \frac{\partial g}{\partial x} dB_t.$$

We regret possible notational confusion in the statement of the lemma. It is common to use the letter t as the time variable, and thus t appears both as the upper limit of integration and as the dummy variable for the function g and its derivative. We trust the reader will safely navigate their way.

Example 9.7 Evaluate $d\left(tB_t^2\right)$.

Solution Let $g(t, x) = tx^2$. Partial derivatives are

$$\frac{\partial g}{\partial t} = x^2, \quad \frac{\partial g}{\partial x} = 2tx, \quad \text{and} \quad \frac{\partial^2 g}{\partial x^2} = 2t.$$

By Ito's Lemma,

$$d(tB_t^2) = (B_t^2 + t)\,dt + 2tB_t\,dB_t.$$

Observe that the usual product rule would give the incorrect answer

$$d(tB_t^2) = B_t^2 dt + 2tB_t\,dB_t.$$

■

Example 9.8 Use Ito's Lemma to evaluate $d(B_t^3)$ and $E(B_t^3)$.

Solution Let $g(t, x) = x^3$. By Ito's Lemma,

$$d(B_t^3) = 3B_t\,dt + 3B_t^2\,dB_t,$$

and

$$B_t^3 = 3\int_0^t B_s\,ds + 3\int_0^t B_s^2\,dB_s.$$

Taking expectations gives

$$E(B_t^3) = 3\int_0^t E(B_s)\,ds + 3E\left(\int_0^t B_s^2\,dB_s\right) = 3(0) + 0 = 0.$$

■

9.3 STOCHASTIC DIFFERENTIAL EQUATIONS

To motivate the discussion, consider an exponential growth process, be it the spread of a disease, the population of a city, or the number of cells in an organism. Let X_t denotes the size of the population at time t. The deterministic exponential growth model is described by an ordinary differential equation

$$\frac{dX_t}{dt} = \alpha X_t, \text{ and } X_0 = x_0,$$

where x_0 is the initial size of the population, and α is the growth rate. The equation says that the population growth rate is proportional to the size of the population, where the constant of proportionality is α. A *solution* of the differential equation is a function X_t, which satisfies the equation. In this case, the solution is uniquely specified

$$X_t = x_0 e^{\alpha t}, \text{ for } t \geq 0.$$

The most common way to incorporate uncertainty into the model is to add a random error term, such as a multiple of white noise W_t, to the growth rate. This gives the *stochastic differential equation*

$$\frac{dX_t}{dt} = (\alpha + \beta W_t)X_t = \alpha X_t + \beta X_t \frac{dB_t}{dt},$$

or

$$dX_t = \alpha X_t \, dt + \beta X_t \, dB_t, \tag{9.6}$$

where α and β are parameters, and $X_0 = x_0$. Equation (9.6) is really a shorthand for the integral form

$$X_t - X_0 = \alpha \int_0^t X_s \, ds + \beta \int_0^t X_s \, dB_s. \tag{9.7}$$

A solution to the SDE is a stochastic process $(X_t)_{t \geq 0}$, which satisfies Equation (9.7).

For the stochastic exponential model, we show that geometric Brownian motion defined by

$$X_t = x_0 e^{\left(\alpha - \frac{\beta^2}{2}\right)t + \beta B_t}, \text{ for } t \geq 0, \tag{9.8}$$

is a solution. Let

$$g(t, x) = x_0 e^{\left(\alpha - \frac{\beta^2}{2}\right)t + \beta x}$$

with partial derivatives

$$\frac{\partial g}{\partial x} = \beta g, \quad \frac{\partial^2 g}{\partial x^2} = \beta^2 g, \quad \text{and} \quad \frac{\partial g}{\partial t} = \left(\alpha - \frac{\beta^2}{2}\right) g.$$

By the extended version of Ito's Lemma,

$$\begin{aligned} g(t, B_t) - g(0, B_0) &= x_0 e^{\left(\alpha - \frac{\beta^2}{2}\right)t + \beta B_t} - x_0 \\ &= \left(\alpha - \frac{\beta^2}{2} + \frac{\beta^2}{2}\right) \int_0^t x_0 e^{\left(\alpha - \frac{\beta^2}{2}\right)s + \beta B_s} \, ds \\ &\quad + \beta \int_0^t x_0 e^{\left(\alpha - \frac{\beta^2}{2}\right)s + \beta B_s} \, dB_s, \end{aligned}$$

which reduces to the solution

$$X_t - X_0 = \alpha \int_0^t X_s \, ds + \beta \int_0^t X_s \, dB_s.$$

Geometric Brownian motion can be thought of as the stochastic analog of the exponential growth function.

Differential equations are the meat and potatoes of applied mathematics. Stochastic differential equations are used in biology, climate science, engineering, economics, physics, ecology, chemistry, and public health.

Example 9.9 (Logistic equation) Unfettered exponential growth is typically unrealistic for biological populations. The *logistic* model describes the growth of a self-limiting population. The standard deterministic model is described by the ordinary differential equation

$$\frac{dP_t}{dt} = rP_t\left(1 - \frac{P_t}{K}\right),$$

where P_t denotes the population size at time t, r is the growth rate, and K is the *carrying capacity*, the maximum population size that the environment can sustain.

If $P_t \approx 0$, then $dP_t/dt \approx rP_t$, and the model exhibits near-exponential growth. On the contrary, if the population size is near carrying capacity and $P_t \approx K$, then $P_t/dt \approx 0$, and the population exhibits little growth.

The solution of the deterministic equation—obtained by separation of variables and partial fractions—is

$$P_t = \frac{KP_0}{P_0 + (K - P_0)e^{-rt}}, \quad \text{for } t \geq 0. \tag{9.9}$$

Observe that $P_t \to K$, as $t \to \infty$; that is, the population size tends to the carrying capacity.

A stochastic logistic equation is described by the SDE

$$dP_t = rP_t\left(1 - \frac{P_t}{K}\right)dt + \sigma P_t\, dB_t,$$

where $\sigma > 0$ is a parameter. Let $(X_t)_{t\geq 0}$ be the geometric Brownian motion process defined by

$$X_t = e^{\left(r-\frac{\sigma^2}{2}\right)t+\sigma B_t}.$$

It can be shown that the solution to the logistic SDE is

$$P_t = \frac{P_0 K X_t}{K + P_0 r \int_0^t X_s\, ds}.$$

When $\sigma = 0$, $X_t = e^{rt}$ and the solution reduces to Equation (9.9). See Figure 9.4 for sample paths of the stochastic logistic process. ■

Example 9.10 Stochastic models are used in climatology to model long-term climate variability. These complex models are typically multidimensional and involve systems of SDEs. They are relevant for our understanding of global warming and climate change.

A simplified system that models the interaction between the atmosphere and the ocean's surface is described in Vallis (2010). Let T_t^A and T_t^S denote the atmosphere and sea surface temperatures, respectively, at time t. The system is

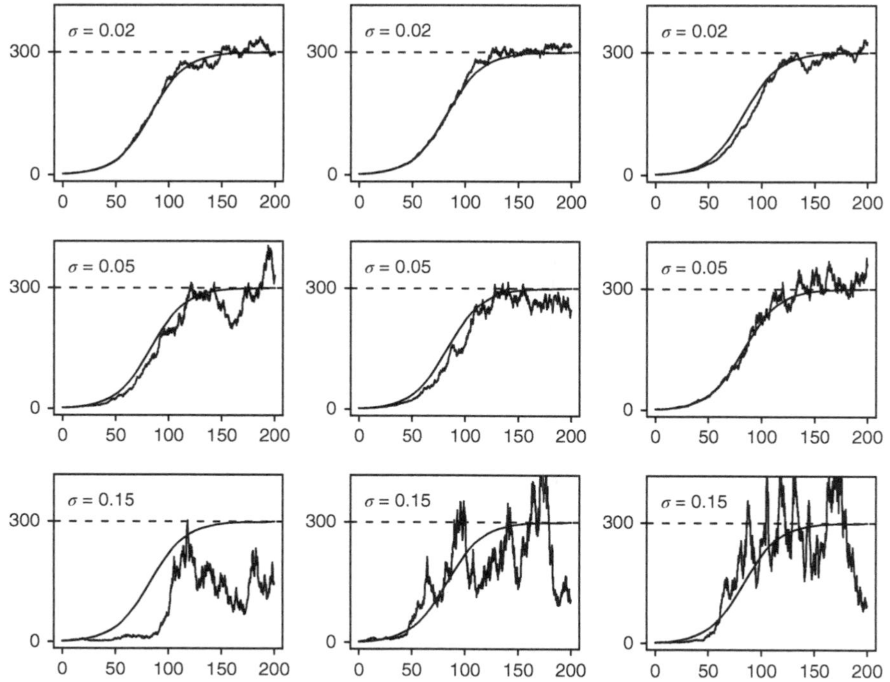

Figure 9.4 Sample paths for the logistic SDE, with $P_0 = 2$, $r = 0.06$, and $K = 300$. Smooth curve is the deterministic logistic function.

$$c_S \frac{dT^S}{dt} = aT^A - bT^S,$$

$$c_A \frac{dT^A}{dt} = cT^S - dT^A + \sigma B_t,$$

where c_A and c_S describe the heat capacity of the atmosphere and sea, respectively, and a, b, c, d, and σ are parameters. The model is based on Newton's law of cooling, by which the rate of heat loss of an object is proportional to the difference in temperature between the object and its surroundings. The Brownian motion term σB_t accounts for random fluctuations that affect the atmosphere.

Assuming the heat capacity of the ocean surface is much greater than that of the atmosphere, the model finds that rapid changes of atmospheric temperatures can affect long-term fluctuations in the ocean temperature, over possibly decades or centuries. The finding has significance in understanding how temperature changes at time scales greater than a year can occur in the earth's climate. ■

Ito's Lemma is an important tool for working with stochastic differential equations. The Lemma can be extended further to include a wide class of stochastic processes, which are solutions to SDEs of the form

$$dX_t = a(t, X_t)dt + b(t, X_t)dB_t, \tag{9.10}$$

where a and b are functions of t and X_t.

The integral form is

$$X_t - X_0 = \int_0^t a(s, X_s)\, ds + \int_0^t b(s, X_s)\, dB_s.$$

Such processes are called *diffusions* or *Ito processes*. A diffusion is a Markov process with continuous sample paths. The functions a and b are called, respectively, the *drift coefficient* and *diffusion coefficient*.

Standard Brownian motion is a diffusion with $a(t, x) = 0$ and $b(t, x) = 1$. The introductory example of this section shows that geometric Brownian motion is a diffusion with $a(t, x) = \alpha x$ and $b(t, x) = \beta x$, for parameters α, β.

Ito's Lemma for Diffusions

Let $g(t, x)$ be a real-valued function whose second-order partial derivatives are continuous. Let $(X_t)_{t\geq 0}$ be an Ito process as defined by Equation (9.10). Then

$$\begin{aligned} g(t, X_t) - g(0, X_0) = &\int_0^t \left(\frac{\partial g}{\partial t} + \frac{\partial g}{\partial x}\alpha(s, X_s) + \frac{1}{2}\frac{\partial^2 g}{\partial x^2}\beta^2(s, X_s) \right) ds \\ &+ \int_0^t \left(\frac{\partial g}{\partial x}\beta(s, X_s) \right) dB_s. \end{aligned}$$

In shorthand differential form,

$$dg = \left(\frac{\partial g}{\partial t} + \frac{\partial g}{\partial x}\alpha(t, X_t) + \frac{1}{2}\frac{\partial^2 g}{\partial x^2}\beta^2(t, X_t) \right) dt + \frac{\partial g}{\partial x}\beta(t, X_t)dB_t.$$

We showed in the introductory example that geometric Brownian motion is a solution to the SDE of Equation (9.6). However, we did not solve the equation directly. Rather, we offered a candidate process and then verified that it was in fact a solution.

In general, solving an SDE may be difficult. A closed-form solution might not exist, and numerical methods are often needed. However, for the stochastic exponential growth model, the SDE can be solved exactly with the help of Ito's Lemma for diffusions.

From Equation (9.6), divide through by X_t to obtain

$$\frac{dX_t}{X_t} = \alpha\ dt + \beta\ dB_t.$$

The left-hand side suggests the function dx/x, whose integral is $\ln x$. This suggests applying Ito's Lemma with $g(t, x) = \ln x$. Derivatives are

$$\frac{\partial g}{\partial t} = 0, \quad \frac{\partial g}{\partial x} = \frac{1}{x}, \quad \text{and} \quad \frac{\partial^2 g}{\partial x^2} = -\frac{1}{x^2}.$$

This gives

$$d \ln X_t = \left(\frac{1}{X_t} \alpha X_t - \frac{1}{2X_t^2} \beta^2 X_t^2 \right) dt + \frac{1}{X_t} \beta X_t \, dB_t = \left(\alpha - \frac{\beta^2}{2} \right) dt + \beta \, dB_t.$$

Integrating gives

$$\ln X_t - \ln x_0 = \left(\alpha - \frac{\beta^2}{2} \right) t + \beta B_t,$$

with solution

$$X_t = x_0 e^{\left(\alpha - \frac{\beta^2}{2} \right) t + \beta B_t}.$$

Example 9.11 (Ornstein–Uhlenbeck process) Mathematical Brownian motion is not necessarily the best model for physical Brownian motion. If B_t denotes the position of a particle, such as a pollen grain, at time t, then the particle's position is changing over time and it must have velocity. The velocity of the grain would be the derivative of the process, which does not exist for mathematical Brownian motion.

The *Ornstein–Uhlenbeck process*, called the Langevin equation in physics, arose as an attempt to model this velocity. In finance, it is known as the Vasicek model and has been used to model interest rates. The process is called *mean-reverting* as there is a tendency, over time, to reach an equilibrium position.

The SDE for the Ornstein–Uhlenbeck process is

$$dX_t = -r(X_t - \mu)dt + \sigma B_t,$$

where r, μ, and $\sigma > 0$ are parameters. The process is a diffusion with

$$a(t, x) = -r(x - \mu) \quad \text{and} \quad b(t, x) = \sigma.$$

If $\sigma = 0$, the equation reduces to an ordinary differential equation, which can be solved by separation of variables. From

$$\frac{dX_t}{X_t - \mu} = -r \, dt,$$

integrating gives

$$\ln (X_t - \mu) = -rt + C,$$

where $C = \ln (X_0 - \mu)$. This gives the deterministic solution

$$X_t = \mu + (X_0 - \mu)e^{-rt}.$$

Observe that $X_t \to \mu$, as $t \to \infty$.

The SDE can be solved using Ito's Lemma by letting $g(t, x) = e^{rt}x$, with partial derivatives

$$\frac{\partial g}{\partial t} = re^{rt}x, \quad \frac{\partial g}{\partial x} = e^{rt}, \quad \text{and} \quad \frac{\partial^2 g}{\partial x^2} = 0.$$

By Ito's Lemma,

$$d\left(e^{rt}X_t\right) = \left(re^{rt}X_t - e^{rt}r(X_t - \mu)\right) dt + e^{rt}\sigma dB_t$$
$$= r\mu e^{rt}dt + e^{rt}\sigma\, dB_t.$$

This gives

$$e^{rt}X_t - X_0 = r\mu \int_0^t e^{rs}\, ds + \sigma \int_0^t e^{rs}\, dB_s = \mu(e^{rt} - 1) + \sigma \int_0^t e^{rs}\, dB_s,$$

with solution

$$X_t = \mu + (X_0 - \mu)e^{-rt} + \sigma \int_0^t e^{-r(t-s)}\, dB_s.$$

See Figure 9.5 for realizations of the Ornstein–Uhlenbeck process.

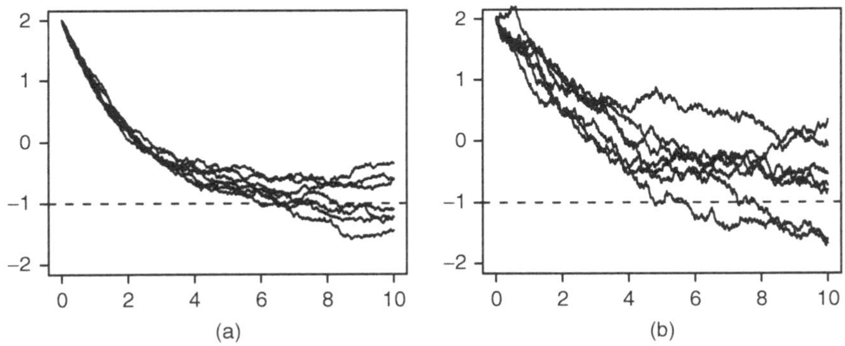

Figure 9.5 Realizations of the Ornstein–Uhlenbeck process with $X_0 = 2$ and $\mu = -1$. (a) $r = 0.5, \sigma = 0.1$. (b) $r = 0.3, \sigma = 0.2$.

If X_0 is constant, then by Equation (9.2), X_t is normally distributed with

$$E(X_t) = \mu + (X_0 - \mu)e^{-rt}$$

and

$$Var(X_t) = \sigma^2 \int_0^t e^{-2r(t-s)}\, ds = \frac{\sigma^2}{2r}\left(1 - e^{-2rt}\right).$$

The limiting distribution, as $t \to \infty$, is normal with mean μ and variance $\sigma^2/2r$. ∎

Numerical Approximation and the Euler–Maruyama Method

The differential form of a stochastic differential equation lends itself to an intuitive method for simulation. Given the SDE

$$dX_t = a(t, X_t)d_t + b(t, X_t)dB_t,$$

the *Euler–Maruyama* method generates a discrete sequence $X_0, X_1, \ldots, X_n$, which approximates the process X_t on an interval $[0, T]$. The method extends the popular Euler method for numerically solving deterministic differential equations.

Partition the interval $[0, T]$ into n equally spaced points

$$0 = t_0 < t_1 < \cdots < t_{n-1} < t_n = T,$$

where $t_i = iT/n$, for $i = 0, 1, \ldots, n$. The differential dt_i is approximated by $t_i - t_{i-1} = T/n$. The stochastic differential dB_{t_i} is approximated by $B_{t_i} - B_{t_{i-1}}$, which is normally distributed with mean 0 and variance $t_i - t_{i-1} = T/n$. Thus, dB_{t_i} can be approximated by $\sqrt{T/n}Z$, where Z is a standard normal random variable. Let

$$X_{i+1} = X_i + a(t_i, X_i)T/n + b(t_i, X_i)\sqrt{T/n}Z_i, \text{ for } i = 0, 1, \ldots, n-1,$$

where $Z_0, Z_1, \ldots, Z_{n-1}$ are independent standard normal random variables. The sequence $X_0, X_1, \ldots, X_n$ is defined recursively and gives a discretized approximate sample path for $(X_t)_{0 \le t \le T}$.

Example 9.12 (Ornstein–Uhlenbeck process) To simulate the solution of the Ornstein–Uhlenbeck SDE

$$dX_t = -r(X_t - \mu)dt + \sigma\, dB_t, \text{ for } 0 \le t \le T,$$

let

$$X_{i+1} = X_i - r(X_i - \mu)T/n + \sigma\sqrt{T/n}Z_i, \text{ for } i = 0, 1, \ldots, n-1.$$

With $n = 1000$, we generate the process with $X_0 = 2$, $\mu = -1$, $r = 0.5$, and $\sigma = 0.1$. Realizations are shown in Figure 9.5(a). ■

R : Ornstein–Uhlenbeck Simulation

```
# ornstein.R
> mu <- -1
> r <- 0.5
> sigma <- 0.1
> T <- 10
> n <- 1000
> xpath <- numeric(n+1)
> xpath[1] <- 2  # initial value
> for (i in 1:n) {
  +   xpath[i+1] <- xpath[i]-r*(xpath[i]-mu)*T/n
  +     + sigma*sqrt(T/n)*rnorm(1) }
> plot(seq(0,T,T/n),xpath,type="l")
```

To simulate the random variable X_T, for fixed T, it is not necessary to store past outcomes X_t, for $t < T$. To generate one outcome of X_T the code simplifies.

```
> x <- 2  # initial value
> for (i in 1:n) {
  + x <- x - r*(x-mu)*T/n + sigma*sqrt(T/n)*rnorm(1) }
> x
[1] -0.9404498
```

Here, we simulate the mean of X_{10} based on 10,000 trials.

```
> trials <- 10000
> simlist <- numeric(trials)
> for (k in 1: trials) {
> x <- 2
> for (i in 1:n) {
  + x <- x - r*(x-mu)*T/n + sigma*sqrt(T/n)*rnorm(1) }
> simlist[k] <- x }
> mean(simlist)
[1] -0.9978892
```

From Example 9.11, the exact mean is

$$E(X_{10}) = \mu + (X_0 - \mu)e^{-r(10)} = -1 + 3e^{-5} = -0.9798.$$

Example 9.13 (Random genetic drift) The SDE

$$dX_t = \sqrt{X_t(1 - X_t)}\, dB_t, \text{ for } 0 \le t \le 1$$

arises as a model for *random genetic drift*. It is a continuous version of the discrete-time Wright–Fisher Markov chain introduced in Example 2.6. The latter is a model for the evolution of a population of N genes consisting of two alleles A and a. In the discrete-time process, the number of A alleles is obtained by drawing from replacement from the gene population. Given i A alleles at time n, the number of A alleles at time $n + 1$ has a binomial distribution with parameters $2N$ and $p = i/2N$. The Markov chain is absorbing with absorbing states 0 and $2N$.

The discrete-time process extends to a continuous-time diffusion $(X_t)_{t\ge 0}$ by a suitable scaling of time and space, where X_t denotes the *proportion* of A alleles in the gene population at time t. The diffusion process is absorbing with absorbing states 0 and 1.

Solving the SDE exactly is beyond the scope of this book. However, numerical methods are used (i) to approximate the sample paths of the process on the time interval $[0, 1.5]$ and (ii) to simulate the probability density function of X_t, for $t = 0.1, 0.2, 0.4, 1$. See Figures 9.6 and 9.7.[2]

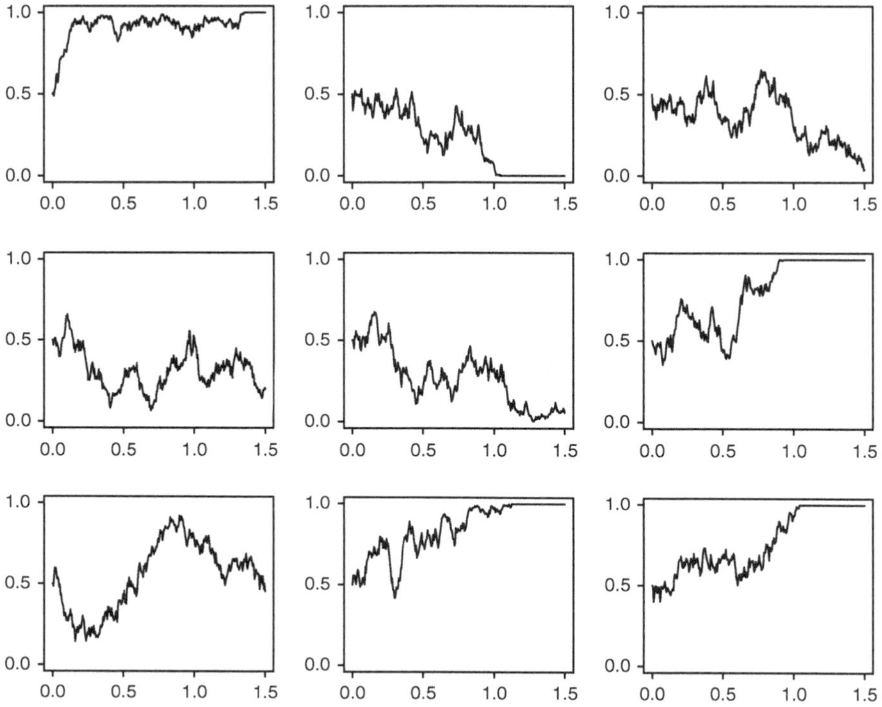

Figure 9.6 Sample paths for the solution of the random genetic drift SDE.

R: Euler–Maruyama Method for Simulating SDE

The following code generates the histograms in Figure 9.7.

```
# drift.R
> par(mfrow=c(2,2))
> times <- c( 0.1, 0.2,0.35,1)
> n = 100  # number of subintervals
> trials <- 10000
> for (k in 1:4) {
> t = times[k]
> simlist <- numeric(trials)
```

[2]In the Euler–Maruyama R code, to insure that the argument to the square foot function is non-negative, the absolute value of $x(1-x)$ is taken. This gives an equivalent model to the original SDE.

```
> for (j in 1:trials) {
> x <- 1/2  # initial state
> for (i in 2:n) {
  x <- x + sqrt(abs(x*(1-x)))*sqrt(t/n)*rnorm(1)   }
> simlist[j] <- x }
> hist(simlist,freq=F,main="", col="gray") }
```

■

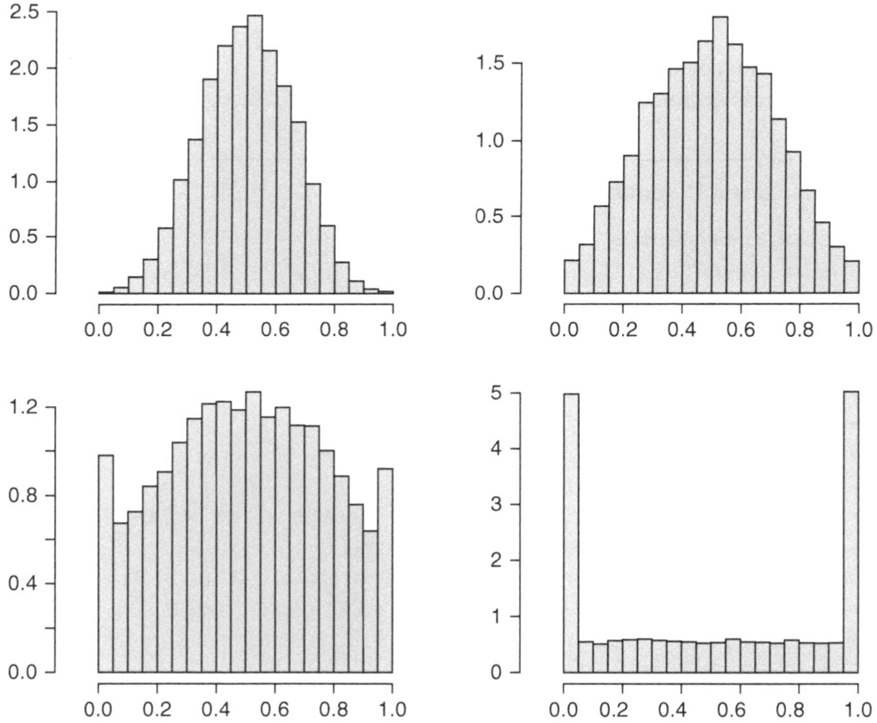

Figure 9.7 Simulating the distribution of X_t in the random genetic drift model, for $t = 0.1, 0.2, 0.35, 1.0$ (top-left to bottom-right).

Example 9.14 (Stochastic resonance) Stochastic resonance is a remarkable phenomenon whereby a signal, which is too weak to be detected, can be boosted by adding noise to the system. The idea is counter-intuitive, since we typically expect that noise (e.g., random error) makes signal detection more difficult. Yet the theory has found numerous applications over the past 25 years in biology, physics, and engineering, and has been demonstrated experimentally in the operation of ring lasers and in the neurons of crayfish.

The phenomenon was first introduced by Roberto Benzi in 1980 in the context of climate research, where it was proposed as a mechanism to explain how dramatic climactic events such as the almost periodic occurrence of the ice ages might be caused by minute changes in the earth's orbit around the sun. The theory has prompted discussions of whether rapid climate change is a hallmark of human impact (e.g., noise in the system).

As explained in Benzi (2010), stochastic resonance can be observed by considering the SDE

$$dX_t = (X_t - X_t^3 + A \sin t)dt + \sigma\, dB_t.$$

Think of the sinusoidal term, called a *periodic forcing*, as representing a weak, external signal, with amplitude A. We are interested in studying the effect of the noise parameter σ on detection of the forcing signal.

The process is simulated using the Euler–Maruyama method.

R : Stochastic Resonance

```
# stochresonance.R
> T <- 100
> n <- 10000
> A <- 0.3
> sigma <- 0.2
> w <- 2*pi/40
> xpath <- numeric(n+1)
> xpath[1]<- 0
> for (i in 2:(n+1)) {
+   x[i]<-x[i-1]+(x[i-1]-x[i-1]^3+A*sin(w*T*(i-1)/n))
+      *T/n+ sigma*sqrt(T/n)*rnorm(1)
> plot(seq(0,T,T/n),x,type="l",ylim=c(-2.8,2.8),
+     xaxt="n",xlab="",ylab="",yaxt="n",lwd=0.5)
> axis(2,c(-1,0,1))
> axis(1,c(0,25,50,75,100))
> curve(A*sin(w*x),0,100,lty=2,add=TRUE)
```

The process $(X_t)_{t\geq 0}$ has two *stable* points, at ± 1. For *small* σ (little noise), paths tend to stay near one of these values, although jumps may occur from one stable point to another. Three sample paths are shown in Figure 9.8 for $\sigma = 0.2$. The periodic forcing function (dashed curve) is not detectable. For this example, the amplitude of the sine function is $A = 0.3$, which is significantly smaller than the distance between the two stable points.

The effect of a relatively *large* random error term, with $\sigma = 2.0$, is apparent in Figure 9.9. The noise swamps any underlying structure. Again, the periodic forcing function is not detectable.

For Figure 9.10, an *optimal* value of σ is chosen at $\sigma = 0.8$. The *hidden* periodic forcing is now apparent. The added noise is sufficient for paths of the process to intersect with the range of the sine wave, which facilitates switching states. The system exhibits stochastic resonance.

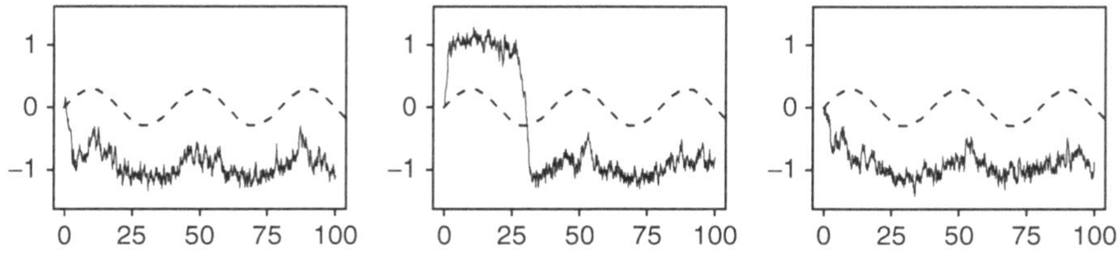

Figure 9.8

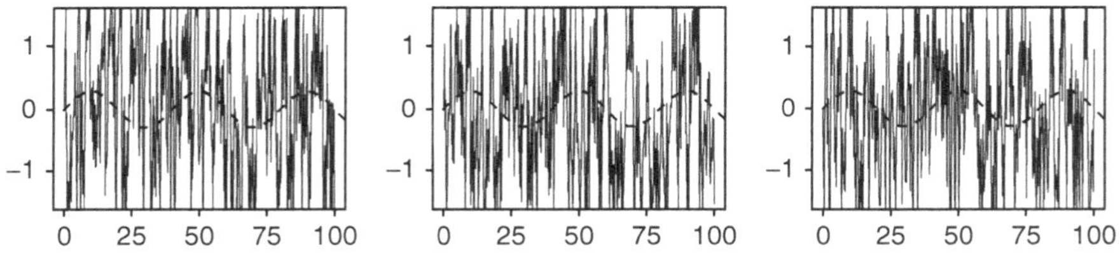

Figure 9.9

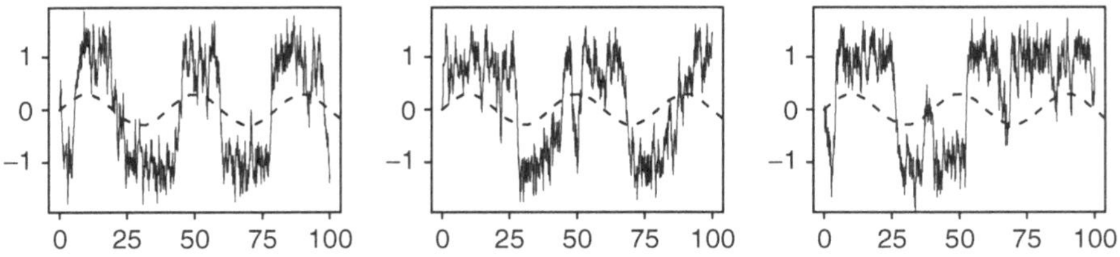

Figure 9.10

■

EXERCISES

9.1 Find the distribution of the stochastic integral $I_t = \int_0^t sB_s \, ds$.

9.2 Show that Brownian motion with drift coefficient μ and variance parameter σ^2 is a diffusion.

9.3 Find $E(B_t^4)$ by using Ito's Lemma to evaluate $d(B_t^4)$.

9.4 Use Ito's Lemma to show that

$$E(B_t^k) = \frac{k(k-1)}{2} \int_0^t E(B_s^{k-2})\, ds, \text{ for } k \geq 2.$$

Use this result to find $E(B_t^k)$, for $k = 1, \ldots, 8$.

9.5 Use the methods of Example 9.6 to derive a martingale that is a fourth-degree polynomial function of Brownian motion.

9.6 Consider the stochastic differential equation

$$dX_t = (1 - 2X_t)dt + 3\, dB_t.$$

(a) Use Ito's Lemma to find $d(e^{rt}X_t)$.

(b) For suitable choice of r, simplify the drift coefficient in the resulting SDE. Solve the SDE and find the mean of X_t, and the asymptotic mean of the process.

9.7 Consider the SDE for the *square root process*

$$dX_t = dt + 2\sqrt{X_t}\, dB_t.$$

With $X_0 = x_0$, show that $X_t = (B_t + x_0)^2$ is a solution.

9.8 R : Show how to use the Euler–Maruyama method to simulate geometric Brownian motion started at $G_0 = 8$, with $\mu = 1$ and $\sigma^2 = 0.25$.

(a) Generate a plot of a sample path on $[0, 2]$.

(b) Simulate the mean and variance of G_2. Compare with the theoretical mean and variance.

9.9 R Use the Euler–Maruyama method to simulate the square root process of Exercise 9.7 with $x_0 = 1$.

(a) Estimate $E(X_3)$, $Var(X_3)$, and $P(X_3 < 5)$.

(b) Using the fact that $X_t = (B_t + x_0)^2$ is a solution to the SDE, compare your simulations in (a) with the exact results.

9.10 R: The random drift model of Example 9.13 is an absorbing process with two absorbing states. Use the Euler-Marayuma method to estimate the expectation and standard deviation of the time until absorption.

9.11 R : The *Cox–Ingersoll–Ross (CIR) model*

$$dX_t = -r(X_t - \mu)dt + \sigma\sqrt{X_t}dB_t$$

has been used to describe the evolution of interest rates. The diffusion has the same drift coefficient as the Ornstein–Uhlenbeck process and is also mean-reverting. The CIR model has the advantage over the Ornstein–Uhlenbeck process as a model for interest rates since, unlike the latter, the process is non-negative. However, unlike that process, the CIR model has no closed-form solution.

With $X_0 = 0$, $\mu = 1.25$, $r = 2$, and $\sigma = 0.2$, simulate the CIR process. Estimate the asymptotic mean and variance by taking $t = 100$. Demonstrate that the process is mean-reverting.

APPENDIX A

GETTING STARTED WITH R

There are many excellent R primers and tutorials on the web. To learn more about R, the best starting place is the homepage of the R Project for Statistical Computing at http://www.r-project.org/. Go there to download the software. The site contains links to books, manuals, demonstrations, and other resources. For this introduction, we assume that you have R up and running on your computer.

1. **R as a calculator**

   When you bring up R the first window you see is the R console. You can type directly on the console, using R like a calculator.

```
> 1+1
[1] 2
> 2-2
[1] 0
> 3*3
[1] 9
> 4/4
[1] 1
> 5^5
[1] 3125
```

Introduction to Stochastic Processes with R, First Edition. Robert P. Dobrow.

R has many built-in math functions. Their names are usually intuitive. R is case sensitive. All commands are in lowercase letters. The # symbol is for comments. Anything appearing after # is ignored by R.

```
> pi
[1] 3.141593
> cos(pi)
[1] -1
> exp(1)    # e^1
[1] 2.718282
> abs(-6)   # |-6|
[1] 6
> factorial(4)
[1] 24
> sqrt(9)
[1] 3
```

Exercises:

1.1. How many orderings are there of a standard 52-card deck?

1.2. An isosceles right triangle has two legs of length 5. Find the length of the hypotenuse.

1.3. Find the tangent of a 60-degree angle.

1.4. The *hyperbolic sine* is the function

$$h(x) = \frac{e^x - e^{-x}}{2}.$$

Evaluate the hyperbolic sine at $x = 1$. The R command for hyperbolic sine is `sinh(x)`. Verify that `sinh(1)` gives the same results.

2. **Navigating the keyboard**

Each command you type in R is stored in history. The up arrow (↑) key scrolls back along this history; the down arrow (↓) scrolls forward. These keystrokes are enormously helpful for navigating your R session.

Suppose you want to calculate log 8 and you type

```
> Log(8)
Error: could not find function "Log"
```

You got an error because R commands start with lower case letters. Rather than retype the command, hit the up arrow key and fix the previously entered entry.

```
> log(8)
[1] 2.079442
```

You can use the <ESC> (escape) key to interrupt a running calculation or output. If R is expecting more input it will preface a line at the console with the + symbol. You can use escape to start anew. In the following, we forget the end parentheses. Repeatedly hitting the return key generated the + symbol as R is expecting the right parentheses to finish the command. By finally typing <ESC>, the command is voided.

```
> exp(1
+
+
+
>
```

3. **Asking for help**

   Prefacing any R command with ? will produce a help file in another window.

```
> ?log
```

Reading the help file for the `log` function you learn that the function has the form `log(x, base = exp(1))`. The `base` argument defaults to exp(1) $= e$. That is, `log(x)` returns the natural logarithm of x. For logarithms in other bases give a second argument to `log`.

```
> log(8,2)
[1] 3
> log(100,10)
[1] 2
```

Exercises:

3.1. Read the help file for the `factorial` command to find the R command for computing the binomial coefficient $\binom{n}{k}$, also called "n choose k." Now compute the number of 5-card Poker hands in a standard 52-card deck $\binom{52}{5}$.

3.2. Use the up arrow key, and replace the 5 in the previous exercise with 13 to find the number of 13-card bridge hands in a standard 52-card deck.

4. **Vectors**

   The real power of R lies in its ability to work with lists of numbers, called vectors. If a and b are integers, the command `a:b` creates a vector of integers from a to b.

```
> 1:10
[1]  1  2  3  4  5  6  7  8  9 10
```

```
> -3:5
[1] -3 -2 -1  0  1  2  3  4  5
```

For more control generating vectors, use the sequence command `seq`.

```
> seq(1,10)
[1]  1  2  3  4  5  6  7  8  9 10
> seq(1,20,2)
[1]  1  3  5  7  9 11 13 15 17 19
> seq(20,1,-4)
[1] 20 16 12  8  4
```

Create and manipulate vectors using the `c` concatenate command.

```
> c(2,3,5,7,11)
[1]  2  3  5  7 11
```

Assign a vector to a variable with the assignment operator `<-`. Here, the first five prime numbers are assigned to a variable called `primes`. The `primes` vector is then lengthened by concatenating three more primes to the original vector.

```
> primes <- c(2,3,5,7,11)
> primes
[1]  2  3  5  7 11
> primes <- c(primes,13,17,19)
> primes
[1]  2  3  5  7 11 13 17 19
```

Elements of vectors are indexed with brackets [ ]. Bracket arguments can be single integers or vectors.

```
> primes[4]
[1] 7
> primes[1:4]
[1] 2 3 5 7
> primes[c(1,4,5,10)]
[1]  2  7 11 NA
```

For the last command we asked for the first, fourth, fifth, and tenth element of `primes`. There is no tenth element so `R` returns an `NA`.

One often wants to find the elements of a vector that satisfy some property, such as those `primes` that are less than 10. This is done with the `which` command, which returns the indices of the vector that satisfy the given property.

```
> primes
[1]  2  3  5  7 11 13 17 19
> which(primes < 10)
[1] 1 2 3 4
> index <- which(primes < 10)
> primes[index]
[1] 2 3 5 7
```

Vectors can consist of numbers, characters, and even strings of characters.

```
> y <- c("Probability", "is", "very", "very",
  "cool")
> y
[1] "Probability" "is" "very" "very" "cool"
> y[c(1,2,5)]
[1] "Probability" "is" "cool"
```

When performing mathematical operations on vectors, the entire vector is treated as a single object.

```
> dog <- seq(0,30,4)
> dog
[1]  0  4  8 12 16 20 24 28
> dog+1
[1]  1  5  9 13 17 21 25 29
> dog*3
[1]  0 12 24 36 48 60 72 84
> 1/dog
[1]  Inf 0.25000 0.12500 0.08333 0.06250 0.05000
[7] 0.041667 0.035714
> cat <- dog+1
> cat
[1]  1  5  9 13 17 21 25 29
> dog*cat
[1]   0  20  72 156 272 420 600 812
```

Notice that when a single number is added to a vector, that number gets added to each element of the vector. But when two vectors of the same length are added together, then corresponding elements are added. This applies to most binary operations.

Many mathematical functions can take vector arguments, returning vectors as output.

```
> factorial(1:7)
[1]    1    2    6   24  120  720 5040
```

```
> sqrt(seq(0,900,100))  # √0, √100, √200, ..., √900
 [1]  0.000 10.000 14.142 17.321 20.000 22.361
 [7] 24.495 26.458 28.284 30.000
```

Here are some common, and intuitive, commands for working with vectors.

```
> x <- c(67.6, 68.7, 66.3, 66.2, 65.5, 70.2, 71.1)
> sum(x)
[1] 475.6
> mean(x)
[1] 67.943
> length(x)
[1] 7
> sort(x)
[1] 65.5 66.2 66.3 67.6 68.7 70.2 71.1
> sort(x,decreasing=T)
[1] 71.1 70.2 68.7 67.6 66.3 66.2 65.5
```

Exercises: Use vector operations:

4.1. Compute the squares of the first ten integers.

4.2. Compute the powers of 2 from 2^1 to 2^{20}.

4.3. For $n = 6$, use the `choose` command to compute the binomial coefficients $\binom{n}{k}$, for $k = 0, \ldots, n$.

4.4. Let $\boldsymbol{x}$ be an n-element vector defined by $x_k = 2\pi k/n$, for $k = 1, \ldots, n$. Find $(\cos x_1, \ldots, \cos x_n)$ for $n = 5$. Repeat for $n = 13$, using the up arrow key rather than retyping the full command.

4.5. Compute the sum of the cubes of the first 100 integers.

4.6. Among the cubes of the first 100 integers, how many are greater than 10,000? Use `which` and `length`.

5. **Generating random numbers**

The `sample` command generates a random sample of a given size from the elements of a vector. The syntax is `sample(vec, size, replace=, prob=)`. Samples can be taken with or without replacement (the default is without). A probability distribution on the elements of `vec` can be specified. The default is that all elements of `vec` are equally likely.

```
> sample(1:10,1)
[1] 8
> sample(1:4,4)
[1] 3 1 4 2
```

```
> sample(c(-8,0,1,4,60),6,replace=T)
[1] 60 -8  1  1  4 60
```

To simulate ten rolls of a six-sided die, type

```
> sample(1:6,10,replace=T)
 [1] 6 3 6 1 3 6 5 3 2 1
```

According to the Red Cross, the distribution of blood types in the United States is O: 44%, A: 42%, B: 10%, and AB: 4%. The following simulates a random sample of 10 people's blood types.

```
> sample(c("O","A","B","AB"),8,replace=T,
   prob=c(0.44,0.42,0.10,0.04))
 [1] "A"   "O"   "O"   "O"   "A"   "A"   "A"   "AB"
```

In a random sample of 50,000 people, how many have blood type B? Use the `which` command and the logical connective ==.

```
> samp <- sample(c("O","A","B","AB"),50000,
    replace=T,prob=c(0.44,0.42,0.10,0.04))
> index <- which(samp=="B")
> length(index)
[1] 5042
```

Other logical connectives are given in Table A.1.

TABLE A.1 Logical Connectives

==	equal to	&	and
!=	not equal to	|	or
>	greater than	!	not
>=	greater than or equal to		
<	less than		
<=	less than or equal to		

In the sample of 50,000 people, how many are either type B or type AB?

```
> length(which(samp=="AB" | samp=="B"))
[1] 7085
```

What proportion of people in the sample are not type O?

```
> length(which(samp != "O"))/50000
[1] 0.56388
```

Exercises:

5.1. Consider a probability distribution on $\{0, 2, 5, 9\}$ with respective probabilities $0.1, 0.2, 0.3, 0.4$. Generate a random sample of size five.

5.2. For the above-mentioned distribution, generate a random sample of size one million and determine the proportion of sample values which are equal to 9.

5.3. Represent the cards of a standard 52-card deck with the numbers 1 to 52. Show how to generate a random five-card Poker hand from a standard 52-card deck. Let the numbers 1, 2, 3, and 4 represent the four aces in the deck. Write a command that will count the number of aces in a random Poker hand. Use `which` and `length`.

5.4. Generate 100,000 integers uniformly distributed between 1 and 20 and count the proportion of samples between 3 and 7.

6. Probability distributions

There are four commands for working with probability distributions in `R`. The commands take the root name of the probability distribution (see Table A.2) and prefix the root with `d`, `p`, `q`, or `r`. These give, respectively, continuous density or discrete probability mass function (`d`), cumulative distribution function (`p`), quantile (`q`), and random variable (`r`).

TABLE A.2 Probability Distributions in R

Distribution	Root	Distribution	Root
beta	beta	log-normal	lnorm
binomial	binom	multinomial	multinom
Cauchy	cauchy	negative binomial	nbinom
chi-squared	chisq	normal	norm
exponential	exp	Poisson	pois
F	f	Student's t	t
gamma	gamma	uniform	unif
geometric	geom	Weibull	weibull
hypergeometric	hyper		

Generate six random numbers uniformly distributed on (0, 1).

```
> runif(6,0,1)
[1] 0.06300 0.44851 0.70231 0.20649 0.14377 0.74398
```

Generate 9 normally distributed variables with mean $\mu = 69$ and standard deviation $\sigma = 2.5$.

```
> rnorm(9,69,2.5)
 [1] 69.548 69.931 68.923 71.153 71.779 68.975
 [7] 67.429 65.621 70.148
```

Find $P(X \leq 2.5)$, where X has an exponential distribution with parameter $\lambda = 1$.

```
> pexp(2.5,1)
[1] 0.91792
```

Suppose SAT scores are normally distributed with mean $\mu = 500$ and standard deviation $\sigma = 100$. Find the 95th quantile of the distribution of SAT scores.

```
> qnorm(0.95,500,100)
[1] 664.49
```

Find $P(X = 10)$, where X has a binomial distribution with parameters $n = 20$ and $p = 0.5$.

```
> dbinom(10,20,0.5)
[1] 0.1762
```

7. **Plots and graphs**

To graph functions and plot data, use `plot`. This workhorse command has enormous versatility. Here is a simple plot comparing two vectors.

```
> radius <- 1:20
> area <- pi*radius^2
> plot(radius,area, main="Area as
function of radius")
```

To graph smooth functions, you can also use the `curve` command. See Figures A.1 and A.2 . The syntax is `curve(expr,from=,to=)`, where `expr` is an expression that is a function of x.

```
> curve(pi*x^2,1,20,xlab="radius",ylab="area",
+ main="Area as function of radius")
```

For displaying data, a histogram is often used, obtained with the command `hist(vec)`. The default is a frequency histogram of counts. A density histogram has bars whose areas sum to one and is obtained using the `hist` command with parameter `freq=F`. A continuous density curve can be overlaid on the histogram by including the parameter `add=T`.

For example, male adult heights in the United States are normally distributed with mean 69 inches and standard deviation 2 inches. Here, we simulate 1,000

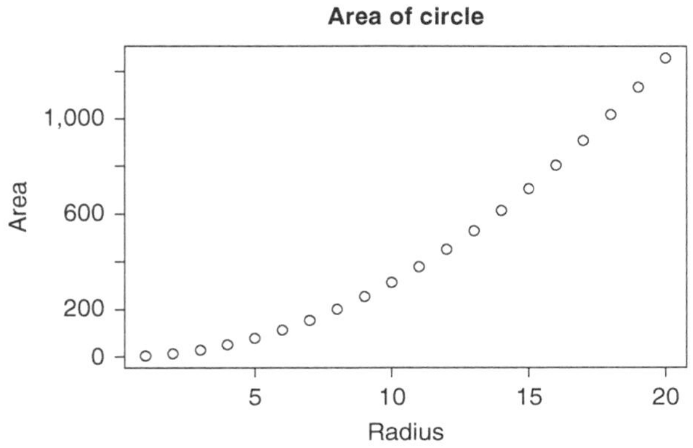

Figure A.1

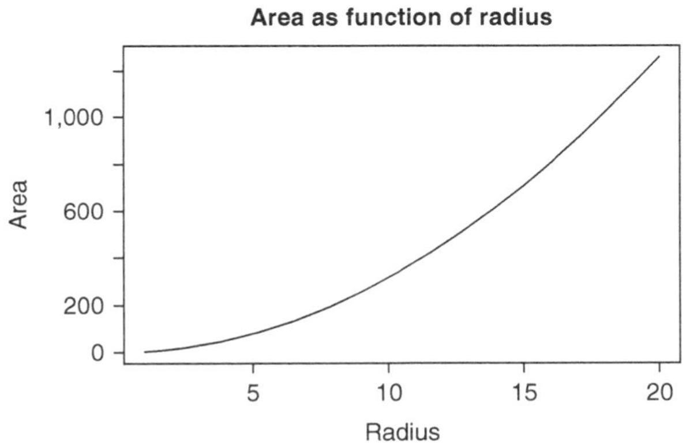

Figure A.2

observations from such a distribution and plot the resulting data. The histogram is first graphed with counts, then with relative frequencies, and then overlaid with a normal density curve. See Figure A.3.

```
> heights <- rnorm(1000,69,2.0)
> hist(heights,main="Distribution of heights")
> hist(heights,main="Distribution of heights",
  freq=F)
> hist(heights,main="Distribution of heights",
  freq=F)
> curve(dnorm(x,69,2),60,80,add=T)
```

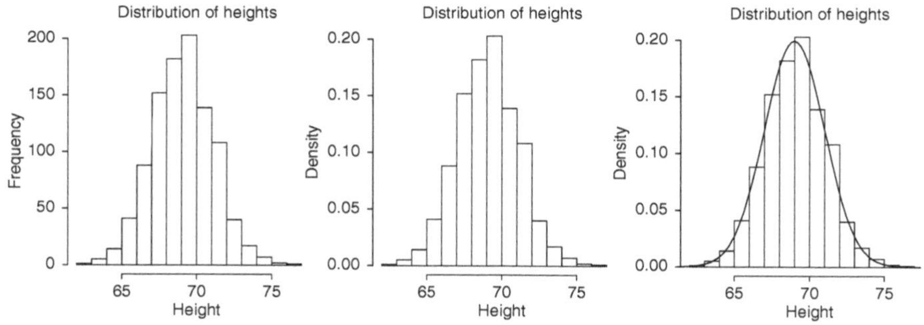

Figure A.3

8. **Script files**

   When you have many R commands to manage, it is useful to keep your work in a script file. A script is a plain text file that contains R commands. The file can be saved and edited and you can execute the entire file or portions of it as you please. Many of the R examples in this book are contained in script files that you can download, execute, and modify.

   Script files can be written in your favorite text editor. They can also be managed directly in your R environment.

   In the Mac environment, click on **File: New Document** to bring up a new window for typing your R commands. To execute a portion of your code, highlight the text that you want to execute, and then hit **Command-Return**. The highlighted code will appear in the R console and be executed.

   In the PC environment, click on **File: New Script** to bring up a new window for a script file. To execute a portion of your code, highlight the text. Under the task bar, press the **Run line or selection** button (third from the left) to execute your code.

   Open the R script file **scriptsample.R**. The file contains two blocks of code. The first block contains commands for plotting the area of a circle as a function of radius for radii from 1 to 20. Highlight the three lines of R code in the first part of the file. Execute the code and you should see a plot similar to that in Figure A.1.

```
# scriptsample.R
### Area of circle
radius <- 1:20
area <- pi*radius^2
plot(radius,area, main="Area as function of radius")
```

```
### Coin flips
n <- 1000 # Number of coin flips
coinflips <- sample(0:1,n,replace=TRUE)
heads <- cumsum(coinflips)
prop <- heads/(1:n)
plot(1:n,prop,type="l",xlab="Number of coins",
  ylab="Running average",
  main="Running proportion of heads in 1000
  coin flips")
abline(h=0.5)
```

The second block of code simulates flipping 1,000 coins, with 1 representing heads and 0 representing tails. The running proportion of heads is plotted from 1 to 1,000. The vector `coinflips` contains the outcome of 1,000 flips. The `cumsum` command generates a cumulative sum and stores the resulting vector in `heads`. The kth element of `heads` gives the number of heads in the first k coin flips. The proportion of heads is then computed and stored in `prop`. Finally, the running proportions are plotted as in Figure A.4. We see that the proportion of heads appears to converge to 1/2, an illustration of the law of large numbers.

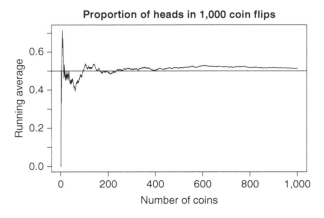

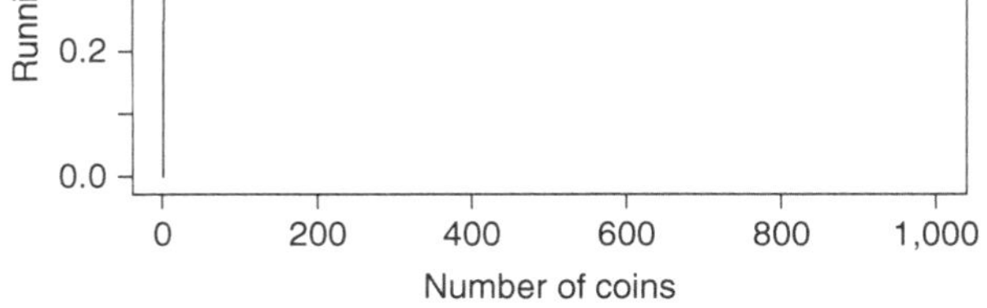

Figure A.4

Exercises:

8.1. Modify the **scriptsample.R** file to compute the volume of a sphere. Plot the volume as a function of radius for radii from 1 to 100.

8.2. Modify the **scriptsample.R** file so that the coin flips are biased and the probability of heads is 0.51. What happens to the resulting plot?

8.3. Given a list of numbers $x_1, \ldots, x_n$, the *sample standard deviation* is

$$s = \sqrt{\frac{1}{n-1}\sum_{i=1}^{n}(x_i - \overline{x})^2},$$

where $\overline{x} = \frac{1}{n}\sum_{i=1}^{n} x_i$ is the average of the x_i, obtained in R with the mean command. Write a command to compute the sample standard deviation for the integers from 1 to n. Use a script file. Compute the standard deviation for $n = 2, 10, 10000$. The R command for the standard deviation is sd(x), where x is a vector. Compare the results of your command with that of the R command.

9. **Functions**

You can create your own functions in R . The syntax is

```
> name <- function(arg1, arg2, . . . ) expression
```

A function can have one, several, or no arguments. Here is a function to compute the area of a circle of given radius.

```
> area <- function(radius) pi*radius^2
```

The name of the function is area. It is a function of one variable.

```
> area(1)
[1] 3.141593
> area(7.5)
[1] 176.7146
```

The function cone computes the volume of a cone of height h and circular base of radius r.

```
> cone <- function(r,h) (1/3)*pi*r^2*h
> cone(1,1)
[1] 1.047198
> cone(3,10)
[1] 94.24778
```

If the function definition contains more than one line of code, enclose the code in curly braces {}. The function allsums takes a vector vec as its input and outputs the sum, sum of squares, and sum of cubes of the elements of vec.

```
> allsums <- function(vec) {
 s1 <- sum(vec)
 s2 <- sum(vec^2)
 s3 <- sum(vec^3)
 c(s1,s2,s3) }
> allsums(1:5)
[1]  15  55 225
> allsums(-5:5)
[1]   0 110   0
```

Exercises:

9.1. Write a function that takes as input the lengths of two sides of a right triangle and returns the length of the hypotenuse.

9.2. A tetrahedron die is a four-sided die with labels $\{1, 2, 3, 4\}$. Write a function with argument n that rolls n tetrahedron dice and computes their average value. Implement your function for $n = 1, 1000$, and 1,000,000.

10. Other useful commands

```
> table(c(0,1,1,1,1,1,2,2,6,6,6,6,6,6,6))
   # create a frequency table
0 1 2 6
1 5 2 7

> min(0:20)    # minimum element
[1] 0

> max(0:20)    # maximum element
[1] 20

> round(pi,3)  # rounds to a given number of
  decimal places
[1] 3.142
```

The `replicate` command is powerful and versatile. The syntax is `replicate(n,expr)`. The expression `expr` is repeated n times and output as a vector.

```
> replicate(6,2)
 [1]  2 2 2 2 2 2
```

Choose 1,000 numbers uniformly distributed between 0 and 1 and find their mean. Then repeat the experiment five times.

```
> replicate(5,mean(runif(1000,0,1)))
```

```
[1] 0.51168 0.50591 0.48371 0.49162 0.50879
```

A Poisson distribution with parameter λ has mean $\mu = \lambda$ and standard deviation $\sigma = \sqrt{\lambda}$. To illustrate the central limit theorem we generate 80 observations $X_1, \ldots, X_{80}$ from a Poisson distribution with parameter $\lambda = 4$ and compute the normalized sum

$$\frac{(X_1 + \cdots + X_n)/n - \mu}{\sigma/\sqrt{n}} = \frac{(X_1 + \cdots + X_{80})/80 - 4}{2/\sqrt{80}}.$$

The simulation is repeated 100,000 times and the resulting data are graphed as a histogram together with a standard normal density curve. See Figure A.5.

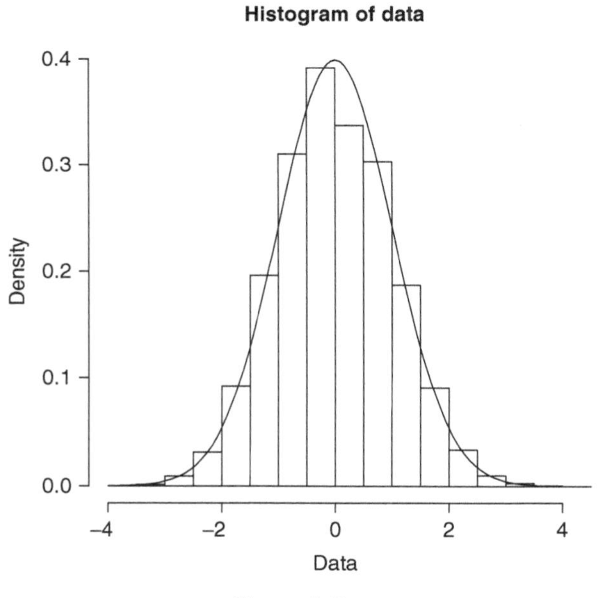

Figure A.5

```
> normsum <- function() (mean(rpois(80,4))-4)/
  (2/sqrt(80))
> normsum()
[1] -0.50312
> normsum()
[1] 1.062132
> data <- replicate(100000,normsum())
> hist(data,freq=F)
> curve(dnorm(x),-4,4,add=T)
```

11. **Working with matrices**

The `matrix` command is used to create matrices from vectors. Specify the number of rows or columns. The default is to fill the matrix by columns. To fill the entries of a matrix from left to right, top to bottom, add the parameter `byrow=T`.

```
> matrix(1:9,nrow=3)
     [,1] [,2] [,3]
[1,]    1    4    7
[2,]    2    5    8
[3,]    3    6    9
> matrix(1:9,nrow=3,byrow=T)
     [,1] [,2] [,3]
[1,]    1    2    3
[2,]    4    5    6
[3,]    7    8    9
```

If x is a scalar, then $x + A$ results in adding x to each component of A. If A and B are matrices, then $A + B$ adds entries componentwise. The same is true for all binary operators.

```
> A <- matrix(c(1,3,0,-4),ncol=2,byrow=T)
> B <- matrix(1:4,ncol=2,byrow=T)
> A
     [,1] [,2]
[1,]    1    3
[2,]    0   -4
> B
     [,1] [,2]
[1,]    1    2
[2,]    3    4
> 2+A
     [,1] [,2]
[1,]    3    5
[2,]    2   -2
> 3*B
     [,1] [,2]
[1,]    3    6
[2,]    9   12
> A+B
     [,1] [,2]
[1,]    2    5
[2,]    3    0
```

```
> A-B
     [,1] [,2]
[1,]    0    1
[2,]   -3   -8
> 2^B
     [,1] [,2]
[1,]    2    4
[2,]    8   16
> A/B
     [,1] [,2]
[1,]    1  1.5
[2,]    0 -1.0
```

For matrix multiplication, or matrix–vector multiplication use the matrix multiplication operator `%*%`.

```
> A %*% A
     [,1] [,2]
[1,]    1   -9
[2,]    0   16
> x <- c(2,1)
> A %*% x
     [,1]
[1,]    5
[2,]   -4
> y <- matrix(c(1,3),nrow=2)
y %*% x
     [,1] [,2]
[1,]    2    1
[2,]    6    3
```

Surprisingly, R does not have a primitive or built-in command for taking matrix powers. Here is a function that will take integer powers of a matrix.

```
matrixpower <- function(mat,k) {
  out <- mat
  for (i in 2:k) {
    out <- out %*% mat
    }
 out
    }

> mat <-matrix(c(0.1,0.9,0.6,0.4),nrow=2,byrow=T)
> mat
```

```
     [,1] [,2]
[1,]  0.1  0.9
[2,]  0.6  0.4
> matrixpower(mat,2)
     [,1] [,2]
[1,] 0.55 0.45
[2,] 0.30 0.70

> matrixpower(mat,10)
          [,1]      [,2]
[1,] 0.4005859 0.5994141
[2,] 0.3996094 0.6003906
> matrixpower(mat,50)
     [,1] [,2]
[1,]  0.4  0.6
[2,]  0.4  0.6
```

For A^T, the transpose of A, type `t(A)`.

```
> A <- matrix(c(3,1,-2,0,0,4),nrow=2,byrow=T)
> A
     [,1] [,2] [,3]
[1,]    3    1   -2
[2,]    0    0    4
> t(A)
     [,1] [,2]
[1,]    3    0
[2,]    1    0
[3,]   -2    4
> A %*% t(A)
     [,1] [,2]
[1,]   14   -8
[2,]   -8   16
```

To solve the linear system $Ax = b$, type `solve(A,b)`. If A is invertible, typing `solve(A)` without the second argument returns the inverse A^{-1}.

```
> A <- matrix(c(2,0,1,1,-1,4,3,1,0),nrow=3,byrow=T)
> A
     [,1] [,2] [,3]
[1,]    2    0    1
[2,]    1   -1    4
[3,]    3    1    0
> b <- c(0,1,1)
```

```
> solve(A,b)
[1] -0.5  2.5  1.0
> A %*% c(-.5,2.5,1)
     [,1]
[1,]    0
[2,]    1
[3,]    1

> solve(A)
     [,1]  [,2]  [,3]
[1,]    1 -0.25 -0.25
[2,]   -3  0.75  1.75
[3,]   -1  0.50  0.50
> A %*% solve(A)
     [,1] [,2] [,3]
[1,]    1    0    0
[2,]    0    1    0
[3,]    0    0    1
```

For integer n, `diag(n)` gives the $n \times n$ identity matrix. If $\boldsymbol{x}$ is a vector, `diag(x)` returns a diagonal matrix whose diagonal elements are $\boldsymbol{x}$. If $\boldsymbol{A}$ is a matrix, `diag(A)` gives the diagonal elements of $\boldsymbol{A}$.

```
> diag(3)
     [,1] [,2] [,3]
[1,]    1    0    0
[2,]    0    1    0
[3,]    0    0    1
> A <- diag(c(1,4,0,2))
> A
     [,1] [,2] [,3] [,4]
[1,]    1    0    0    0
[2,]    0    4    0    0
[3,]    0    0    0    0
[4,]    0    0    0    2
> diag(A)
[1] 1 4 0 2
```

For an $n \times n$ matrix A, the command `eigen(A)` returns the eigenvalues and eigenvectors of A in a two-component list. The first component of the list `eigen(A)$values` is a vector containing the n eigenvalues. The second component `eigen(A)$vectors` is an $n \times n$ matrix whose columns contain the corresponding eigenvectors. To illustrate, we diagonalize the matrix $A = \begin{bmatrix} 1 & 1 & 1 \\ 1 & 1 & 1 \\ 1 & 1 & 1 \end{bmatrix}$. That is, we find an invertible matrix S, and a diagonal matrix D

such that $A = SDS^{-1}$. The diagonal elements of D are the eigenvalues of A. The columns of S are the corresponding eigenvectors.

```
> A <- matrix(replicate(9,1),nrow=3)
> A
     [,1] [,2] [,3]
[1,]    1    1    1
[2,]    1    1    1
[3,]    1    1    1
> eigen(A)$values
[1] 3 0 0
> eigen(A)$vectors
           [,1]        [,2]        [,3]
[1,] 0.57735027  0.70710678  0.40824829
[2,] 0.57735027 -0.70710678  0.40824829
[3,] 0.57735027  0.00000000 -0.81649658
> D <- diag(eigen(A)$values)
> D
     [,1] [,2] [,3]
[1,]    3    0    0
[2,]    0    0    0
[3,]    0    0    0
> S <- eigen(A)$vectors
> S
           [,1]        [,2]        [,3]
[1,] 0.57735027  0.70710678  0.40824829
[2,] 0.57735027 -0.70710678  0.40824829
[3,] 0.57735027  0.00000000 -0.81649658
> S %*% D %*% solve(S)   # SDS^-1 = A
     [,1] [,2] [,3]
[1,]    1    1    1
[2,]    1    1    1
[3,]    1    1    1
```

Exercises:

11.1. Let $A = \begin{pmatrix} 1 & 3 & -1 \\ 2 & 1 & 0 \\ 4 & -2 & 3 \end{pmatrix}$. Find the following:

(i) A^3
(ii) A^T
(iii) A^{-1}
(iv) $p(A)$, where $p(x) = x^5 - 3x^2 + 7$.

11.2. Construct a random 4×4 matrix each of whose elements is uniformly distributed on $(0, 1)$. Find the eigenvalues.

11.3. Solve the linear system

$$\begin{aligned} 2x + 3y - z &= 1, \\ x + y + z &= 2, \\ x + 2y - 3z &= 3. \end{aligned}$$

TABLE A.3 Summary of R Commands for Matrix Algebra

Command	Description
`matrix(vec)`	Creates a matrix from a vector
`A+B`	Matrix addition
`s*A`	Scalar multiplication
`A*B`	Elementwise multiplication
`A %*% B`	Matrix multiplication
`x %*% y`	If $\boldsymbol{x}$ and $\boldsymbol{y}$ are vectors, returns dot product
`solve(A)`	Inverse $\boldsymbol{A}^{-1}$
`solve(A,b)`	Solves the linear system $\boldsymbol{Ax} = \boldsymbol{b}$.
`t(A)`	Transpose $\boldsymbol{A}^T$
`det(A)`	Determinant of A
`diag(s)`	If s is a scalar, creates the $s \times s$ identity matrix
`diag(x)`	If $\boldsymbol{x}$ is a vector, creates diagonal matrix whose diagonal elements are $\boldsymbol{x}$
`diag(A)`	If $\boldsymbol{A}$ is a matrix, returns the diagonal elements
`eigen(A)`	Eigenvalues and eigenvectors of $\boldsymbol{A}$ `eigen(A)$values` is vector of eigenvalues `eigen(A)$vectors` is matrix of eigenvectors

APPENDIX B

PROBABILITY REVIEW

This review of the basics of probability theory focuses on random variables and their distributions. Few results are proven and we refer the reader to a standard undergraduate probability textbook for more complete results. Conditional probability and conditional distribution are discussed in Chapter 1.

Probability begins with a *random experiment*, which is loosely defined as an experiment for which the outcome is uncertain. Given such an experiment, the *sample space* Ω is the set of all possible outcomes. Individual outcomes, that is, the elements of the sample space, are denoted by ω. An *event* A is a subset of the sample space. Say that A occurs if the outcome of the experiment is contained in A.

A probability P is a function that assigns to each event a number between 0 and 1 in such a way that the following conditions are satisfied:

1. $0 \leq P(A) \leq 1$, for all $A \subseteq \Omega$.
2. $P(\Omega) = 1$.
3. Given a sequence of disjoint events $A_1, A_2, \ldots,$

$$P\left(\bigcup_{i=1}^{\infty} A_i\right) = \sum_{i=1}^{\infty} P(A_i).$$

We interpret $P(A)$ to mean *the probability that A occurs*.

A *random variable* is a real-valued function defined on a sample space. That is, the outcomes of a random variable are determined by a random experiment. For instance,

Introduction to Stochastic Processes with R, First Edition. Robert P. Dobrow.

assume that three coins are flipped. Let X be the number of heads. Then, X is a random variable that takes values 0, 1, 2, or 3, depending on the outcome of the coin flips.

Write $P(X = x)$ for the probability that X takes the value x, and $P(X \leq x)$ for the probability that X takes a value less than or equal to x. More generally, for $R \subseteq \mathbb{R}$, write $P(X \in R)$ for the probability that X takes a value that is contained in R. The notation $\{X \in R\}$ is shorthand for $\{\omega : X(\omega) \in R\}$, which is the set of all outcomes ω with the property that $X(\omega)$ is contained in R.

The *distribution* of a random variable X describes the set of values of X and their corresponding probabilities.

The function $F(x) = P(X \leq x)$ is the *cumulative distribution function (cdf) of X*. The cdf takes values between 0 and 1 and is defined for all real numbers. The cdf gives complete probabilistic information about a random variable in the sense that knowing the cdf is equivalent to knowing the distribution of the random variable.

Example B.1 For the random experiment of flipping a fair coin three times, let X denote the number of heads that occur. Letting H denote heads, T denote tails, and keeping track of the order of coin flips, the sample space is

$$\Omega = \{HHH, HHT, HTH, HTT, THH, THT, TTH, TTT\}.$$

If all outcomes are equally likely, then each of the eight outcomes occurs with probability $1/8$. This gives

$$P(X = x) = \begin{cases} P(\{TTT\}) & = 1/8, \quad \text{if } x = 0, \\ P(\{HTT, THT, TTH\}) & = 3/8, \quad \text{if } x = 1, \\ P(\{HHT, HTH, THH\}) & = 3/8, \quad \text{if } x = 2, \\ P(\{HHH\}) & = 1/8, \quad \text{if } x = 3, \end{cases}$$

and cdf

$$F(x) = P(X \leq x) = \begin{cases} 0, & \text{if } x < 0, \\ 1/8, & \text{if } 0 \leq x < 1, \\ 4/8, & \text{if } 1 \leq x < 2, \\ 7/8, & \text{if } 2 \leq x < 3, \\ 1, & \text{if } x \geq 3. \end{cases}$$

■

B.1 DISCRETE RANDOM VARIABLES

A random variable that takes values in a finite or countably infinite set is called a *discrete random variable*. As a function of x, the function $P(X = x)$ is the *probability mass function (pmf) of X*. The pmf describes the distribution of a discrete random variable. For $R \subseteq \mathbb{R}$,

$$P(X \in R) = \sum_{x \in R} P(X = x).$$

The *expectation*, or mean, of a discrete random variable X is defined as

$$E(X) = \sum_x xP(X = x).$$

The expectation is a *weighted average* of the values of X, with weights given by the pmf. Intuitively, the expectation of X is the long-run average value of X over repeated trials.

If g is a function and X is a random variable, then $Y = g(X)$ is a *function of a random variable*, which itself is a random variable that takes the value $g(x)$ whenever X takes the value x. A useful formula for computing the expectation of a function of a random variable is

$$E(Y) = E(g(X)) = \sum_x g(x)P(X = x). \tag{B.1}$$

The expectation is also computed as $E(Y) = \sum_y yP(Y = y)$, which requires knowledge of the distribution of Y.

Example B.2 The radius R of a circle is a random variable that takes values 1, 2, 4, and 8 with respective probabilities 0.1, 0.2, 0.3, and 0.4. Find the expected area of the circle.

Solution Let Y be the area of the circle. The expected area is

$$\begin{aligned} E(Y) &= E\left(\pi R^2\right) \\ &= \sum_{r=1}^{4} \pi r^2 P(R = r) \\ &= \pi\left(1(0.1) + 4(0.2) + 16(0.3) + 64(0.4)\right) \\ &= 31.3\pi. \end{aligned}$$

■

The expectation of a linear function of a random variable is a linear function of the expectation. From Equation (B.1), for constants a and b,

$$\begin{aligned} E(aX + b) &= \sum_x (ax + b)P(X = x) \\ &= a\sum_x xP(X = x) + b\sum_x P(X = x) \\ &= aE(X) + b, \end{aligned}$$

The *variance* of a random variable is a measure of variability or discrepancy from the mean. It is defined as

$$Var(X) = E\left((X - E(X))^2\right) = \sum_x (x - E(X))^2 P(X = x).$$

A computationally useful formula is $Var(X) = E\left(X^2\right) - (E(X))^2$. For constants a and b,

$$\begin{aligned} Var(aX + b) &= E\left(((aX + b) - (aE(X) + b))^2\right) \\ &= E\left(a^2(X - E(X))^2\right) \\ &= a^2 Var(X). \end{aligned}$$

The *standard deviation* of a random variable is defined as $SD(X) = \sqrt{Var(X)}$.

B.2 JOINT DISTRIBUTION

The *joint probability mass function* of X and Y is $P(X = x, Y = y)$, which is a function of x and y. The *joint cumulative distribution function* is

$$F(x, y) = P(X \leq x, Y \leq y) = \sum_{i \leq x} \sum_{j \leq y} P(X = i, Y = j).$$

From the joint pmf of X and Y one can obtain the individual, or marginal, distributions of each random variable. For instance,

$$P(X = x) = P(X = x, -\infty < Y < \infty) = \sum_y P(X = x, Y = y).$$

Similarly, $P(Y = y) = \sum_x P(X = x, Y = y)$.

The *covariance* is a measure of linear association between two random variables. It is defined as

$$\text{Cov}(X, Y) = E((X - E(X))(Y - E(Y))) = E(XY) - E(X)E(Y).$$

The *correlation* between X and Y is

$$\text{Corr}(X, Y) = \frac{\text{Cov}(X, Y)}{SD(X)\, SD(Y)}.$$

The correlation satisfies $-1 \leq \text{Corr}(X, Y) \leq 1$ and is equal to ± 1 if one random variable is a linear function of the other.

If $g(x, y)$ is a function of two variables, and X and Y are random variables, then $g(X, Y)$ is a random variable whose expectation is

$$E(g(X, Y)) = \sum_x \sum_y g(x, y) P(X = x, Y = y).$$

In the case when $g(x, y) = x + y$,

$$\begin{aligned} E(X+Y) &= \sum_x \sum_y (x+y)P(X=x, Y=y) \\ &= \sum_x x \sum_y P(X=x, Y=y) + \sum_y y \sum_x P(X=x, Y=y) \\ &= \sum_x xP(X=x) + \sum_y yP(Y=y) \\ &= E(X) + E(Y), \end{aligned}$$

which gives the important *linearity* property of expectation. For random variables $X_1, \dots, X_n$,

$$E(X_1 + \cdots + X_n) = E(X_1) + \cdots + E(X_n).$$

Example B.3 The solution of the classic *matching problem* is an elegant application of the linearity of expectation.

At the baseball stadium a group of n people wearing baseball hats all throw their hats into the air when their favorite player hits a home run. If the hats are mixed up at random when they fall to the ground, and each person picks up one hat, how many people, on average, will get their own hat back?

Solution Let $I_1, \dots, I_n$ be a sequence of random variables where

$$I_k = \begin{cases} 1, & \text{if the } k\text{th person gets their hat back,} \\ 0, & \text{otherwise,} \end{cases}$$

for $k = 1, \dots, n$. Then, $X = I_1 + \cdots + I_n$ is the number of people who get their hat back. For each k,

$$\begin{aligned} E(I_k) &= (1)P(k\text{th person gets their hat}) \\ &\quad + (0)P(k\text{th person does not get their hat}) \\ &= P(k\text{th person gets their hat}) \\ &= \frac{1}{n}, \end{aligned}$$

since there are n hats to choose from and exactly one belongs to the kth person. By linearity of expectation,

$$E(X) = E(I_1 + \cdots + I_n) = E(I_1) + \cdots + E(I_n) = \sum_{k=1}^{n} \frac{1}{n} = 1.$$

On average, one person gets their hat back. Remarkably, the solution does not depend on n. ■

For the variance of a sum of random variables,

$$Var(X+Y) = Var(X) + Var(Y) + 2\,\mathrm{Cov}(X,Y).$$

More generally,

$$Var\left(\sum_{i=1}^{n} X_i\right) = \sum_{i=1}^{n} Var(X_i) + 2\sum_{i<j} \mathrm{Cov}(X_i, X_j).$$

Events A and B are *independent* if $P(A \cap B) = P(A)P(B)$. Intuitively, events are independent if knowledge of whether or not one occurs has no influence on the probability of whether or not the other occurs.

We say that discrete random variables X and Y are *independent* if

$$P(X = x, Y = y) = P(X = x)P(Y = y) \text{ for all } x, y.$$

Equivalently,

$$P(X \in R, Y \in S) = P(X \in R)P(Y \in S) \text{ for all } R, S \subseteq \mathbb{R}.$$

If X and Y are independent random variables, then

$$\begin{aligned} E(XY) &= \sum_x \sum_y xyP(X = x, Y = y) \\ &= \sum_x \sum_y xyP(X = x)P(Y = y) \\ &= \sum_x xP(X = x) \sum_y yP(Y = y) = E(X)E(Y), \end{aligned}$$

and thus $\mathrm{Cov}(X, Y) = 0$. Hence, for independent random variables, $Var(X + Y) = Var(X) + Var(Y)$.

Sequences of independent and identically distributed (i.i.d.) random variables are common models. For instance, an infinite sequence of fair coin flips can be modeled as an independent sequence $X_1, X_2, \ldots,$ where for each k, $X_k = 1$, if the kth flip is heads, and 0, if the kth flip is tails. The random variable $S_n = X_1 + \cdots + X_n$ is the number of heads in the first n coin flips.

In statistics, one often models a simple random sample as an i.i.d. sequence of random variables from a common population.

B.3 CONTINUOUS RANDOM VARIABLES

A continuous random variable takes values in an uncountable set, most commonly $\mathbb{R}$, $(0, \infty)$ or (a, b), with $a < b$. For continuous random variables $P(X = x) = 0$ for all x, and probabilities are computed by integrating the *probability density function*. The density function plays a role analogous to the pmf for discrete variables for computing probabilities.

A function f is a probability density function of X if

1. $f(x) \geq 0$, for all x
2. $\int_{-\infty}^{\infty} f(x)\,dx = 1$,
3. For all $R \subseteq \mathbb{R}$, $P(X \in R) = \int_R f(x)\,dx$.

The cumulative distribution function of X is

$$F(x) = P(X \leq x) = \int_{-\infty}^{x} f(t)\,dt.$$

Differentiating with respect to x gives

$$\frac{d}{dx}F(x) = f(x).$$

That is, the density is the derivative of the cdf.

Example B.4 Let X be a continuous random variable with density function

$$f(x) = cx^2, \text{ for } 0 < x < 3.$$

(i) Find the constant c. (ii) Find $P(1 < X < 2)$. (iii) Find the density function of $Y = X^2$.

Solution

(i) To find c, solve

$$1 = \int_{-\infty}^{\infty} f(x)\,dx = \int_0^3 cx^2\,dx = 9c,$$

which gives $c = 1/9$.

(ii) The desired probability is

$$P(1 < X < 2) = \int_1^2 f(x)\,dx = \int_1^2 \frac{x^2}{9}\,dx = \frac{7}{27}.$$

(iii) To find the density of Y first find the cdf of Y and then differentiate. Since X takes values between 0 and 3, $Y = X^2$ takes values between 0 and 9. For $0 < y < 9$,

$$P(Y \leq y) = P(X^2 \leq y) = P(X \leq \sqrt{y}).$$

Taking the derivative with respect to y and applying the chain rule gives

$$f_Y(y) = \frac{d}{dy}P(X \leq \sqrt{y}) = \frac{1}{2\sqrt{y}}f_X(\sqrt{y}) = \frac{1}{2\sqrt{y}}\frac{y}{9} = \frac{\sqrt{y}}{18},$$

for $0 < y < 9$. ■

Formulas for expectation and variance are analogous to the discrete formulas with density function replacing pmf and integrals replacing sums. Thus, for a continuous random variable X with density function f,

$$E(X) = \int_{-\infty}^{\infty} xf(x)\,dx \text{ and } Var(X) = \int_{-\infty}^{\infty} (x - E(X))^2 f(x)\,dx.$$

If g is a function, then $E(g(X)) = \int_{-\infty}^{\infty} g(x)f(x)\,dx$.

For continuous random variables, the joint density of X and Y is the function $f(x, y)$ that satisfies

$$P((X, Y) \in R) = \iint_R f(x, y)\,dx\,dy, \text{ for all } R \subseteq \mathbb{R}^2.$$

The joint cdf of X and Y is the function

$$F(x, y) = P(X \le x, Y \le y) = \int_{-\infty}^{x} \int_{-\infty}^{y} f(s, t)\,dt\,ds.$$

If X and Y are independent, then $P(X \le x, Y \le y) = P(X \le x)P(Y \le y)$ for all x and y. Equivalently, the joint density function factors into the product of the marginal densities. That is,

$$f(x, y) = f_X(x)\,f_Y(y), \text{ for all } x, y.$$

B.4 COMMON PROBABILITY DISTRIBUTIONS

Uniform (Discrete) Distribution

The simplest model for a random variable X taking values in a finite set is that all outcomes are equally likely. We say that X has a *uniform distribution*. Historically, probability began by considering equally likely outcomes, mostly in games of chance.

For a finite set R, let $|R|$ denote the number of elements of R. If X is uniformly distributed on $S = \{s_1, \ldots, s_k\}$, then

$$P(X = s_i) = \frac{1}{k}, \text{ for } i = 1, \ldots, k$$

and

$$P(X \in R) = \frac{|R|}{|S|} = \frac{|R|}{k}, \text{ for } R \subseteq S.$$

In the discrete uniform case, probability reduces to counting. The probability that X is contained in R is the number of elements of R divided by the number of elements of the sample space.

Mean and variance are

$$E(X) = \frac{s_1 + \cdots + s_k}{k} \text{ and } Var(X) = \left(\frac{s_1^2 + \cdots + s_k^2}{k}\right) - \left(\frac{s_1 + \cdots + s_k}{k}\right)^2.$$

For the case $S = \{1, \ldots, k\}$, this gives

$$E(X) = \frac{k+1}{2} \text{ and } Var(X) = \frac{k^2-1}{12}.$$

Bernoulli Distribution

A *Bernoulli* random variables takes values 1 and 0, with probabilities p and $1 - p$, respectively. It is common to refer to the dichotomous values of a Bernoulli variable as *success* and *failure*.

If X has a Bernoulli distribution with parameter $0 < p < 1$, then

$$E(X) = p \text{ and } Var(X) = p(1-p).$$

Binomial Distribution

Assume that $X_1, \ldots, X_n$ is an i.i.d. sequence of Bernoulli random variables with common parameter p, where each X_i represents success or failure on the ith trial. Let $X = X_1 + \cdots + X_n$. Then, X counts the number of successes in n trials, and has a *binomial distribution with parameters n and p.*

The pmf of the binomial distribution is derived by a counting argument using the fact that the event $\{X = k\}$ can be expressed as the set of all 0-1 sequences of length n with exactly k 1s. The number of such sequences is counted by the binomial coefficient $\binom{n}{k}$. It follows that

$$P(X = k) = \binom{n}{k} p^k (1-p)^{n-k}, \text{ for } k = 0, 1, \ldots, n,$$

with

$$E(X) = np \text{ and } Var(X) = np(1-p).$$

Example B.5 Whether or not tomato seeds germinate in Angel's garden is modeled as independent Bernoulli random variables with germination (success) probability $p = 0.8$. If 100 seeds are planted, find the probability that at least 75 germinate.

Solution Let X be the number of seeds that germinate. Then, X has a binomial distribution with parameters $n = 100$ and $p = 0.8$. The desired probability is

$$P(X \geq 75) = \sum_{k=75}^{100} \binom{100}{k} (0.8)^k (0.2)^{100-k} = 0.9125.$$

Since $P(X \geq 75) = 1 - P(X < 75) = 1 - P(X \leq 74)$, the probability is obtained in R by typing

```
> 1-pbinom(74,100,0.8)
[1] 0.9125246
```

■

Geometric Distribution

Given a sequence $X_1, X_2, \ldots$ of i.i.d. Bernoulli variables with parameter p, with X_i representing success or failure on the ith trial, let N be the index of the first trial in which success occurs. That is $\{N = k\}$ if and only if $X_i = 0$ for all $i < k$ and $X_k = 1$. Then, N has a *geometric distribution with parameter p* and

$$P(N = k) = (1 - p)^{k-1} p, \text{ for } k = 1, 2, \ldots,$$

with

$$E(N) = \frac{1}{p} \text{ and } Var(N) = \frac{1-p}{p^2}.$$

Example B.6 In Texas hold 'em poker, players are initially dealt two cards. If the deck is reshuffled after each play, find the expected number of deals until a player gets at least one ace.

Solution The outcome of each deal is modeled as a Bernoulli variable with parameter

$$\begin{aligned} p &= P(\text{Player gets at least one ace}) \\ &= 1 - P(\text{Player gets no aces}) \\ &= 1 - \left(\frac{48}{52}\right)\left(\frac{47}{51}\right) = \frac{33}{221} = 0.1493. \end{aligned}$$

The number of deals required for a player to get at least one ace has a geometric distribution with parameter p. The expected number of required deals is $1/p = 221/33 = 6.697$. ■

Poisson Distribution

The Poisson distribution arises as a model for counts of independent events that occur in some fixed region of time or space. Examples include the number of traffic accidents along a stretch of highway, the number of births on a hospital ward, and the number of wrong numbers to your cell phone. The distribution is sometimes called the *law of rare events* and arises when the chance that some event occurs in a small interval of time or space is small.

The distribution depends on a parameter $\lambda > 0$, which can be interpreted as the *rate* of occurrence in a unit interval. A random variable X has a *Poisson distribution with parameter* λ, if

$$P(X = k) = \frac{e^{-\lambda}\lambda^k}{k!}, \text{ for } k = 0, 1, \ldots$$

The distribution has the property that mean and variance are both equal to λ. For the mean,

$$E(X) = \sum_{k=0}^{\infty} kP(X = k) = \sum_{k=0}^{\infty} k\frac{e^{-\lambda}\lambda^k}{k!} = \sum_{k=1}^{\infty} \frac{e^{-\lambda}\lambda^k}{(k-1)\,!}$$

$$= e^{-\lambda}\lambda \sum_{k=1}^{\infty} \frac{\lambda^{k-1}}{(k-1)\,!} = e^{-\lambda}\lambda \sum_{k=0}^{\infty} \frac{\lambda^k}{k!} = e^{-\lambda}\lambda e^{\lambda} = \lambda.$$

Example B.7 During the peak month of May, tornados hit Oklahoma at the rate of about 21.7 per month, according to the National Weather Service. Find the probability that there will be fewer than 15 tornados next May.

Solution If it is assumed that successive tornado hits are independent events, then the Poisson distribution is a reasonable model. Let X be the number of tornados which hit Oklahoma next May. Then,

$$P(X < 15) = P(X \leq 14) = \sum_{k=0}^{14} \frac{e^{-21.7}(21.7)^k}{k!} = 0.054.$$

In R, type

```
> ppois(14,21.7)
[1] 0.05400056
```

■

The Poisson distribution is closely related to the binomial distribution and arises as a limiting distribution when the number of trials is large and the success probability is small. If X has a binomial distribution with large n and small p, then the distribution of X will be approximately equal to a Poisson distribution with parameter $\lambda = np$.

Poisson Approximation of Binomial Distribution

Consider the binomial distribution with parameters n and p_n. Assume that

$$\lim_{n\to\infty} p_n = 0 \text{ and } \lim_{n\to\infty} np_n = \lambda > 0.$$

Then, the binomial pmf converges to the pmf of a Poisson distribution with parameter λ. That is, for $k = 0, 1, \ldots,$

$$\lim_{n\to\infty} \frac{n!}{k!(n-k)!} p_n^k (1-p_n)^{n-k} = \frac{e^{-\lambda}\lambda^k}{k!}.$$

Proof. Consider the binomial probability with $p = \lambda/n$, which gives

$$\begin{aligned}
\frac{n!}{k!(n-k)!} p^k (1-p)^{n-k} &= \frac{n!}{k!(n-k)!} \left(\frac{\lambda}{n}\right)^k \left(1-\frac{\lambda}{n}\right)^{n-k} \\
&= \frac{n(n-1)\cdots(n-k+1)}{k!} \left(\frac{\lambda}{n}\right)^k \left(1-\frac{\lambda}{n}\right)^{n-k} \\
&= \frac{n^k\left(1-\frac{1}{n}\right)\cdots\left(1-\frac{k-1}{n}\right)}{k!} \left(\frac{\lambda}{n}\right)^k \left(1-\frac{\lambda}{n}\right)^{-k} \left(1-\frac{\lambda}{n}\right)^{n} \\
&= \frac{\lambda^k}{k!} \left[\left(1-\frac{1}{n}\right)\cdots\left(1-\frac{k-1}{n}\right)\right] \left(1-\frac{\lambda}{n}\right)^{-k} \left(1-\frac{\lambda}{n}\right)^{n}. \qquad \text{(B.2)}
\end{aligned}$$

Take the limit as $n \to \infty$, and consider the four factors on the right-hand side of Equation (B.2).

(i) Since λ and k are constants, $\lambda^k/k!$ stays unchanged in the limit.

(ii) For fixed k,

$$\lim_{n\to\infty} \left(1-\frac{1}{n}\right)\cdots\left(1-\frac{k-1}{n}\right) = 1^{k-1} = 1.$$

(iii)

$$\lim_{n\to\infty} \left(1-\frac{\lambda}{n}\right)^{-k} = 1^{-k} = 1.$$

(iv) Recall that the constant $e = 2.71827 \ldots$ is defined as the limit

$$\lim_{x\to\infty} \left(1+\frac{1}{x}\right)^x = e.$$

Make the substitution $1/x = -\lambda/n$, so that $n = -\lambda x$. This gives

$$\lim_{n\to\infty} \left(1-\frac{\lambda}{n}\right)^n = \lim_{x\to\infty} \left(1+\frac{1}{x}\right)^{-\lambda x} = \left[\lim_{x\to\infty} \left(1+\frac{1}{x}\right)^x\right]^{-\lambda} = e^{-\lambda}.$$

Plugging in the four limits in Equation (B.2) gives the result. ∎

Example B.8 Mutations in DNA sequences occur from environmental factors and mistakes when a cell copies its DNA in preparation for cell division. The mutation rate per nucleotide of human DNA has been estimated at about 2.5×10^{-8}. There are about 3.3×10^9 nucleotide bases in the human DNA genome. Find the probability that exactly 80 DNA sites mutate in a random person's DNA.

Solution Let X be the number of mutations. If successive mutations are independent, then X has a binomial distribution with $n = 3.3 \times 10^9$ and $p = 2.5 \times 10^{-8}$. The distribution is approximated by a Poisson distribution with parameter

$$\lambda = np = (33 \times 10^9)(2.5 \times 10^{-8}) = 82.5.$$

The desired probability is

$$P(X = 80) \approx \frac{e^{-82.5}(82.5)^{80}}{80!} = 0.043.$$

Note that the exact probability using the binomial distribution is

$$P(X = 80) = \binom{3.3 \times 10^9}{80} (2.5 \times 10^{-8})^{80}(1 - 2.5 \times 10^{-8})^{3.3\times 10^9 - 80}.$$

The approximate probability using the Poisson distribution can be compared with the exact probability in R. We display 12 significant digits, and see that the approximation is good to nine digits.

```
> options(digits=12)
> dpois(80,82.5)
[1] 0.0428838140788
> dbinom(80,3.3*10^9,2.5*10^(-8))
[1] 0.042883814558
```

■

Multinomial Distribution

The multinomial distribution generalizes the binomial distribution. Consider a sequence of n i.i.d. random variables, where each variable takes one of k possible values. Assume that the ith value occurs with probability p_i, with $p_1 + \cdots + p_k = 1$. For $i = 1, \ldots, k$, let X_i denote the number of times outcome i occurs. Then, $(X_1, \ldots, X_k)$ has a *multinomial distribution with parameters* $n, p_1, \ldots, p_k$. The joint pmf is

$$P(X_1 = x_1, \ldots, X_k = x_k) = \frac{n!}{x_1! \cdots x_k!} p_1^{x_1} \cdots p_k^{x_k}, \text{ for } x_1 + \cdots + x_k = n.$$

Marginally, each X_i has a binomial distribution with parameters n and p_i.

Example B.9 According to the Red Cross, the distribution of blood types in the United States is as follows: O: 44%, A: 42%, B: 10%, and AB: 4%. In a sample of six people, find the probability that three are type O, two are A, one is B, and none are AB.

Solution Let X_O, X_A, X_B, and X_{AB} denote the number of people in the sample with the respective blood types. Then, (X_O, X_A, X_B, X_{AB}) has a multinomial distribution with parameters $6, 0.44, 0.42, 0.10, 0.04$. The desired probability is

$$\begin{aligned} P(X_O = 3, X_A = 2, X_B = 1, X_{AB} = 0) \\ &= \frac{6!}{3!2!1!0!}(0.44)^3(0.42)^2(0.10)^1(0.04)^0 \\ &= 0.09016. \end{aligned}$$

■

Uniform (Continuous) Distribution

For $a < b$, the uniform distribution on (a, b) is a continuous model for equally likely outcomes on a bounded interval. The density function is constant. A random variable X is *uniformly distributed on* (a, b) if the density of X is

$$f(x) = \frac{1}{b-a}, \quad \text{for } a < x < b.$$

Mean and variance are

$$E(X) = \frac{a+b}{2} \quad \text{and} \quad Var(X) = \frac{(b-a)^2}{12}.$$

For $a < c < d < b$,

$$P(c < X < d) = \frac{d-c}{b-a} = \frac{\text{Length of } (c, d)}{\text{Length of } (a, b)}.$$

Exponential Distribution

The exponential distribution is a positive continuous distribution which often arises as a model for arrival, or waiting, times. Applications include the time when customers arrive at a queue, when electronic components fail, and when calls come in to a call center. The distribution depends on a parameter $\lambda > 0$, which has the interpretation of the arrival rate.

A random variable X has an *exponential distribution with parameter* λ, if the density function of X is

$$f(x) = \lambda e^{-\lambda x}, \quad \text{for } x > 0.$$

The cdf is

$$F(x) = P(X \leq x) = \int_0^x \lambda e^{-\lambda t}\,dt = 1 - e^{-\lambda x}, \text{ for } x > 0.$$

Mean and variance are

$$E(X) = \frac{1}{\lambda} \text{ and } Var\,(X) = \frac{1}{\lambda^2}.$$

The exponential distribution plays a prominent role in probability models because of its *memoryless* property. A random variable X is memoryless if

$$P(X > s + t | X > s) = P(X > t)\,, \text{ for all } s, t > 0.$$

The exponential distribution is the only continuous distribution which is memoryless.

Example B.10 The lifetime of an electronic component is modeled with an exponential distribution. If components fail on average after 1,200 hours, find the probability that the component lasts more than 1,300 hours.

Solution Let X be the time until the component fails. Since the parameter of an exponential distribution is the reciprocal of the mean, model X with an exponential distribution with parameter $\lambda = 1/1200$. The desired probability is

$$\begin{aligned} P(X > 1300) &= \int_{1300}^{\infty} f(x)\,dx \\ &= \int_{1300}^{\infty} \frac{1}{1200} e^{-x/1200}\,dx \\ &= e^{-1300/1200} = 0.3385. \end{aligned}$$

In R, type

```
> 1-pexp(1300,1/1200)
[1] 0.3384654
```

■

Normal Distribution

The normal distribution plays a central role in statistics, is a common model for numerous natural and biological phenomenon, and arises as the limit for many random processes and distributions. It is also called the Gaussian distribution after Carl Friedrich Gauss who discovered its utility as a model for astronomical measurement errors. The distribution is parameterized by two numbers μ and σ^2, which are the mean and variance, respectively, of the distribution.

A random variable X has a *normal distribution with parameters μ and σ^2*, if the density function of X is

$$f(x) = \frac{1}{\sqrt{2\pi\sigma^2}} e^{-\frac{(x-\mu)^2}{2\sigma^2}}, \text{ for } -\infty < x < \infty.$$

The shape of the density is the famous *bell curve*. The density is symmetric about the line $x = \mu$. The inflection points, where the curvature of the density changes sign, occur one standard deviation from the mean, at the points $\mu \pm \sigma$.

The normal distribution has the property that a linear function of a normal random variable is normal. If X is normally distributed with mean μ and variance σ^2, then $Y = aX + b$ is normally distributed with mean

$$E(Y) = E(aX + b) = aE(X) + b = a\mu + b$$

and variance

$$Var(Y) = Var(aX + b) = a^2 Var(X) = a^2\sigma^2.$$

A common heuristic for working with the normal distribution is the *68–95–99.7 rule*, which says that for a normal distribution, the probability of being one, two, and three standard deviations from the mean is, respectively about 0.68, 0.95, and 0.997.

The *standard normal distribution* is a normal distribution with mean 0 and variance 1.

Example B.11 Assume that babies' birth weights are normally distributed with mean 120 ounces and standard deviation 20 ounces. (i) Find the approximate probability that a random baby's birth weight is between 100 and 140 ounces. (ii) Find the exact probability that a baby's birth weight is less than 136 ounces. (iii) *Low birth weight* is defined as the 5th percentile of the birth weight distribution. At what weight is a baby's birth weight considered low?

Solution

(i) Let X be a random baby's birth weight. The desired probability is $P(100 < X < 140)$. Observe that 100 is one standard deviation below the mean and 140 is one standard deviation about the mean. The event $\{100 < X < 140\}$ is the event that the birth weight is within one standard deviation of the mean. By the 68–95–99.7 rule, the probability is about 0.68.

(ii) The desired probability is

$$P(X < 136) = \int_{-\infty}^{136} f(x)\, dx = \int_{-\infty}^{136} \frac{1}{\sqrt{2\pi(20^2)}} e^{-\frac{(x-120)^2}{2(20)^2}}\, dx.$$

There is no elementary closed form for the cdf of the normal distribution, and numerical methods are needed to solve integrals such as this. In R, type

```
> pnorm(136,120,20)
[1] 0.78814460
```

(iii) The problem asks for the 5th percentile of the normal distribution. Equivalently, we seek the number q such that $P(X \leq q) = 0.05$. In R, use the quantile function.

```
> qnorm(0.05,120,20)
[1] 87.102927461
```

Babies with birth weights less than 87 ounces are considered to have low birth weight. ■

Bivariate Normal Distribution

The bivariate normal distribution for X and Y, and the multivariate normal distribution for $X_1, \ldots, X_m$, are generalizations of the normal distribution to higher dimensions. The bivariate normal distribution is specified by five parameters: $\mu_X, \mu_Y, \sigma_X^2, \sigma_Y^2, \rho$. These are the means and variances of X and Y, and their correlation. If $\mu_X = \mu_Y = 0$ and $\sigma_X^2 = \sigma_Y^2 = 1$, this gives the bivariate standard normal distribution with joint density

$$f(x, y) = \frac{1}{2\pi\sqrt{1-\rho^2}} e^{-\frac{x^2-2\rho xy+y^2}{2(1-\rho^2)}},$$

for $-\infty < x, y < \infty$ and $-1 < \rho < 1$.

The marginal and conditional distributions of a bivariate normal distribution are normal. In particular, the conditional distribution of X given $Y = y$ is normal with mean ρy and variance $1 - \rho^2$. Similarly, the conditional distribution of Y given $X = x$ is normal with mean ρx and variance $1 - \rho^2$.

If X and Y have a bivariate normal distribution, then $aX + bY$ is normally distributed for all nonzero constants a and b.

If $\rho = 0$, we say that X and Y are *uncorrelated*. If normal random variables are uncorrelated, then they are independent. Note that this is not true for random variables in general. If two random variables X and Y are uncorrelated it does not necessarily mean that they are independent.

Gamma Distribution

The gamma distribution is a nonnegative continuous distribution which depends on two parameters. The distribution encompasses a large family of unimodal, skewed and symmetric distribution shapes and is a popular distribution for modeling positive continuous processes. It also arises naturally as the distribution of a sum of i.i.d. exponential random variables.

A random variable X has a *gamma distribution with parameters* $r > 0$ *and* $\lambda > 0$, if the density function of X is

$$f(x) = \frac{1}{\Gamma(r)} \lambda^r x^{r-1} e^{-\lambda x}, \text{ for } x > 0,$$

where $\Gamma(r)$ is the *gamma function*

$$\Gamma(r) = \int_0^\infty t^{r-1} e^{-t}\, dt.$$

If r is a positive integer, then $\Gamma(r) = (r-1)!$ For a gamma random variable,

$$E(X) = \frac{r}{\lambda} \text{ and } Var\,(X) = \frac{r}{\lambda^2}.$$

A sum of n independent exponential random variables with common parameter λ has a gamma distribution with parameters n and λ.

Example B.12 Bella is fishing and the time it takes to catch a fish is exponentially distributed with mean 20 minutes. Every time she catches a fish, she throws it back in the water, and continues fishing. Find the probability that she will catch five fish in the first hour.

Solution Let X_1 denote how long it takes for Bella to catch her first fish. For $k = 2, \ldots, 5$, let X_k be the time from when she throws her $(k-1)$th fish in the water to when she catches her kth fish. Then, $X_1, \ldots, X_5$ are i.i.d. exponential random variables with parameter $\lambda = 1/20$, and $S = X_1 + \cdots + X_5$ is the total time it takes for Bella to catch five fish. The random variable S has a gamma distribution with parameters $r = 5$ and $\lambda = 1/20$. The desired probability is

$$P(S \le 60) = \int_0^{60} \frac{1}{\Gamma(5)} \left(\frac{1}{20}\right)^5 x^4 e^{-x/20}\, dx = 0.185.$$

■

Beta Distribution

The beta distribution is a continuous distribution on (0, 1) that depends on two parameters. It is a generalization of the uniform distribution. A random variable X *has a beta distribution with parameters* $a > 0$ *and* $b > 0$, if the density function of X is

$$f(x) = \frac{1}{B(a,b)} x^{a-1}(1-x)^{b-1}, \text{ for } 0 < x < 1,$$

where

$$B(a,b) = \int_0^1 x^{a-1}(1-x)^{b-1}\,dx = \frac{\Gamma(a)\Gamma(b)}{\Gamma(a+b)}.$$

For a beta random variable,

$$E(X) = \frac{a}{a+b} \quad \text{and} \quad Var(X) = \frac{ab}{(a+b+1)(a+b)^2}.$$

The uniform distribution on (0, 1) is obtained for $a = b = 1$.

B.5 LIMIT THEOREMS

The classic limit theorems of probability are concerned with sequences of i.i.d. random variables. If $X_1, X_2, \ldots$ is such a sequence with common mean $\mu = E(X_1) < \infty$, let $S_n = X_1 + \cdots + X_n$. The *law of large numbers* says that the sequence of averages S_n/n converges to μ, as $n \to \infty$. There are two versions of the law—weak and strong.

Theorem B.1. (Weak law of large numbers). *For any* $\epsilon > 0$,

$$\lim_{n\to\infty} P\left(\left|\frac{S_n}{n} - \mu\right| < \epsilon\right) = 1.$$

While the weak law asserts that for large n, the average S_n/n is with high probability close to μ, it does not say that having come close to μ, the sequence of averages will always stay close to μ.

If a sequence of numbers $x_1, x_2, \ldots$ converges to a limit x then eventually, for n sufficiently large, the terms $x_n, x_{n+1}, x_{n+2}, \ldots$ will all be arbitrarily close to x. That is, for any $\epsilon > 0$, there is some index N such that $|x_n - x| \le \epsilon$, for all $n \ge N$.

The strong law of large numbers asserts that with probability 1 the sequence of averages $S_1/1, S_2/2, S_3/3, \ldots$ behaves precisely in this way.

Theorem B.2. (Strong law of large numbers).

$$P\left(\lim_{n\to\infty} \frac{S_n}{n} = \mu\right) = 1.$$

By the law of large numbers, $S_n/n - \mu$ converges to 0, as $n \to \infty$. The *central limit theorem* asserts that $(S_n/n - \mu)/(\sigma/\sqrt{n})$ converges to a normally distributed random variable. Remarkably, this is true for *any* i.i.d. sequence $X_1, X_2, \ldots$ with finite mean and variance.

Theorem B.3. (Central limit theorem). *For all t,*

$$\lim_{n\to\infty} P\left(\frac{S_n/n-\mu}{\sigma/\sqrt{n}} \le t\right) = P(Z \le t),$$

where Z has a standard normal distribution.

The central limit theorem gives that for large n, $X_1 + \cdots + X_n$ has an approximate normal distribution with mean $n\mu$ and variance $n\sigma^2$.

Example B.13 The number of accidents per week in a stretch of highway has a Poisson distribution with parameter $\lambda = 2$. If the number of accidents is independent from week to week, what is the probability that over a year's time there will be more than 100 accidents?

Solution Let $X_1, \ldots, X_{52}$ be the number of accidents, respectively, during each week of the year. Then, $S_{52} = X_1 + \cdots + X_{52}$ is the total number of accidents in a year. The X_i are i.i.d. with common mean and variance $\lambda = 2$. By the central limit theorem,

$$\begin{aligned} P(S_{52} > 100) &= P\left(\frac{S_{52}/52-\mu}{\sigma/\sqrt{52}} > \frac{100/52-2}{\sqrt{2}/\sqrt{52}}\right) \\ &\approx P(Z > -0.392) = 0.652. \end{aligned}$$

We can compare the central limit approximation with the exact result. The sum of independent Poisson random variables has a Poisson distribution, and $X_1 + \cdots + X_{52}$ has a Poisson distribution with parameter $52\lambda = 104$. The exact probability is

$$P(S_{52} > 100) = 1 - P(S_{52} \le 100) = 1 - \sum_{k=0}^{100} \frac{e^{-104}104^k}{k!} = 0.629. \qquad \blacksquare$$

B.6 MOMENT-GENERATING FUNCTIONS

Moment-generating functions are remarkably versatile tools for proving results involving sums and limits of random variables. Let X be a random variable. The *moment-generating function (mgf) of* X is the function $m(t) = E\left(e^{tX}\right)$, defined for all t for which the expectation exists.

The name *moment-generating* comes from the fact that the moments of X can be derived by taking successive derivatives of the mgf. In particular,

$$m'(t) = \frac{d}{dt}E\left(e^{tX}\right) = E\left(\frac{d}{dt}e^{tX}\right) = E\left(Xe^{tX}\right),$$

and $m'(0) = E(X)$. In general, the kth derivative of the mgf gives

$$m^{(k)}(0) = E\left(X^k\right), \text{ for } k = 1, 2, \ldots$$

Example B.14 Let X be a Bernoulli random variable with success parameter p. Find the mgf of X.

Solution The mgf of X is

$$m(t) = E\left(e^{tX}\right) = e^{t(1)}p + e^{t(0)}(1-p) = 1 - p + pe^t. \qquad \blacksquare$$

Example B.15 Let X have a standard normal distribution. Find the mgf.

Solution The mgf is

$$\begin{aligned} m(t) &= \int_{-\infty}^{\infty} e^{tx} \frac{1}{\sqrt{2\pi}} e^{-x^2/2}\, dx = \frac{1}{\sqrt{2\pi}} \int_{-\infty}^{\infty} e^{-(x^2-2tx)/2}\, dx \\ &= e^{t^2/2} \frac{1}{\sqrt{2\pi}} \int_{-\infty}^{\infty} e^{-(x-t)^2/2}\, dx = e^{t^2/2}, \end{aligned}$$

as the last integral gives the density of a normal distribution with mean t and variance 1 which integrates to 1. Check that $m'(0) = 0 = E(X)$ and $m''(0) = 1 = E\left(X^2\right)$. ■

Here are four key properties of the mgf.

Properties of Moment-Generating Functions

1. If X and Y are independent, then the mgf of $X + Y$ is the product of their respective mgfs. That is,

$$\begin{aligned} m_{X+Y}(t) &= E\left(e^{t(X+Y)}\right) = E\left(e^{tX}e^{tY}\right) = E\left(e^{tX}\right)E\left(e^{tY}\right) \\ &= m_X(t)m_Y(t). \end{aligned}$$

2. For constant c,

$$m_{cX}(t) = m_X(ct).$$

3. Moment-generating functions uniquely determine the underlying probability distribution. That is, if $m_X(t) = m_Y(t)$ for all t, then the distributions of X and Y are the same.

4. *Continuity Theorem:* Let $X_1, X_2, \ldots$ be a sequence of random variables with corresponding mgfs $m_{X_1}, m_{X_2}, \ldots$ Assume that X is a random variable such that for all t, $m_{X_n}(t) \to m_X(t)$, as $n \to \infty$. Then,

$$\lim_{n\to\infty} P(X_n \le x) = P(X \le x),$$

at each x for which $P(X \le x)$ is continuous.

Example B.16 Let X and Y be independent Poisson random variables with respective parameters λ_1 and λ_2. Use moment-generating functions to show that $X + Y$ has a Poisson distribution with parameter $\lambda_1 + \lambda_2$.

Solution The mgf of a Poisson random variable with parameter λ is

$$m(t) = \sum_{k=0}^{\infty} e^{tk} \frac{e^{-\lambda}\lambda^k}{k!} = e^{-\lambda} \sum_{k=0}^{\infty} \frac{(\lambda e^t)^k}{k!} = e^{-\lambda} e^{\lambda e^t} = e^{\lambda(e^t - 1)}.$$

Hence, the mgf of $X + Y$ is

$$m_{X+Y}(t) = m_X(t) m_Y(t) = e^{\lambda_1(e^t-1)} e^{\lambda_2(e^t-1)} = e^{(\lambda_1+\lambda_2)(e^t-1)},$$

which is the mgf of a Poisson random variable with parameter $\lambda_1 + \lambda_2$. ■

APPENDIX C

SUMMARY OF COMMON PROBABILITY DISTRIBUTIONS

DISCRETE DISTRIBUTIONS	
Distribution	**PMF, Expectation, Variance**
Uniform $(1, \ldots, n)$	$P(X = k) = \frac{1}{n}, \quad k = 1, \ldots, n$ $E(X) = \frac{n+1}{2} \quad Var(X) = \frac{n^2-1}{12}$
Binomial	$P(X = k) = \binom{n}{k} p^k (1-p)^{n-k}, \quad k = 0, 1, \ldots, n$ $E(X) = np \quad Var(X) = np(1-p)$
Poisson	$P(X = k) = \frac{e^{-\lambda}\lambda^k}{k!}, \quad k = 0, 1, 2, \ldots$ $E(X) = \lambda \quad Var(X) = \lambda$
Geometric	$P(X = k) = (1-p)^{k-1}p, \quad k = 1, 2, \ldots$ $E(X) = \frac{1}{p} \quad Var(X) = \frac{1-p}{p^2}$ $P(X > t) = (1-p)^t$
Negative binomial	$P(X = k) = \binom{k-1}{r-1} p^r (1-p)^{k-r},$ $k = r, r+1, \ldots, \quad r = 1, 2, \ldots$ $E(X) = \frac{r}{p} \quad Var(X) = \frac{r(1-p)}{p^2}$
Hypergeometric	$P(X = k) = \frac{\binom{D}{k}\binom{N-D}{n-k}}{\binom{N}{n}}, k = 0, 1, \ldots, n$ $E(X) = \frac{nD}{N} \quad Var(X) = \frac{nD(N-D)}{N^2}\left(1 - \frac{n-1}{N-1}\right)$

Introduction to Stochastic Processes with R, First Edition. Robert P. Dobrow.

CONTINUOUS DISTRIBUTIONS	
Distribution	**Density, CDF, Expectation, Variance**
Uniform (a, b)	$f(x) = \frac{1}{b-a}, \quad a < x < b$ $F(x) = \frac{x-a}{b-a}, \quad a < x < b$ $E(X) = \frac{a+b}{2} \quad Var(X) = \frac{(b-a)^2}{12}$
Exponential	$f(x) = \lambda e^{-\lambda x}, \quad x > 0$ $F(x) = 1 - e^{-\lambda x}, \quad x > 0$ $E(X) = \frac{1}{\lambda} \quad Var(X) = \frac{1}{\lambda^2}$
Normal	$f(x) = \frac{1}{\sigma\sqrt{2\pi}} \exp\left(\frac{-(x-\mu)^2}{2\sigma^2}\right), \quad -\infty < x < \infty$ $E(X) = \mu \quad Var(X) = \sigma^2$
Gamma	$f(x) = \lambda e^{-\lambda x} \frac{(\lambda x)^{r-1}}{\Gamma(r)}, \quad x > 0,$ where $\Gamma(r) = \int_0^\infty t^{r-1} e^{-t}\, dt$ $E(X) = \frac{r}{\lambda} \quad Var(X) = \frac{r}{\lambda^2}$
Beta	$f(x) = \frac{1}{B(\alpha, \beta)} x^{\alpha-1}(1-x)^{\beta-1}, \quad 0 < x < 1,$ where $B(\alpha, \beta) = \frac{\Gamma(\alpha)\Gamma(\beta)}{\Gamma(\alpha+\beta)}.$ $E(X) = \frac{\alpha}{\alpha+\beta} \quad Var(X) = \frac{\alpha\beta}{(\alpha+\beta+1)(\alpha+\beta)^2}$

APPENDIX D

MATRIX ALGEBRA REVIEW

D.1 BASIC OPERATIONS

A *matrix* is a rectangular array of elements. Rectangles are arranged in rows and columns. The general form of an $m \times n$ matrix (m rows and n columns) is

$$\boldsymbol{A} = \begin{pmatrix} a_{11} & a_{12} & \cdots & a_{1n} \\ a_{21} & a_{22} & \cdots & a_{2n} \\ \vdots & \vdots & \ddots & \vdots \\ a_{m1} & a_{m2} & \cdots & a_{mn} \end{pmatrix}.$$

The element a_{ij} is the entry in the ith row and jth column of $\boldsymbol{A}$. An $n \times n$ matrix is said to be *square*.

A *row vector* is a $1 \times n$ matrix. A *column vector* is an $n \times 1$ matrix. Matrices are denoted with bold uppercase letters. Vectors are denoted with bold lowercase letters. The ith component of the vector $\boldsymbol{x}$ is denoted x_i. The Euclidean space $\mathbb{R}^n$ consists of all n-element vectors of real numbers.

Given n-element vectors $\boldsymbol{x}$ and $\boldsymbol{y}$, the *dot product*, or *inner product*, of $\boldsymbol{x}$ and $\boldsymbol{y}$ is the number

$$\boldsymbol{x} \cdot \boldsymbol{y} = x_1 y_1 + \cdots + x_n y_n = \sum_{i=1}^{n} x_i y_i.$$

Introduction to Stochastic Processes with R, First Edition. Robert P. Dobrow.

A real number is called a *scalar*. If s is a scalar and $\boldsymbol{A}$ a matrix, then *scalar multiplication* is defined as the product $s\boldsymbol{A}$, which is the matrix obtained by multiplying each of the elements of $\boldsymbol{A}$ by s. For example,

$$(-3)\begin{pmatrix} 4 & 0 \\ -1 & 1 \end{pmatrix} = \begin{pmatrix} -12 & 0 \\ 3 & -3 \end{pmatrix}.$$

Matrix addition is the operation of adding two matrices of the same dimension. Corresponding elements are added. For example,

$$\begin{pmatrix} 3 & 1 & 0 \\ 2 & -4 & 1 \end{pmatrix} + \begin{pmatrix} 2 & 1 & -2 \\ 0 & -4 & 5 \end{pmatrix} = \begin{pmatrix} 5 & 2 & -2 \\ 2 & -8 & 6 \end{pmatrix}.$$

Linear Combination

A *linear combination* of vectors $\boldsymbol{x}_1, \ldots, \boldsymbol{x}_k$ is a vector of the form

$$s_1\boldsymbol{x}_1 + \cdots + s_k\boldsymbol{x}_k,$$

where $s_1, \ldots, s_k$ are scalars. The s_i are called the *coefficients* of the linear combination.

Observe that $\begin{pmatrix} 8 & -1 & -3 \end{pmatrix}$ is a linear combination of $\begin{pmatrix} 1 & 1 & 0 \end{pmatrix}$, $\begin{pmatrix} 2 & -1 & 1 \end{pmatrix}$, and $\begin{pmatrix} 3 & 3 & 3 \end{pmatrix}$ since

$$\begin{pmatrix} 8 & -1 & -3 \end{pmatrix} = (2)\begin{pmatrix} 1 & 1 & 0 \end{pmatrix} + (3)\begin{pmatrix} 2 & -1 & 1 \end{pmatrix} + (0)\begin{pmatrix} 3 & 3 & 3 \end{pmatrix}.$$

The coefficients of this linear combination are 2, 3, and 0.

We define the matrix–vector product $\boldsymbol{Ax}$ of an $m \times n$ matrix $\boldsymbol{A}$ and an $n \times 1$ column vector $\boldsymbol{x}$. Write the columns of $\boldsymbol{A}$ as $\boldsymbol{a}_1, \ldots, \boldsymbol{a}_n$. Then, $\boldsymbol{Ax}$ is defined as the $m \times 1$ column vector, which is the linear combination of the columns of $\boldsymbol{A}$ whose coefficients are the components of $\boldsymbol{x}$. That is,

$$\boldsymbol{Ax} = \begin{pmatrix} | & | & \cdots & | \\ \boldsymbol{a}_1 & \boldsymbol{a}_2 & \cdots & \boldsymbol{a}_n \\ | & | & \cdots & | \end{pmatrix}\begin{pmatrix} x_1 \\ x_2 \\ \vdots \\ x_n \end{pmatrix} = x_1\boldsymbol{a}_1 + x_2\boldsymbol{a}_2 + \cdots + x_n\boldsymbol{a}_n.$$

The ith component of $\boldsymbol{Ax}$ is

$$(\boldsymbol{Ax})_i = \sum_{j=1}^{n} a_{ij}x_j, \text{ for } i = 1, \ldots, m.$$

Equivalently, the ith component of $\boldsymbol{Ax}$ is the dot product of the ith row of $\boldsymbol{A}$ and the vector $\boldsymbol{x}$. For example,

$$\begin{pmatrix} 2 & 1 & 1 \\ 3 & -1 & 2 \\ 1 & 0 & -4 \\ 4 & 2 & 1 \end{pmatrix} \begin{pmatrix} 1 \\ 1 \\ -2 \end{pmatrix} = \begin{pmatrix} 2(1) + 1(1) + 1(-2) \\ 3(1) + (-1)(1) + 2(-2) \\ 1(1) + 0(1) + (-4)(-2) \\ 4(1) + 2(1) + 1(-2) \end{pmatrix} = \begin{pmatrix} 1 \\ -2 \\ 9 \\ 4 \end{pmatrix}$$
$$= (1)\begin{pmatrix} 2 \\ 3 \\ 1 \\ 4 \end{pmatrix} + (1)\begin{pmatrix} 1 \\ -1 \\ 0 \\ 2 \end{pmatrix} + (-2)\begin{pmatrix} 1 \\ 2 \\ -4 \\ 1 \end{pmatrix}.$$

D.2 LINEAR SYSTEM

A *linear system* of m equations in n unknowns is a collection of linear equations of the form

$$\begin{array}{ccccccccc} a_{11}x_1 & + & a_{12}x_2 & + & \cdots & + & a_{1n}x_n & = & b_1 \\ a_{21}x_1 & + & a_{22}x_2 & + & \cdots & + & a_{2n}x_n & = & b_2 \\ \vdots & & \vdots & & & & \vdots & & \vdots \\ a_{m1}x_1 & + & a_{m2}x_2 & + & \cdots & + & a_{mn}x_n & = & b_m \end{array}$$

Such a system can be written succinctly in matrix form as

$$\boldsymbol{Ax} = \boldsymbol{b},$$

where

$$\boldsymbol{A} = \begin{pmatrix} a_{11} & a_{12} & \cdots & a_{1n} \\ a_{21} & a_{22} & \cdots & a_{2n} \\ \vdots & \vdots & \ddots & \vdots \\ a_{m1} & a_{m2} & \cdots & a_{mn} \end{pmatrix}, \quad \boldsymbol{x} = \begin{pmatrix} x_1 \\ x_2 \\ \vdots \\ x_n \end{pmatrix}, \quad \text{and} \quad \boldsymbol{b} = \begin{pmatrix} b_1 \\ b_2 \\ \vdots \\ b_m \end{pmatrix}.$$

Linear systems can have no solutions, infinitely many solutions, or exactly one solution. For instance, the system

$$\begin{aligned} 2x_1 + x_2 &= 5 \\ 2x_1 + x_2 &= 4 \end{aligned}$$

has no solutions. The system

$$\begin{aligned} 2x_1 + x_2 &= 5 \\ 4x_1 + 2x_2 &= 10 \end{aligned}$$

has infinitely many solutions of the form

$$\boldsymbol{x} = \begin{pmatrix} x_1 \\ x_2 \end{pmatrix} = \begin{pmatrix} a \\ 5 - 2a \end{pmatrix}, \quad \text{for all real } a.$$

And the system

$$\begin{aligned} 2x_1 + x_2 &= 5 \\ x_1 - x_2 &= -2 \end{aligned}$$

has the unique solution $\boldsymbol{x} = \begin{pmatrix} 1 \\ 3 \end{pmatrix}$.

D.3 MATRIX MULTIPLICATION

If $\boldsymbol{A}$ is an $m \times n$ matrix and $\boldsymbol{B}$ is an $n \times p$ matrix, then the matrix product $\boldsymbol{AB}$ is defined as the $m \times p$ matrix whose ith column is the matrix–vector product of $\boldsymbol{A}$ and the ith column of $\boldsymbol{B}$. Writing

$$\boldsymbol{B} = \begin{pmatrix} | & | & \cdots & | \\ \boldsymbol{b}_1 & \boldsymbol{b}_2 & \cdots & \boldsymbol{b}_p \\ | & | & \cdots & | \end{pmatrix}$$

gives

$$\boldsymbol{AB} = \boldsymbol{A} \begin{pmatrix} | & | & \cdots & | \\ \boldsymbol{b}_1 & \boldsymbol{b}_2 & \cdots & \boldsymbol{b}_p \\ | & | & \cdots & | \end{pmatrix} = \begin{pmatrix} | & | & \cdots & | \\ \boldsymbol{Ab}_1 & \boldsymbol{Ab}_2 & \cdots & \boldsymbol{Ab}_p \\ | & | & \cdots & | \end{pmatrix}.$$

The ijth element of $\boldsymbol{AB}$ is

$$(\boldsymbol{AB})_{ij} = \sum_{k=1}^{n} a_{ik} b_{kj}.$$

Equivalently, the ijth element of $\boldsymbol{AB}$ is the dot product of the ith row of $\boldsymbol{A}$ and the jth column of $\boldsymbol{B}$. For example,

$$\begin{pmatrix} 0 & 1 & 4 \\ 1 & 2 & 0 \end{pmatrix} \begin{pmatrix} 2 & 3 \\ 1 & 0 \\ 1 & 3 \end{pmatrix} = \begin{pmatrix} 5 & 12 \\ 4 & 3 \end{pmatrix}.$$

Matrix multiplication is not commutative That is, $\boldsymbol{AB}$ does not necessarily equal $\boldsymbol{BA}$.

D.4 DIAGONAL, IDENTITY MATRIX, POLYNOMIALS

Given an $n \times n$ matrix $\boldsymbol{A}$, the entries $a_{11}, \ldots, a_{nn}$ are called the diagonal elements of $\boldsymbol{A}$. An $n \times n$ matrix $\boldsymbol{A}$ is a *diagonal matrix* if $a_{ij} = 0$, for all $i \neq j$.

The $n \times n$ *identity matrix*, denoted $\boldsymbol{I}_n$, is the diagonal matrix all of whose diagonal elements are 1. The columns of the $n \times n$ identity matrix are called the *standard basis vectors of* $\mathbb{R}^n$, denoted $\boldsymbol{e}_1, \ldots, \boldsymbol{e}_n$. That is, $\boldsymbol{e}_k$ is the n-element column vector of all 0s except for a 1 in the kth position.

If $\boldsymbol{A}$ is an $n \times n$ matrix, then $\boldsymbol{AI}_n = \boldsymbol{I}_n\boldsymbol{A} = \boldsymbol{A}$.

If $\boldsymbol{A}$ is a square matrix, then $\boldsymbol{AA} = \boldsymbol{A}^2$ is also a square matrix. Similarly, $\boldsymbol{A}^k$ is well-defined for all integer k. It follows that if $p(x)$ is a polynomial function and $\boldsymbol{A}$ is a square matrix, then $p(\boldsymbol{A})$ is well-defined. For instance, let $p(x) = x^3 - 5x + 6$ and $\boldsymbol{A} = \begin{pmatrix} 1 & 1 \\ 1 & 0 \end{pmatrix}$. Then,

$$p(\boldsymbol{A}) = \boldsymbol{A}^3 - 5\boldsymbol{A} + 6\boldsymbol{I} = \begin{pmatrix} 3 & 2 \\ 2 & 1 \end{pmatrix} - (5)\begin{pmatrix} 1 & 1 \\ 1 & 0 \end{pmatrix} + (6)\begin{pmatrix} 1 & 0 \\ 0 & 1 \end{pmatrix} = \begin{pmatrix} 4 & -3 \\ -3 & 7 \end{pmatrix}.$$

D.5 TRANSPOSE

Given an $m \times n$ matrix $\boldsymbol{A}$, the *transpose* $\boldsymbol{A}^T$ is the $n \times m$ matrix whose ijth element is the jith element of $\boldsymbol{A}$.

A matrix $\boldsymbol{A}$ is *symmetric* if $\boldsymbol{A} = \boldsymbol{A}^T$. That is, $a_{ij} = a_{ji}$ for all i, j. A symmetric matrix is necessarily square.

D.6 INVERTIBILITY

A square matrix $\boldsymbol{A}$ is *invertible* if there exists a matrix $\boldsymbol{B}$ such that $\boldsymbol{AB} = \boldsymbol{BA} = \boldsymbol{I}$. The matrix $\boldsymbol{B}$ is denoted $\boldsymbol{A}^{-1}$ and is called *the inverse of* $\boldsymbol{A}$.

Since $\boldsymbol{AA}^{-1} = \boldsymbol{I}$, it follows that the ith column of $\boldsymbol{A}^{-1}$ is the solution of the linear system $\boldsymbol{Ax} = \boldsymbol{e}_i$, where $\boldsymbol{e}_i$ is the ith standard basis vector.

The solution of a general linear system $\boldsymbol{Ax} = \boldsymbol{b}$ is unique if and only if $\boldsymbol{A}$ is invertible. In that case, the solution is $\boldsymbol{x} = \boldsymbol{A}^{-1}\boldsymbol{b}$.

Properties of Inverse, Transpose

1. $(\boldsymbol{A}^T)^T = \boldsymbol{A}$
2. $(\boldsymbol{A}^{-1})^{-1} = \boldsymbol{A}$
3. $(\boldsymbol{A}^T)^{-1} = (\boldsymbol{A}^{-1})^T$
4. $(\boldsymbol{AB})^T = \boldsymbol{B}^T\boldsymbol{A}^T$
5. $(\boldsymbol{AB})^{-1} = \boldsymbol{B}^{-1}\boldsymbol{A}^{-1}$

D.7 BLOCK MATRICES

It is sometimes convenient to partition a matrix $\boldsymbol{A}$ into smaller *blocks*, such as

$$\boldsymbol{A} = \left(\begin{array}{ccc|c} 1 & 2 & 3 & 4 \\ 5 & 6 & 7 & 8 \\ 9 & 10 & 11 & 12 \\ \hline 13 & 14 & 15 & 16 \end{array}\right) = \left(\begin{array}{c|c} \boldsymbol{B} & \boldsymbol{C} \\ \hline \boldsymbol{D} & \boldsymbol{E} \end{array}\right),$$

where

$$\boldsymbol{B} = \begin{pmatrix} 1 & 2 & 3 \\ 5 & 6 & 7 \\ 9 & 10 & 11 \end{pmatrix}, \quad \boldsymbol{C} = \begin{pmatrix} 4 \\ 8 \\ 12 \end{pmatrix}, \quad \boldsymbol{D} = \begin{pmatrix} 13 & 14 & 15 \end{pmatrix}, \text{ and } \quad \boldsymbol{E} = \begin{pmatrix} 16 \end{pmatrix}.$$

Matrix operations on block matrices can be carried out by treating the blocks as matrix elements. Thus,

$$\begin{aligned} A^2 &= \left(\begin{array}{c|c} \boldsymbol{B} & \boldsymbol{C} \\ \hline \boldsymbol{D} & \boldsymbol{E} \end{array}\right) \left(\begin{array}{c|c} \boldsymbol{B} & \boldsymbol{C} \\ \hline \boldsymbol{D} & \boldsymbol{E} \end{array}\right) \\ &= \left(\begin{array}{c|c} \boldsymbol{B}^2 + \boldsymbol{CD} & \boldsymbol{BC} + \boldsymbol{CE} \\ \hline \boldsymbol{DB} + \boldsymbol{ED} & \boldsymbol{DC} + \boldsymbol{E}^2 \end{array}\right) \\ &= \left(\begin{array}{ccc|c} 90 & 100 & 110 & 120 \\ 202 & 228 & 254 & 280 \\ 314 & 356 & 398 & 440 \\ \hline 426 & 484 & 542 & 600 \end{array}\right). \end{aligned}$$

D.8 LINEAR INDEPENDENCE AND SPAN

Given a set of vectors, if at least one vector can be written as a linear combination of the others, then the vectors are called *linearly dependent*. If none of the vectors in the set can be written as a linear combination of the other vectors, then the vectors are called *linearly independent*.

The vectors $\left\{ \begin{pmatrix} 2 \\ -1 \\ 3 \end{pmatrix}, \begin{pmatrix} 0 \\ 1 \\ -1 \end{pmatrix}, \begin{pmatrix} 4 \\ -3 \\ 7 \end{pmatrix} \right\}$ are linearly dependent, as

$$\begin{pmatrix} 4 \\ -3 \\ 7 \end{pmatrix} = (2) \begin{pmatrix} 2 \\ -1 \\ 3 \end{pmatrix} + (-1) \begin{pmatrix} 0 \\ 1 \\ -1 \end{pmatrix}.$$

The vectors $\{\boldsymbol{e}_1, \boldsymbol{e}_2, \boldsymbol{e}_3\} = \left\{ \begin{pmatrix} 1 \\ 0 \\ 0 \end{pmatrix}, \begin{pmatrix} 0 \\ 1 \\ 0 \end{pmatrix}, \begin{pmatrix} 0 \\ 0 \\ 1 \end{pmatrix} \right\}$ are linearly independent.

The *span* of a set of vectors is the set of all linear combinations of those vectors.

The span of $\left\{ \begin{pmatrix} 1 \\ 0 \\ 0 \end{pmatrix}, \begin{pmatrix} 0 \\ 1 \\ 0 \end{pmatrix} \right\}$ is the set of all vectors of the form

$$a \begin{pmatrix} 1 \\ 0 \\ 0 \end{pmatrix} + b \begin{pmatrix} 0 \\ 1 \\ 0 \end{pmatrix} = \begin{pmatrix} a \\ b \\ 0 \end{pmatrix}, \quad \text{for scalars } a \text{ and } b.$$

Geometrically, this set is the x–y plane in $\mathbb{R}^3$.

D.9 BASIS

A *basis* for $\mathbb{R}^n$ is a set of n vectors $\{\boldsymbol{b}_1, \ldots, \boldsymbol{b}_n\}$ in $\mathbb{R}^n$, which are linearly independent and span $\mathbb{R}^n$. The fact that $\{\boldsymbol{b}_1, \ldots, \boldsymbol{b}_n\}$ span $\mathbb{R}^n$ means that every vector in $\mathbb{R}^n$ can be written as a linear combination of the $\boldsymbol{b}_i$. Together with linear independence we obtain the following result.

Theorem D.1. *If $\{\boldsymbol{b}_1, \ldots, \boldsymbol{b}_n\}$ is a basis for $\mathbb{R}^n$, then every vector in $\mathbb{R}^n$ can be written uniquely as a linear combination of the $\boldsymbol{b}_i$.*

To obtain this unique representation for a given set of vectors $\{\boldsymbol{b}_1, \ldots, \boldsymbol{b}_n\}$, let $\boldsymbol{B}$ be the square matrix obtained by making the $\boldsymbol{b}_i$ the columns of $\boldsymbol{B}$. That is,

$$\boldsymbol{B} = \begin{pmatrix} | & | & \cdots & | \\ \boldsymbol{b}_1 & \boldsymbol{b}_2 & \cdots & \boldsymbol{b}_n \\ | & | & \cdots & | \end{pmatrix}.$$

The matrix $\boldsymbol{B}$ is invertible. If $\boldsymbol{x} = s_1\boldsymbol{b}_1 + \cdots + s_n\boldsymbol{b}_n$ for some choice of $s_1, \ldots, s_n$, then

$$\boldsymbol{x} = \begin{pmatrix} | & | & \cdots & | \\ \boldsymbol{b}_1 & \boldsymbol{b}_2 & \cdots & \boldsymbol{b}_n \\ | & | & \cdots & | \end{pmatrix} \begin{pmatrix} s_1 \\ \vdots \\ s_n \end{pmatrix} = \boldsymbol{B} \begin{pmatrix} s_1 \\ \vdots \\ s_n \end{pmatrix}$$

and thus

$$\begin{pmatrix} s_1 \\ \vdots \\ s_n \end{pmatrix} = \boldsymbol{B}^{-1}\boldsymbol{x}.$$

D.10 VECTOR LENGTH

The *length* (also magnitude or norm) of a vector $\boldsymbol{x}$ is defined as

$$\| \boldsymbol{x} \| = \sqrt{x_1^2 + \cdots + x_n^2} = \sqrt{\boldsymbol{x} \cdot \boldsymbol{x}}.$$

A vector $\boldsymbol{x}$ has unit length if $\| \boldsymbol{x} \| = 1$. For any nonzero vector $\boldsymbol{v}$, the vector $\left(\frac{1}{\|\boldsymbol{v}\|}\right)\boldsymbol{v}$ has unit length, since

$$\left\| \left(\frac{1}{\| \boldsymbol{v} \|}\right) \boldsymbol{v} \right\| = \left(\frac{1}{\| \boldsymbol{v} \|}\right) \| \boldsymbol{v} \| = 1.$$

A most important inequality in linear algebra is the Cauchy–Schwarz inequality, which says that for any n-element vectors $\boldsymbol{x}$ and $\boldsymbol{y}$,

$$|\boldsymbol{x} \cdot \boldsymbol{y}| \leq \| \boldsymbol{x} \| \| \boldsymbol{y} \| .$$

For example, letting $\boldsymbol{x} = \begin{pmatrix} 1 & \cdots & 1 \end{pmatrix}$ and $\boldsymbol{y} = \begin{pmatrix} y_1 & \cdots & y_n \end{pmatrix}$, the inequality yields

$$|y_1 + \cdots + y_n| \leq \sqrt{n}\sqrt{y_1^2 + \cdots + y_n^2}.$$

Equivalently,

$$(y_1 + \cdots + y_n)^2 \leq n(y_1^2 + \cdots + y_n^2).$$

D.11 ORTHOGONALITY

Vectors $\boldsymbol{x}$ and $\boldsymbol{y}$ are said to be *orthogonal* if $\boldsymbol{x} \cdot \boldsymbol{y} = 0$. Geometrically, orthogonal vectors are perpendicular.

A set of vectors $\{\boldsymbol{v}_1, \ldots, \boldsymbol{v}_k\}$ is *orthonormal* if the vectors have unit length and are pairwise orthogonal. That is,

$$\boldsymbol{v}_i \cdot \boldsymbol{v}_j = \begin{cases} 1, & \text{if } i = j, \\ 0, & \text{if } i \neq j. \end{cases}$$

A square matrix whose columns are orthonormal is called an *orthogonal matrix*. If $\boldsymbol{U}$ is an orthogonal matrix, then $\boldsymbol{U}$ is invertible and $\boldsymbol{U}^{-1} = \boldsymbol{U}^T$. The latter follows from the definition of orthonormal vectors since the ijth element of $\boldsymbol{U}^T\boldsymbol{U}$ is the dot product of the ith column of $\boldsymbol{U}$ and the jth column of $\boldsymbol{U}$.

The matrix

$$\boldsymbol{U} = \begin{pmatrix} 1/\sqrt{3} & 1/\sqrt{3} & 1/\sqrt{3} \\ 1/\sqrt{6} & 1/\sqrt{6} & -\sqrt{2/3} \\ -1/\sqrt{2} & 1/\sqrt{2} & 0 \end{pmatrix}$$

is an orthogonal matrix as

$$\begin{pmatrix} 1/\sqrt{3} & 1/\sqrt{3} & 1/\sqrt{3} \\ 1/\sqrt{6} & 1/\sqrt{6} & -\sqrt{2/3} \\ -1/\sqrt{2} & 1/\sqrt{2} & 0 \end{pmatrix} \begin{pmatrix} 1/\sqrt{3} & 1/\sqrt{6} & -1/\sqrt{2} \\ 1/\sqrt{3} & 1/\sqrt{6} & 1/\sqrt{2} \\ 1/\sqrt{3} & -\sqrt{2/3} & 0 \end{pmatrix} = \begin{pmatrix} 1 & 0 & 0 \\ 0 & 1 & 0 \\ 0 & 0 & 1 \end{pmatrix}.$$

D.12 EIGENVALUE, EIGENVECTOR

Let $\boldsymbol{A}$ be a square matrix. If there exists a scalar λ and nonzero (column) vector $\boldsymbol{x}$ such that $\boldsymbol{A}\boldsymbol{x} = \lambda\boldsymbol{x}$, we say that λ is an *eigenvalue* of $\boldsymbol{A}$ with corresponding *eigenvector* $\boldsymbol{x}$.

For example, let $\boldsymbol{A} = \begin{pmatrix} 3 & 0 \\ 1 & -4 \end{pmatrix}$. Observe that

$$\begin{pmatrix} 3 & 0 \\ 1 & -4 \end{pmatrix} \begin{pmatrix} 0 \\ 1 \end{pmatrix} = \begin{pmatrix} 0 \\ -4 \end{pmatrix} = (-4) \begin{pmatrix} 0 \\ 1 \end{pmatrix},$$

which shows that $\lambda = -4$ is an eigenvalue of $\boldsymbol{A}$ with corresponding eigenvector $\begin{pmatrix} 0 \\ 1 \end{pmatrix}$. Further observe that

$$\begin{pmatrix} 3 & 0 \\ 1 & -4 \end{pmatrix}\begin{pmatrix} 7 \\ 1 \end{pmatrix} = \begin{pmatrix} 21 \\ 3 \end{pmatrix} = (3)\begin{pmatrix} 7 \\ 1 \end{pmatrix},$$

which shows that $\lambda = 3$ is an eigenvalue of $\boldsymbol{A}$ with corresponding eigenvector $\begin{pmatrix} 7 \\ 1 \end{pmatrix}$.

One can show that $\lambda = -4$ and $\lambda = 3$ are the only eigenvalues of $\boldsymbol{A}$.

D.13 DIAGONALIZATION

A square matrix $\boldsymbol{A}$ is *diagonalizable* if there exists an invertible matrix $\boldsymbol{S}$ and a diagonal matrix $\boldsymbol{D}$ such that $\boldsymbol{A} = \boldsymbol{SDS}^{-1}$. If $\boldsymbol{A}$ is diagonalizable then the entries of $\boldsymbol{D}$ are the eigenvalues of $\boldsymbol{A}$ and the columns of $\boldsymbol{S}$ are corresponding eigenvectors.

For example, let $\boldsymbol{A} = \begin{pmatrix} 6 & -3 & 7 \\ 6 & -3 & 6 \\ 0 & 0 & -1 \end{pmatrix}$. The eigenvalues of $\boldsymbol{A}$ are $\lambda = 3, -1, 0$, with respective eigenvectors $\left\{\begin{pmatrix} 1 \\ 1 \\ 0 \end{pmatrix}, \begin{pmatrix} -1 \\ 0 \\ 1 \end{pmatrix}, \begin{pmatrix} 1 \\ 2 \\ 0 \end{pmatrix}\right\}$.

Let $\boldsymbol{S} = \begin{pmatrix} 1 & -1 & 1 \\ 1 & 0 & 2 \\ 0 & 1 & 0 \end{pmatrix}$, with $\boldsymbol{S}^{-1} = \begin{pmatrix} 2 & -1 & 2 \\ 0 & 0 & 1 \\ -1 & 1 & -1 \end{pmatrix}$. This gives

$$\begin{aligned} \boldsymbol{SDS}^{-1} &= \begin{pmatrix} 1 & -1 & 1 \\ 1 & 0 & 2 \\ 0 & 1 & 0 \end{pmatrix}\begin{pmatrix} 3 & 0 & 0 \\ 0 & -1 & 0 \\ 0 & 0 & 0 \end{pmatrix}\begin{pmatrix} 2 & -1 & 2 \\ 0 & 0 & 1 \\ -1 & 1 & -1 \end{pmatrix} \\ &= \begin{pmatrix} 6 & -3 & 7 \\ 6 & -3 & 6 \\ 0 & 0 & -1 \end{pmatrix} = \boldsymbol{A}. \end{aligned}$$

That is, $\boldsymbol{A}$ is diagonalizable.

A sufficient condition for an $n \times n$ matrix to be diagonalizable is that there exists n distinct eigenvalues. The following theorem gives a necessary and sufficient condition for diagonalizability.

Theorem D.2. *An $n \times n$ matrix $\boldsymbol{A}$ is diagonalizable if and only if there exists a basis for $\mathbb{R}^n$ consisting of eigenvectors of $\boldsymbol{A}$.*

One advantage of diagonalizability is that it simplifies matrix products. If $\boldsymbol{A}$ is diagonalizable, then

$$\boldsymbol{A}^k = (\boldsymbol{S}\boldsymbol{D}\boldsymbol{S}^{-1})^k = \boldsymbol{S}\boldsymbol{D}^k\boldsymbol{S}^{-1}.$$

In the previous example, for $k \geq 1$,

$$\boldsymbol{A}^k = \boldsymbol{S}\boldsymbol{D}^k\boldsymbol{S}^{-1} = \begin{pmatrix} 1 & -1 & 1 \\ 1 & 0 & 2 \\ 0 & 1 & 0 \end{pmatrix} \begin{pmatrix} 3^k & 0 & 0 \\ 0 & (-1)^k & 0 \\ 0 & 0 & 0 \end{pmatrix} \begin{pmatrix} 2 & -1 & 2 \\ 0 & 0 & 1 \\ -1 & 1 & -1 \end{pmatrix}$$

$$= \begin{pmatrix} (2)3^k & -3^k & (2)3^k - (-1)^k \\ (2)3^k & -3^k & (2)3^k \\ 0 & 0 & (-1)^k \end{pmatrix}.$$

A square matrix $\boldsymbol{A}$ is *orthogonally diagonalizable* if there exists an orthogonal matrix $\boldsymbol{U}$ and a diagonal matrix $\boldsymbol{D}$ such that $\boldsymbol{A} = \boldsymbol{U}\boldsymbol{D}\boldsymbol{U}^T$.

Theorem D.3 *(Spectral Theorem).* A matrix is orthogonally diagonalizable if and only if it is symmetric.

ANSWERS TO SELECTED ODD-NUMBERED EXERCISES

Solutions for Chapter 1

1.1 (b) X_t is the student's status at the end of year t.
State space (discrete): $S = \{$Drop Out, Frosh, Sophomore, Junior, Senior, Graduate$\}$.
Index set (discrete): $I = \{0, 1, 2, \dots \}$.
(e) X_t is the arrival time of student t.
State space (continuous): $[0, 60]$
Index set (discrete): $\{1, 2, \dots, 30\}$.
(f) X_t is the order of the deck of cards after t shuffles.
State space (discrete): Set of all orderings of the deck (52! elements).
Index set (discrete): $\{0, 1, 2, \dots \}$

1.3

$$\sum_{i=1}^{k} P(A|B_i \cap C)P(B_i|C) = \sum_{i=1}^{k} \left(\frac{P(A \cap B_i \cap C)}{P(B_i \cap C)}\right)\left(\frac{P(B_i \cap C)}{P(C)}\right)$$

$$= \sum_{i=1}^{k} \frac{P(A \cap B_i \cap C)}{P(C)} = \frac{1}{P(C)} \sum_{i=1}^{k} P(A \cap B_i \cap C)$$

$$= \frac{P(A \cap C)}{P(C)} = P(A|C).$$

Introduction to Stochastic Processes with R, First Edition. Robert P. Dobrow.

1.5 (a) Uniform on {1, 2, 3, 4, 5, 6 }.
(b) Uniform on {2, 3, 4, 5, 6 }.

1.7 Let X denote the time until the rat finds the cheese. Let 1, 2, and 3 denote each door, respectively. Then,

$$E(X) = E(X|1)P(1) + E(X|2)P(2) + E(X|3)P(3)$$

$$= (2 + E(X))\frac{1}{3} + (3 + E(X))\frac{1}{3} + (1)\frac{1}{3}$$

$$= 2 + E(X)\frac{2}{3}.$$

Thus, $E(X) = 6$ minutes.

1.11 Let x_k be the probability of reaching n when the gambler's fortune is k. As in Example 1.10.

$$x_k = x_{k+1}p + x_{k-1}q, \text{ for } 1 \leq k \leq n-1,$$

with $x_0 = 0$ and $x_n = 1$, which gives

$$x_{k+1} - x_k = (x_k - x_{k-1})\frac{p}{q}, \text{ for } 1 \leq k \leq n-1.$$

It follows that

$$x_k - x_{k-1} = \cdots = (x_1 - x_0)(p/q)^{k-1} = x_1(p/q)^{k-1}, \text{ for all } k.$$

This gives $x_k - x_1 = \sum_{i=2}^{k} x_1(p/q)^{k-1}$, or

$$x_k = \sum_{i=1}^{k} x_1(p/q)^{k-1} = x_1\frac{1-(p/q)^k}{1-p/q}.$$

For $k = n$, this gives

$$1 = x_n = x_1\frac{1-(p/q)^n}{1-p/q}.$$

Thus, $x_1 = (1 - p/q)/(1 - (p/q)^n)$, which gives

$$x_k = \frac{1-(p/q)^k}{1-(p/q)^n}, \text{ for } k = 0, \ldots, n.$$

1.13 (a) $f_{Y|X}(y|x) = 2y/(1 - x^2)$, for $x < y < 1$.
(b) The conditional distribution is uniform on $(0, y)$.

1.15 The area of the circle is π. The equation of the circle is $x^2 + y^2 = 1$. The joint density is

$$f(x, y) = \frac{1}{\pi}, \quad \text{for } -1 < x < 1, -\sqrt{1-x^2} < y < \sqrt{1-x^2}$$

Integrating out the y term gives the marginal density

$$f_X(x) = \int_{-\sqrt{1-x^2}}^{\sqrt{1-x^2}} \frac{1}{\pi}\, dy = \frac{2\sqrt{1-x^2}}{\pi}, \text{ for } -1 < x < 1.$$

The conditional density is

$$f_{Y|X}(y|x) = \frac{f(x,y)}{f_X(x)} = \frac{1/\pi}{2\sqrt{1-x^2}/\pi} = \frac{1}{2\sqrt{1-x^2}},$$

for $-1\sqrt{1-x^2} < y < \sqrt{1-x^2}$. The conditional distribution of Y given $X = x$ is uniform on $\left(-\sqrt{1-x^2}, \sqrt{1-x^2}\right)$.

1.17 $E(X|X > 2) = 4.16525$.

1.21

$$\begin{aligned}\int_0^\infty P(T > t)\, dt &= \int_0^\infty \int_t^\infty f(s)\, ds\, dt = \int_0^\infty \int_0^s f(s)\, dt\, ds \\ &= \int_0^\infty sf(s)\, ds = E(T).\end{aligned}$$

1.23 (b) For $m > n$,

$$\begin{aligned}E(S_m|S_n) &= E(S_n + X_{n+1} + \cdots + X_m|S_n) \\ &= E(S_n|S_n) + E(X_{n+1} + \cdots + X_m|S_n) \\ &= S_n + \sum_{i=n+1}^{m} E(X_i|S_n) = S_n + \sum_{i=n+1}^{m} E(X_i) \\ &= S_n + (m-n)\mu.\end{aligned}$$

1.27 Let T be the total amount spent at the restaurant. Then,

$$E(T) = 200(15) = \$3000,$$

and

$$Var(T) = 9(200) + 15^2(40^2) = 361800, \qquad SD(T) = \$601.50.$$

1.29 Yes.

Solutions for Chapter 2

2.1 (a) 0.6;
(b) $P^2_{32} = 0.27$;
(c) $P_{31}\alpha_3/(\alpha P)_1 = (0.3)(0.5)/(0.17) = 15/17 = 0.882$;
(d) $(0.182, 0.273, 0.545) \cdot (1, 2, 3) = 2.363$.

2.3 $P^3_{10} = 0.517$.

2.5 (a)

$$P = \begin{array}{c} \\ 0 \\ 1 \\ 2 \\ 3 \end{array} \begin{array}{c} \begin{array}{cccc} 0 & 1 & 2 & 3 \end{array} \\ \begin{pmatrix} 0 & 1 & 0 & 0 \\ 1/4 & 0 & 3/4 & 0 \\ 0 & 1/4 & 0 & 3/4 \\ 0 & 0 & 1 & 0 \end{pmatrix} \end{array}.$$

(b) 19/64.
(c) 0.103.

2.9

$$P = \begin{array}{c} \\ a \\ b \\ c \\ d \\ e \end{array} \begin{array}{c} \begin{array}{ccccc} a & b & c & d & e \end{array} \\ \begin{pmatrix} 0 & 0 & 3/5 & 0 & 2/5 \\ 1/7 & 2/7 & 0 & 0 & 4/7 \\ 0 & 2/9 & 2/3 & 1/9 & 0 \\ 0 & 1 & 0 & 0 & 0 \\ 3/4 & 0 & 0 & 1/4 & 0 \end{pmatrix} \end{array}.$$

2.11 (b) $P^3_{0,5} = 0.01327$.

2.25

Socializing	Traveling	Milling	Feeding	Resting
0.148	0.415	0.096	0.216	0.125

Solutions for Chapter 3

3.1

$$\pi = \left(\frac{1}{5}, \frac{1}{3}, \frac{2}{5}, \frac{1}{15}\right).$$

3.3 P and R are regular. Q is regular for $0 < p < 1$.

3.5 (a) All non-negative vectors of the form (a, a, b, c, a), where $3a + b + c = 1$.

3.7 The transition matrix is doubly stochastic. The stationary distribution is uniform.

3.11 (a)

$$\pi_j = \begin{cases} 1/(2k), & \text{if } j = 0, k, \\ 1/k, & \text{if } j = 1, \ldots, k-1. \end{cases}$$

(b) 2,000 steps.

3.13 Communication classes are {4} (recurrent, absorbing); {1, 5} (recurrent); {2, 3} (transient). All states have period 1.

$$\boldsymbol{P} = \begin{array}{c} \\ 2 \\ 3 \\ 1 \\ 5 \\ 4 \end{array} \begin{array}{c} \begin{array}{ccccc} 2 & 3 & 1 & 5 & 4 \end{array} \\ \begin{pmatrix} 1/2 & 1/6 & 1/3 & 0 & 0 \\ 1/4 & 0 & 0 & 1/4 & 1/2 \\ 0 & 0 & 1/2 & 1/2 & 0 \\ 0 & 0 & 1 & 0 & 0 \\ 0 & 0 & 0 & 0 & 1 \end{pmatrix} \end{array}.$$

3.17
$$\boldsymbol{P}^n = \begin{pmatrix} 1/2^n & 1 - 1/2^n \\ 0 & 1 \end{pmatrix} \quad \text{and} \quad \lim_{n\to\infty} \boldsymbol{P}^n = \begin{pmatrix} 2 & +\infty \\ 0 & +\infty \end{pmatrix}.$$

3.19 Let p_x be the expected time to hit d for the walk started in x. By symmetry, $p_b = p_e$ and $p_c = p_e$. Solve

$$p_a = \frac{1}{2}(1 + p_b) + \frac{1}{2}(1 + p_c),$$

$$p_b = \frac{1}{2}(1 + p_a) + \frac{1}{2}(1 + p_c),$$

$$p_c = \frac{1}{4}(1 + p_b) + \frac{1}{4} + \frac{1}{4}(1 + p_a) + \frac{1}{4}(1 + p_c).$$

This gives $p_a = 10$.

3.23
$$\pi_i = 2(k + 1 - i)/(k(k + 1)), \quad \text{for } i = 1, \dots, k.$$

3.25 (a) For $k = 2$, $\boldsymbol{\pi} = (1/6, 2/3, 1/6)$. For $k = 3$, $\boldsymbol{\pi} = (1/20, 9/20, 9/20, 1/20)$.

3.29 Communication classes are: (i) $\{a\}$ transient; (ii) $\{e\}$ recurrent; (iii) $\{c, d\}$ transient; and (iv) $\{b, f, g\}$ recurrent. The latter class has period 2. All other states have period 1.

3.33 For all states i and j, and $m > 0$,

$$P^{N+m}_{ij} = \sum_k P^m_{ik} P^N_{kj}.$$

Since $P^N_{kj} > 0$ for all k, the only way the expression above could be zero is if $P^m_{ik} = 0$ for all k, which is not possible since $\boldsymbol{P}^m$ is a stochastic matrix whose rows sum to 1.

3.43 (a) The chain is ergodic for all $0 \le p, q \le 1$, except $p = q = 0$ and $p = q = 1$. (b) The chain is reversible for all $p = q$, with $0 < p < 1$.

3.47

$$P = \begin{array}{c} \\ 1 \\ 2 \\ 3 \end{array}\begin{array}{c} \begin{array}{ccc} 1 & 2 & 3 \end{array} \\ \begin{pmatrix} 1/3 & 4/9 & 2/9 \\ 0 & 1/2 & 1/2 \\ 1/2 & 0 & 1/2 \end{pmatrix} \end{array}.$$

3.53 (a) The probability that A wins is $1/(2-p)$.
(c) $\alpha = 1270/6049 \approx 0.210$ and $\beta = 737/6049 \approx 0.122$. For the first method, A wins with probability 0.599. For the second method, A wins with probability 0.565.

3.61 Yes. T is a stopping time.

3.63

$$\pi = (0.325, 0.207, 0.304, 0.132, 0.030, .003, .0003).$$

3.67

```
> allsixes <- function() {
+ i <- 0
+ ct <- 0
+ while (ct < 5)
+ { x <- sample(1:6,5-ct,replace=T)
+ sixes <- sum(x==6)
+ ct <- ct + sixes
+ i <- i+1 }
+ i
+ }
> sim <- replicate(10000, allsixes())
> mean(sim)
[1] 13.0873
```

Solutions for Chapter 4

4.1 $P_{0,j} = 1$, if $j = 0$, and 0, otherwise.

$$P_{1,j} = \begin{cases} a, & \text{if } j = 0, \\ b, & \text{if } j = 1, \\ c, & \text{if } j = 2. \end{cases}$$

For the second row of $\boldsymbol{P}$,

$$\begin{array}{c} \\ 2 \end{array}\begin{array}{c} \begin{array}{cccccc} 0 & 1 & 2 & 3 & 4 & 5\ \cdots \end{array} \\ \left(\begin{array}{cccccc} a^2 & 2ab & 2ac+b^2 & 2bc & c^2 & 0\ \cdots \end{array} \right) \end{array}$$

4.3 From Exercise 4.2, $G_X(s) = e^{\lambda(s-1)}$ and $G_Y(s) = e^{\mu(s-1)}$. Then,

$$G_{X+Y}(s) = G_X(s)G_Y(s) = e^{\lambda(s-1)}e^{\mu(s-1)} = e^{(\lambda+\mu)s-1}.$$

Thus, $X + Y$ has a Poisson distribution with parameter $\lambda + \mu$.

4.7

$$\begin{aligned} G(s) &= (1-p) + ps^3. \\ \mu &= G'(1) = 3p. \\ \sigma^2 &= G''(1) + G'(1) - G'(1)^2 = 6p + 3p - (3p)^2 = 9p(1-p). \\ E\left(Z^4\right) &= \mu^4 = (3p)^4 = 81p^4. \\ Var\left(Z^4\right) &= 9p(1-p)(3p)^3(81p^4 - 1)/(3p - 1). \end{aligned}$$

4.11

$$\begin{aligned} E(Z_n) &= G'1_n(1) = G'(G_{n-1}(1))G_{n-1}'(1) = G'(1)G_{n-1}'(1) \\ &= \mu G_{n-1}'(1) = \mu E(Z_{n-1}). \end{aligned}$$

The result follows by induction.

4.15 Solve $s = 1/4 + s/4 + s^2/4 + s^3/4$. $e = 0.414$.

4.17 (a)

$$e = \begin{cases} (1-p)/p, & \text{if } p > 1/2, \\ 1, & \text{if } p \leq 1/2. \end{cases}$$

(b)

$$P_{2i,2j} = \binom{2i}{2j} p^{2j}(1-p)^{2i}.$$

4.21 (a) $\mu = c/(1-p)^2$
(c) 0.693, 0.803

4.23

$$G_Z(s) = \frac{1}{a_0}(G(s) - a_0).$$

4.29 (a)

```
> pgf <- function(s) {
+   0.8 + 0.1*s^4 + 0.1*s^9
+ }
> x <- 0.5  # initial value
> e <- pgf(x)
> for (i in 1:100) {
+   e <- pgf(e)
+ }
> e
[1] 0.9152025
```

(b) 0.101138

4.33

```
> branch <- function(n,lam) {  ## Poisson
+   z <- c(1,rep(0,n))
+   for (i in 2:(n+1)) {
+   z[i] <- sum(rpois(z[i-1],lam))
+   }
+   return(z)    }
    # Assume extinction occurs by 50th generation
> n <- 10000
> simlist <- replicate(n, sum(branch(50,0.60)))
> mean(simlist)
[1] 2.5308
> var(simlist)
[1] 9.594811
```

Solutions for Chapter 5

5.1 Let

$$\boldsymbol{P} = \begin{array}{c} \\ \text{Truck} \\ \text{Car} \end{array} \begin{array}{c} \begin{array}{cc} \text{Truck} & \text{Car} \end{array} \\ \begin{pmatrix} 1/5 & 4/5 \\ 1/4 & 3/4 \end{pmatrix} \end{array},$$

with stationary distribution $\boldsymbol{\pi} = (5/21, 16/21)$. By the strong law of large numbers the toll collected is about

$$1000\left(5\left(\frac{5}{21}\right) + 1.5\left(\frac{16}{21}\right)\right) = \$2333.33.$$

5.3 (a) Compute $(10000) \times \boldsymbol{\lambda P}^2$, with $\boldsymbol{\lambda} = (0.6, 0.3, 0.1)$. This gives Car: 3,645, Bus: 4,165, Bike: 2,190.

For long-term totals, find the stationary distribution and compute $(10000) \times \boldsymbol{\pi}$ to get Car: 2083, Bus: 4583, Bike: 3333.

(b) Current: $271(0.6) + 101(0.3) + 21(0.1) = 195$ g. Long-term: $271(0.208) + 101(0.458) + 21(0.333) = 109.75$ g.

5.7 Assume that the chain is currently at state i. Let j be the proposal state, chosen uniformly on $\{0, 1, \ldots, n\}$. Let $U \sim \text{Uniform}(0, 1)$. Accept j as the next state of the chain if

$$U < \frac{\binom{n}{j} p^j (1-p)^{n-j}}{\binom{n}{i} p^i (1-p)^{n-i}} = \frac{i!(n-i)!}{j!(n-j)!}\left(\frac{p}{1-p}\right)^{j-i}.$$

Otherwise, stay at state i.

5.15

$$P = \begin{array}{c} \\ 123 \\ 132 \\ 213 \\ 231 \\ 312 \\ 321 \end{array} \begin{array}{c} \begin{array}{cccccc} 123 & 132 & 213 & 231 & 312 & 321 \end{array} \\ \begin{pmatrix} 1/2 & 1/8 & 1/8 & 1/8 & 1/8 & 0 \\ 1/8 & 1/2 & 1/8 & 0 & 1/8 & 1/8 \\ 1/8 & 1/8 & 1/2 & 1/8 & 0 & 1/8 \\ 1/8 & 0 & 1/8 & 1/2 & 1/8 & 1/8 \\ 1/8 & 1/8 & 0 & 1/8 & 1/2 & 1/8 \\ 0 & 1/8 & 1/8 & 1/8 & 1/8 & 1/2 \end{pmatrix} \end{array}.$$

5.19

```
> trials <- 20000
> n <- 50
> p <- 1/4
> sim <- numeric(trials)
> for (k in 1:trials) {
+ state <- 0
+ # run chain for 60 steps to be near stationarity
+ for (i in 1:60) {
+ y <- sample(0:n,1)
+ acc <- factorial(state)*factorial(n-state)/
  (factorial(y)*factorial(n-y))
+   *(p/(1-p))^(y-state)
+ if (runif(1) < acc) state <- y
+ }
+ sim[k] <- if (state >= 10 & state <= 15) 1 else 0
+ }
> mean(sim)  # estimate of P(10 <= X <= 15)
[1] 0.6712
# exact probability
> pbinom(15,n,p)-pbinom(9,n,p)
[1] 0.6732328
```

Solutions for Chapter 6

6.1 (a) 0.048;
(b) 0.1898;
(c) 0.297.

6.3 (a) 0.082;
(b) 0.0257;
(c) 0.01299.

6.7 (a) 1/2;
(b) 1/4;
(c) 1/6.

6.9 Let X be geometrically distributed with parameter p. The cumulative distribution function for X is

$$P(X \leq x) = \sum_{k=1}^{x} P(X = k) = \sum_{k=1}^{x} (1-p)^{k-1}p = p\frac{1-(1-p)^x}{1-(1-p)} = 1-(1-p)^x.$$

This gives,

$$P(X > s+t | X > s) = \frac{P(X > s+t)}{P(X > s)} = \frac{(1-p)^{s+t}}{(1-p)^s} = (1-p)^t = P(X > t).$$

6.11 Let X be a memoryless, continuous random variable. Let $g(t) = P(X > t)$. By memorylessness,

$$P(X > t) = P(X > s+t | X > s) = \frac{P(X > s+t)}{P(X > s)}.$$

Thus, $g(s+t) = g(s)g(t)$. It follows that

$$g(t_1 + \cdots + t_n) = g(t_1)\cdots g(t_n).$$

Let $r = p/q = \sum_{i=1}^{p}(1/q)$. Then, $g(r) = (g(1/q))^p$. Also,

$$g(1) = g\left(\sum_{i=1}^{q} \frac{1}{q}\right) = g\left(\frac{1}{q}\right)^q,$$

or $g(1/q) = g(1)^{1/q}$. This gives

$$g(r) = g(1)^{p/q} = g(1)^r = e^{r \ln g(1)},$$

for all rational r. By continuity, for all $t > 0$, $g(t) = e^{-\lambda t}$, where

$$\lambda = -\ln g(1) = -\ln P(X > 1).$$

6.15 (a) 0.112.
(b) 0.472.
(c) 0.997.

6.17

$$E\left(\sum_{n=1}^{N_t} S_n^2\right) = \frac{\lambda t^3}{3}.$$

6.19 $E(T) = 88.74$.

6.25 The expected time of failure was 8:43 a.m. on the last day of the week.

6.27 (a) 0.747;
(b) 0.632;
(c) 0.20.

6.29 0.77.

6.35 $P(N_C = 0) = e^{(1-e^{-1})\pi} = 0.137.$

6.39 $P(N_1 = 1) = (e-2)/e = 0.264.$

6.41 The goal scoring Poisson process has parameter $\lambda = 2.68/90$. Consider two independent thinned processes, each with parameter $p\lambda$, where $p = 1/2$. By conditioning on the number of goals scored in a 90-minute match, the desired probability is

$$\sum_{k=0}^{\infty} \left(\frac{e^{-90\lambda/2}(90\lambda/2)^k}{k!} \right)^2 = \sum_{k=0}^{\infty} \left(\frac{e^{-1.34}1.34^k}{k!} \right)^2 = 0.259.$$

6.43 Mean and variance are 41.89.

Solutions for Chapter 7

7.3

$$Q = \begin{pmatrix} -a & a/2 & a/2 \\ b/2 & -b & b/2 \\ c/2 & c/2 & -c \end{pmatrix}.$$

$$\pi = \left(\frac{bc}{ac+bc+ab}, \frac{ac}{ac+bc+ab}, \frac{ab}{ac+bc+ab} \right).$$

7.7 (a)

$$\begin{aligned}
P'_{11}(t) &= -P_{11}(t) + 3P_{13}(t) \\
P'_{12}(t) &= -2P_{12}(t) + P_{11}(t) \\
P'_{13}(t) &= -3P_{13}(t) + 2P_{12}(t) \\
P'_{21}(t) &= -P_{21}(t) + 3P_{23}(t) \\
P'_{22}(t) &= -2P_{22}(t) + P_{21}(t) \\
P'_{23}(t) &= -3P_{23}(t) + 2P_{22}(t) \\
P'_{31}(t) &= -P_{31}(t) + 3P_{33}(t) \\
P'_{32}(t) &= -2P_{32}(t) + P_{31}(t) \\
P'_{33}(t) &= -3P_{33}(t) + 2P_{32}(t)
\end{aligned}$$

(b)

$$\boldsymbol{P}(t) = \begin{pmatrix} -1 & 0 & 1 \\ -3 & 1 & 1 \\ 3 & 0 & 1 \end{pmatrix} \begin{pmatrix} e^{-4t} & 0 & 0 \\ 0 & e^{-2t} & 0 \\ 0 & 0 & 1 \end{pmatrix} \begin{pmatrix} -1/4 & 0 & 1/4 \\ -3/2 & 1 & 1/2 \\ 3/4 & 0 & 1/4 \end{pmatrix}$$

$$= \frac{1}{4} \begin{pmatrix} 3 + e^{-4t} & 0 & 1 - e^{-4t} \\ 3 + 3e^{-4t} - 6e^{-2t} & 4e^{-2t} & 1 - 3e^{-4t} + 2e^{-2t} \\ 3 - 3e^{-4t} & 0 & 1 + 3e^{-4t}. \end{pmatrix}$$

7.11

$$\frac{d}{dt} e^{tA} = \frac{d}{dt} \sum_{n=0}^{\infty} \frac{t^n}{n!} \boldsymbol{A}^n = \sum_{n=0}^{\infty} \frac{1}{n!} \boldsymbol{A}^n \frac{d}{dt} t^n = \sum_{n=1}^{\infty} \frac{1}{n!} \boldsymbol{A}^n n t^{n-1}$$

$$= \boldsymbol{A} \sum_{n=1}^{\infty} \frac{t^{n-1}}{(n-1)} \boldsymbol{A}^{n-1} = \boldsymbol{A} \sum_{n=0}^{\infty} \frac{t^n}{n!} \boldsymbol{A}^n = \boldsymbol{A} e^{tA}.$$

The second equality is done similarly.

7.13 Taking limits on both sides of $\boldsymbol{P}'(t) = \boldsymbol{P}(t)\boldsymbol{Q}$ gives that $\boldsymbol{0} = \boldsymbol{\pi}\boldsymbol{Q}$. This uses the fact that if a differentiable function $f(t)$ converges to a constant then the derivative $f'(t)$ converges to 0.

7.15 (a)

$$\boldsymbol{Q} = \begin{array}{c} \\ 0 \\ 1 \\ 2 \\ 3 \end{array} \begin{array}{c} \begin{array}{cccc} 0 & 1 & 2 & 3 \end{array} \\ \begin{pmatrix} -1 & 1 & 0 & 0 \\ 0 & -1 & 1 & 0 \\ 0 & 0 & -1 & 1 \\ 1 & 0 & 0 & -1 \end{pmatrix} \end{array}.$$

7.17 The population process is a Yule process. The distribution of X_8, the size of the population at $t = 8$, is negative binomial, with mean and variance

$$E(X_8) = 651,019 \quad \text{and } SD(X_8) = 325,509.$$

7.21 Make 4 an absorbing state. We have

$$(-\boldsymbol{V})^{-1} = \left(- \begin{pmatrix} -1 & 1 & 0 & 0 \\ 1 & -2 & 1 & 0 \\ 0 & 2 & -3 & 1 \\ 0 & 0 & 3 & -4 \end{pmatrix} \right)^{-1} = \begin{pmatrix} 10 & 9 & 4 & 1 \\ 9 & 9 & 4 & 1 \\ 8 & 8 & 4 & 1 \\ 6 & 6 & 3 & 1 \end{pmatrix},$$

with row sums $(24, 23, 21, 16)$. The desired mean time is 23.

7.25 (a) $\boldsymbol{\psi} = (0.1, 0.3, 0.3, 0.3)$.
(b) We have that $(q_1, q_2, q_3, q_4) = (1, 1/2, 1/3, 1/4)$. The stationary distribution $\boldsymbol{\pi}$ is proportional to $(0.1, 0.3(2), 0.3(3), 0.3(4))$. This gives

$$\boldsymbol{\psi} = \frac{1}{2.8}(0.1, 0.6, 0.9, 1.2) = (0.036, 0.214, 0.321, 0.428).$$

7.27 (a) If the first dog has i fleas, then the number of fleas on the dog increases by one the first time that one of the $N - i$ fleas on the other dog jumps. The time of that jump is the minimum of $N - i$ independent exponential random variables with parameter λ. Similarly, the number of fleas on the first dog decreases by one when one of the i fleas on that dog first jumps.
(b) The local balance equations are $\pi_i(N - i)\lambda = \pi_{i+1}(i + 1)\lambda$. The equations are satisfied by the stationary distribution

$$\pi_k = \binom{N}{k}\left(\frac{1}{2}\right)^k, \text{ for } k = 0, 1, \dots, N,$$

which is a binomial distribution with parameters N and $p = 1/2$.
(c) 0.45 minutes.

7.29 (b) The embedded chain transition matrix, in canonical form, is

$$\tilde{\boldsymbol{P}} = \begin{array}{c} \\ 1 \\ 5 \\ 2 \\ 3 \\ 4 \end{array}\begin{array}{c} \begin{array}{ccccc} 1 & 5 & 2 & 3 & 4 \end{array} \\ \begin{pmatrix} 1 & 0 & 0 & 0 & 0 \\ 0 & 1 & 0 & 0 & 0 \\ 1/3 & 0 & 0 & 2/3 & 0 \\ 0 & 0 & 1/2 & 0 & 1/2 \\ 0 & 3/5 & 0 & 2/5 & 0 \end{pmatrix} \end{array}.$$

By the discrete-time theory for absorbing Markov chains, write

$$\tilde{\boldsymbol{Q}} = \begin{pmatrix} 0 & 2/3 & 0 \\ 1/2 & 0 & 1/2 \\ 0 & 2/5 & 0 \end{pmatrix} \quad \text{and} \quad \tilde{\boldsymbol{R}} = \begin{pmatrix} 1/3 & 0 \\ 0 & 0 \\ 0 & 3/5 \end{pmatrix}.$$

The matrix of absorption probabilities is

$$(\boldsymbol{I} - \tilde{\boldsymbol{Q}})^{-1}\tilde{\boldsymbol{R}} = \begin{array}{c} \\ 2 \\ 3 \\ 4 \end{array}\begin{array}{c} \begin{array}{cc} 1 & 5 \end{array} \\ \begin{pmatrix} 4/7 & 3/7 \\ 5/14 & 9/14 \\ 1/7 & 6/7 \end{pmatrix} \end{array}.$$

The desired probability is $3/7$.

7.31 The process is an M/M/2 queue with $\lambda = 2$, $\mu = 3$, and $c = 2$. The desired probability is

$$\pi_0 = \left(1 + \frac{2}{3} + \frac{1}{3}\right)^{-1} = \frac{1}{2}$$

7.33 (a) The long-term expected number of customers in the queue L is the mean of a geometric distribution on $0, 1, 2, \dots$, with parameter $1 - \lambda/\mu$, which is $\lambda/(\mu - \lambda)$. If both λ and μ increase by a factor of k, this does not change the value of L.
(b) The expected waiting time is $W = L/\lambda$. The new waiting time is $L/(k\lambda) = W/k$.

7.37 (c) Choose N such that $P(Y > N) < 0.5 \times 10^{-3}$, where Y is a Poisson random variable with parameter $9 \times 0.8 = 7.2$. This gives $N = 17$.

Solutions for Chapter 8

8.3 (a) 0.013.
(b) $f_{X_2|X_1}(x|0) = \frac{1}{\sqrt{2\pi}} e^{-x^2/2}$, for $-\infty < x < \infty$.
(c) 3.
(d) X_1.

8.5 (a) For the joint density of B_s and B_t, since

$$\{B_s = x, B_t = y\} = \{B_s = x, B_t - B_s = y - x\},$$

it follows that

$$f_{B_s,B_t}(x,y) = f_{B_s}(x) f_{B_{t-s}}(y-x) = \frac{1}{\sqrt{2\pi s}} e^{-x^2/2s} \frac{1}{\sqrt{2\pi(t-s)}} e^{-(y-x)^2/2(t-s)}.$$

(b) $E(B_s|B_t = y) = sy/t$ and $Var(B_s|B_t = y) = s(t-s)/t$.

8.7 One checks that the reflection is a Gaussian process with continuous paths. Furthermore, the mean function is $E(-B_t) = 0$ and the covariance function is

$$E((-B_s)(-B_t)) = E(B_s B_t) = \min\{s, t\}$$

8.11 0.11.

8.13 $E(X_s X_t) = st\mu^2 + \sigma^2 s$.

8.19 $E(M_t) = \sqrt{\frac{2t}{\pi}}$ and $Var(M_t) = (1 - 2/\pi)t$.

8.23 (a) $\arcsin(\sqrt{r/t}) / \arcsin(\sqrt{r/s})$.
(b) $\sqrt{s}/\sqrt{t}$.

8.25 0.1688.

8.27 Write $Z_{n+1} = \sum_{i=1}^{Z_n} X_i$. Then,

$$\begin{aligned} E\left(\frac{Z_{n+1}}{\mu^{n+1}} | Z_n, \dots, Z_0\right) &= \frac{1}{\mu^{n+1}} E\left(\sum_{i=1}^{Z_n} X_i | Z_n, \dots, Z_0\right) \\ &= \frac{1}{\mu^{n+1}} E\left(\sum_{i=1}^{Z_n} X_i | Z_n\right) \\ &= \frac{1}{\mu^{n+1}} Z^n \mu = \frac{Z_n}{\mu^n}. \end{aligned}$$

8.31 (a) 4/9.
(b) 20.

8.35 $SD(T) = \sqrt{2/3}a^2$.

8.43 (a) Black–Scholes price is \$35.32.
(b) Price is increasing in each of the parameters, except strike price, which is decreasing.
(c) $\sigma^2 \approx 0.211$.

Solutions for Chapter 9

9.1 The distribution is normal, with

$$E\left(\int_0^t sB_s\, ds\right) = \int_0^t sE(B_s)\, ds = 0,$$

and

$$\begin{aligned}
Var\left(\int_0^t sB_s\, ds\right) &= E\left(\left(\int_0^t sB_s\, ds\right)^2\right) = \int_{x=0}^t \int_{y=0}^t E(xB_x yB_y)\, dy\, dx \\
&= \int_{x=0}^t \int_{y=0}^x xyE(B_x B_y)\, dy\, dx + \int_{x=0}^t \int_{y=x}^t xyE(B_x B_y)\, dy\, dx \\
&= \int_{x=0}^t \int_{y=0}^x xy^2\, dy\, dx + \int_{x=0}^t \int_{y=x}^t x^2 y\, dy\, dx \\
&= \int_{x=0}^t \frac{x^4}{3}\, dx + \int_{x=0}^t x^2\left(\frac{t^2}{2} - \frac{x^2}{2}\right)\, dx \\
&= \frac{t^5}{15} + \frac{t^5}{6} - \frac{t^5}{10} = \frac{2t^5}{15}.
\end{aligned}$$

9.3 By Ito's Lemma, with $g(t, x) = x^4$,

$$d(B_t^4) = 6B_t^2 dt + 4B_t^3\, dB_t,$$

which gives

$$B_t^4 = 6\int_0^t B_s^2\, ds + 4\int_0^t B_s^3\, dB_s,$$

and

$$E(B_t^4) = 6\int_0^t E(B_s^2)\, ds = 6\int_0^t s\, ds = 3t^2.$$

9.5 The desired martingale is $B_t^4 - 6tB_t^2 + 3t^2$.

9.9 (b) $E(X_3) = 4$; $Var(X_3) = 30$; $P(X_3 < 5) = 0.7314$.

REFERENCES

S. D. Abbott and M. Richey. Take a walk on the Boardwalk. *College Mathematics Journal*, 28(3):162–171, 1997.

D. Aldous and P. Diaconis. Shuffling cards and stopping times. *American Mathematical Monthly*, 93:333–348, 1986.

D. Aldous and J. A. Fill. Reversible Markov Chains and Random Walks on Graphs, 2002. Unfinished monograph, recompiled 2014, available at http://www.stat.berkeley.edu/aldous/RWG/book.html.

K.B. Athreya and P.E. Ney. *Branching Processes*. Springer-Verlag, Berlin, 1972.

N. Bartolomeo, P. Trerotoli, and G. Serio. Progression of liver cirrhosis to HCC: an application of hidden Markov model. *BMC Medical Research Methodology*, 11(38):1–8, 2011. http://www.biomedcentral.com/1471-2288/11/38. Open access.

D. Bayer and P. Diaconis. Trailing the dovetail shuffle to its lair. *Annals of Applied Probability*, 2(2):294–313, 1992.

N. Becker. On parametric estimation for mortal branching processes. *Biometrika*, 61:393–399, 1974.

R. Benzi. Stochastic resonance: from climate to biology. *Nonlinear Processes in Geophysics*, 17:431–441, 2010.

G. Blom and L. Holst. Embedding procedures for discrete problems in probability. *Mathematical Scientist*, 16:29–40, 1991.

M. C. Bove et al. Effect of El Niño on U.S. landfalling hurricanes, revisited. *Bulletin of the American Meteorological Society*, 79(11):2477–2482, 1998.

S. Brin and L. Page. The anatomy of a large-scale hypertextual web search engine. *Computer Networks and ISDN Systems*, 30(1–7):107–117, 1998.

Introduction to Stochastic Processes with R, First Edition. Robert P. Dobrow.

M. Broadie and D. Joneja. An application of Markov chain analysis to the game of squash. *Decision Sciences*, 24(5):1023–1035, 1993.

G. Casella and E. I. George. Explaining the Gibbs sampler. *The American Statistician*, 46:167–174, 1992.

G. W. Cobb and Y.-P. Chen. An application of Markov chain Monte Carlo to community ecology. *American Mathematical Monthly*, 110(4):265–288, 2003.

S. L. Dans et al. Effects of tour boats on dolphin activity examined with sensitivity analysis of Markov chains. *Conservation Biology*, 26:708–716, 2012. doi: 10.1111/j.1523-1739.2012.01844.x.

P. Diaconis. Dynamical bias in coin tossing. *SIAM Review*, 49(2):211–235, 2007.

P. Diaconis. The Markov chain Monte Carlo revolution. *Bulletin of the American Mathematical Society*, 46(2):179–205, 2009.

P. Diaconis and F. Mosteller. Methods for studying coincidences. *Journal of the American Statistical Association*, 84(408):853–861, 1989.

P. M. Dixon. Nearest neighbor methods, 2012. http://www.public.iastate.edu/pdixon/stat406/NearestNeighbor.pdf, visited 2014-08-11.

R. Dobrow. *Probability with Applications and R*. John Wiley & Sons, Inc., 2013.

J. Dongarra and F. Sullivan. Top ten algorithms of the century. *Computing in Science & Engineering*, 2:22–23, 2000.

R. Durrett. *Probability Models for DNA Sequence Evolution*. Springer-Verlag, 2002.

A. Einstein. On the movement of small particles suspended in stationary liquids required by the molecular-kinetic theory of heat. *Annalen der Physik*, 17:549–560, 1905.

C. P. Farrington and A. D. Grant. The distribution of time to extinction in subcritical branching processes: applications to outbreaks of infectious disease. *Journal of Applied Probability*, 36:771–779, 1999.

W. Feller. *An Introduction to Probability Theory and Its Applications*. John Wiley & Sons, Inc., 1968.

S. Geman and D. Geman. Stochastic relaxation, Gibbs distributions, and the Bayesian restoration of images. *IEEE Transactions on Pattern Analysis and Machine Intelligence*, 6:721–741, 1984.

D. Gross, J. F. Shortle, J. M. Thompson, and C. M. Harris. *Fundamentals of Queueing Theory*. John Wiley & Sons, Inc., 2008.

P. Guttorp. *Stochastic Modeling of Scientific Data*. Chapman and Hall, 1995.

J. C. Hendricks et al. Use of immunological markers and continuous-time Markov models to estimate progression of HIV infection in homosexual men. *AIDS*, 10:649–656, 1996.

P. Hoel, S. Port, and C. Stone. *Introduction to Stochastic Processes*. Waveland Press, 1986.

F. M. Hoppe. Branching processes and the effect of parlaying bets on lottery odds. In *Optimal Play: Mathematical Studies of Games and Gambling*, pages 361–373. University of Nevada, 2007.

R. Horn and C. Johnson. *Matrix Analysis*. Cambridge University Press, 1990.

J. S. Horne et al. Analyzing animal movements using Brownian bridges. *Ecology*, 88(9): 2354–2363, 2007.

D. M. Hull. A reconsideration of Lotka's extinction probability using a bisexual branching process. *Journal of Applied Probability*, 38(3):776–780, 2001.

M. A. Jones. Win, lose or draw: a Markov chain analysis of overtime in the National Football League. *College Mathematics Journal*, 35(5):330–336, 2004.

R. H. Jones et al. Tree population dynamics in seven South Carolina mixed-species forests. *Bulletin of the Torrey Botanical Club*, 121:360–368, 1994.

S. Karlin and H.M. Taylor. *A First Course in Stochastic Processes*. Academic Press, 1975.

F. P. Kelly. *Reversibility and Stochastic Networks*. John Wiley & Sons, Inc., 1994.

K. P. Kimou et al. An efficient analysis of honeypot data based on Markov chain. *Journal of Applied Sciences*, 10:196–202, 2010.

G. Kolata. In shuffling cards, 7 is winning number. New York Times, January 9, 1990.

G.-Y. Lin. Simple Markov chain model of smog probability in the South Coast Air Basin of California. *Professional Geographer*, 33(2):228–236, 1981.

J. S. Liu. *Monte Carlo Strategies in Scientific Computing*. Springer-Verlag, 2001.

A. J. Lotka. The extinction of families. *Journal of the Washington Academy of Sciences*, 21:377–380, 1931.

B. Mann. How many times should you shuffle a deck of cards? *UMAP Journal*, 15(4): 303–331, 1994.

R. Mansuy. The origins of the word "martingale". *Electronic Journal for History of Probability and Statistics*, 5(1), 2009. http://www.jehps.net/juin2009/Mansuy.pdf.

D. L. Martell. A Markov chain model of day to day changes in the Canadian Forest Fire Weather Index. *International Journal of Wildland Fire*, 9(4):265–273, 1999.

D. G. Morrison. On the optimal time to pull the goalie: a Poisson model applied to a common strategy used in ice hockey. *TIMS Studies in Management Science*, 4:67–68, 1976.

J. R. Norris. *Markov Chains*. Cambridge University Press, 1998.

P. K. Newton, J. Mason, K. Bethel, L. A. Bazhenova, J. Nieva, et al. A stochastic Markov chain model to describe lung cancer growth and metastasis. *PLoS ONE*, 7(43):e34637, 2012. http://doi:10.1371/journal.pone.0034637. Open access.

H. Poincaré. *Calcul des Probabilités*. Gauthier Villars, Paris, 1912.

J. G. Propp and D. B. Wilson. Exact sampling with coupled Markov chains and applications to statistical mechanics. *Random Structures and Algorithms*, 9:223–252, 1996.

H. J. Ryser. Combinatorial properties of matrices of zeros and ones. *Canadian Journal of Mathematics*, 9:371–377, 1957.

C. Robert and G. Casella. A short history of Markov chain Monte Carlo: subjective recollections from incomplete data. *Statistical Science*, 26:102–115, 2011.

J. S. Rosenthal. Convergence rates for Markov chains. *SIAM Review*, 37:387–405, 1995.

S. Ross. *Stochastic Processes*. John Wiley & Sons, Inc., 1996.

A. D. Sahin and Z. Sen. First-order Markov chain approach to wind speed modeling. *Journal of Wind Engineering and Industrial Aerodynamics*, 89:263–269, 2001.

Z. Schechner et al. A Poisson process model for hip fracture risk. *Medical and Biological Engineering and Computing*, 48(8):799–810, 2010.

S. H. Sellke et al. Modeling and automated containment of worms. *IEEE Transactions on Dependable and Secure Computing*, 5(2):71–86, 2008.

B. Sinclair. Discrete-time Markov chains: state classifications, 2005 http://cnx.org/content/m10852/latest/.

H. S. Stern. A Brownian motion model for the progress of sports scores. *Journal of the American Statistical Association*, 89(427):1128–1134, 1994.

I. Stewart. Monopoly revisited. *Scientific American*, pages 116–119, October 1996.

I. Stewart. The mathematical equation that caused the banks to crash. http://www.theguardian.com/science/2012/feb/12/black - scholes - equation - credit - crunch, 2012. [The Guardian/The Observer online; posted 11-February-2012].

N. Strigul et al. Modelling of forest stand dynamics using Markov chains. *Environmental Modelling and Software*, 31:64–75, 2012.

D. Teets and K. Whitehead. The discovery of Ceres: how Gauss became famous. *Mathematics Magazine*, 72:83–93, 1999.

H. Tsai. Estimates of earthquake recurrences in the Chiayi-Tainan area, Taiwan. *Engineering Geology*, 63:157–168, 2002.

G. K. Vallis. Mechanisms of climate variability from years to decades. In T. Palmer and P. Williams, editors, *Stochastic Physics and Climate Modelling*. Cambridge University Press, Cambridge, 2010.

B. S. Van der Laan and A. S. Louter. A statistical model for the costs of passenger car traffic accidents. *Journal of the Royal Statistical Society, Series D*, 35(2):163–174, 1986.

N. Wiener. *I am a Mathematician*. MIT Press, 1956.

Wolfram Alpha LLC. 2015. Wolfram Alpha. http://www.wolframalpha.com/input/?i=MatrixExp%5Bt%7B%7B3r%2Cr%2Cr%2Cr%7D%2C%7Br%2C3r%2Cr%2Cr%7D%2C%7Br%2Cr%2C3r%2Cr%7D%2C%7Br%2Cr%2Cr%2C3r%7D%7D%5D+%2F%2F+Simplify (access Sept. 30, 2015).

R. Zhang et al. Using Markov chains to analyze changes in wetland trends in arid Yinchuan Plain, China. *Mathematical and Computer Modelling*, 54:924–930, 2011.

INDEX

Introduction to Stochastic Processes with R, First Edition. Robert P. Dobrow.